EFFECTS OF GASEOUS AIR POLLUTION IN AGRICULTURE AND HORTICULTURE

Proceedings of Previous Easter Schools in Agricultural Science, published by Butterworths, London

*SOIL ZOOLOGY   Edited by D. K. McE. Kevan (1955)
*THE GROWTH OF LEAVES   Edited by F. L. Milthorpe (1956)
*CONTROL OF THE PLANT ENVIRONMENT   Edited by J. P. Hudson (1957)
*NUTRITION OF THE LEGUMES   Edited by E. G. Hallsworth (1958)
*THE MEASUREMENT OF GRASSLAND PRODUCTIVITY   Edited by J. D. Ivins (1959)
*DIGESTIVE PHYSIOLOGY AND NUTRITION OF THE RUMINANT   Edited by D. Lewis (1960)
*NUTRITION OF PIGS AND POULTRY   Edited by J. T. Morgan and D. Lewis (1961)
*ANTIBIOTICS IN AGRICULTURE   Edited by M. Woodbine (1962)
*THE GROWTH OF THE POTATO   Edited by J. D. Ivins and F. L. Milthorpe (1963)
*EXPERIMENTAL PEDOLOGY   Edited by E. G. Hallsworth and D. V. Crawford (1964)
*THE GROWTH OF CEREALS AND GRASSES   Edited by F. L. Milthorpe and J. D. Ivins (1965)
*REPRODUCTION IN THE FEMALE MAMMAL   Edited by G. E. Lamming and E. C. Amoroso (1967)
*GROWTH AND DEVELOPMENT OF MAMMALS   Edited by G. A Lodge and G. E. Lamming (1968)
*ROOT GROWTH   Edited by W. J. Whittington (1968)
*PROTEINS AS HUMAN FOOD   Edited by R. A. Lawrie (1970)
*LACTATION   Edited by I. R. Falconer (1971)
*PIG PRODUCTION   Edited by D. J. A. Cole (1972)
*SEED ECOLOGY   Edited by W. Heydecker (1973)
 HEAT LOSS FROM ANIMALS AND MAN: ASSESSMENT AND CONTROL   Edited by J. L. Monteith and L. E. Mount (1974)
*MEAT   Edited by D. J. A. Cole and R. A. Lawrie (1975)
*PRINCIPLES OF CATTLE PRODUCTION   Edited by Henry Swan and W. H. Broster (1976)
*LIGHT AND PLANT DEVELOPMENT   Edited by H. Smith (1976)
 PLANT PROTEINS   Edited by G. Norton (1977)
 ANTIBIOTICS AND ANTIBIOSIS IN AGRICULTURE   Edited by M. Woodbine (1977)
 CONTROL OF OVULATION   Edited by D. B. Crighton, N. B. Haynes, G. R. Foxcroft and G. E. Lamming (1978)
 POLYSACCHARIDES IN FOOD   Edited by J. M. V. Blanshard and J. R. Mitchell (1979)
 SEED PRODUCTION   Edited by P. D. Hebblethwaite (1980)
 PROTEIN DEPOSITION IN ANIMALS   Edited by P. J. Buttery and D. B. Lindsay (1981)
 PHYSIOLOGICAL PROCESSES LIMITING PLANT PRODUCTIVITY   Edited by C. Johnson (1981)
 ENVIRONMENTAL ASPECTS OF HOUSING FOR ANIMAL PRODUCTION   Edited by J. A. Clark (1981)
 CHEMICAL MANIPULATION OF CROP GROWTH AND DEVELOPMENT   Edited by J. S. McLaren (1982)

These titles are now out of print but are available in microfiche editions

Effects of Gaseous Air Pollution in Agriculture and Horticulture

M. H. UNSWORTH, PhD
University of Nottingham, England

D. P. ORMROD, PhD
University of Guelph, Ontario

BUTTERWORTH SCIENTIFIC
London Boston Sydney Wellington Durban Toronto

First published 1982

© The several contributors named in the list of contents 1982

British Library Cataloguing in Publication Data

Effects of gaseous air pollution in agriculture
and horticulture.
1. Plant, Effects of pollution on—Congresses
I. Unsworth, M.H.   II. Ormrod, D.P.
581.3′1   QK751

ISBN 0–408–10705–7

Typeset by Scribe Design, Gillingham, Kent
Printed in England by the University Press, Cambridge

PREFACE

Everyone is aware of the numerous discussions concerning the more or less pernicious influence of the gases given off by certain manufactories. The ruin now of a manufacturer, now of a horticulturist, may result from the declaration of an expert; hence, it is incumbent on scientific men not to pronounce on these delicate questions without substantial proof.

(From an address by A. de Candolle, 1866, quoted in First Report of the Relations between Climate and Crops, p. 24, Cleveland Abbe, Government Printing Office, Washington D.C., 1905.)

Of the previous 31 Easter Schools in Agricultural Science, many have been concerned with factors restricting the productivity of crops. However, the proceedings of this 32nd School are the first to be devoted entirely to the impact of air pollution on agriculture and horticulture. There are several reasons why the topic was selected: improved instrumentation and monitoring networks have shown the existence of significant concentrations of gaseous pollutants in many agricultural regions of industrialized countries, and transport of pollutants across national boundaries has become a topic of concern; the past decade has seen a substantial increase in investigations of responses of crop plants to pollutants both in controlled environments and in the field; there has been increasing recognition of the benefits of establishing stronger links between research groups in different parts of the world to help in understanding the similarities and differences in approaches to evaluating the influence of air quality on crops.

In structuring the meeting, we sought to begin by reviewing progress in physical, physiological and biochemical aspects of the subject before moving to discussions of effects of pollutants on developmental processes such as flowering and fruiting, and on growth and yield. In practice, responses to pollutants interact with nutrition and disease, and effects of mixtures of pollutant gases often differ from effects of single gases: these topics are particularly important in assessing economic effects of air pollution and are considered in separate chapters. Finally we asked for three reviews to place the topic of air pollution and its effects in a broader perspective, considering, for example, the genetic basis of sensitivity to air-pollutant stress, and the directions future research should take.

Reflecting the international interest in responses to air pollution, the School attracted participants from 16 countries, and throughout the meeting there were wide-ranging and open discussions, both formal and informal. An indication of the range of interests represented at the meeting can be gained not only from the review papers in this volume but also from the summaries of the recent research presented at the Poster Session. We are especially grateful to all participants at the conference for their contributions towards the friendly and relaxed atmosphere and we hope that the professional and social links they established at Sutton Bonington will continue to flourish.

The success of the conference owed much to the experience and efficiency of the Conference Secretary, Edna Lord and to Lynda Hardy who assisted her. Particular thanks are also due to Valerie Black for organizing the Poster Session. Many other members of the Department of Physiology and Environmental Science helped to ensure the smooth running of the meeting. During the formal sessions the experience of the chairmen Professor J.L. Monteith, Professor J.K.A. Bleasdale, Professor S.V. Krupa, Professor A.J. Rutter, Dr P.J. Saunders and Professor K. Mellanby contributed much to the flow of discussion from participants.

Finally, on behalf of the University of Nottingham we thank all the organizations which gave financial assistance to the School.

M.H. UNSWORTH
D.P. ORMROD

Postscript

During the preparation of this book we were saddened to hear of the sudden death of one of the contributors, Derek Cowling. Derek pioneered some of the earliest work in Britain to assess responses of ryegrass to sulphur dioxide and to sulphur nutrition, and his enthusiasm and productivity were a great stimulus for air-pollution research. He will be greatly missed by his many friends throughout the scientific community.

ACKNOWLEDGEMENTS

The financial assistance given to the Conference by the following companies is gratefully acknowledged:

Analysis Automation
Analytical Development Company
Associated Octel Company
Boots Company Limited
Crump Scientific Instruments
Delta-T Devices
London Brick Company Limited
Shell International Petroleum Company Limited
Solartron Electronic Group
Techmation Limited

The following organizations also supported a Trade Exhibition during the Conference:

Analysis Automation Ltd, Cherwell Boathouse, Bradwell Rd, Oxford.
Analytical Development Co. Ltd, Pindar Rd, Hoddesdon, Herts.
Crump Scientific Instruments, 166 High St. Rayleigh, Essex.
Delta-T Devices, 128 Low Rd, Burwell, Cambridge.
Solartron Electronic Group, Farnborough, Hants.
Techmation Ltd, 58 Edgware Way, Edgware, Middlesex.

CONTENTS

I

Defining the Polluted Environment

1

AIR POLLUTANTS IN AGRICULTURE AND HORTICULTURE

D. FOWLER
J.N. CAPE
*Institute of Terrestrial Ecology, Bush Estate, Penicuik, Midlothian,
Scotland*

Introduction

Important changes in the characteristics of major sources of air pollutants
have taken place in the last two decades. In general these changes have led
to more effective dispersion and consequently smaller concentrations of
pollutant gases close to sources. For effects on plants, current interest in
gaseous pollution is centred on the influence on crop production of
moderate concentrations, 20–50 ppb of each of the gases SO_2 and NO_2 and
60–150 ppb of O_3, which occur over large areas of agricultural land in the
industrialized countries of the world.

Concentrations of gas pollutants to which crops are exposed must be
defined in terms of their spatial distribution to assess the scale of potential
effects, and in terms of their frequency of occurrence to allow controlled-
environment studies of the effects of 'typical' gas mixtures on plant
performance. In this chapter, these two aspects of distribution of gas
pollutants are considered in turn. Our attempt to define those agricultural
areas in Europe and North America that are exposed to potentially
damaging concentrations of SO_2, NO_2 and O_3 is speculative (particularly
for NO_2 and O_3) because little monitoring information is available for
these gases in agricultural areas. The distribution of gas concentrations in
time is considered in detail and two alternative methods for predicting the
frequency of occurrence of high concentrations are described. This information, although seldom incorporated in earlier studies, is now being
recognized as a vital component of experimental design where effects on
crop yield rather than on physiological mechanisms are being investigated.

To link controlled-environment studies of gas effects with field exposure
of agricultural crops requires a method for defining the dose of pollutant
received by field crops. The final section of this chapter considers a method
for estimation of dose from measurements of gas concentration and
stomatal conductance.

Pollutant-gas concentrations

The pollutant gases we are considering are not restricted to Europe and
North America, but are a characteristic of the atmosphere over these and

other industrialized countries. The areas under consideration are those for which most monitoring information is available, although for the tasks we have set ourselves this information is far from adequate. To put into perspective the problem of sulphur dioxide and nitrogen oxides, annual emissions in North America and Western Europe together amount to about 50×10^6 tonnes SO_2 and 30×10^6 tonnes NO_x*, in comparison with natural emissions globally of gases containing 60×10^6 tonnes S (natural emissions of NO_x have not been quantified). It is clear that the air chemistry of sulphur and nitrogen oxides is strongly influenced by anthropogenic emissions in North America and Western Europe.

Primary pollutants such as nitric oxide (NO) and sulphur dioxide in large concentrations are associated with major source areas, whereas secondary pollutants such as photochemical ozone and acidity in rain may have a much wider distribution, and in the latter case may influence ecosystems thousands of kilometres from the major source areas. Although the emphasis in this volume is on pollutant gases, the association between gaseous pollutants, particulate material and the chemical composition of rainfall must not be forgotten.

It is difficult to generalize in a useful way on the relative distribution of agricultural and horticultural areas and of major sources of pollution. Most productive cropland is situated between areas of high population density and areas of very low ambient pollutant-gas concentrations. Consequently, crops are exposed to a mixture of primary and secondary pollutants, in contrast to the vegetation in remote areas which is affected mainly by secondary pollutants resulting from chemical transformation of emitted gases.

With the sources of primary gas pollutants being the industrial areas, where the regional *emission density* (the sum of all known measured sources over a specified area) approaches $30\,g\,SO_2\,m^{-2}$ per year, the zones with very large concentrations are well defined and contain a significant proportion of urban land area. In Western Europe the main areas are the Ruhr in West Germany and the industrial Midlands and North of England, where annual *arithmetic mean concentrations* of SO_2 may exceed $100\,\mu g\,m^{-3}$ and NO_2 concentrations, although less well determined, are of the order of $50\,\mu g\,m^{-3}$. For the region as a whole (area $1.1 \times 10^8\,ha$, including the UK, West Germany, the Netherlands, Belgium and France) the amount of agricultural land exposed to annual arithmetic mean concentrations exceeding $100\,\mu g\,SO_2\,m^{-3}$ and $50\,\mu g\,NO_2\,m^{-3}$ has been computed as $1.3 \times 10^6\,ha$, or about 1% of the total area (*Figure 1.1*). The map in *Figure 1.1* is based on predicted concentrations from the Longe Range Transport of Air Pollution (LRTAP) model (OECD) for 1974 emissions and meteorology, with the isopleths of concentration interpolated by hand. For Great Britain, rural monitoring data (Warren Spring Laboratory, 1979) have been used to help define the zones (*Figure 1.2*) although the data are not very accurate in rural areas (Bailey, Barrett and Cooper 1978).

The $50\,\mu g\,SO_2\,m^{-3}$ isopleth encloses an area of $5.7 \times 10^6\,ha$ or 5% of the total area of Europe considered (*Figure 1.1*), although in the UK this concentration is exceeded for 10% of the land area (*Figure 1.2*). The

---

* (NO_x is expressed as NO_2)

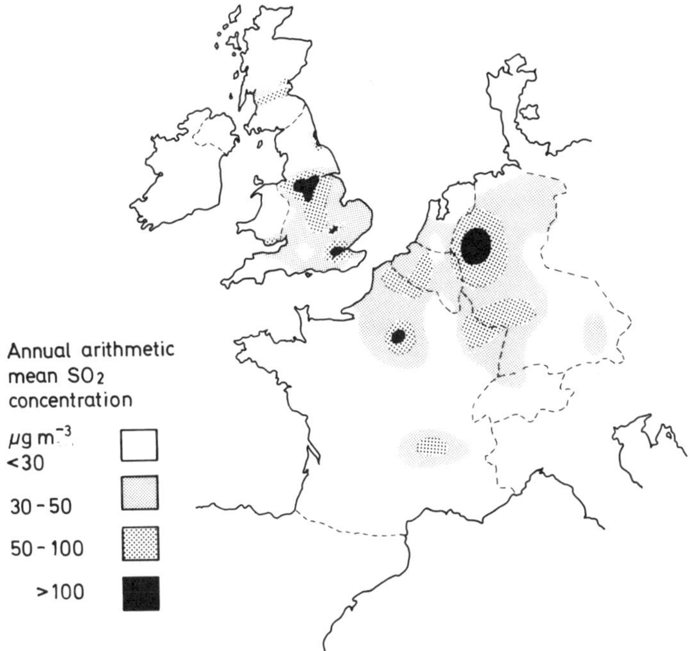

Figure 1.1   Distribution of sulphur dioxide over Western Europe (based on 1974 emission data)

30 µg SO$_2$ m^{-3} contour encloses 23% of the European area and over 40% of the UK, including most of the agricultural land. (*Table 1.1*). No significant area of agricultural land outside the 30 µg SO$_2$ m^{-3} contour is exposed to potentially harmful SO$_2$ gas concentrations. An alternative map for the UK (Martin, 1980), which specifically excludes the influences of local sources and urban areas, shows smaller overall concentrations for SO$_2$ but with a similar pattern. Part of the difference is attributable to the different years to which the data relate.

The agricultural area of the United States exposed to high concentrations of SO$_2$ cannot be estimated from monitoring stations because these are (as in the UK and Europe) nearly all situated in towns or around major emitters. We were unable to make appropriate predictions from modelling studies. However, the areas in which the emission density of SO$_2$ approaches the average value for the area within the 30 µg SO$_2$ m^{-3} isopleth in Europe (30 g SO$_2$ m^{-2} y^{-1}) are the Ohio river states of Indiana, Ohio and Pennsylvania, in which half the US emissions of SO$_2$ occur (NAS, 1978). The emission density in these states is about 25 µg SO$_2$ m^{-2} y^{-1}, and this is the only region in which elevated average SO$_2$ concentrations are probable over very large areas (except in the vicinity of large towns and the Washington DC–Philadelphia–New York region of the East Coast). This region amounts to 3.2 × 10^7 ha and is comparable in size with the area in W. Europe within the 30 µg m^{-3} isopleth. Extrapolating further to predict the area in the US within the 50 µg m^{-3} and the 100 µg m^{-3} contours is not

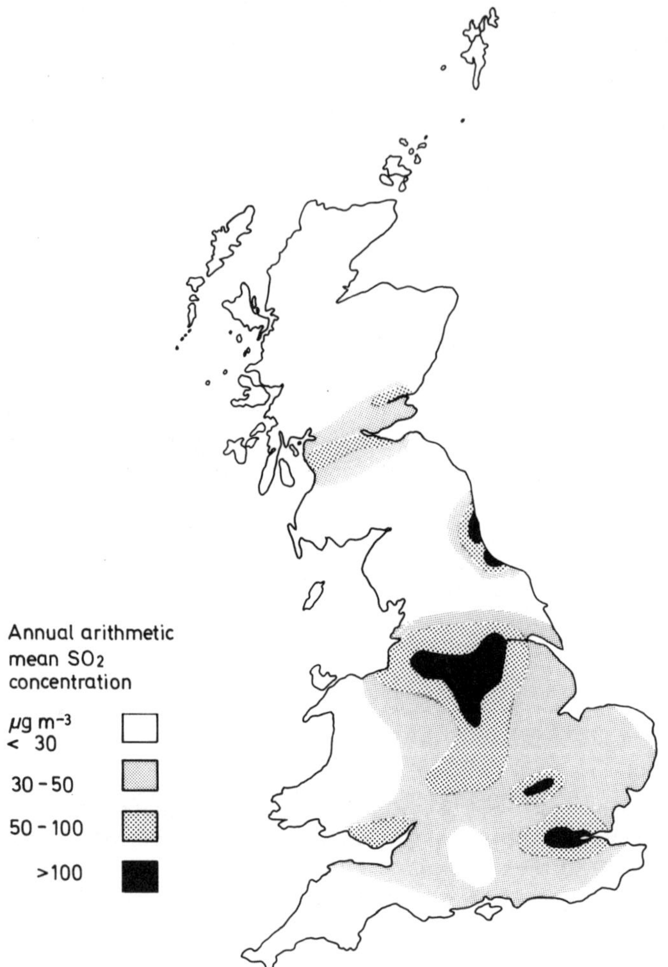

Figure 1.2   Distribution of sulphur dioxide over Great Britain (based on 1974 emission data and measured concentrations in 1978–1979 (Warren Spring, 1979))

justified on the basis of analogy with Europe, because these areas are much more sensitive to the distribution of major sources within the defined region. For Europe and N. America, the lack of monitoring data for rural and remote areas is a great drawback for those engaged in quantifying effects of air pollutants on crop production. Redistribution of monitoring stations to provide information for (1) urban areas, (2) specific major sources and (3) rural/remote areas would help redress the current imbalance. For Canada, the agricultural areas exposed to large concentrations of SO_2 are localized and do not constitute a large area. The areas affected by an annual mean SO_2 concentration in excess of $30\,\mu g\,m^{-3}$ are restricted to the major cities of Ontario and Quebec and industrialized sections of the St Lawrence Seaway, although here again there are few rural monitoring stations.

Table 1.1 RURAL AREAS OF WESTERN EUROPE AND NORTH AMERICA EXPOSED TO GASEOUS POLLUTANTS (SPAIN AND ITALY OMITTED)

Region	Total area (ha)	Annual arithmetic mean SO_2			Annual arithmetic mean NO_2		
		$>30\,\mu g\,m^{-3}$	$>50\,\mu g\,m^{-3}$	$>100\,\mu g\,m^{-3}$	$>25\,\mu g\,m^{-3}$	$>40\,\mu g\,m^{-3}$	$>80\,\mu g\,m^{-3}$
Western Europe	1.1×10^8	2.6×10^7	5.7×10^6	1.3×10^6	2.6×10^7	5.7×10^6	1.3×10^6
Great Britain	2.3×10^7	1.0×10^7	2.3×10^6	2.4×10^5	1.0×10^7	2.3×10^6	2.4×10^5
Canada	6.8×10^7		Not significant			Significant but insufficient data	
USA	4.4×10^8	3.2×10^7	Data not available		3.2×10^7	Data not available	

The total area of agricultural land in W. Europe and N. America exposed to annual average SO_2 concentrations exceeding $30 \,\mu g \, m^{-3}$ is therefore about $6.0 \times 10^7 \, ha$.

NITROGEN OXIDES

Concentrations of NO and NO_2 for the areas being considered are difficult to estimate because there are so few stations which monitor these gases and, as with SO_2, most of the monitoring instruments are located in cities. For the UK the mass ratio NO_2/SO_2 at rural stations is about 0.8 (A. Apling, personal communication), which seems large when one compares the ratio of emissions for NO_x and SO_2 (for Europe this is about 0.5) and may imply a rather longer residence time for NO_x in the atmosphere than for SO_2. However, Martin and Barber (1980) have reported an NO_2/SO_2 ratio of 0.5 for a remote site in S.W. Ireland. An estimate of the zones of NO_2 pollution may be obtained for Europe by assuming that the distribution of this gas follows the same pattern as that for SO_2 and that, on average, the NO_2/SO_2 ratio of 0.8 applies. Given these assumptions, *Table 1.1* shows the agricultural areas of Europe that are likely to be exposed to annual mean NO_2 concentrations of 25, 40 and $80 \,\mu g \, m^{-3}$. These projections are, of course, poorly supported by field measurements and the predictions here should not be seen as an alternative to monitoring but conversely, as justification for appropriate monitoring.

Although based on limited measurements, the large NO_2/SO_2 ratio (and the implied longer NO_2 residence time) may result from less efficient removal mechanisms for nitrogen compounds. Deposition velocities for dry deposition of NO on to vegetation are smaller than those for SO_2 by a factor of about five ($1\text{–}2 \, mm \, s^{-1}$ for NO and $8 \, mm \, s^{-1}$ for SO_2), and NO_2-deposition velocities seem generally to be smaller than those for SO_2 ($5 \, mm \, s^{-1}$ for NO_2; Bengtson, Skarby and Grennfelt, 1980). Rates of wet removal of NO and NO_2 have not attracted much attention but the poorer solubilities of these gases would again lead to a less efficient direct rain scavenging for NO and NO_2 than for SO_2. Lastly, the annual cycle in SO_2 concentration at rural sites (*Figure 1.3*) is not as well defined for NO_2 (*Figure 1.4*) which may suggest that there are summertime natural sources of NO_x (Söderlund and Svensson, 1976; Klepper, 1979; Tingey, 1979).

Concentrations of NO_x in agricultural areas of the USA should be of considerable interest because emissions of $25 \times 10^6 \, t \, NO_2 \, y^{-1}$ represent a much larger relative source than the $6 \times 10^6 \, t \, NO_2 \, y^{-1}$ emissions for W. Europe, the NO_x/SO_2 emission ratios being 0.7 and 0.5 for these areas respectively. With the longer residence times already implied, this would lead to a larger contribution of nitrate-related acid to the total rain acidity in remote regions downwind of major sources. Assuming that the ratio of emission may be used to gauge ambient concentrations, by analogy with European data the rural area of the US exposed to annual average NO_2 concentrations exceeding $25 \,\mu g \, m^{-3}$ is $3.2 \times 10^7 \, ha$.

Ozone exhibits quite different properties as a pollutant because it is strictly secondary in nature; it may be distributed very widely after its photochemical production in polluted air that is advected over generally clean-air regions. There is also a significant natural background concentration which is usually between 10 and 40 ppb, although natural concentrations may occasionally reach 80 ppb or more after a localized incursion of stratospheric air (Derwent *et al.*, 1978). These properties of tropospheric ozone concentrations limit the usefulness of annual arithmetic mean concentrations for characterizing ozone pollution at a given site. The frequency distribution for ozone concentration is also complex in form and will be considered in a later section. Episodes of ozone pollution at many sites occur with a much less predictable frequency because the conditions for ozone formation require a combination of several factors. The basic chemical ingredients for the process are often present over large areas, but without the necessary temperature and radiation levels.

Taking the British Isles as an example, the potential for the formation of photochemical ozone in phytotoxic concentrations occurs throughout most of the country (Ashmore, Bell and Reily, 1980). The frequency of occurrence is very variable but is of the order of $100\,\mathrm{h\,y^{-1}}$ during which the concentration exceeds 50 ppb. The mean duration of a single event is about 6 h and the maximum concentration is of the order of 100 ppb. To illustrate the variability, this year (1980) with a relatively poor summer has yielded a total time of 30 h when the ozone concentration was above 50 ppb at the Institute of Terrestrial Ecology monitoring site in Central Scotland (Nicholson *et al.*, 1980). Martin and Barber (1980; *see also* page 449) also report the frequency of occurrence of photochemical ozone in the Midlands of England. The area of Europe potentially influenced by phytotoxic ozone levels is very much larger than the area within the $30\,\mu\mathrm{g\,m^{-3}}$ isopleth for SO_2 and includes most of the central, southern and eastern areas of W. Europe. The main problem in estimating frequency of occurrence and geographical extent of ozone events is that, in appropriate conditions, a polluted air mass may travel long distances (1000 km).

In contrast to the situation in Europe, in N. America the presence of large concentrations of photochemical ozone has been recognized for a long time. The Los Angeles basin is now internationally renowned for very large and persistent concentrations of ozone and peroxyacetyl nitrate (PAN) and a considerable area of agricultural land in California is influenced by these gases. The other areas of interest lie principally to the east of the Mississippi and extend to most of the land between the Mississippi river and the East coast (W.W. Heck, personal communication). In Canada the only areas attracting concern are the southern regions of Ontario and S.W. Quebec where the frequency of occurrence of photochemical ozone events is of the order of $30\,\mathrm{d\,y^{-1}}$ with average daily concentrations in excess of 50 ppb (H.H. Neumann, personal communication).

Cyclic variations in concentrations of pollutant gases

There are two major time-scales for cyclic variations which, to a greater or lesser degree, are applicable to all pollutant gases. These are the annual cycle, which for primary pollutants reflects the demand for energy throughout the year, and the diurnal cycle which depends more on meteorological and chemical factors. For Europe, energy demand results in winter maxima for SO_2 concentrations and probably for nitric oxide concentrations, whereas for the US the greatest emissions occur in summer, although this peak is not pronounced (there are wide regional differences). The annual variation in SO_2 concentration at rural sites in the UK is given in *Figure 1.3* where the winter maximum exceeds the annual arithmetic mean

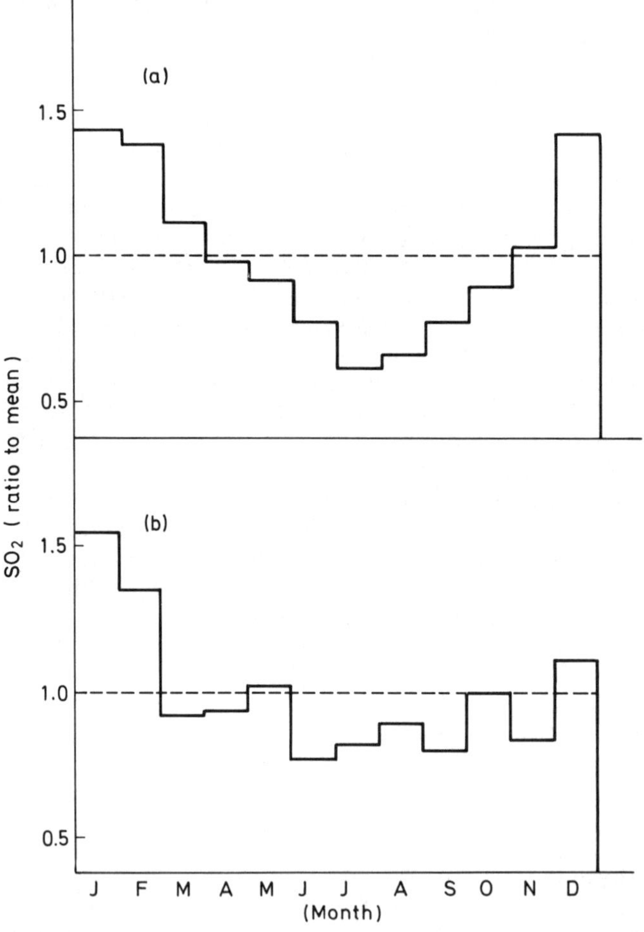

Figure 1.3   Annual cycle of sulphur dioxide concentration: (a) Canada 'residential' 1977 + 1978; (b) UK 'rural' 1978–79 (Canada: H.H. Neumann, personal communication; UK: Warren Spring, 1979)

by 50% (taken from 31 rural monitoring stations – Warren Spring Laboratory, 1979). A similar set of data has been obtained from Canadian monitoring stations with an 'r' classification (the least-polluted sites, which are residential rather than strictly rural). The mean annual variation in SO_2 concentration at these 12 Canadian sites, given in *Figure 1.3*, is very similar to the pattern shown for the UK. No comparable figure is given for the US because there are insufficient data available for rural sites, but the much less pronounced annual emission pattern with a small summer maximum is likely to produce only a weak annual cycle in ambient SO_2 concentrations in remote areas.

For NO_x, although emissions are approximately proportional to those of sulphur for many of the large sources (vehicles excepted), no clear annual cycle in concentration at the Canadian 'r' stations is evident (*Figure 1.4*).

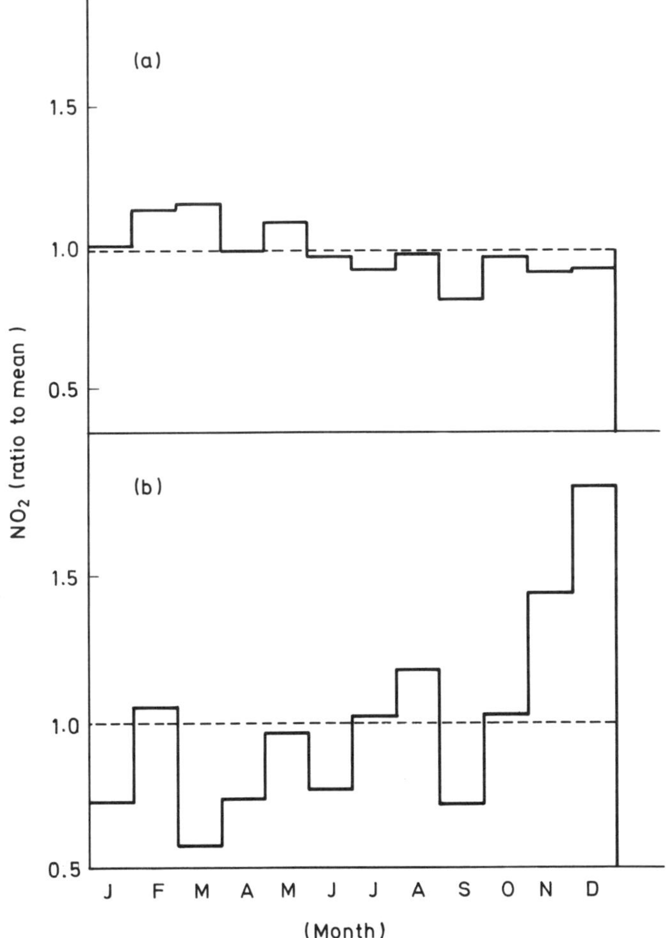

Figure 1.4   Annual cycle of nitrogen dioxide concentration: (a) Canada 'residential' 1977 + 1978; (b) Devilla (Central Scotland) 1978 (Canada: H.H. Neumann, personal communication; Devilla: Nicholson *et al.*, 1980)

Monitoring data for a rural site in the Central Lowlands of Scotland (Devilla Forest) show a small winter maximum for NO_2 (*Figure 1.4*). It is interesting to note the lack of a pronounced annual cycle in NO_2 concentration: we can offer no immediate explanation for this, but it may be the result of natural NO and NO_2 production in soils (Nelson and

Bremner, 1970) and by plants (Klepper, 1979; Tingey, 1979). An annual cycle in NO_2 concentration *has* been observed by Martin and Barber (1980) in the Midlands of England.

For ozone, the origin of the annual cycle in concentration is a matter of current debate (*e.g.* Fishman and Crutzen, 1978), although it is generally held that the spring maximum is the result of the peak in stratosphere/troposphere exchange early in the year. The annual cycle in ozone concentration from the work of Singh, Ludwig and Johnson (1978) is shown in *Figure 1.5*, in generalized form. Two examples from specific monitoring networks are also provided on the same figure.

Diurnal cycles are influenced more strongly by meteorological and deposition factors but, here again, there is considerable debate over the relative importance of air chemistry and deposition processes in producing diurnal cycles in concentration (Hov, Isaksen and Hesstvedt, 1978; Garland and Derwent, 1979; Eastman and Stedman, 1980). In practice the two

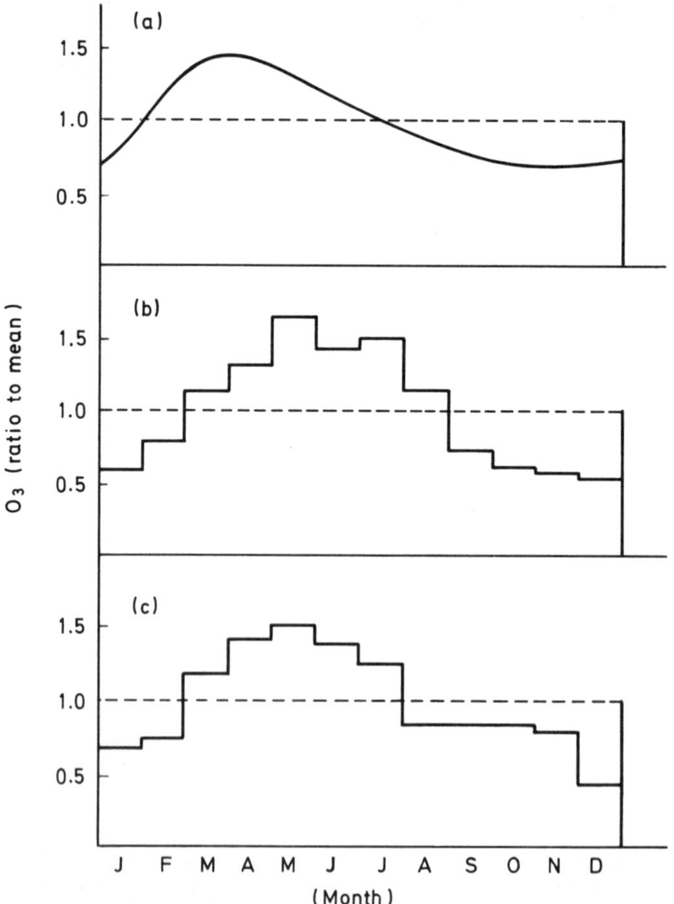

Figure 1.5 Annual cycle of ozone concentration: (a) Schematic (from Singh *et al.*, 1978); (b) Canada 'residential' 1977–78; (c) Devilla (Central Scotland) 1978 (Canada: H.H. Neumann, personal communication; Devilla: Nicholson *et al.*, 1980)

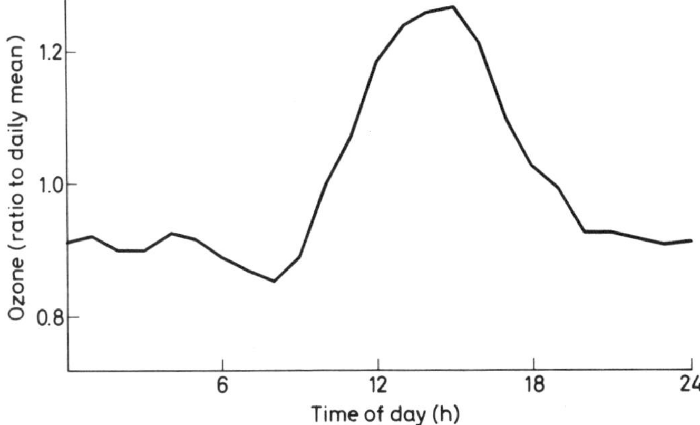

Figure 1.6 Diurnal cycle of ozone concentration, Devilla (Central Scotland) 1978 (Nicholson *et al.*, 1980)

processes are somewhat confounded. Considering the mean diurnal cycle in ozone concentration from one year's monitoring at a site in central Scotland (*Figure 1.6*), the daytime maximum is rather late in the day to be explained entirely as the result of the break-up of a nocturnal inversion and is probably due in part to photochemical ozone production.

The daily cycle in SO_2 and NO_x concentrations shows a similar pattern to ozone except that the maxima occur at about 10.00 h solar time (*Figure 1.7*) rather than mid-afternoon, as in the case of ozone. The nocturnal minimum for NO_x and SO_2 concentrations may be explained in terms of depletion by dry deposition beneath a low-level inversion layer, and the steep increase in concentration during the morning may be explained by the break-up of the nocturnal inversion. The data in *Figure 1.7* show a small mid-evening peak in NO_x and SO_2, which is probably not a general feature of NO_x and SO_2 concentrations and seems to be due more to local sources close to the monitoring site in question.

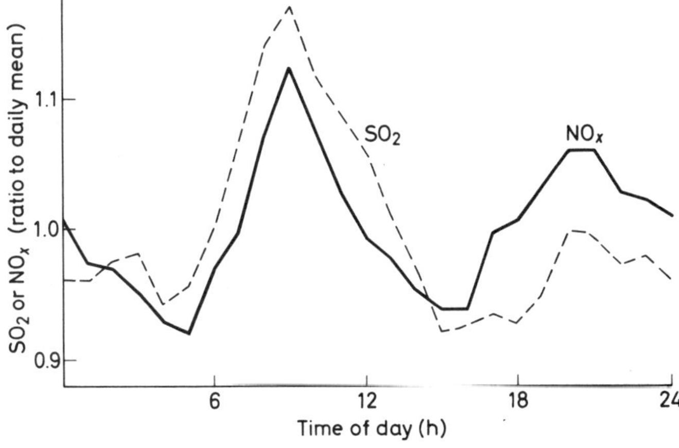

Figure 1.7 Diurnal cycle of NO_x and SO_2, Devilla (Central Scotland) 1978 (Nicholson *et al.*, 1980)

Statistical description of pollutant-gas concentrations

Although it is convenient to describe polluted regions in terms of annual average concentrations, effects on vegetation are strongly dependent on the variation of concentration with time. Potentially damaging concentrations occur only at infrequent intervals in most rural areas and any assessment of likely adverse effects must take explicit account of the statistical distribution of pollutant concentrations, particularly if a threshold value for an effect has been identified. The same considerations apply to all studies of effects, and considerable effort has been expended on the continuous monitoring of pollutant-gas concentrations in an attempt to characterize distributions, thereby enabling predictions to be made of the occurrence of extreme values. However, this monitoring relates to sites which are very seldom typical of agricultural areas, and the distributions may differ from those experienced in rural areas. This is particularly true for monitoring sites close to sources, which generally show greater variability (related to wind direction) than sites exposed to high concentrations of pollutant transported from surrounding source areas.

Accurate descriptions and predictions require that observed concentrations be fitted to a statistical model, and there may be several such models which give equivalent results. Some possible models, and their application to pollutant-gas concentration, have been discussed elsewhere (Bencala and Seinfeld, 1976). For most practical purposes the simplest model is often the best, particularly when all that is required is a good description of the observed values. If predictions based on extrapolation from a limited data set are required, then more care must be taken in selecting the statistical model employed.

There are two methods of analysing pollutant-concentration data: by classification within periods of time, or by levels of concentration. The time-averaging method is simpler to use, where average concentrations are recorded over consecutive periods, but relies on the assumption that a given frequency distribution (the log-normal distribution) is a good fit to the actual variations in concentration. This method, developed by Larsen (1973), allows for prediction of median and maximum values for a given averaging time if data are available for another averaging time. For example, if daily mean values are available over a period of several years, then the hourly median and maximum and the annual median and maximum may be predicted. The expressions take the simple form (Larsen, 1974)

$$\text{median concentration} = (\text{averaging time})^x$$
$$\text{maximum concentration} = (\text{averaging time})^{x'}$$

Such methods are particularly suited to the design of air-quality criteria, as they may be expressed in terms of average concentrations which should not be exceeded for more than one hour or one day per year. Air-quality criteria in the USA are based on the idea of time-averaging.

The alternative method for analysing concentration data is to calculate the frequency of occurrence as a function of concentration. The raw data may be hourly or daily averages, or 'instantaneous' measurements of ambient concentrations. The calculations are more lengthy, as a complete set of data must be considered in order to represent frequency of occurrence as a fraction of the total, but the interpretation is more straightforward, and statistical methods for fitting the data to a model are better-developed. There is a wide variety of models available for description or prediction, and the most widely used are discussed below. An excellent account, with worked examples, of the models and statistical techniques appropriate to the analysis of frequency distributions may be found in the *Flood Studies Report* (NERC, 1975).

DESCRIPTION OF FREQUENCY DISTRIBUTION OF POLLUTANT
CONCENTRATIONS

If concentrations of a pollutant gas are plotted as a histogram of frequencies, a typical distribution (*Figure 1.8a*) is seen to be strongly skewed with a long 'tail' at high values. If the frequency classes are plotted in terms of the

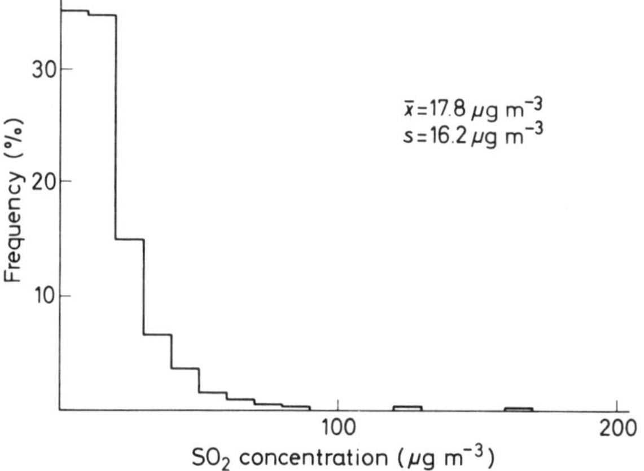

Figure 1.8(a)   Frequency distribution of daily average sulphur dioxide concentrations at Bush (Midlothian) during 1977 and 1978

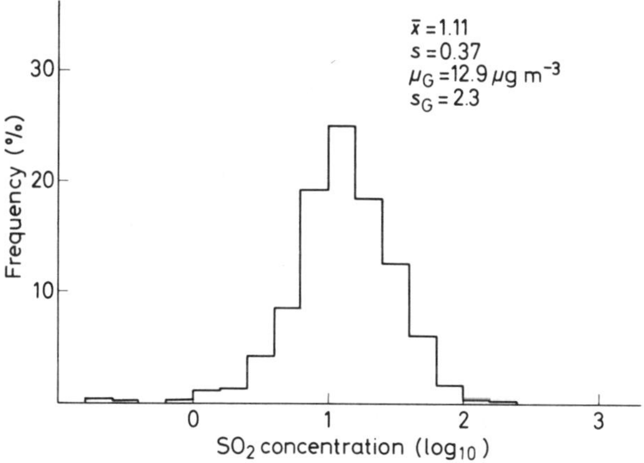

Figure 1.8(b)   Frequency distribution of logarithms of daily average sulphur dioxide concentrations (data as for *Figure 1.8(a)*)

logarithm of concentration, however, then a more symmetrical distribution is obtained (*Figure 1.8b*). This illustrates the approximately log-normal distribution of much pollutant-gas concentration data. The log-normal distribution of pollutant concentration is related to the same forms of distribution of wind speed (Mage, 1980a) and the dispersion of pollutants from sources. The primary pollutants (SO_2, NO and primary NO_2) fit a log-normal distribution reasonably well, especially at sites close to sources, but secondary pollutants (O_3 and secondary NO_2) are less well described, because the concentrations may be dependent upon localized chemical processes rather than atmospheric dispersion. For data from rural and remote sites the distributions may deviate from log-normality, reflecting the influence of secondary processes.

THE LOG-NORMAL DISTRIBUTION (TWO-PARAMETER)

If this distribution is chosen as a model for a given set of data, the mean and standard deviation must be obtained for the logarithm values (*Figure 1.8b*). In principle it is possible to calculate the geometric mean (μ_G) and standard deviation (s_G) from

$$\mu_G = \left(\prod_{i=1}^{n} x_i \right)^{1/n} = \exp \left[\frac{1}{n} \sum_{i=1}^{n} \ln x_i \right] \tag{1.1}$$

$$s_G = \exp \left\{ \left[\frac{\frac{1}{n} \sum_{i=1}^{n} (\ln x_i)^2 - (\ln \mu_G)^2}{(n-1)} \right]^{1/2} \right\} \tag{1.2}$$

However, this may give rise to misleading values for the two parameters, and may not provide the best fit for the data. All monitoring methods have lower limits of detection, which means that values recorded at, or close to, this lower limit are subject to large uncertainties. There may also be uncertainties due to rounding errors. These uncertainties are magnified when the logarithm values are considered. There is also the problem of assigning logarithms to values falling below the limit of detection and recorded as zero. Analytical methods for the direct estimation of the two parameters are available (Kushner, 1976) for such data sets, but in many cases a graphical solution may be satisfactory. The functional form of the distribution is given in *Table 1.2*.

   The usual graphical solution makes use of the cumulative frequency distribution, which is the integral of the frequency distribution. This may be plotted as percentile points against concentration class, a percentile point being the fraction (expressed as a percentage) of the data with concentration less than a given value. The percentile axis is scaled in normal probability units and the concentration axis is logarithmic. Rounding errors can be removed by using the upper bound of the concentration

Table 1.2   FUNCTIONAL FORMS OF FREQUENCY DISTRIBUTIONS

Log-normal (2-parameter)
$$\phi(x) = \frac{1}{x\sigma\sqrt{2\pi}}\exp\left\{-\tfrac{1}{2}\left[\frac{\ln x - \mu}{\sigma}\right]^2\right\}$$

parameters to be fitted : σ, μ

Log-normal (3-parameter)
$$\phi(x) = \frac{1}{\sigma(x-\epsilon)\sqrt{2\pi}}\exp\left\{-\tfrac{1}{2}\left[\frac{(\ln(x-\epsilon)-\mu)}{\sigma}\right]^2\right\}$$

parameters to be fitted : σ, μ, ϵ

Log-normal (4-parameter)
(S_B distribution)
$$\phi(Z) = \frac{1}{\sigma\sqrt{2\pi}}\exp\left\{-\tfrac{1}{2}\left[\frac{(Z-\mu)}{\sigma}\right]^2\right\}$$

where $Z = \gamma + \eta\ln[(x-\epsilon)/(\epsilon+\lambda-x)]$
parameters to be fitted : $\gamma, \eta, \epsilon, \lambda$

classes, which also removes 'zero' values. The uncertainty associated with small values can be realized by showing error limits on the plot. A log-normal distribution is then represented by a straight line (*Figure 1.9*). The median (50th percentile) concentration then gives the geometric mean, and the mean of the log distribution. The standard deviation of the log distribution is obtained from the slope of the line, for example by

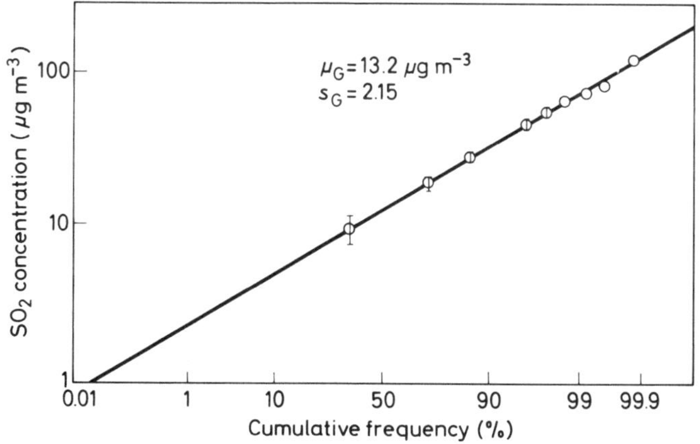

Figure 1.9   Log-probability plot of data from *Figure 1.8*

taking the difference in logarithms of concentrations at the 90th percentile (the point which gives the concentration exceeded 10% of the time) and the 50th percentile, and dividing by the standard normal variate $Z = 1.28$. The values obtained in this way for the data of *Figure 1.9* give

$$\mu_G = 13.2\ \mu g\,m^{-3}$$

$$s_G = 2.15$$

The arithmetic mean may be calculated directly (Equation 1.4 below) or may be estimated from the geometric mean and standard deviation. Assuming a perfectly log-normal distribution

$$\bar{x} = \frac{2\,\mu_G\,\exp[(\ln s_G)^2/2]}{\sqrt{\pi}} \tag{1.3}$$

$$where\ \bar{x} = \ \frac{1}{n}\sum_{i=1}^{n} x_i \tag{1.4}$$

THE LOG-NORMAL DISTRIBUTION (THREE-PARAMETER)

In some cases a plot of the data distribution on log probability graph paper shows a marked deviation from linearity, particularly at low concentrations. An expedient method for obtaining a straight-line plot is to increase or decrease all the concentration values by a constant (small) amount. Although this appears to be an arbitrary approach to fitting a simple statistical model, there may be a sound physical basis for the procedure. For example, if sulphur dioxide concentration is estimated from the acidity of a 'trap' solution, then any other gas contributing to acidity will lead to an overestimate, and ammonia will lead to an underestimate of the amount of sulphur dioxide present. Alternatively, a three-parameter model as described here may be justified from first principles. Ott and Mage (1976) have considered the physical and chemical processes involved in the dispersion of a pollutant gas from source, including homogenous and heterogenous chemical reactions. It is assumed that the various processes are independently linked to meteorological factors such as wind speed and direction, and are thereby strongly correlated with each other. For physical reality in the model the distribution is truncated (censored) at zero concentration. The functional form is given in *Table 1.2*. This three-parameter model, determined by the geometric mean and standard deviation and additive constant, has been used with considerable success in describing pollutant-gas concentrations.

THE LOG-NORMAL DISTRIBUTION (FOUR-PARAMETER)

Although a three-parameter model may be justified on theoretical grounds, the problem of selecting an appropriate model may be treated in terms of a distribution which is required to satisfy reality. If the model is constrained to have an upper and lower limit to concentration which is physically meaningful, and a log-normal distribution between these limits, then the functional form of the frequency distribution is that described by Johnson as an S_B distribution (Johnson, 1949) and given in *Table 1.2*. The lower limit, as determined by fitting the model to the data, then gives a value for the 'background' concentration, while the dependence of the fit is usually insensitive to the value of the upper limit. However, the problem of fixing the four parameters from a given set of data is not simple. Mage

(1980b) has described an algebraic method for calculating the four parameters from the data, using four equally spaced percentile points. These four percentile points need not be placed symmetrically about the median, although in theory this gives the best estimate of the four parameters. In practice, a large proportion of the data may be at small concentrations where the measurements are relatively inaccurate, and it may be advantageous to use only the data at the 'top end' of the distribution. This will be particularly advantageous if it is the distribution of large concentrations that is required for prediction of extreme values (*see next section*). The four parameters obtained using this method will not be unique, but the distribution obtained and predictions made from it are generally not too sensitive to the choice of percentile points. This is probably the best 'simple' distribution based on the whole data set for prediction of extreme values.

Extreme-value statistics

In many cases it is found that the small fraction of data at the largest concentrations does not fit the model distribution very well. This may be because the physical processes responsible for these concentrations differ from those which control the rest of the data. If one wishes to predict the occurrence of large concentrations, it may be more appropriate to base the prediction on the distribution of the extreme values only, rather than on the data as a whole. Such methods have been used extensively in hydrology to predict flood levels. The *Flood Studies Report* (NERC, 1975) describes a number of possible models for the analysis of extreme values, of which two will be discussed here and are fully described by Gumbel (1959).

The data set of extreme values from which predictions are to be made must contain only independent values. It is, therefore, not possible to consider the maximum 10-minute average per hour as a suitable extreme value, as concentration levels are likely to be strongly correlated from one hour to the next. The shortest practicable period for pollutant data is 12 hours but, even so, correlations may occur if a maximum value overlaps the boundary between two time periods. The longer the periods of observation, the more likely it is that extreme values within those periods will be independent.

Once the set of extreme values has been obtained, it may be examined using a graphical technique. The frequency distribution of extreme values is assumed to have the form $\phi(x)$, where

$$\phi(x) = \exp\left\{ -e^{-\alpha(x-\mu)} \right\} \tag{1.5}$$

where x is the concentration

   $\phi(x)$ is the frequency of occurrence

   α and μ are two parameters to be fitted

Concentration values are plotted on the vertical axis of a graph which has the horizontal axis in units of y, where

$$y = \alpha(x-\mu) \tag{1.6}$$

The corresponding probability function

$$\phi(y) = \exp\{-e^{-y}\} \tag{1.7}$$

then allows calculation of the return period $T(x)$ for any concentration x as

$$T(x) = [1 - \phi(y)]^{-1} \tag{1.8}$$

and return periods may also be noted on the y axis.

The y values must now be assigned to the observed extreme values, and this is achieved by placing the N observations (x) in order of increasing magnitude and assigning a rank (m) to the ith value. The y value corresponding to x_i is then

$$y_i = -\ln\{-\ln[m/(N+1)]\} \tag{1.9}$$

If the plotted points fall on, or close to, a straight line, then predictions may be made based on the probability function given by Equation 1.5, where the parameters α and μ may be estimated from the graph, or by an analytical method such as least-squares fitting of a straight line to the data. The expected return period for a given concentration (x) may then be calculated from Equations 1.5 and 1.8. Gumbel (1959) also gives expressions which allow the calculation of 'control curves'. These define the limits between which the predictions are likely to be valid.

If the data when plotted do not follow a straight line, and if there is a pronounced curvature (concave upwards) then it may still be possible to make predictions by using a suitable transformation of the original data. This allows the graphical technique to be used for asymptotic probability functions of the form

$$\phi(x) = \exp\left[-\left(\frac{\mu-E}{x-E}\right)^{\alpha}\right] \tag{1.10}$$

where E is a positive lower limit to concentration values.

Applying the transformation

$$x^1 = \ln(x-E) \tag{1.11}$$

to the data will allow the transformed data to be plotted against y as a straight line if the probability function defined in Equation 1.10 describes the distribution of the extreme values.

As an example, an extreme-value analysis may be applied to the data of *Figures 1.8* and *1.9*, taking the maximum daily value in consecutive 10-day periods as the extreme values. The Gumbel plot is shown in *Figure 1.10* and is nonlinear, suggesting a transformation of the form given in Equation 1.11. When transformed ($E = 0$) and replotted, (*Figure 1.11*), the data form a straight line with parameters $\mu = 26.2$, $\alpha = 2.10$. Substituting for μ, α and E, Equation 1.10 and Equation 1.8 allow the

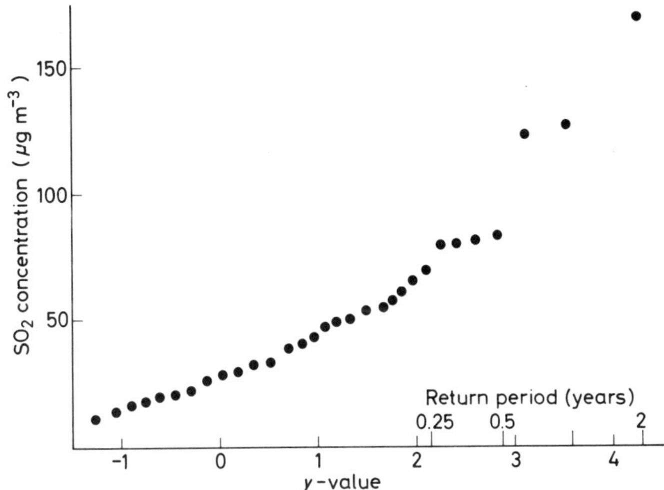

Figure 1.10   Extreme values of daily average concentration of sulphur dioxide (highest in consecutive 10-day periods). Units on horizontal axis are values of y calculated from equation (1.9). Data as for *Figure 1.8*

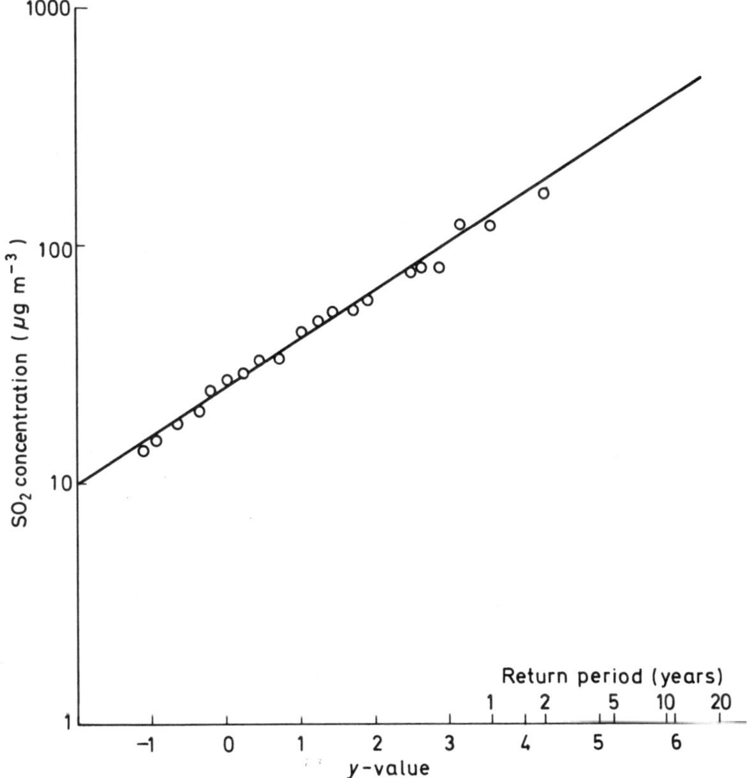

Figure 1.11   As *Figure 1.10* but with concentration values plotted on a logarithmic scale, following equation (1.11)

Table 1.3 PREDICTION OF HIGHEST DAILY AVERAGE
CONCENTRATIONS OF SO$_2$ AT BUSH BASED ON DATA FOR
1977 AND 1978 (μg m^{-3})

	Return period (years)			
	1	2	5	10
Observed (1977/1978)	84/169	–	–	–
Observed (1977–1978)	–	169	–	–
Predicted: log-normal (*Figure 1.9*)	135	160	210	240
Predicted: extreme values (*Figure 1.11*)	144	200	310	440

prediction of the highest daily value to be expected in a given period. The
results of the predictions are given in *Table 1.3*, and compared with
predictions based on the log-normal distribution of *Figure 1.9*. The marked
differences emphasize the need to choose an appropriate statistical model
before attempting to predict extreme values, but as both methods seem to
be well supported by the same set of data, the use of a single method to
predict the occurrence of large concentrations may not be very reliable.
This is of particular relevance for the calculation of air-quality criteria
standards.

Classification of sites

If data from a large number of sites are considered, a striking pattern in
SO$_2$ distribution emerges. The daily mean value (geometric or arithmetic)
may vary over an order of magnitude, but the geometric standard deviation
is approximately constant for all sites, whether urban or rural. In order to
remove the bias of interference at small concentrations, the concentration
ratio of the 99th and 50th percentiles was used to estimate the geometric
standard deviation. The results are shown in *Table 1.4*. It is clear from the

Table 1.4 GEOMETRIC STANDARD DEVIATION FOR
SULPHUR DIOXIDE DISTRIBUTIONS

Region	s_G	*No. of sites*
Canada residential 1978*	2.2 ± 0.4	14
UK rural 1978–79†	1.9 ± 0.2	31
Canada industrial and commercial 1977*	1.9 ± 0.4	27
UK industrial (C1 and C2) 1978–79†	1.9 ± 0.2	50

*H.H. Neumann, personal communication
†Warren Spring Laboratory (1979)

error bars that there is some variation between sites, as might be expected,
but what is interesting is the relative constancy of the geometric standard
deviation at rural and urban sites. This means, in effect, that average
annual values may be used to estimate the whole distribution over a wide
area. Taking s_G for sulphur dioxide as 2.0 allows the calculation of the
geometric mean from an arithmetic mean (Equation 1.3) $\mu_G \simeq 0.7\bar{x}$, and
estimation of the whole distribution. On this basis it is possible to estimate
the fraction of time that certain values are expected to be exceeded for the
regions described in the first part of this chapter (*Table 1.5*). Note that

Table 1.5 ESTIMATES OF TIME (%) THAT SO₂ CONCENTRATIONS WILL BE EXCEEDED

| *Annual arithmetic mean* | *% time during which concentrations exceed:* | | | |
(μg m^{-3})	50 μg m^{-3}	100 μg m^{-3}	200 μg m^{-3}	300 μg m^{-3}
< 30	< 11	< 1.3	< 0.06	< 0.01
30–50	11–31	1.3–7.0	0.06–0.62	0.01–0.10
50–100	31–69	7.0–30	0.62–7.0	0.1–1.8
> 100	> 69	> 30	> 7	> 1.8

these estimates give an annual average and do not indicate how the large values are distributed in time. In order to estimate this distribution we would need information on the average duration of large concentration 'events', and this is likely to be strongly site-dependent. There are less data available for NO_2, and as the distributions are less well fitted by a simple log-normal distribution the estimates obtained by using such a distribution will be less reliable. Canadian data for 1978 (H.H. Neumann, personal communication) have been used to obtain a geometric standard deviation of 1.5 ± 0.2 for 17 residential sites and 1.4 ± 0.1 for 27 industrial and commercial sites. Again, the values do not appear to be strongly site-dependent, but what is of interest is that the standard deviation is smaller than that for SO_2. This may in part be a reflection of the lack of annual

Table 1.6 ESTIMATES OF TIME (%) THAT NO₂ CONCENTRATIONS WILL BE EXCEEDED

| *Annual arithmetic mean* | *% time during which concentrations exceed:* | | | |
(μg m^{-3})	50 μg m^{-3}	100 μg m^{-3}	150 μg m^{-3}	200 μg m^{-3}
< 25	< 0.5	< 0.01	–	–
25–40	0.5–12	0.01–0.05	< 0.01	–
40–80	12–80	0.05–11	< 0.01–0.8	0.05
> 80	> 80	> 11	> 0.8	> 0.05

cycle in NO_2 concentration at Canadian residential sites and the consequent small variations in average concentration from season to season, or it may indicate that there is a relatively constant background concentration on which the anthropogenic variations are superimposed. A longer residence time for NO_2 in the troposphere would also lead to the suppression of large fluctuations in concentration and a reduced standard deviation. Estimates of exposure time for NO_2 based on a value of $s_G = 1.4$ are given in *Table 1.6* for different annual mean values. The geometric mean is now $\mu_G \simeq 0.84\bar{x}$ (Equation 1.3).

Pollutant dose

The purpose of this section is to attempt to construct a useable bridge between the average concentrations in the first section, the daily and frequency distribution aspects of variations in concentration, and effects of specific exposures to pollutants that will be described in subsequent chapters. Although this subject has received consideration by a number of workers (*see* Runeckles, 1974; Heck and Brandt, 1977) there seems little prospect of devising a unit of dose that incorporates all factors that influence plant sensitivity. The list of such factors is so long that the unit produced would be too unwieldy to be workable, and in many situations too many variables would remain unmeasured. To make progress, one way of considering dose is to remove as many of the effect-related (i.e. biological) factors as possible so that a simple unit that may be evaluated for given experimental or field conditions is obtained.

If we consider the medical usage of dose for ionizing radiation we find an array of different units (roentgen, rad, rem, dose equivalent rate) are used, but the rad (radiation absorbed dose, $J\,kg^{-1}$) is the basic unit of absorbed dose. Following Runeckles (1974), air-pollutant 'dosages' may be best described as the amount of pollutant absorbed. Taking this process one stage further, the pollutant absorbed dose (pad) may be obtained as the product of concentration, time and stomatal (or canopy) conductance for the gas in question. This quantity, in addition to providing an estimate of the amount of gas absorbed by the plant through stomata, also provides a dose in units that may readily be appreciated, i.e. $g\,m^{-2}$ (the area may be ground area or leaf area), in contrast to the conventional and much-criticized unit of dose, the product of concentration and time (units $g\,m^{-3}\,s$).

Taking as an example the agricultural areas exposed to annual arithmetic mean SO_2 concentrations of $30\,\mu g\,m^{-3}$ we may determine the pollutant absorbed dose for a given period. In the example provided (*Table 1.7*), a

Table 1.7 AN EXAMPLE OF POLLUTANT ABSORBED DOSAGE CALCULATED FOR A CEREAL CROP DURING MAY AND JUNE

Assumptions:
  8-hour growing day
  $5\,mm\,s^{-1}$ canopy conductance for SO_2 (stomatal uptake only)
  $30\,\mu g\,m^{-3}$ annual mean SO_2 concentration, which corrected for annual and diurnal variations corresponds to $26\,\mu g\,m^{-3}$ as the growing-day average SO_2 concentration

Pollutant absorbed dose
  PAD = mean concentration × exposure time × canopy conductance
  (units = $\mu g\,m^{-3}$ s $m\,s^{-1}$)

For one 8-hour day the dose would be $3.7\,mg\,SO_2\,m^{-2}$ ground area (per unit leaf area estimates could be obtained simply by dividing the result by leaf area index). For a 60-day growing period the dose would be $225\,mg\,SO_2\,m^{-2}$ ground area.

---

single 8-hour growing day, and a 60-day period of the growing season for a cereal crop are considered, and the dose obtained may be expressed as the quantity absorbed per unit ground or leaf area. This technique may be used for studying short- or long-term effects for which gas monitoring data are available and for which the dose is required to enable comparisons to be made between experiments.

One of the problems of such a simple concept of dose is that the same dose may in different conditions lead to different effects, for example because of the time over which the dose is received. The very convenient

Table 1.8   ESTIMATE OF NUMBER OF DAYLIGHT HOURS PER DAY DURING THE GROWING SEASON THAT SO_2 CONCENTRATIONS EXCEED A GIVEN VALUE

Annual arithmetic mean	*Hours during which concentrations exceed:*			
$(\mu g\ m^{-3})$	$50\ \mu g\ m^{-3}$	$100\ \mu g\ m^{-3}$	$200\ \mu g\ m^{-3}$	$300\ \mu g\ m^{-3}$
< 30	< 0.6	< 0.06	< 0.01	–
30–50	0.6–1.8	0.06–0.36	0.01–0.09	< 0.01
50–100	1.8–5.0	0.36–1.9	0.09–0.78	0.01–0.09
> 100	> 5.0	> 1.9	> 0.78	> 0.09

Assumes 8 hours of daylight over 6-month period, summer concentrations = 0.8 × annual mean, daytime concentrations = 1.1 × daily mean and log-normal distribution with $s_G = 2$.

finding that the geometric standard deviation of SO_2 concentration in rural areas is fairly constant (at about 2.0) enables the prediction of the duration of exposure to concentrations within defined thresholds from the monitoring data in the first section of this chapter (*Table 1.8*). This enables estimates to be made of dose during specific large concentrations in field conditions, which may then be related to the short-term effects of a similar dose over a similar period in a controlled environment. Predictions of dry matter lost, using such work, may then be compared with the results from a long-term integrating type of experiment where a crop is grown in open-top chambers or in a field fumigation experiment.

Acknowledgements

The authors wish to acknowledge the many helpful comments of ITE colleagues, in particular those of Mr I.A. Nicholson, and from Dr C.E. Jeffree, University of Edinburgh and Mr A Martin, CEGB. This work was supported financially by the UK Department of the Environment, the EEC and NERC.

References

ASHMORE, M.R., BELL, J.N.B. and REILY, C.L. (1980). *Environmental Pollution* (B), **1**, 195–216

BAILEY, D.L.R., BARRETT, D.F. and COOPER, I.C. (1978). *Warren Spring Report WSL 270 (AP)*. Stevenage, UK

BENCALA, K.E. and SEINFELD, J.H. (1976). *Atmospheric Environment*, **10**, 941–950

BENGTSON. C., SKÄRBY, L. and GRENNFELT, P. (1980). In *Proceedings of International Conference on the Ecological Impact of Acid Precipitation*, pp. 154–155 (Drabløs, D. and Tollan, A., Eds). SNSF Project, Ås, Norway

DERWENT, R.G., EGGLETON, A.E.J., WILLIAMS, M.L. and BELL, C.A. (1978). *Atmospheric Environment*, **12**, 2173–2177

EASTMAN, J.E. and STEDMAN, D.H. (1980). *Atmospheric Environment*, **14**, 731–732

FISHMAN, J. and CRUTZEN, P.J. (1978). *Nature,* **274,** 855–858

GARLAND, J.A. and DERWENT, R.G. (1979). *Quarterly Journal of the Royal Meteorological Society,* **105,** 169–183

GUMBEL, E.J. (1959). *Journal of the Institute of Water Engineers,* **12,** 157–184

HECK, W.W. and BRANDT, C.S. (1977). In *Air Pollution,* Vol II, Ch. 4, pp. 157–229. (Stern, A.C., Ed.). Academic Press, New York

HOV, O., ISAKSEN, I.S.A. and HESSTVEDT, E. (1978). *Atmospheric Environment,* **12,** 2469–2479

JOHNSON, N.L. (1949). *Biometrika,* **36,** 149–176

KLEPPER, L.A. (1979). *Atmospheric Environment,* **13,** 537–542; 1475

KUSHNER, E.J. (1976). *Atmospheric Environment,* **10,** 975–980

LARSEN, R.I. (1973). *Journal of the Air Pollution Control Association,* **23,** 933–940

LARSEN, R.I. (1974). *Journal of the Air Pollution Control Association,* **24,** 551–558

MAGE, D.T. (1980a). *Atmospheric Environment,* **14,** 367–374

MAGE, D.T. (1980b). *Technometrics* **22,** 247–251

MARTIN, A. (1980). *Environmental Pollution* (B) **1**(3), 177–193

MARTIN, A. and BARBER, F.R. (1981). *Atmospheric Environment,* **15,** 567 –578

NATIONAL ACADEMY OF SCIENCES (NAS) (1978). *Sulfur Oxides.* NAS, Washington, DC

NATURAL ENVIRONMENTAL RESEARCH COUNCIL (NERC) (1975). *Flood Studies Report, Vol. 1. Hydrological Studies*

NICHOLSON, I.A., FOWLER, D., PATERSON, I.S., CAPE, J.N. and KINNAIRD, J.W. (1980). In *Proceedings of International Conference on the Ecological Impact of Acid Precipitation,* pp. 144–145 (Drabløs, D. and Tollan, A., Eds). SNSF Project, Ås, Norway

NELSON, D.W. and BREMNER, J.M. (1970). *Soil Biology and Biochemistry,* **2,** 203–215

OTT, W.R. and MAGE, D.T. (1976). *Computers & Operations Research,* **3,** 209–216

RUNECKLES, V.C. (1974). *Environmental Conservation,* **1,** 305–308

SINGH, H.B., LUDWIG, F.L. and JOHNSON, W.B. (1978). *Atmospheric Environment,* **12,** 2185–2196

SÖDERLUND, R. and SVENSSON, B.H. (1976). *SCOPE Report 7, Ecological Bulletin 22, Swedish Natural Science Research Council (Stockholm)* 23–73

TINGEY, D.T. (1979). *Atmospheric Environment,* **13,** 1475

WARREN SPRING LABORATORY (1979). *National Survey on Smoke and Sulphur Dioxide, April 1978–March 1979*

2

BIOLOGICAL INDICATORS OF AIR POLLUTION

A.C. POSTHUMUS
Research Institute for Plant Protection, Wageningen, The Netherlands

Introduction

Air pollution in relation to vegetation, including agricultural and horticultural crops, may be characterized in different ways. First of all, the
concentrations of different pollutants in the air may be measured and their
possible effects may be estimated if the exposure–effect relationships in the
prevailing conditions are known. But a second, more direct approach is the
measurement of the actual effects of the ambient air pollutants on the
plants in question.

For the first approach, several physical and chemical methods have been
developed already, and very sophisticated techniques with semi-automatic
or fully automatic apparatus are available. However, exposure–effect
relationships for many plant species and varieties with differing degrees of
sensitivity to the different air pollutants are seldom available or are
inadequate, and the influence of the external and internal environmental
conditions of the plants is generally poorly understood. For these reasons it
is easier in the present circumstances to study and measure the air-
pollution effects themselves on the naturally growing plants or cultivated
crop plants, and it is even better to expose standardized indicator and/or
accumulator plants in well-known conditions.

Monitoring, meaning continuous measurement, of effects of air pollution on vegetation may be very important, for example to warn of possible
damage to natural or cultivated plants and to raise the alarm concerning
possible risks to men, animals and materials from specific air pollutants to
which plants are (much) more sensitive. In addition, continuous effect
measurements with the same indicator plants over areas of reasonable
extent may help to study the presence and distribution of air pollutants in
time and space, to check air-pollution abatement measures and to analyse
air-pollution trends. It should be clear, however, that the emphasis of the
studies is on the occurrence and intensity of effects and that these studies
do not replace measurements of the concentrations of air pollutants.

In this chapter, examples of monitoring of air-pollution effects on plants
will be discussed and it will be stressed that both concentrations and effects
of air pollutants should be monitored simultaneously, as they complement
each other. Measurements of concentrations are reasonably accurate and
objective and may be helpful in predicting and explaining the occurrence of

effects on plants, but they cannot replace measurements of effects, which integrate the influences of intrinsic plant properties, exposure times, concentrations of pollutants and external and internal conditions of the plants.

As an example of a national programme, the national monitoring network for air-pollution effects on plants in the Netherlands will be reviewed. Existing international cooperation will be mentioned, and the need for truly international air-pollution effect monitoring networks will be stressed.

Monitoring of biological effects of air pollutants by the use of plants as indicators and/or accumulators

Because of the relatively high sensitivity of plants in general to several air pollutants, and the occasional fairly specific symptoms accompanying the effects of different air pollutants on plants, certain plant species may be used as indicators for the detection, recognition and monitoring of air-pollution effects. When the plants accumulate the polluting compounds without changing their chemical nature by metabolism, and the pollutants are easily analysed in samples of plant material, such plants may also be used as accumulators. If the accumulation of air pollutants by plants is also considered to be an effect of the atmospheric pollution, plants are very suitable for detecting, recognizing and monitoring air-pollution effects.

Several types of air-pollution effects are known and these may be divided into acute effects of exposures to high concentrations over short periods and chronic effects of exposures to low concentrations over long periods. Examples of acute effects are clearly visible chlorosis and necrosis of leaf tissue; leaf, flower or fruit abscission; and epinastic curvatures of leaves and leaf stems. Chronic effects may appear as retardation or disturbance of normal growth and development (resulting in reduction of growth, yield or quality of agricultural and horticultural crop plants), or slow discoloration (chlorosis), leaf-tip necrosis and, ultimately, total die-back of plant organs may be caused. In some cases the symptoms of acute and chronic effects may be fairly specific for a particular air pollutant or for a combination of different pollutants.

Many different plant species may be useful as indicators and/or accumulators of air pollutants by showing special symptoms or effects. For example, species of lichens (mostly epiphytic), mosses, ferns and higher vascular plants (mostly phanerogames) have been used for this purpose. Biological-effect monitoring may be performed by using the natural vegetation and the crop plants present in the area studied, but differences in soil, water and other (for example climatic) conditions may influence the effects and diminish the comparability of results between sites. It is therefore better to use selected indicator and/or accumulator plants, cultivated in conditions of soil and watering that are standardized as far as possible.

Higher plants have been mainly used for this purpose, for example the tobacco variety Bel W_3 in the Netherlands (Floor and Posthumus, 1977) and in the United Kingdom (Ashmore, Bell and Reily, 1978), although transplanted lichens have been used in the Ruhr area of the Federal Republic of Germany (Schönbeck, 1969; Schönbeck *et al*; 1970). Several species and varieties of natural and cultivated plants, which have been shown to be sensitive to one or more air pollutants, are in current use in effect-monitoring networks. *Table 2.1* shows an example of a set of indicator and accumulator plants of differing degrees of sensitivity, based on the experience of the national monitoring network for air-pollution effects on plants in the Netherlands (Posthumus, 1976).

Table 2.1 REVIEW OF INDICATOR AND ACCUMULATOR PLANTS USED FOR EFFECT MEASUREMENTS OF DIFFERENT AIR POLLUTANTS IN THE NETHERLANDS

Plant species and cultivar	Air pollutant	Symptoms/effects

Gladiolus (*Gladiolus gandavensis* L.)
    cvs Snow Princess and
    Flowersong
Tulip
(*Tulipa gesneriana* L.)
    cvs Blue Parrot and
    Preludium
	Hydrogen fluoride (HF)	Leaf-tip and marginal necrosis (Snow Princess and Blue Parrot) and fluoride concentrations in dry matter (Flowersong and Preludium)

Tobacco
(*Nicotiana tabacum* L.)
    cv. Bel W$_3$
Spinach
(*Spinacia oleracea* L.)
    cvs Subito and Dynamo
	Ozone (O$_3$)	Leaf upper-surface speckle necrosis

Small nettle (*Urtica urens* L.) Annual meadow grass (*Poa annua* L.)	Peroxyacetyl nitrate (PAN)	Leaf under-surface band forming necrosis
Lucerne (*Medicago sativa* L.) cv. Du Puits Buckwheat (*Fagopyrum esculentum* Mönch)	Sulphur dioxide (SO$_2$)	Interveinal chlorosis and necrosis
Petunia (*Petunia nyctaginiflora* Juss.) cv. White Joy	Ethylene (C$_2$H$_4$)	Flower-bud abortion, small flowers
Italian ryegrass (*Lolium multiflorum* Lam.) cv. Optima	Fluoride and metal ions (F, Cd, Mn, Pb, Zn)	Ion concentrations in dry matter

Indicator plants in more-or-less standardized conditions in the open air may be used to detect, recognize and measure the acute effects of exposures to air pollutants and to study the presence, intensity and distribution of these effects in place and time. Chronic effects may be determined by comparing growth, development, production and quality of indicator plants in pairs of small greenhouses, one of which is ventilated with charcoal-filtered air and one with unfiltered ambient air. Growth retardations and yield reductions (fresh- or dry-matter production) have been found in the unfiltered compared with the charcoal-filtered air. In this way, possible crop losses in polluted areas may be indicated, although the effects of air pollution on crop yield in practice are often quite different because of the different internal and external environments of the plants. The use of open-top chambers (Heagle, Body and Heck, 1973; Mandl *et al.*, 1973) instead of closed greenhouses is an improvement, but still different from the open-air situation. Plants grown in the open air are much more resistant to air pollution than plants cultivated in greenhouses. As a consequence, greenhouse crops are generally more vulnerable than crops in the field, if exposed to outdoor concentrations of pollutants.

Accumulator plants may also be cultivated in a standardized way at several sites to study the presence and distribution of air pollution. In particular, the nonconvertible, persistent air pollutants such as fluoride and heavy metals may be measured relatively easily by physical/chemical analysis. This method is applied in the Federal Republic of Germany using standardized grass cultures (Scholl, 1974), and has produced a considerable amount of information (Prinz and Scholl, 1975). It must be stressed, however, that the accumulation of the pollutants in the grass is strongly dependent on the growth of the plants and on the climatic conditions. In order to make comparisons between samples from different sites or over various periods, it is therefore necessary to know the internal and external conditions of the plants. Other plant species, for example mosses, have also been used as accumulators, but with the same problems. Only tree bark may be used for accumulation of pollutants without the problem of interference by rapid growth of the material (Lötschert and Köhm, 1978).

For a better understanding of biological-effect monitoring research with plants, some definitions and descriptions may be useful.

Indicator plants are plants which may show clear symptoms of effects, indicating the possible presence of some pollutant(s). These symptoms may be fairly specific and lead to the qualitative determination of a pollutant, but usually they do not provide a definite identification, and the presence of a specific pollutant must be proved by other methods. The indicator plants serve to detect and identify the effects of the pollutants, but these effects may also be measured quantitatively to monitor the effect intensities. Permanent measurements of effect intensities may be used for surveys of air quality in relation to plants. The influences of the internal and external conditions of the plants are already included in the effects.

Accumulator plants are plants which readily accumulate specific air-polluting compounds. These compounds may be analysed in the plant material after some time by physical/chemical methods to identify the pollutants and to obtain a quantitative measurement of the pollution burden (total amount of pollutants accumulated over a specific period). As the uptake of pollutants by the plants (determined in washed samples of plant material) is regarded as an effect of the air pollution, this may provide a method for air-pollution effect monitoring by accumulator plants. The total pollution burden may also be monitored by measuring the pollutant concentrations in unwashed samples of plant material.

Sometimes the same plant species may act as both indicator and accumulator for a special pollutant, for example tulips and gladioli for hydrogen fluoride. These species are also examples of plants reacting to the pollutant (HF) with both acute and chronic responses, depending on the concentration and the exposure period. In the long run, enough fluoride may accumulate to cause symptoms of acute response (leaf-tip and marginal necrosis) and, in addition, reductions in bulb or tuber yield may result (Spierings and Wolting, 1971).

Standardization of indicator and accumulator plants for air-borne pollution is a very important prerequisite to eliminate unnecessary variation in the effects studied. As the effects of air pollutants on plants are (apart from the influences of the nature and concentration of the pollutant and of the exposure time) dependent on plant species and variety, developmental and physiological stage of the plants and physical environmental conditions, it is quite understandable that the selection of the plant material and the growth conditions are very important. Selected uniform seeds or other plant material which is as genetically homogeneous as possible should be used (for example, clones). Growth conditions should be optimal and identical at all monitoring sites to be compared. In principle, this is possible only when totally artificial growth cabinets are used for the cultivation and exposure of the plants so that only differences in the quality of the air passing through the cabinets will influence the plants. In practice, this procedure has not been used in air-pollution effect monitoring, because of its high costs. The use of plants grown in soil at different sites (with different soils and different preceding crops) has been practised frequently, but is not to be encouraged. Standardization, as far as possible, of the soil and water conditions of the plants, is advisable and costs relatively little. Differences in climatic factors cannot be excluded in this way, but these characteristics can be measured on the spot to be included in the comparison of the effects. Possible influences or effects of biotic and abiotic pathogens should be excluded as far as possible by proper treatments, or should be known exactly.

Monitoring of biological effects of air pollutants by the use of plants as indicators and/or accumulators has been applied on a local, regional or national scale. On a local scale, indicator and accumulator plants have been used to survey the effects of air pollution from a single source or cluster of sources on special horticultural, agricultural or forestry crops, or on natural vegetation (van Raay, 1969). The same indicators and accumulators may be used to sustain possible claims for compensation of economic losses. On a larger, regional or even national scale, the results of biological-effect monitoring networks have been used to estimate the distribution of these effects in place and time (Floor and Posthumus, 1977). This may lead to conclusions about differences in the extent of air-pollution effects between different regions of a country. In the long run it is also possible to study trends in incidences of air-pollution effects, for example the occurrence of maximal values of ozone-effect intensities in summer. On an international scale it would be possible to compare the air-pollution effects burden between adjacent regions of different countries and eventually to trace sources of pollutants, if necessary across national frontiers. It could be a worthwhile source of information to have an extensive biological-effects monitoring network spread over Western European countries to show the distribution, according to place and time, of specific effects of air pollution on plants.

In every case, the selection of plants and conditions should be adapted to the specific aim of the study. When the pollutant to be studied is known, selection of specific sensitive indicator plants is the first step towards adequate results. In cases of unknown pollutants, a series of plant species, sensitive to different compounds, should be used simultaneously. For comparison of the effects of air pollutants on plants at all locations of a regional monitoring network, it would be preferable to use identical indicator and accumulator plants in standardized conditions. In all cases it would be advisable to expose more tolerant or insensitive varieties of the indicator plant species next to the sensitive ones, to distinguish possible effects of other biotic or abiotic factors such as pathogens (viruses, bacteria and fungi), insect pests, frost and possible nutritional deficiencies.

A major problem in biological-effects monitoring with plants is the possibility of interactions when more than one pollutant is present (which is often the case). Two or more different air-polluting compounds may act on the plants additively, synergistically or antagonistically (*see* Ormrod, this volume, Chapter 15). This means that the intensity of the effect of a combination of air pollutants may be respectively equal to, higher than, or lower than the sum of the effect intensities of the pollutants when these are applied at the same concentrations and conditions, but separately. Moreover, combination effects may be different for mixtures of the same pollutants at different exposure times, concentration levels and ratios of the components. Combined effects of mixtures of pollutants are therefore very difficult to interpret when no information about the nature and concentration of the pollutants is available. The ideal indicator plant, for this reason, should be a plant sensitive to only one air-polluting compound. As this is unlikely for all the different pollutants, it is worth while to seek indicator plants, the sensitivity of which is as specific as possible to one component of the total complex of air pollution. It is, nevertheless, advisable to expose several indicator plants with different sensitivities at the same place and time in order to discern the effects of different airborne pollutants.

Example of a national monitoring network for air-pollution effects

Under the auspices of the Directorate-General for Environmental Hygiene of the Netherlands Ministry of Public Health and Environmental Hygiene, the Research Institute for Plant Protection at Wageningen has determined, since 1973, the effects of several air pollutants on indicator plants, initially in one part of the country. By 1976 the effects-monitoring network was extended to 40 experimental fields, spread all over the country with a concentration in the area west of Rotterdam (*Figure 2.1*). This effects-monitoring network is operated in close cooperation with the nationwide automated monitoring network for air-pollution concentrations, directed by the National Institute of Public Health at Bilthoven. Results for the years from 1976 to 1979 have been published in Dutch in reports of the Bilthoven Institute (Anonymous, 1976; 1978a,b,c; 1979a,b).

Biological-effect monitoring is performed only during the vegetation period of the indicator and/or accumulator plants (from April until November) at 37 stations of the national monitoring network for air-pollution concentrations and at three other locations. The methods and

Figure 2.1 Location of the experimental sites of the national monitoring network for effects of air pollution on plants in the Netherlands during 1978. The numbers refer to regional codes

materials are the same at all 40 experimental fields and these fields are visited weekly in the monitoring period, but not all on the same day. This restriction should be taken into account when comparing the results of the concentration measurements and the biological-effect measurements resulting from exposures for one week.

The basic concept of effect measurements is the exposure of sensitive indicator and/or accumulator plants in a standardized way. Use is made in the Netherlands of a special plant cultivation set, derived from a similar system developed and used in the Federal Republic of Germany (Scholl, 1969; van Haut, Scholl and van Haut, 1972). Plants are grown in standard soil in plastic containers, with ceramic filter candles in the soil for automatic watering. Three of these containers are placed in a larger container, which serves as a water store. In these standardized cultivation sets (*Figure 2.2*), all species of indicator and accumulator plants may be

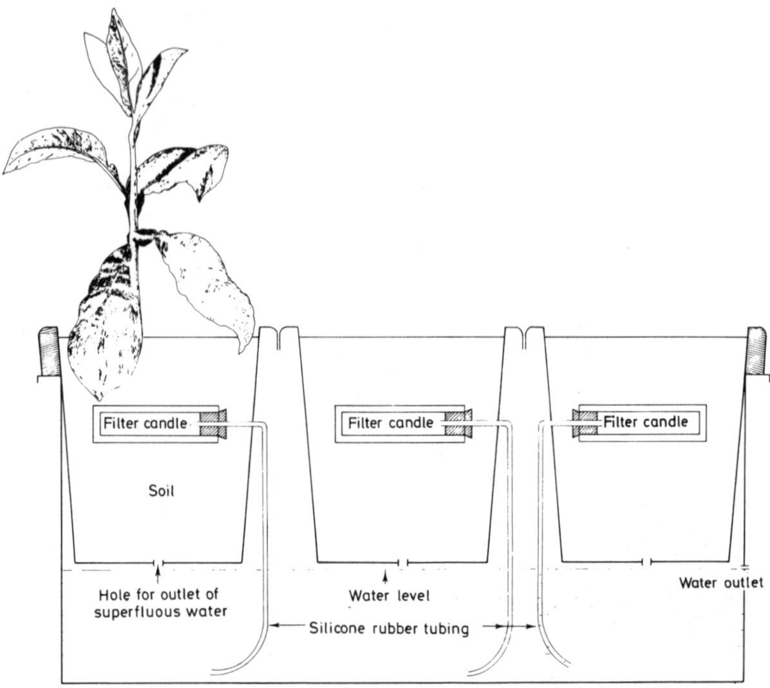

Figure 2.2   A standardized cultivation set for indicator and accumulator plants

cultivated by adapting the number of filter candles per container to the water requirement of the plants.

Both acute and chronic effects of exposures to airborne pollutants are monitored. Acute effects are measured weekly on sensitive indicator plants in cultivation sets in the open air. Chronic effects are studied and measured by comparing the growth and productivity of sensitive plants after long-term exposures in cultivation sets outdoors and, at some locations, also in pairs of small greenhouses, one of which is ventilated with charcoal-filtered air and one with unfiltered air. In addition, standardized cultures of Italian ryegrass in the open air have been used to accumulate several air-polluting compounds during 14-day periods for analyses of these compounds in the harvested plant material. Examples will be given of these three types of effect monitoring. A list of indicator and accumulator plants used in the Netherlands' national network for air-pollution effects is presented in *Table 2.1*, and could be extended with other plants (for example red clover and pea for monitoring the effects of sulphur dioxide (Posthumus, 1978)).

Acute effects of ozone on the sensitive tobacco variety Bel W_3 are determined by estimating, for all leaves of four plants per location, the percentage of leaf area damaged by the specific speckle necrosis, according to the following ratings: 0%, 0–5%, 5–10%, 10–25%, 25–50%, 50–75% and 75–100%. From these weekly assessments the ultimate mean value for all leaves is calculated for every location. Thus a distribution of the intensity of the ozone effect in place and time during the vegetation period is obtained. From these results it appears that the mean intensity of the

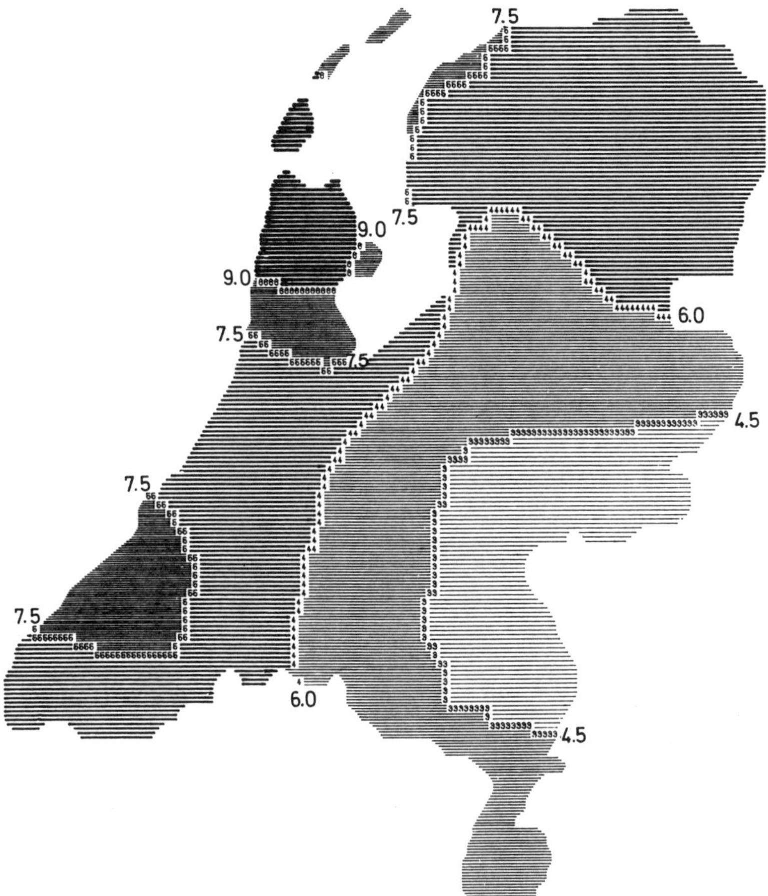

Figure 2.3   Distribution of the mean ozone-effect intensity on tobacco variety Bel W_3 (as percentage of leaf area damaged, indicated by the large numbers), measured weekly during the vegetation period from 6 June until 28 October 1977, in the Netherlands

ozone effect is higher in the western half of the Netherlands than in the eastern half (*Figure 2.3*), and that there are weeks every year, correlated with sunny weather, when this intensity is maximal (Floor and Posthumus, 1977). *Figure 2.4* shows the weekly variations of the mean leaf damage (in % leaf area) for tobacco Bel W_3 at the experimental fields of the network in the northern and southern, and in the eastern and western halves of the Netherlands for the monitoring periods from 1976 to 1978.

Acute and chronic effects of HF are studied on sensitive monocotyledonous ornamental plants such as tulips and gladioli. In the spring the tulip variety Blue Parrot and in the summer the gladiolus variety Snow Princess are grown in the cultivation sets outdoors. The necrosis of leaf tips and margins caused by hydrogen fluoride is fairly specific and may be assessed quantitatively by measuring the mean length of the necrotic tips, as an estimate of the necrotic leaf area, after exposures for weeks or months.

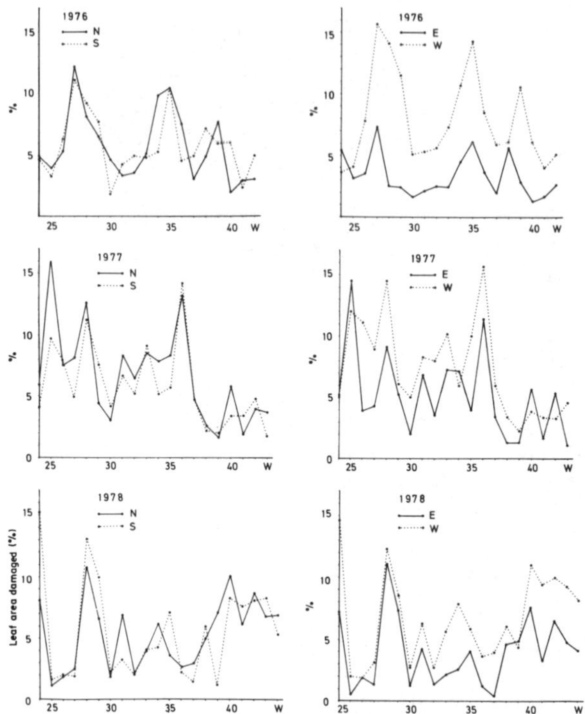

Figure 2.4   Variation of the weekly mean ozone-effect intensities on tobacco variety Bel W$_3$ (as percentage of leaf area damaged) at the experimental fields of the national monitoring network for air-pollution effects in the northern (N) and southern (S), and in the eastern (E) and western (W) halves of the Netherlands during the vegetation periods of the years from 1976 to 1978 (W is week number from the beginning of each year)

The fluoride content of leaf tips of tulips (5.0 cm long) and gladioli (7.5 cm) appeared to be very well correlated with leaf necrosis. From the fluoride-effects monitoring network it has been possible to show the distribution of fluoride pollution over the Netherlands, the south-western part of the country being most polluted (*Figure 2.5*).

Chronic effects of air pollution, expressed as growth and yield reductions of horticultural crop plants, have been studied at several sites in the industrial area west of Rotterdam. Pairs of small greenhouses, ventilated with filtered and unfiltered air, have been used to grow tomato, lettuce and other crops. Yield reductions of up to 20% in the unfiltered air, compared with the filtered air, have been found. Differences in these reductions for separate locations indicated the differences in air-pollution effect. These types of results might be used for selection of areas where the air quality is suitable for horticulture.

Physicochemical analyses of samples of standardized grass cultures for lead, cadmium, zinc and fluoride have revealed possible differences in these air-polluting compounds in different regions of the Netherlands. This may be important for the productivity and/or quality of some crops, even when no acute effects are produced on the plants.

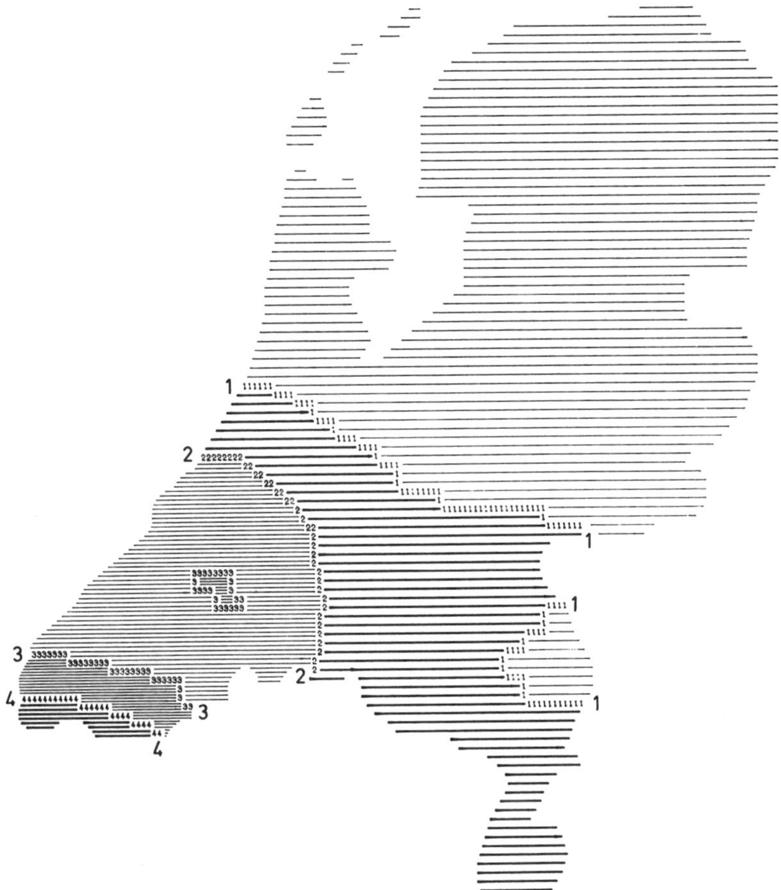

Figure 2.5   Distribution of the intensity of the fluoride effect on the sensitive gladiolus cultivar Snow Princess (in length (cm) of necrotic leaf-tip, indicated by the large numbers), at the end of the vegetation period from 6 May until 26 August 1977 in the Netherlands

In the national monitoring network for air pollution in the Netherlands it should be possible to compare results of concentration measurements and biological-effect measurements. However, up to now this has not been done intensively, for several reasons. Sulphur dioxide concentrations have been measured at more than 200 sites very frequently, but the concentrations have been rather low during the vegetation period, and no visible effects of SO_2 have been measured on the indicator plants, except during a short air-pollution episode in 1978. Clear-cut effects of hydrogen fluoride have been found and measured quantitatively, but HF is not measured in the automated chemical/physical measuring network. For ozone, very good evidence for effects on indicator plants has been obtained, but O_3 concentration measurements in the network up to 1979 were sparse and unreliable. For 1979 a fairly good qualitative correlation was found between the intensities of the ozone effect on tobacco Bel W_3 and the peak O_3 concentration values (*Figure 2.6*). In addition, the spatial distribution

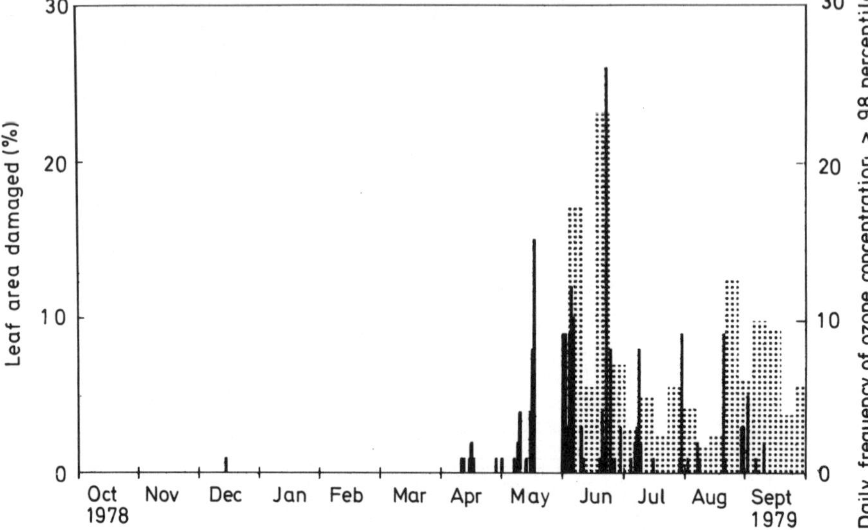

Figure 2.6   Average weekly ozone-effect intensities on tobacco variety Bel W_3 (as percentage of leaf area damaged, ▩) and total daily frequencies of ozone peak concentration values (> 98-percentile values, ■) in 1979 within the national monitoring network for air pollution in the Netherlands

of the mean values of effect intensities and peak concentrations of O_3 during the summer of 1979 were well correlated (*Figure 2.7*). In general, no more quantitative correlations between concentration and effect measurements have been established so far. This will undoubtedly be the aim of future research, but other factors such as climate will also have to be taken into account.

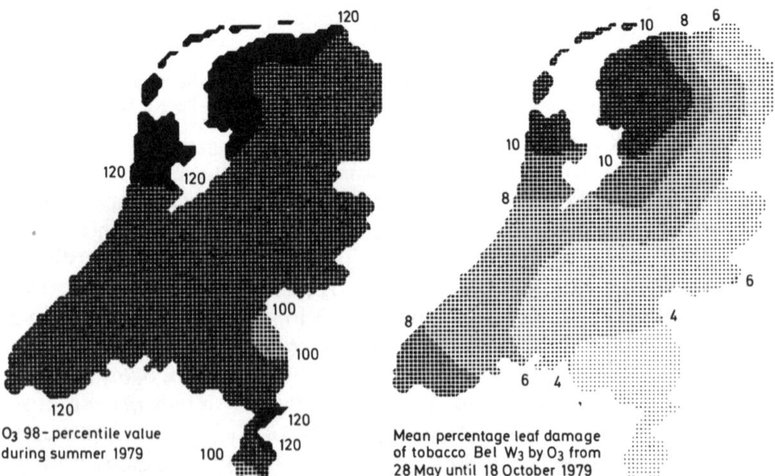

Figure 2.7   Distribution of the 98-percentile daily ozone concentration values (in $\mu g/m^3$) and the mean ozone-effect intensities on tobacco variety Bel W_3 (as percentage of leaf area damaged), measured weekly during the vegetation period from 28 May until 18 October 1979 in the Netherlands

Possible international programmes for monitoring networks for air-pollution effects on plants

There has been some international cooperation in Europe to monitor air-pollution effects on plants. Methods, materials and results of this type of biological-effect monitoring have been exchanged between researchers, for example between the United Kingdom, the Federal Republic of Germany, Switzerland, Belgium, Denmark, Sweden, Finland, Poland and the Netherlands. In addition to such forms of cooperation, the establishment of international (for example European) programmes for the monitoring of effects of airborne pollution on vegetation should be promoted. The need for this type of cooperation is quite clear, when transfrontier air pollution accounts for a great deal of the total pollution burden, as is the case in the Netherlands.

Real international monitoring networks set up and maintained under the auspices of some international authority or organization would be highly advantageous to the participating countries. The aim should be to achieve comparable results by using standardized methods and materials. For the purpose of comparison of the effects of air pollution on plants at all points of an international air-quality surveillance network, it is perhaps necessary to use identical indicator and accumulator plants at all sites. To this end, standardized methods for cultivation, exposure, handling and observation of selected plants must be developed, as has already been done for the European Economic Community (Posthumus, 1980). The personnel involved should also receive adequate relevant training.

Differences in climate between the countries participating in such international programmes pose problems relating to the selection of uniform collections of indicator and accumulator plants. It should be borne in mind, however, that the effects measured after a period of exposure of the plants are the integrated results of the influences of air pollutants and other (for example climatic) factors. As all these factors also act on natural vegetation, effects on indicator plants may help to predict possible general risks. These risks may be compared for the plants in different regions or countries, and threatened zones may be delimited.

Discussion and conclusions

The first step towards controlling airborne pollutants is the monitoring of their concentrations and effects. Without data on concentrations and effects, there is no possibility of studying the occurrence and distribution of air pollutants and their effects in place and time, of setting standards for the protection of man and his environment, and of checking the results of abatement measures. We need an alarm system in order to prevent the occurrence of excessive concentrations, resulting in hazardous effects on men, animals, plants and materials.

For these purposes both physicochemical concentration measurements and biological-effect measurements with plants are needed. The concentrations measured by monitors will not be sufficient to predict all possible effects of air pollutants, and indicator plants will never be able to give information about the identity and concentration of polluting agents in the air. There is no question of replacing ambient-air monitoring by physico-chemical methods with effect monitoring with plants: both should be used jointly. Immission concentration measurements (i.e. concentration at the 'living' level), effect-intensity measurements on plants (including physicochemical leaf analysis), and measurements of meteorological para-meters may produce the total picture of the pollution situation.

Plants may be used as indicators and accumulators of air pollutants for detection, recognition and monitoring purposes. An important problem remains: how representative are the effects on the indicator and accumula-tor plants for the vegetation as a whole and for the separate species of natural and crop plants? The comparability of effects on indicator and accumulator plants with effects on other plants is not very well known, and much information about the exposure–effect relationships for many plant species under all possible conditions is still inadequate, or even nonexis-tent. The occurrence of special effects of combinations of different air-polluting compounds is also a great problem. Additive, synergistic, and antagonistic combination effects have to be taken into account as far as possible (for example synergistic effects of combinations of O_3 and SO_2, or of SO_2 and NO_2). In the meantime the specificity of indicator plants for particular air-polluting compounds has become a great problem. Some-times, although there may not be any really specific sensitivity to any one particular pollutant, it may be possible to discriminate between different pollution situations by using sets of indicator plants with different sensitivi-ties.

When plants are used for the monitoring of air-pollution effects, a high degree of standardization of the plant material and of physical and chemical environmental conditions is a prerequisite. In international monitoring networks, geographical and climatic differences should be taken into account, but it seems to be better to use uniform material initially and to accept an integrated effect of air pollutants and environ-mental conditions, than to use different and uncomparable plant material adapted to different local conditions. The risk of effects of air pollutants on vegetation is largely determined by environmental factors, so these should not be excluded from effect-monitoring research.

In the belief that there are many problems and uncertainties in the use of plants as biological indicators and accumulators of air pollution, it is worth while summarizing the advantages of this approach:

1. It provides a direct method of studying the effects of the prevailing air pollution on living organisms.
2. It provides a measure of the integrated effects of all environmental factors, including air pollutants and weather conditions.
3. It is possible to study the relationships between concentrations and effects when both are measured at the same sites.
4. It provides possibilities of determining spatial and temporal trends in the occurrence and intensity of effects of several air pollutants on natural and cultivated plants.
5. It sometimes enables the analysis of polluting compounds by measur-ing accumulation within plants.

6. It acts as a sensitive early-warning system which may stimulate prophylactic measures to prevent or diminish disastrous effects of air pollution.

Acknowledgements

The author thanks Mr H. Floor of the Research Institute for Plant Protection at Wageningen and Mr N.D. van Egmond of the National Institute of Public Health at Bilthoven for kindly providing illustrative material and for their very helpful cooperation.

References

ANONYMOUS (1976). *Nationaal Meetnet voor Luchtverontreiniging, Rijksinstituut voor de Volksgezondheid, Bilthoven, Rapport 212 LMO*
ANONYMOUS (1978a). *Nationaal Meetnet voor Luchtverontreiniging, Rijksinstituut voor de Volksgezondheid, Bilthoven, Rapport 4/78 LMO*
ANONYMOUS (1978b). *Nationaal Meetnet voor Luchtverontreininging, Rijksinstituut voor de Volksgezondheid, Bilthoven, Rapport 105/78*
ANONYMOUS (1978c). *Nationaal Meetnet voor Luchtverontreiniging, Rijksinstituut voor de Volksgezondheid, Bilthoven, Rapport nr. 241/78*
ANONYMOUS (1979a). *Nationaal Meetnet voor Luchtverontreiniging, Rijksinstituut voor de Volksgezondheid, Bilthoven, Rapport nr. 114/79*
ANONYMOUS (1979b). *Nationaal Meetnet voor Luchtverontreiniging, Rijksinstituut voor de Volksgezondheid, Bilthoven, Rapport nr. 236/79*
ASHMORE, M.R., BELL, J.N.B. and REILY, C.L. (1978). *Nature*, **276**, 813–815
FLOOR, H. and POSTHUMUS, A.C. (1977). *VDI-Berichte*, **270**, 183–190
HEAGLE, A.S., BODY, D.E. and HECK, W.W. (1973). *Journal of Environmental Quality*, **2**, 365–368
LÖTSCHERT, W. and KÖHM, H.J. (1978). *Oecologia*, **37**, 121–132
MANDL, R.H., WEINSTEIN, L.H., McCUNE, D.C. and KEVENY, M. (1973). *Journal of Environmental Quality*, **2**, 371–376
POSTHUMUS, A.C. (1976). In *Proceedings of the Kuopio Meeting on Plant Damages Caused by Air Pollution*, pp.115–120 (Kärenlampi, L., Ed.). University of Kuopio, Finland
POSTHUMUS, A.C. (1978). *VDI–Berichte*, **314**, 225–230
POSTHUMUS, A.C. (1980). *Report EUR 66422 EN*. Commission of the European Communities, Brussels
PRINZ, B. and SCHOLL, G. (1975). *Schriftenreihe der Landesanstalt für Immissionsschutz des Landes Nordrhein Westfalen, Essen*, **36**, 62–86
SCHOLL, G. (1969). *Zeitschrift für Pflanzenernahrung ünd Bodenkunde*, **124**, 126–129
SCHOLL, G. (1974). *Staub-Reinhaltung der Luft*, **34**, 89–92
SCHÖNBECK, H. (1969). *Staub-Reinhaltung der Luft*, **29**, 14–18
SCHÖNBECK, H., BUCK, M., VAN HAUT, H. and SCHOLL, G. (1970). *VDI–Berichte*, **149**, 225–236
SPIERINGS, F.H. and WOLTING, H.G. (1971). *VDI–Berichte*, **164**, 19–21

VAN HAUT, H., SCHOLL, G. and VAN HAUT, G. (1972). *Landwirtschaftliche Forschung*, **25**, 42–47
VAN RAAY, A. (1969). In *Proceedings of the 1st European Congress on the Influence of Air Pollution on Plants and Animals, Wageningen, 1968*, pp.319–328. PUDOC, Wageningen

3

EXPOSURE TO GASEOUS POLLUTANTS AND UPTAKE BY PLANTS

M.H. UNSWORTH
Department of Physiology and Environmental Science, University of Nottingham School of Agriculture

Introduction

To understand the responses of plants to air pollution, and to compare results from different experiments, it is necessary to define the exposure that the plants receive. But how should we define that exposure? And what other features of the plant's environment should be measured? These questions are not confined to air-pollution research; they apply equally to other research fields where stresses are applied and responses elicited. To avoid raising the reader's hopes I should make it clear that this paper does not give unique answers to such questions, but it is an attempt to show how plants are linked to their atmospheric environment and how design criteria for exposure systems may be established.

In air-pollution research, the aims of experimentation with plants fall broadly into three categories:

1. To investigate physiological or biochemical mechanisms of response.
2. To study the action of pollutants on growth and development.
3. To quantify the impact of pollutants on crop productivity.

The first two categories often require controlled environments; the third ideally allows crops to grow normally in the field, with only air-pollution concentration being modified. Categories (1) and (2) are well established and many examples may be found in this book. Category (3) has received surprisingly little attention from experimentalists until recently, especially in Britain, although there have long been speculators willing to extrapolate from laboratory to the field!

Exposure to pollution is usually defined in terms of the ambient concentration or the 'ambient dose', the product of concentration and time. This concept of dose has much to commend it from the viewpoint of air-quality legislation but is probably the main factor responsible for confusion in the literature when comparing plant responses in different experiments. Just as it would be uncharitable to assume that everyone in a bar was equally inebriated because all were surrounded by an equal number of bottles, so is it futile to attempt to relate responses of plants to air pollution merely to the concentration of pollutant surrounding them. The important intermediate step of relating response to uptake (or perhaps following the alcoholic analogy it should be 'intake') has been increasingly taken in recent years, as exemplified by several contributions to this book. Ultimately such studies will lead to clearer understanding of ambient dose-response relationships.

In this chapter I will briefly review various designs of exposure systems for air-pollution research, and will show how systems may be analysed to predict pollutant uptake, carbon dioxide exchange and heat balance of plants.

Designs of exposure systems

This review is not intended to be exhaustive, but to give examples of types of systems commonly used. Heck, Krupa and Linzon (1979) give more detail of several systems.

CUVETTES

Cuvettes designed to contain single leaves or parts of intact plants are commonly used in studying CO_2 exchange and water relations. Šesták, Čatský and Jarvis (1971) give detailed designs. Examples of cuvettes for studying responses to pollutants and pollutant uptake have been given by Srivastava, Jolliffe and Runeckles (1975), Taylor and Tingey (1979), Winner and Mooney (1980), Taylor, McLaughlin and Shriner (page 458) and Bengtson, Grennfelt and Skarby (page 461 this volume). For pollutant applications, care is required to select materials for construction that minimize absorption and adsorption; teflon is often used. One difficulty with cuvettes is the possibility that responses when only part of a plant is exposed to a pollutant may differ from those when the pollutant surrounds the whole plant.

CHAMBERS FOR SEVERAL PLANTS

This type of system is the most common. Early versions were described in detail by Thomas and Hill (1935) and by Setterstrom and Zimmerman (1938). Designs developed by Heck, Dunning and Johnson (1968) have been adapted and used by many workers (e.g. Lockyer, Cowling and Jones, 1976). Use of materials with low uptake of pollutants is important. Recent studies have also shown the importance of adequate air movement (Ashenden and Mansfield, 1977; Unsworth and Mansfield, 1980) and this has stimulated development of wind-tunnel exposure systems (Horsman and Wellburn, 1977; Horsman, Roberts and Bradshaw, 1979) and continuously stirred tank reactors (Rogers *et al.*, 1977; Heck, Philbeck and Dunning, 1978). It is becoming clear that, even when there is generally good air movement in chambers, the relations between plant community structure, air flow and response are complex. Many years ago it was recognized that water use by isolated plants could be considerably enhanced when dry air blew across them—the 'clothes-line' effect. In a similar manner, plants exposed to pollutants in chambers may experience enhanced uptake in comparison with typical field crops, because in chambers there is more air movement through the canopy and because concentration gradients around the plants are smaller than in the field.

OPEN-TOP CHAMBERS

Because the environment of closed chambers often differs radically from that in the field, chambers without tops were developed to allow modification of the gaseous environment with minimal interference with normal growth (Heagle, Body and Heck, 1973; Mandl *et al.*, 1973). Open-top chambers are most effective when used in pairs, filtering air to one chamber and comparing plant growth with that in a second chamber ventilated with ambient air (e.g. Thompson, Kats and Cameron, 1976; Buckenham, Parry and Whittingham, this volume page 479). Alternatively, pollutants may be injected into previously filtered air to obtain a range of treatments (Heagle *et al.*, 1979). Open-top chambers work least effectively when strong winds produce turbulent incursion of ambient air into 'filtered' chambers; this may limit their use for studying responses to SO_2 which may occur in high concentrations in windy weather. There is disagreement over whether crop growth and yields are similar in chambers ventilated with unfiltered air and in the field. Heagle *et al.* (1979) found good agreement for spinach (cv. Winter Bloomsdale) but substantial differences in growth and development rates were found by Olszyk, Tibbits and Hertzberg (1980) for alfalfa and by Buckenham, Parry and Whittingham (pages 479–480) for barley.

FIELD EXPOSURE SYSTEMS

To attempt further to reduce differences between the field environment and a pollutant treatment, several groups have developed fumigation systems sometimes described as topless or bottomless or both! Shinn, Clegg and Stuart (1977) developed a linear gradient system where lines of polyethylene tubing with holes in the walls were used to distribute polluted air along rows of a crop. By changing the frequency of holes with distance from the source, a linear gradient of pollution concentration was established. Such systems seem to rely on a well-established prevailing wind direction. A variant of this idea is the exposure system described by Greenwood *et al.* (page 452) in which SO_2 was released from line sources forming sides of a 30 m square. By controlling sources in terms of wind direction, a central region of the square can be fumigated continuously at a preset concentration.

More complex zonal air-pollution systems (ZAPS) (Lee, Preston and Lewis, 1978) for exposing large areas of fruit trees, natural grass or crops, use networks of pipes over the area to be treated (de Cormis, Bonte and Tisne, 1975; Heitschmidt, Lauenroth and Dodd, 1978; Muller, Miller and Sprugel, 1979).

Modifications of the open-top idea for treating smaller areas have been described by Roberts *et al.* (1979), who used short open-top chambers to study growth of grasses, and Runeckles *et al.* (1978) who described a bottomless 'downdraft field chamber' in an attempt to get realistic air movement around plants.

Analysis of pollutant uptake by leaves and canopies

LEAVES

The analysis of pollutant uptake (flux) by leaves and vegetated surfaces has followed two main paths (a third approach has recently been suggested and will be discussed later). The approach most commonly used in atmospheric modelling derives from Chamberlain, who developed the concept of a deposition velocity v_g first for particles then for gases, radioactive and otherwise (*for reviews see* Chamberlain, 1975; 1980). Deposition velocity v_g (m s^{-1}) is the ratio

(flux density to a surface (μg m^{-2} s^{-1}))/(airborne concentration at a reference height (μg m^{-3}))

The alternative approach treats flux density as analogous to electrical current and, following Ohm's Law, expresses

$$\text{Flux density } (\mu\text{g m}^{-2}\,\text{s}^{-1}) = \frac{\text{Concentration difference } (\mu\text{g m}^{-3})}{\text{Resistance } (\text{s m}^{-1})}$$

When the concentration difference is that between a reference height and a sink where concentration is zero it will be seen that

Deposition velocity = 1/resistance

The resistance analogy has the advantage that the total resistance may be partitioned into component resistances corresponding to paths through boundary layers, stomata, intercellular spaces, and to chemical sinks (Bennett, Hill and Gates, 1973; Black and Unsworth, 1979; Unsworth, 1981), and thus the influence of physiological, biochemical and physical factors on flux may be identified. Some authors prefer to quote component conductances

(conductance = 1/resistance)

e.g. Tingey and Taylor (this volume, Chapter 6), but the formulae for combining conductances are a little more awkward than those for resistance.

Rogers *et al.* (1977) proposed an analysis of pollutant fluxes to vegetation using a chemical engineering analogy to derive rate constants. Their principal rate constant for flux density to a leaf can be shown to be v_g/V where v_g is the deposition velocity and V is the chamber volume. Inherently an analysis which yields a 'constant' depending on chamber dimensions seems undesirable.

CANOPIES

Monteith (1963, 1965) extended the resistance analogies for single leaves to apply to analysis of evaporation and CO_2 exchange from canopies of field crops. Fowler and Unsworth (1974, 1979) used the same approach to analyse SO_2 fluxes to wheat, and similar analyses for ozone have been reported by Wesely *et al.* (1978), and Leuning *et al.* (1979). Unsworth (1981) has reviewed the subject. For a canopy where the main sinks are in leaves and the resistance to uptake by a single leaf is r_p, the 'canopy resistance' r_c is approximately r_p/L where L is the leaf area index.

Analysis of the environment of exposure chambers

The most common objective in designing an exposure chamber is to create a controlled environment in which pollution concentrations may be maintained. Questions which should be asked, but usually are not, are:

1. Will plants grow in this chamber? Will they have enough light, carbon dioxide, nutrients?
2. How hot will the plants get? Will leaf temperatures influence development, respiration, enzyme activity?
3. At what rate will plants transpire? Could there be water stress, leading to restrictions on cell expansion or stomatal closure?
4. At what rate will pollutant be absorbed by the plants? How will plant responses influence pollution concentration in the chamber?
5. Can a planned concentration be maintained?

In illustrating how physical principles may be used to define the atmospheric environments of exposure chambers, I will suggest answers to some of these questions, using published details of chambers.

MASS BALANCE OF ENCLOSED CHAMBERS

Figure 3.1 shows the mass balance of a typical exposure chamber. Pollutant enters the chamber at flow rate f $(m^3 s^{-1})$ and concentration C_i $(\mu g\,m^{-3})$ giving a mass flux $Q_i = fC_i$ $(\mu g\,s^{-1})$. Within the chamber there are fluxes Q_w $(\mu g\,s^{-1})$ to the walls, area A_w (m^2) and Q_p $(\mu g\,s^{-1})$ to the plants, leaf area A_p (m^2). It is convenient to express Q_p and Q_w as products of flux densities q_p, q_w $(\mu g\,m^{-2}\,s^{-1})$ respectively and appropriate areas, i.e.

$$Q_w = q_w A_w \quad \text{and} \quad Q_p = q_p A_p$$

Within the chamber, uptake by walls (and ceiling) and plants reduces the pollutant concentration to C_c $(\mu g\,m^{-3})$. The flux leaving the chamber is

Q_o $(\mu g\,s^{-1})$, at a flow rate f, i.e. $Q_o = fC_c$

The mass balance of the chamber is

$$Q_i = Q_w + Q_p + Q_o \tag{3.1}$$

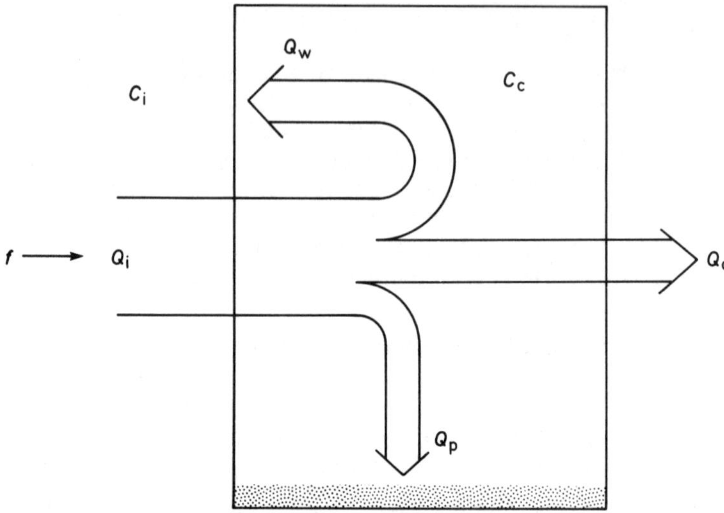

Figure 3.1   The mass balance of an exposure chamber. Air entering at flow rate f and concentration C_i gives an input flux Q_i which is balanced by fluxes Q_w and Q_p to walls and plants respectively and by outlet flux Q_o at concentration C_c

which may be expanded and rearranged to give

$$f(C_i - C_c) = q_w A_w + q_p A_p \tag{3.2}$$

Alternatively the flow rate f may be written

$$f = NV$$

where N is the rate of air changes (s^{-1}) and V is the chamber volume (m^3), the product of base area A_b and height h.

Substituting $f = NA_b h$ in Equation 3.2 and dividing,

$$Nh\,(C_i - C_c) = q_w \frac{A_w}{A_b} + q_p \frac{A_p}{A_b} \tag{3.3}$$

The ratio A_p/A_b is leaf area per unit base area, which is leaf area index L. Similarly A_w/A_b may be defined as a wall area index W, hence

$$Nh\,(C_i - C_c) = q_w\,W + q_p\,L \tag{3.4}$$

The flux density on the left side of Equation 3.4 can be expressed as a concentration difference divided by a ventilation resistance r_v, thus defining

$$r_v = (Nh)^{-1} = A_b/f \tag{3.5}$$

Similarly the flux densities q_w and q_p may be written

$$q_w = (C_c - 0)/r_w \text{ and } q_p = (C_o - 0)/r_p \tag{3.6}$$

where r_w and r_p are the total resistances controlling uptake to walls and leaves respectively from chamber concentration C_c to sinks where concentration is zero. Both resistances are the sum of boundary layer resistances r_{bw}, r_{bp} depending on the shape and structure of the surface and on air movement in the chamber, and surface resistances r_{sw}, r_{sp} which, for leaves, may have stomatal, cuticular and internal components. Details and typical values were given by Black and Unsworth (1979) and Unsworth (1981). The resistance analogue of Equation 3.4 therefore is

$$\frac{C_i - C_c}{r_v} = \frac{C_c\,W}{r_{bw} + r_{sw}} + \frac{C_c\,L}{r_{bp} + r_{sp}} \tag{3.7}$$

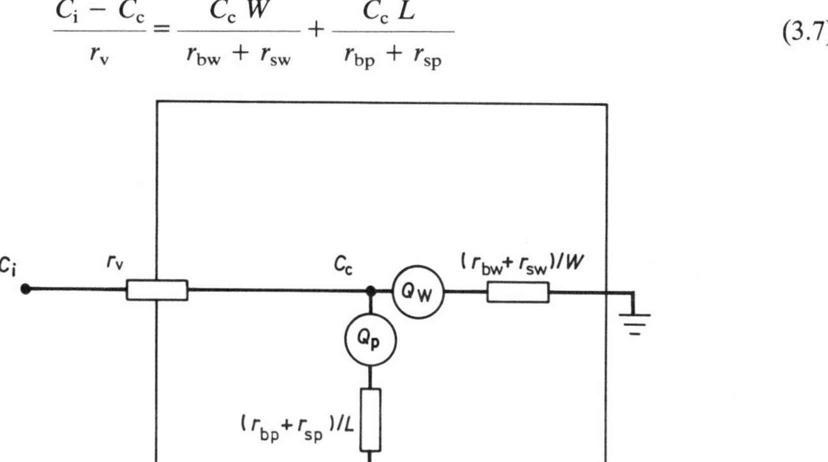

Figure 3.2   Resistance analogue corresponding to *Figure 3.1*. C_i and C_c are the gas concentrations at the inlet and in the chamber, suffix b refers to resistances of boundary layers and suffix s refers to surface resistances of walls (w) and plants (p) respectively. W is the 'wall area index' and L the 'leaf area index'. The ventilation resistance is r_v

Figure 3.2 illustrates Equation 3.7 in electrical terms. Three cases show the implications of this analogue: if r_v is very large (low ventilation rates) so that it dominates the uptake resistances, $C_i - C_c$ will be large, i.e. there will be substantial depletion of pollutant concentrations in the chamber; if r_v and C_i are constant, but r_{sp} varies significantly, there may be large changes in C_c as a result of plant uptake; if r_{bp} and r_{bw} are large (low air movement in the chamber), uptake by plants and walls will be restricted.

Carbon dioxide depletion

Equations 3.3–3.7 may be applied to analyse the mass balance of chambers described in the literature if adequate information is given or if informed guesses can be made. As a relevant application, carbon dioxide concentration can be estimated: it is generally recognized that CO_2 concentration in experimental chambers should differ by less than $\pm 10\%$ of the ambient value which is about 330 ppm ($= 0.60 \, \text{g m}^{-3}$ at 20°C) if physiological responses to CO_2 are to be avoided.

Thomas and Hill (1935) pioneered the use of enclosures in the field to study CO_2 exchange of crops in relation to SO_2 exposure. They covered plots of alfalfa with transparent chambers, floor area $A_b \simeq 4 \, \text{m}^2$, height $\simeq$ 1.5 m. Maximum rates of net photosynthesis per unit ground area q_b were about 6 g $CO_2 \, \text{m}^{-2} \, \text{h}^{-1}$ (1.7 mg m^{-2} s^{-1}). The flow rate f of air through a chamber could be varied from about $4 \, \text{m}^3 \, \text{min}^{-1}$ to more than $12 \, \text{m}^3 \, \text{min}^{-1}$ (70 to $200 \times 10^{-3} \, \text{m}^3 \, \text{s}^{-1}$). Since uptake of CO_2 by walls was zero, the mass balance is

$$f\,(C_i - C_c) = q_b A_b \qquad\qquad (3.8)$$

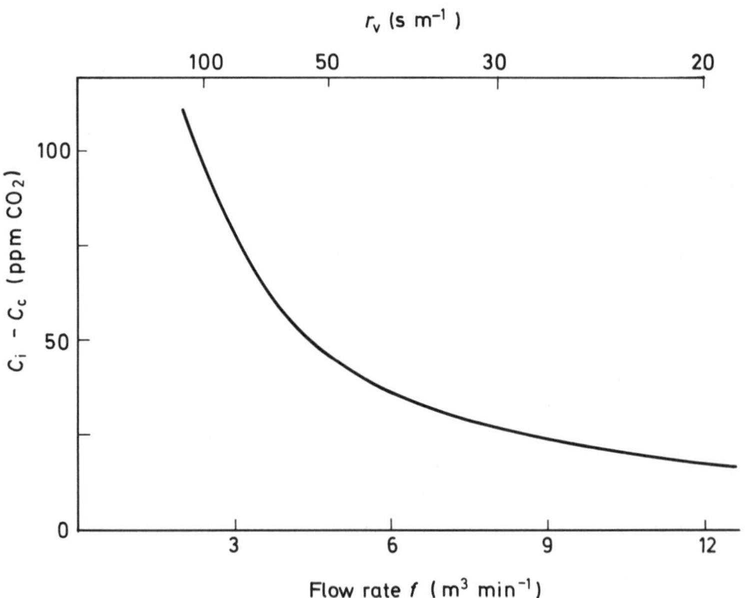

Figure 3.3   Dependence on flow rate of the CO_2 concentration gradient ($C_i - C_c$) between the inlet air and the chamber for an exposure chamber of dimensions given by Thomas and Hill (1935). Equivalent ventilation resistances r_v are calculated from Equation 3.5

Figure 3.3 shows depletion of CO_2 for crops photosynthesizing at the maximum rate as a function of ventilation rate of the chamber. Clearly, flow rates below about $8 \, \text{m}^3 \, \text{min}^{-1}$ would have led to unacceptable depletion of CO_2. Thomas recognized this, using high flow rates during the day and low flow rates at night, thus optimizing the precision of his CO_2 analysis. For comparison with the later part of this chapter, *Figure 3.3* also shows that ventilation resistances r_v (Equation 3.5) in this system range from 57 s m^{-1} at $4 \, \text{m}^3 \, \text{min}^{-1}$ to 19 s m^{-1} at $12 \, \text{m}^3 \, \text{min}^{-1}$.

A more recent example to illustrate CO_2 depletion comes from the original description of a continuously stirred tank reactor (CSTR) (Rogers *et al.*, 1977). The chamber was cylindrical, base area A_b, $28 \times 10^{-2}\,m^2$, height h, 0.71 m, volume V, $200 \times 10^{-3}\,m^3$. Flow rate f was $10\,\ell\,min^{-1}$ ($167 \times 10^{-6}\,m^3\,s^{-1}$), giving a ventilation resistance $r_v = 1690\,s\,m^{-1}$. This chamber was designed primarily to test the characteristics of the CSTR with small numbers of plants, but it is interesting to show the implications for CO_2 concentration if large areas of plant material had been exposed to pollutants in the chamber. Suppose plants had a net photosynthetic rate per unit leaf area q_p of 0.8 mg $CO_2\,m^{-2}\,s^{-1}$ (typical of light saturation) and

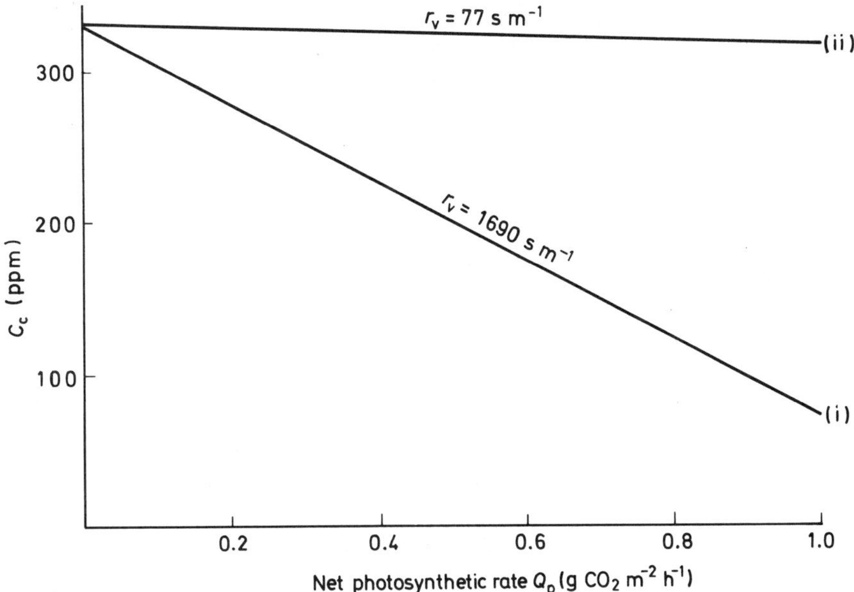

Figure 3.4 Dependence of the CO_2 concentration C_c inside two designs of continuously stirred tank reactor (CSTR) on the photosynthetic rate of plant material. (i) Design of Rogers *et al.* (1977), $r_v = 1690\,s\,m^{-1}$. (ii) Design of Heck, Philbeck and Dunning (1978), $r_v = 77\,s\,m^{-1}$

that increasing areas of leaf were added to make L vary from zero to 0.3, i.e. $Q_p = q_p L$ varied from zero to 0.24 mg $CO_2\,m^{-2}\,s^{-1}$. Alternatively this variation in Q_p could be achieved with $L = 0.3$ by altering irradiance. *Figure 3.4* shows the dependence of C_c on Q_p, calculated from the mass balance

$$(C_i - C_c)/r_v = Q_p \qquad\qquad (3.9)$$

Clearly, the chamber could accommodate very little plant material without severe depletion of CO_2. It should be stressed that the 'production-type' CSTR described by Heck, Philbeck and Dunning (1978) and used by many others has much higher ventilation rates ($100–700\,\ell\,min^{-1}$) and different dimensions, giving a minimum r_v of $77\,s\,m^{-1}$. *Figure 3.4* shows that CO_2 depletion is unlikely in this design. A further caveat is that when CO_2 is depleted, photosynthetic rates would fall: consequently, calculations leading to *Figure 3.4* represent a 'worst case' approximation.

Pollutant depletion

Concentrations of pollutants in chambers are usually monitored and controlled by sampling from within the chamber. However, in some cases the concentration C_i of pollutant in the incoming air is maintained constant, and it is useful to calculate how the concentration C_c around the plants would vary as a result of changing uptake by plants or walls. As an example the measurements of Cowling and Koziol (1978) are analysed. The chambers used for this work were described by Lockyer, Cowling and Jones (1976), but to measure photosynthetic responses to SO_2, Cowling and Koziol reduced the flow rate of air through each chamber from 700 ℓ min^{-1} to 200 ℓ min^{-1}. For the chamber dimensions quoted this increased r_v from about 30 s m^{-1} to 110 s m^{-1}. In the experiments the concentration of SO_2 in the incoming air C_i was maintained constant and was quoted as the exposure concentration. Estimates of the true concentration C_c surrounding the plants may be made from Equation 3.7 with appropriate values of resistances, and of L and W.

From the dimensions of the chambers, $W = 7.0$ and base area $= 0.37$ m^2. The leaf area is not given, and changed during and between experiments as plants grew. To illustrate principles, the leaf area index L is assumed to vary from zero to 3 (A_p varies from 0 to 1.1 m^2). A value of 250 s m^{-1} is assumed for $(r_{bp} + r_{sp})$, based on measurements of SO_2 uptake by barley leaves (Spedding, 1969). To illustrate what might happen at night when stomata close, calculations are repeated with $(r_{bp} + r_{sp}) = 2500$ s m^{-1}. Spedding also measured SO_2 uptake rates of artificial surfaces, and found that rates varied with humidity and with time, presumably as sites for absorption or adsorption became saturated. Values equivalent to $(r_{bw} + r_{sw})$ in Equation 3.7 ranged from about 100 s m^{-1} to 5000 s m^{-1}. We will therefore assume two values of $(r_{bw} + r_{sw})$: infinity (corresponding to zero uptake by walls) and 3000 s m^{-1}, a likely value at humidities employed in these experiments.

Rearranging Equation 3.7 to give the ratio of chamber concentration to inlet concentration, yields

$$\frac{C_c}{C_i} = \left[1 + \frac{r_v W}{r_{bw} + r_{sw}} + \frac{r_v L}{r_{bp} + r_{sp}} \right]^{-1} \tag{3.10}$$

Figure 3.5 shows values of C_c/C_i estimated for the experiments of Koziol and Cowling (1978) and demonstrates that, at the low flow rate employed in these experiments, concentrations of SO_2 around the plants could have been substantially lower than those measured at the inlet and quoted in the publication. Unsworth and Mansfield (1980) did similar calculations for the same chambers used at the higher flow rate (700 ℓ min^{-1}) and showed that significant depletion of SO_2 could occur when leaf areas of grasses were large. Subsequent measurements by Koziol (1980) supported these calculations. It seems likely that some of the reported experiments using this apparatus (e.g. Cowling and Lockyer, 1976, 1978) should be re-evaluated because the true exposure concentrations were lower than those quoted.

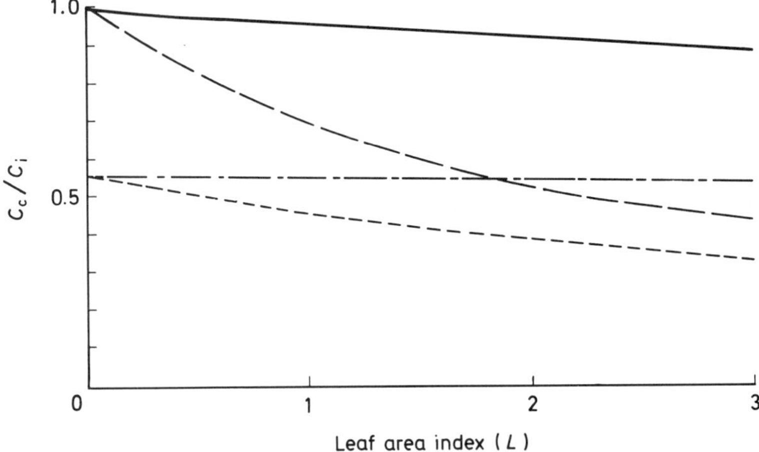

Figure 3.5   Dependence of the ratio C_c/C_i of SO_2 concentration inside and at the inlet of the exposure chamber used by Koziol and Cowling (1978) on leaf area index for several values of wall and leaf resistances

Symbol	$(r_{bw} + r_{sw})$ $s\,m^{-1}$	$(r_{bp} + r_{sp})$ $s\,m^{-1}$
———————	∞	2500
— — — —	∞	250
—·—·—	3000	2500
- - - - - - - -	3000	250

Although specific examples have been analysed here, there are probably many more cases hidden in the literature where inadequate flow rates in exposure chambers have resulted in depletion of carbon dioxide and air pollutants in the atmosphere around plants.

MASS BALANCE OF OPEN-TOP CHAMBERS

When an open-top chamber is used to expose plants there is always some incursion of polluted air in the top of the chamber. *Figure 3.6a* shows the mass balance schematically, and *Figure 3.6b* shows the appropriate resistance analogue. The mass flux Q_i entering the chamber

$$Q_i = f_1 C_i + f_2 C_e$$

where f_1 and f_2 are flow rates through the inlet and by incursion and C_i and C_e are the concentrations of pollutant in inlet air and in the external air respectively. The flux out of the chamber is

$$Q_o = (f_1 + f_2)C_c$$

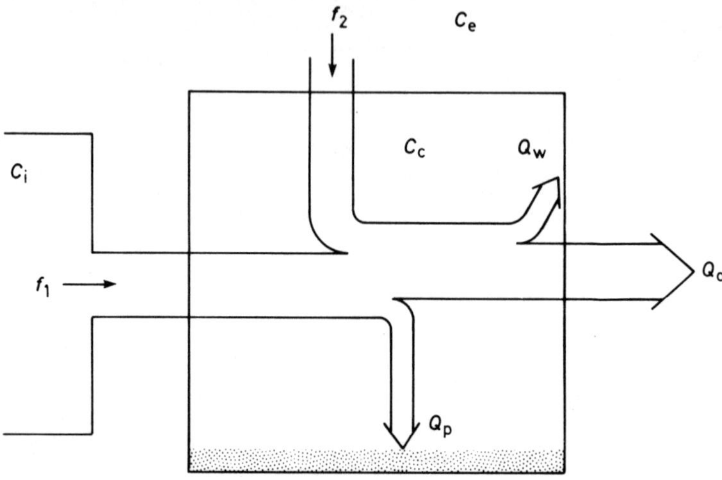

Figure 3.6(a)   The mass balance of an open-top chamber (compare with *Figure 3.1*). Air enters the chamber at rate f_1, concentration C_i, from the forced ventilation system and at rate f_2, concentration C_e, from natural incursion at the open top

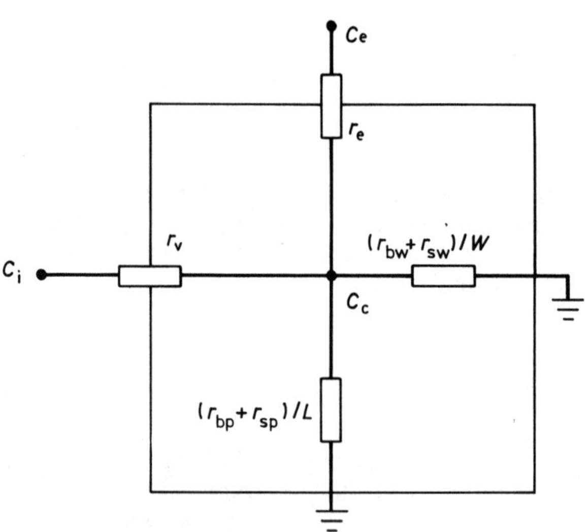

Figure 3.6(b)   Resistance analogue corresponding to *Figure 3.6(a)* (compare with *Figure 3.2*). Incursion is described by resistance r_e

Following the development of Equations 3.1–3.7, the resistance analogue may be written

$$\frac{C_i - C_c}{r_v} + \frac{C_e - C_c}{r_e} = \frac{C_c W}{r_{bw} + r_{sw}} + \frac{C_c L}{r_{bp} + r_{sp}} \tag{3.11}$$

Values of r_v may be estimated from published dimensions and flow rates. The chambers of Mandl *et al.* (1973) were cylindrical (diameter 2.2 m, height 2.4 m) and the ventilation rate was about two air changes min^{-1}, giving $r_v \simeq 13 \text{ s m}^{-1}$. Chambers described by Heagle, Body and Heck (1973), which were 3.1 m in diameter and 2.4 m high, operated at four air changes min^{-1}, giving $r_v \simeq 6 \text{ s m}^{-1}$. Note that in both cases ventilation resistances were much smaller than likely values for $(r_{bw} + r_{sw})/W$ or $(r_{bp} + r_{sp})/L$, so that C_c was likely to be relatively insensitive to plant and wall uptake.

Estimates of the incursion resistance r_e can be made from observations of C_c when the air inlet is filtered to remove pollutants. Putting $C_i = 0$ and neglecting wall and plant uptake in Equation 3.11 gives

$$r_e/r_v \simeq (C_c/(C_e - C_c))^{-1} \tag{3.12}$$

Heagle, Body and Heck (1973) published mean values of C_c and C_e (ozone) at various external wind speeds, and *Table 3.1* shows calculated values of r_e/r_v and r_e, assuming $r_v = 6 \text{ s m}^{-1}$. M.A.J. Parry (personal communication) modified open-top chambers to minimize incursion for experiments in an area polluted with SO_2 and fluoride (*see* Buckenham, Parry and Whittingham, this volume, page 479). The mean value of r_e/r_v, based on 18 days of observation of SO_2 concentrations in winter with empty chambers was 1.6 ± 0.5, and was not significantly related to wind speed, which ranged from 1 to 5 m s^{-1}.

Table 3.1 THE RATIO r_e/r_v, AND r_e AT
VARIOUS WIND SPEEDS, CALCULATED FROM
EQUATION 3.12 FROM MEASUREMENTS BY
HEAGLE, BODY AND HECK (1973). VALUES OF
r_e ARE CALCULATED ASSUMING $r_v = 6 \text{ s m}^{-1}$

Wind speed (m s^{-1})	2.3	3.6	4.9	6.3
r_e/r_v	3.0	1.9	1.0	1.3
r_e (s m^{-1})	18	11	6	8

It is often reported that pollutant concentrations in open-top chambers used without filters (e.g. as controls in filtered-versus-nonfiltered experiments) are lower than in the field. This has been attributed to 'degradation in the chamber air handling system' (Heagle *et al.*, 1979) and will be caused, in part, by differences in ozone uptake by plants inside and outside the chamber. Estimates of pollutant depletion or of CO_2 concentration in open-top chambers may be made by calculating the 'effective ventilation resistance' r_{eff} as the resultant of r_v and r_e acting in parallel,

$$r_{eff} = [r_v^{-1} + r_e^{-1}]^{-1} \tag{3.13}$$

and then applying the methods described earlier for closed chambers.

Heat balance in exposure chambers

To answer questions concerning temperatures of leaves and air, evaporation rates and humidity in chambers, it is necessary to consider the heat balance of plants in chambers and the heat balance of the chambers themselves. The principles of environmental physics concerned are well established in texts such as Monteith (1975) and Campbell (1977). As an example, we will calculate how the atmospheric environment inside an open-top chamber may differ from its surroundings, and how transpiration rates and leaf temperatures would differ.

LEAF HEAT BALANCE

Figure 3.7 shows the basic principles of heat balance. A net radiative energy flux density R_N (W m^{-2}) at the vegetated surface is balanced by the sum of a latent heat flux λE (W m^{-2}) where λ is the latent heat of vaporization of water (Jg^{-1}) and E is the evaporation rate (g m^{-2} s^{-1}), and a convective heat flux C (W m^{-2}). Minor terms associated with conduction,

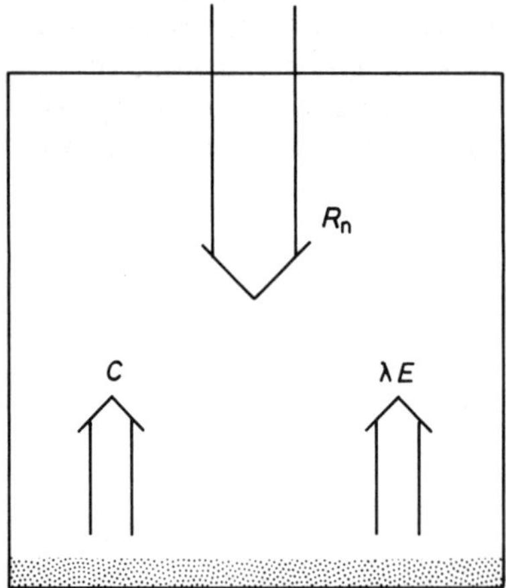

Figure 3.7   The heat balance of vegetation in an exposure chamber. The net radiative flux R_n is balanced by convective flux C and latent heat flux λE

storage and advection are ignored here. The net radiation is the difference between the income of radiation from the sun, atmosphere and walls of the chamber and the loss of radiation by reflection and emission from the vegetation. Emission, calculated as 'black-body' radiation, depends on leaf temperature T_p which is variable. It is convenient to work in terms of isothermal net radiation R_{ni} (Monteith, 1975) defined by

$$R_{ni} = R_{abs} - \sigma T_c^4$$

where R_{abs} is the absorbed radiation, T_c is air temperature (Kelvin) in the chamber and σ is Stefan's constant, 56.7 nW m^{-2} K^{-4}. Heat can be lost from leaves not only by convection but by radiation, and this may be allowed for by assuming that a radiative resistance r_{rp} acts in parallel with the boundary-layer resistance r_{bp} (Monteith, 1975) to give a resultant 'environmental' resistance to heat transfer r_{ep}. Water vapour is transferred through a resistance r_{vp}, composed of stomatal resistance r_{sp} and r_{bp} in series. Analysis of the heat balance (Monteith, 1975) enables latent and convective heat flux densities to be calculated as

$$\lambda E = \frac{1}{\Delta + \gamma^*}(\Delta R_{ni} + \frac{\rho c_p (e_s - e_c)}{r_{ep}}) \tag{3.12}$$

and

$$C = \frac{1}{\Delta + \gamma^*}(\gamma^* R_{ni} - \rho c_p \frac{(e_s - e_c)}{r_{ep}}) \tag{3.13}$$

where the meaning of symbols is given in *Table 3.2*. The Table also shows values assumed for a simple example in which the base of an open-top chamber is completely covered with a single layer of horizontal leaves

Table 3.2 MEANING OF SYMBOLS IN EQUATIONS AND IN TEXT AND ESTIMATES OF THEIR VALUES WHEN A SINGLE LEAF LAYER COVERS THE BASE OF AN OPEN-TOP CHAMBER IN BRIGHT SUNSHINE. Values based on examples and tables in Monteith (1975)

Symbol	Value	Details
R_{ni}	$640\ \mathrm{W\ m^{-2}}$	Isothermal net radiation, assuming leaf reflection coefficient $= 0.2$ and wall radiative temperature $= T_c$
T_c	$30\,°\mathrm{C}$	Air temperature in chamber
e_c	$2.00\ \mathrm{kPa}$	Water vapour pressure in chamber
e_s	$4.25\ \mathrm{kPa}$	Saturation vapour pressure at T_c
γ	$0.066\ \mathrm{kPa\ K^{-1}}$	Psychrometric constant
γ^*	$0.27\ \mathrm{kPa\ K^{-1}}$	Modified psychrometric constant, $\gamma r_{vp}/r_{ep}$
Δ	$0.244\ \mathrm{kPa\ K^{-1}}$	$\delta e_s/\delta T$
ρc_p	$1.18 \times 10^3\ \mathrm{J\ m^{-3}\ K^{-1}}$	Volumetric heat capacity of air at T_c
λ	$2.43\ \mathrm{kJ\ g^{-1}}$	Latent heat of vaporization of water
r_{bp}	$100\ \mathrm{s\ m^{-1}}$	Leaf boundary-layer resistance
r_{sp}	$100\ \mathrm{s\ m^{-1}}$	Leaf stomatal resistance
r_{vp}	$200\ \mathrm{s\ m^{-1}}$	Leaf water vapour transfer resistance $(r_{bp} + r_{sp})$
r_{rp}	$95\ \mathrm{s\ m^{-1}}$	Leaf radiative resistance, $0.5\ \rho c_p/4\sigma T_c^3$ (T_c in °K)
r_{ep}	$49\ \mathrm{s\ m^{-1}}$	Leaf 'environmental' resistance to convective transfer, $(r_{bp}^{-1} + r_{rp}^{-1})^{-1}$

(LAI = 1) exposed to bright sunshine. For the values in *Table 3.2*, Equations 3.12 and 3.13 give $\lambda E = 410\ \mathrm{W\ m^{-2}}$ ($E = 0.17\ \mathrm{g\ m^{-2}\ s^{-1}}$) and $C = 230\ \mathrm{W\ m^{-2}}$. Knowing C, leaf temperature T_p can be calculated from

$$C = \rho c_p \frac{(T_p - T_c)}{r_{ep}} \tag{3.14}$$

which gives $T_p = 39.6\,°\mathrm{C}$. This high value for leaf temperature occurs for two reasons: the isothermal net radiation is greater than would occur in the open because the walls of the chamber radiate at a much higher temperature than the effective temperature of the sky; the leaf boundary-layer resistance has been chosen as about double that commonly observed in the open, to account for the smaller amount of turbulent mixing which might occur in an enclosure. At a leaf temperature of nearly 40°C, some degree of stomatal closure would be expected and the value $r_{sp} = 100\,\mathrm{s\,m^{-1}}$ was deliberately fixed at about twice the minimum value commonly observed in the field.

There do not seem to be reports of maximum leaf temperatures in open-top chambers, but the high value calculated here suggests that such measurements, together with observations of net radiation, should be made. If large differences between leaf temperatures of plants inside and outside open-top chambers *do* occur, there could be important differences in water relations, photosynthesis and respiration.

CHAMBER HEAT BALANCE

When the leaf heat balance has been solved for a given set of conditions, values of the flux densities C and λE may be used to calculate the differences between temperature and humidity inside and outside the chambers as a function of ventilation rate. (In fact ventilation would also alter boundary-layer resistances of leaves, but this is ignored in this treatment.) The situation is analogous to Equation 3.6 where a potential difference divided by a ventilation resistance balances fluxes inside the chamber. In appropriate units the equations are

$$\rho c_p\,(T_i - T_c)/r_{eff} = -\,C \tag{3.15}$$

and

$$\rho c_p\,(e_i - e_c)/\gamma r_{eff} = -\,\lambda E \tag{3.16}$$

where e_i and T_i are the vapour pressure and temperature of the incoming external air, r_{eff} is the effective ventilation resistance defined in Equation 3.13 and other terms are given in *Table 3.2*. The negative signs in Equations 3.15 and 3.16 are because the chamber is a source of latent and convective heat rather than a sink, as with pollutants.

Figure 3.8 illustrates, for the conditions in *Table 3.2*, the resistance analogues of the heat balance for convective and latent heat of an open-top chamber of the Heagle, Body and Heck (1973) type with $r_e = 18\,\mathrm{s\,m^{-1}}$ and r_v variable. *Figure 3.9* shows the amounts by which temperature and vapour pressure in the chamber exceed the external values as a function of ventilation rate.

Maximum values of excess temperature reported in open-top chambers ventilated at two to four air changes $\mathrm{min^{-1}}$ in still, sunny weather are 1–2°C (Mandl *et al.*, 1973; Heagle *et al.*, 1979; Olszyk *et al.*, 1980), consistent with these calculations. There do not appear to be any reports of vapour pressure differences, but relative humidity has been noted to differ little between air inside and outside chambers. *Figure 3.10* shows calculations of

(a)

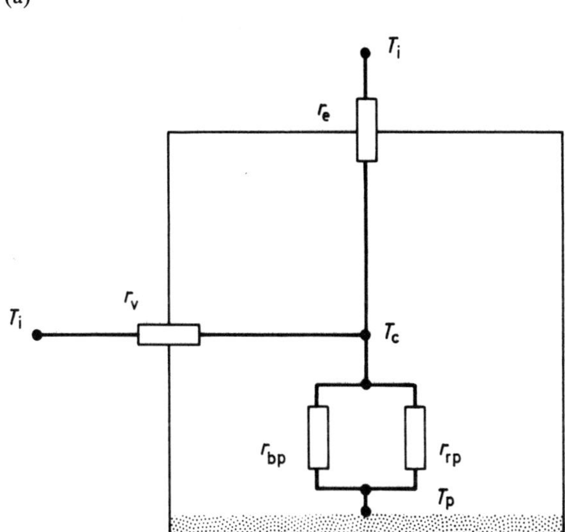

(b)

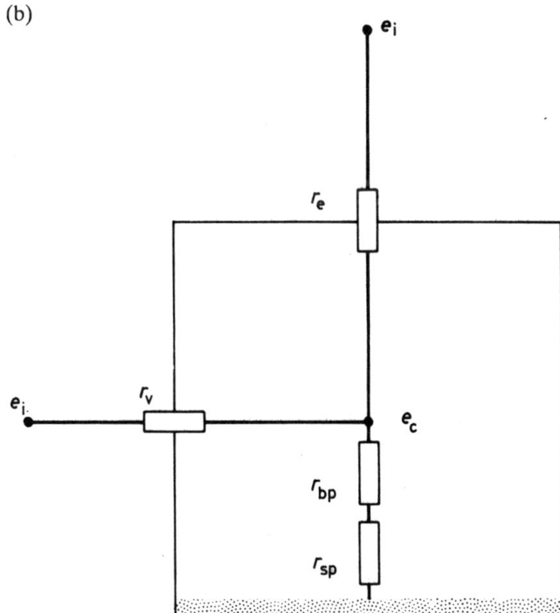

Figure 3.8   Resistance analogues for convective (a) and latent (b) heat fluxes in an open-top chamber. The chamber contains a single layer of leaves (leaf area index = 1). Temperature and vapour pressure inside and outside the chamber are T_c, e_c and T_i, e_i respectively, and the leaf temperature is T_p. The resistance r_{rp} in *Figure 3.8(a)* is a radiative resistance described in the text

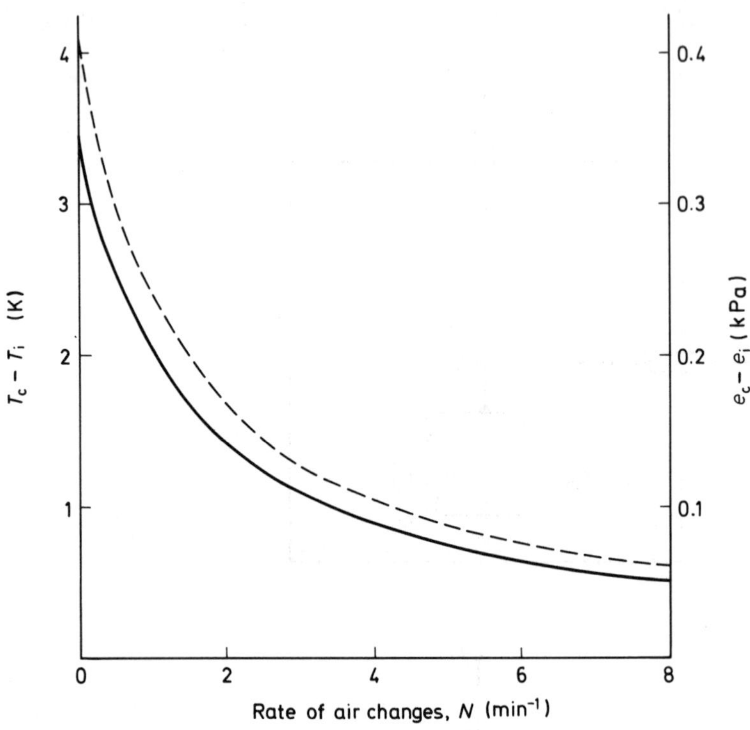

Figure 3.9   Dependence on chamber ventilation rate N of the temperature difference $(T_c - T_i$:———) and vapour pressure difference $(e_c - e_i$: ———) between air inside and outside an open-top chamber for conditions given in *Table 3.2*

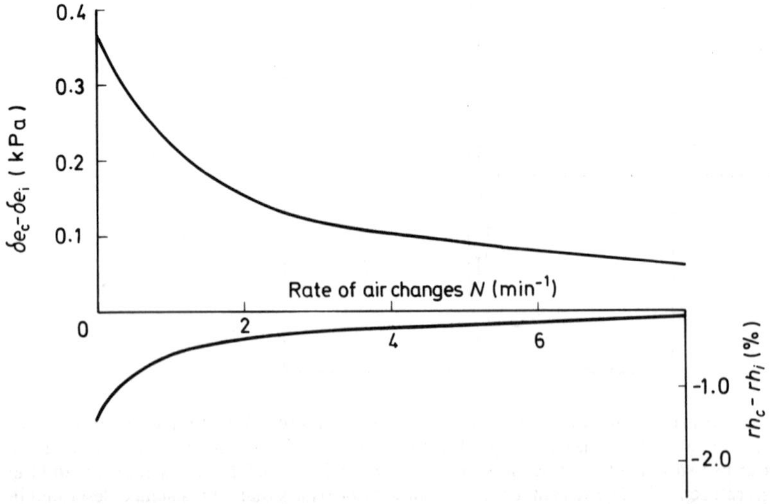

Figure 3.10   Dependence on chamber ventilation rate N of the difference in saturation deficit $(\delta e_c - \delta e_i)$ and relative humidity $(rh_c - rh_i)$ between air inside and outside an open-top chamber for conditions given in *Table 3.2*

relative humidity differences based on *Figure 3.9* and shows that < 1% difference would be expected at ventilation rates typically employed. Probably more important are the differences in vapour pressure deficit in the air (saturation vapour pressure − vapour pressure) also shown in *Figure 3.10*. Vapour pressure deficit can modify stomatal responses to pollutants in sensitive species (Black and Unsworth, 1980). This feature of chamber environment and the possibility of large leaf–air temperature differences may be responsible for some of the differences in growth and development noted between plants in chambers and in the field.

Conclusion

Methods of exposing plants to pollutants are generally much better defined than in the past, but anomalies between responses observed in different systems may be due to unobserved differences in chamber environments. There is a need for better quantification of resistances to gas transfer in exposure systems because of the importance of resistances in the exchange of pollutants, carbon dioxide and water vapour. Much progress has been made in the last few years in measuring and interpreting the transfer of pollutants from the atmosphere to plants in the laboratory and in the field. We are entering an exciting new era in pollution research where responses can be related to a true 'dose' rather than to external exposure. It remains to be seen whether such an approach will increase or decrease our understanding of plant responses! As a realist, I suspect that Murphy's Law will apply and confusion will reign in the initial stages. But, ultimately, better understanding of how responses relate to the environment in which plants are exposed must lead to sounder and more generally applicable dose–response relationships.

References

ASHENDEN, T.W. and MANSFIELD, T.A. (1977). *Journal of Experimental Botany*, **28**, 729–735

BENNETT, J.H., HILL, A.C. and GATES, D.M. (1973). *Journal of the Air Pollution Control Association*, **23**, 203–206

BLACK, V.J. and UNSWORTH, M.H. (1979). *Nature*, **282**, 68–69

BLACK, V.J. and UNSWORTH, M.H. (1980). *Journal of Experimental Botany*, **31**, 667–677

CAMPBELL, G.S. (1977). *An Introduction to Environmental Biophysics*. Springer Verlag, New York

CHAMBERLAIN, A.C. (1975). In *Heat and Mass Transfer in the Biosphere*, pp. 561–582 (DeVries, D.A. and Afgan, N., Eds), Scripta Book Co, Washington, DC

CHAMBERLAIN, A.C. (1980). In *Atmospheric Sulphur Deposition*, pp. 185–197 (Shriner, D.S., Richmond, C.R. and Lindberg, S.E., Eds), Ann Arbor Science

COWLING, D.W. and LOCKYER, D.R. (1976). *Journal of Experimental Botany*, **27**, 411–417

COWLING, D.W. and LOCKYER, D.R. (1978). *Journal of Experimental Botany,* **29**, 257–265

COWLING, D.W. and KOZIOL, M.J. (1978). *Journal of Experimental Botany,* **29**, 1029–1036

DE CORMIS, L., BONTE, J. and TISNE, A. (1975). *Pollution Atmospherique,* **17**, 103–107

FOWLER, D. and UNSWORTH, M.H. (1974). *Nature,* **249**, 389–390

FOWLER, D. AND UNSWORTH, M.H. (1979). *Quarterly Journal of the Royal Meteorological Society,* **105**, 767–783

HEAGLE, A.S., BODY, D.E. and HECK, W.W. (1973). *Journal of Environmental Quality,* **2**, 365–368

HEAGLE, A.S., PHILBECK, R.B., ROGERS, H.H. and LETCHWORTH, M.B. (1979). *Phytopathology,* **69**, 15–20

HECK, W.W., DUNNING, J.A. and JOHNSON, H. (1968). *Pub. APTD-68-6,* US Department of Health, Education and Welfare, National Centre for Air Pollution Control, Cincinnatti

HECK, W.W., KRUPA, S.V. and LINZON, S.N. (1979). (Eds) *Methodology for the Assessment of Air Pollution Effects on Vegetation.* Air Pollution Control Association, Pittsburgh

HECK, W.W., PHILBECK, R.B. and DUNNING, J.A. (1978). *ARS-S-181.* Agricultural Research Service, USDA

HEITSCHMIDT, R.K., LAUENROTH, W.K. and DODD, J.L. (1978). *Journal of Applied Ecology,* **15**, 859–868

HORSMAN, D.C. and WELLBURN, A. (1977). *Environmental Pollution,* **13**, 33–39

HORSMAN, D.C., ROBERTS, T.M. and BRADSHAW, A.D. (1979). *Journal of Experimental Botany,* **30**, 485–493

KOZIOL, M.J. (1980). *Journal of Experimental Botany,* **31**, 1413–1423

KOZIOL, M.J. and COWLING, D.W. (1978). *Journal of Experimental Botany,* **29**, 1431–1439

LEE, J.J., PRESTON, E.M. and LEWIS, R.A. (1978). In *Proceedings of the 4th Joint Conference on Sensing of Environmental Pollutants,* pp. 49–53. American Chemical Society, Washington DC

LEUNING, R., UNSWORTH, M.H., NEUMANN, H.H. and KING, K.M. (1979). *Atmospheric Environment,* **13**, 1155–1163

LOCKYER, D.R., COWLING, D.W. and JONES, L.H.P. (1976). *Journal of Experimental Botany,* **27**, 397–409

MANDL, R.L., WEINSTEIN, L.H., McCUNE, D.C. and KEVENY, M. (1973). *Journal of Environmental Quality,* **2**, 371–376

MONTEITH, J.L. (1963). In *Environmental Control of Plant Growth,* pp. 95–112 (Evans, L.T., Ed.), Academic Press, New York

MONTEITH, J.L. (1965). In *Symposium of the Society of Experimental Biology,* **19**, 205–234

MONTEITH, J.L. (1975). *Principles of Environmental Physics,* Revised Edition. Edward Arnold, London

MULLER, R.N., MILLER, J.E. and SPRUGEL, D.G. (1979). *Journal of Applied Ecology,* **16**, 567–576

OLSZYK, D.M., TIBBITTS, T.W. and HERTZBERG, W.M. (1980). *Journal of Environmental Quality,* **9**, 610–615

ROBERTS, T.M., BELL, R., HORSMAN, D.C. and BRADSHAW, A.D. (1979). In *Proceedings of Society for Chemical Industry Symposium, London, 1979*

ROGERS, H.H., JEFFRIES, H.E., STAHEL, E.P., HECK, W.W., RIPPERTON, L.A. and WITHERSPOON, A.M. (1977). *Journal of the Air Pollution Control Association,* **27**, 1192–1197

RUNECKLES, V.C., STALEY, L.M., BULLEY, N.R. and BLACK, T.A. (1978). *Canadian Journal of Botany,* **56**, 768–778

ŠESTÁK, Z., ČATSKÝ, J. and JARVIS, P.G. (1971) (Eds). *Plant Photosynthetic Production: Manual of Methods.* Junk N.V., The Hague

SETTERSTROM, C. and ZIMMERMAN, P.W. (1938). *Contributions from Boyce Thompson Institute,* **9**, 161–169

SHINN, J.H., CLEGG, B.R. and STUART, M.L. (1977). *UCRL Reprint 804 11,* Lawrence Livermore Laboratory, California

SPEDDING, D.J. (1969). *Nature,* **224**, 1229–1230

SRIVASTAVA, H.S., JOLLIFFE, P.A. and RUNECKLES, V.C. (1975). *Canadian Journal of Botany,* **53**, 466–474

TAYLOR, G.E. Jr. and TINGEY, D.T. (1979). *Research Report EPA-600/ 3-79-108.* Corvallis Environmental Research Laboratory, Corvallis, Oregon

THOMAS, M.D. and HILL, G.R. (1935). *Plant Physiology,* **10**, 291–307

THOMPSON, C.R., KATS, G. and CAMERON, J.W. (1976). *Journal of Environmental Quality,* **5**, 410–412

UNSWORTH, M.H. (1981). In *Plants and their Atmospheric Environment,* Ch. 7. (Grace, J., Ford, E.D. and Jarvis, P.G., Eds), Blackwells, Oxford

UNSWORTH, M.H. and MANSFIELD, T.A. (1980). *Environmental Pollution, (Series A)* **23**, 115–120

WESELY, M.L., EASTMAN, J.A., COOK, D.R. and HICKS, B.B. (1978). *Boundary Layer Meteorology,* **15**, 361–373

WINNER, W.E. and MOONEY, H.A. (1980). *Oecologia,* **44**, 290–295

II

Physiological and Biochemical Responses to Pollutants

4

EFFECTS OF SULPHUR DIOXIDE ON PHYSIOLOGICAL PROCESSES IN PLANTS

V.J. BLACK*
Department of Physiology and Environmental Science, University of Nottingham School of Agriculture

Introduction

Sulphur dioxide, SO_2, which is a byproduct of fossil fuel consumption, is one of the major pollutants occurring in the UK. Concentrations in the atmosphere range from short-term peaks of a few ppm near point sources, to average concentrations of 70 ppb in industrial areas, and a background of 10 ppb in many parts of rural Britain (Fowler and Cape, this volume, Chapter 1). Whether these low concentrations of SO_2 are sufficient to have an adverse effect on agricultural and horticultural productivity has been debated for many years. Wislicenus (1901) and Stoklasa (1923) suggested that plant yields could be depressed by low concentrations of SO_2 in the absence of visible injury. In contrast, Katz (1949) concluded that there was no evidence to support this concept of 'invisible injury'. Even as recently as 1972, Davis found yield losses in SO_2-fumigated soybeans only when visible damage had occurred. In addition, it has been suggested that when the sulphur status of the soil is poor, low concentrations of SO_2 may act as a plant nutrient and improve the yield of some forest ecosystems (Smith, 1974) and agricultural crops (Cowling and Lockyer, 1978). The majority of recent reports, however, have suggested that SO_2 may indeed reduce several components of growth and yield in a range of species; these components include dry-matter production and tillering, number of spikelets, percentage ripened grains, relative growth rate, shoot dry weight, root:shoot ratio, mean tiller weight and leaf area (Taniyama, 1972; Taniyama *et al.*, 1972; Bell and Clough, 1973; Lockyer, Cowling and Jones, 1976; Ashenden, 1978, 1979; Bell, Rutter and Relton, 1979; Crittenden and Read, 1979; Ayazloo, Bell and Garsed, 1980; Davies, 1980). Nevertheless it has proved difficult to define the magnitude of these SO_2-induced effects on growth and yield of major agricultural crops.

Definition of the magnitude of these effects on grasses has proved to be particularly difficult. For instance, Bell, Rutter and Relton (1979) reported that shoot dry weight of *Lolium perenne* was reduced after 173 days' exposure to 43 μg SO m^{-3} (15 ppb) but there was no effect after 144 days' exposure to 66 μg m^{-3} (23 ppb). In contrast, no reduction in the growth of the same species after 51 days at 419 μg m^{-3} (147 ppb) was observed by Cowling and Koziol (1978). Reasons for these apparent discrepancies may lie in reports that the degree of depression of yield and growth may be

*Present address: Ecology Section, Department of Human Sciences, Loughborough University of Technology

strongly influenced by many plant and environmental factors. Katz and Ledingham (1939), Crittenden and Read (1979) and Bell, Rutter and Relton (1979), for example, found that sensitivity to SO_2 varied with plant age. Heck, Dunning and Hindawi (1965) concluded that plants were more sensitive to pollutants when they were grown in soils of low fertility than in well-fertilized soils. Similarly Ayazloo, Bell and Garsed (1980) indicate that high nitrogen decreased the sensitivity to SO_2 injury. Certain species or cultivars possess innate or evolved resistance to SO_2 (Roose, Bradshaw and Roberts, this volume, Chapter 18). Prevailing environmental conditions during exposure have been shown to be important. Bell (this volume, Chapter 11) suggested that *Lolium perenne* is more susceptible in winter than in summer. Greater yield reductions have been reported also at higher wind speeds (Ashenden and Mansfield, 1977) and when irradiance is low and photoperiod short (Davies, 1980). Thus any relationship between exposure to SO_2 and reductions in growth or yield depends not only on pollutant concentration but also on plant status and the multivariate microclimate.

Why do low concentrations of SO_2 affect growth and yield in some species and in certain environmental conditions, but not in others? Clearly, any reduction in growth or yield must reflect the interference of the pollutant with cellular metabolism, carbon dioxide exchange or water loss. However, the numerous attempts to define and elucidate these SO_2-induced changes in physiological processes have highlighted the complexity and lack of mechanistic understanding of these responses (*for reviews see* Mudd, 1975; Mansfield, 1976; Hallgren, 1978; Heath, 1980). In this review, therefore, examples have been selected from the literature which illustrate the range of physiological responses to SO_2 and the environmental factors known to influence these responses, and which suggest possible mechanisms to explain these responses. In the field it is difficult to separate responses to a pollutant from responses to ever-changing environmental conditions, and so most research has been carried out in chambers to compare the stomatal behaviour, transpiration, respiration and photsynthesis of whole plants or leaves exposed to filtered air with those exposed to a range of pollutant concentrations.

Responses of stomata and transpiration

A pollutant-induced change in stomatal aperture will have important consequences. First, there will be an enhancement or depression in photosynthetic carbon dioxide uptake and transpirational water loss. Secondly, the quantity or rate at which pollutant enters the plant and arrives at the metabolic sites in the underlying mesophyll tissue will be altered. Stomatal responses may thus have an important role in determining the magnitude of effects of pollutants on gas exchange and, ultimately, growth and yield.

There are, however, few definitive pieces of research aimed at elucidating and understanding stomatal responses to SO_2. Stomatal conductance is often inferred from measurements of transpiration in uncontrolled and ill-defined conditions and is rarely studied throughout the experimental period. Also, too often responses are determined after unrealistic exposure to high concentrations (> 1 ppm) for long periods (> 1 h), a situation very rare in the field. At these concentrations, any direct stomatal responses are often compounded by concomitant changes in whole-leaf water potential, cell collapse and visible injury.

MAGNITUDE AND DIRECTIONS OF RESPONSES

Stomatal effects induced by SO_2 are varied in magnitude and direction. The literature indicates that stomata may be induced either to open or close in response to SO_2, depending on the species examined, the concentration and length of exposure to SO_2, and the prevailing environmental conditions. Some of the first observations of SO_2-induced stomatal responses were made by Mansfield and Majernik (1970), who demonstrated an enhanced stomatal opening in *Vicia faba* plants exposed to concentrations of SO_2 greater than 0.25 ppm. Other species exhibiting a similar response include *Zea mays* (Unsworth, Biscoe and Pinckney, 1972), pine (Farrar, Relton and Rutter, 1977), *Phaseolus vulgaris* (Ashenden, 1978; Rist and Davis, 1979), pea and corn (Klein *et al.*, 1978), grapevine (Shertz, Kender and Musselman, 1980), radish, sunflower and tobacco (Black and Unsworth, 1980), navy beans, cucumber, soybeans and white bean (Beckerson and Hofstra, 1979) and *Atriplex triangularis* and *A. sabulosa* (Winner and Mooney, 1980c). These responses were reported to occur within a few minutes of exposure to SO_2 and resulted in a 10–20% increase in stomatal conductance in several C_4 species and a maximum increase of 200% in the C_3 species *Atriplex triangularis* (Winner and Mooney, 1980c). In contrast, stomatal closure or depressed transpiration rates in response to SO_2 have been reported in many species. These include pinto beans (Sij and Swanson, 1974), *Pelargonium hortorum* (Bonte, 1975; Bonte, De Cormis and Louguet, 1977), pine (Caput *et al.*, 1978), *Diplacus aurantiacus* and *Heteromeles arbutifolia* (Winner and Mooney, 1980a,b), peanut, tomato, radish, perilla and spinach (Kondo and Sugahara, 1978), castor oil, Swiss chard, rice, poplar, plane, sunflower, cucumber (Furukawa *et al.*, 1980), alday, wheat, corn, sorghum and bean (Kondo, Maruta and Sugahara, 1980), apple (Shertz, Kender and Musselman, 1980) and birch (Biggs and Davis, 1980). The maximum inhibition of transpiration rates observed ranged from 35–75% and occurred within ten minutes to four hours following exposure, depending on the species examined. However, in the majority of these investigations stomatal responses to only high concentrations of SO_2 were examined. It is not known, therefore, whether these species also show enhanced opening at lower SO_2 concentrations.

INFLUENCE OF ENVIRONMENT

Environmental conditions during exposure may also modify stomatal responses to low concentrations of SO_2. For instance, light is an important factor in eliciting a response. As the stomatal pore is the major pathway for the uptake of gaseous pollutants (Spedding, 1969; Black and Unsworth,

1979c), stomata must be open before responses to SO_2 can be demonstrated. For example, stomata are not usually affected by SO_2 if exposure occurs in the dark when stomata are closed (Black and Unsworth, 1980), although Garsed and Mochrie (1980) have demonstrated uptake of radioactivity from droplets containing $^{35}Na_2SO_3$ applied to the leaf surface of *Vicia faba*.

Swain (1923) first proposed that relative humidity (r.h.) was an important factor in influencing responses to SO_2 after observing that plants were more susceptible when r.h. was high, conditions which often give large stomatal apertures. However, Majernik and Mansfield (1970) were the first to demonstrate that r.h. also influenced the direct response of the stomata themselves to SO_2. They found that stomatal apertures of *Vicia faba* increased following exposure to 0.25–1.0 ppm SO_2 at relative humidities above 40% at 18°C but decreased when humidity was below 40%. Similar responses to humidity or vapour-pressure deficit have been demonstrated by Black and Unsworth (1980) in *Vicia faba*, *Helianthus annuus* and *Nicotiana tabacum*. These are species which are able to exert some control over water loss by reducing stomatal apertures as evaporative demand increases, i.e. they are sensitive to vapour-pressure deficit, whereas species which are not sensitive in this way, e.g. *Phaseolus vulgaris*, showed only enhanced stomatal opening over a range of relative humidities (*Figure 4.1*). It has been suggested (Aston, 1976) that in species sensitive to

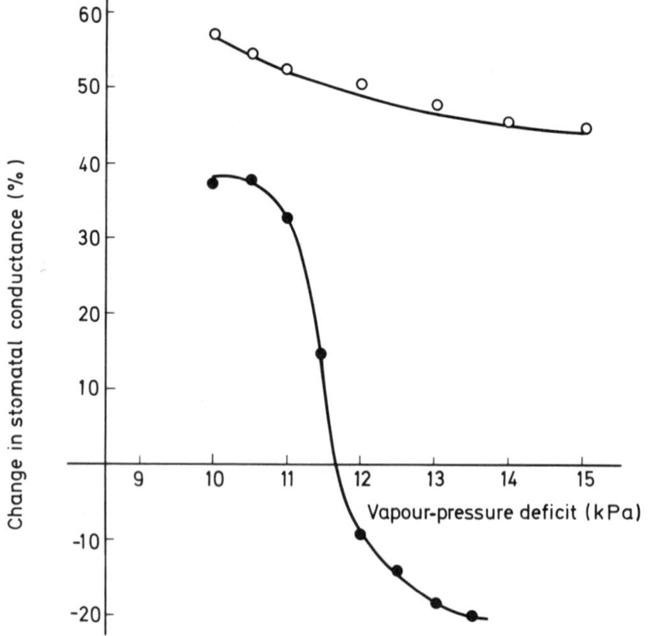

Figure 4.1   Variation of the % change in stomatal conductance,
$$\frac{[\text{conductance }(SO_2) - \text{conductance (control)}]}{\text{conductance (control)}} \times 100,$$
with vapour-pressure deficit in plants exposed to 35 ppb SO_2: (○) *Phaseolus vulgaris*; (●) *Helianthus annuus* (after Black and Unsworth, 1980)

vapour-pressure deficit, as deficits become large, transpirational water loss from either the guard cells themselves, or the epidermal cells adjacent to the guard cells, will increase and loss of guard-cell turgor with stomatal closure will ensue. Stomatal responses to SO_2 exposure could interact with this postulated mechanism. The increased stomatal conductance induced by the action of SO_2 will result in higher transpiration rates and thus greater water loss from the stomatal cells. Therefore, in species sensitive to vapour-pressure deficit, as gradients of evaporative demand increase, loss of guard-cell turgor in polluted stomata will be more rapid than in controls, and enhanced stomatal closure will result. This closure in response to large deficits in polluted plants may be accentuated by the presence of damaged epidermal cells, adjacent to the guard cells; such damage has been observed in species such as *V. faba* (Black and Black, 1979a,b). On exposure to low concentrations of SO_2, this enhanced closure at large vapour-pressure deficits was reversible. Stomatal conductances increased again when deficits became smaller. In contrast, the stomatal closure observed when *Vicia faba* was exposed to high concentrations of SO_2 was neither dependent on vapour-pressure deficit nor reversible.

MECHANISMS OF ACTION ON STOMATA

Once inside the plant, how does SO_2 effect changes in stomatal aperture? Are these responses mediated by changes in whole-leaf water potential? Although, after exposure of plants to high concentrations of SO_2, we have observed significant changes in leaf-water potential of *Vicia faba* which may effect changes in stomatal aperture, no measurable changes in leaf-water potential have been observed in *Vicia faba* and petunia (Elkiey and Ormrod, 1979) after exposure to low concentrations of SO_2.

It has also been suggested that stomatal responses are mediated through changes in carbon dioxide concentration in the intercellular species. For instance, Koziol and Jordan (1978) propose that the increases in stomatal conductance reported by Mansfield and Majernik (1970) may be the result of enhanced internal CO_2 concentrations caused by SO_2-induced increases in respiration. Alternatively, it has been suggested that CO_2 concentrations in the substomatal cavity will increase as SO_2 depresses photosynthetic CO_2 fixation, resulting in enhanced stomatal closure. However, this hypothesis cannot explain the enhanced stomatal opening exhibited by many species in response to low SO_2 concentrations or when plants are polluted in the dark with stomata open. There is also a difference between the concentration and time required to effect the stomatal and photosynthetic respiratory responses, stomata tending to respond more rapidly and at lower concentrations. In addition, photosynthetic responses are often concentration-dependent and readily reversible, whereas stomatal responses of several species are less so. Thus, there is considerable evidence to discount this theory as the major cause of changes in stomatal aperture induced by low SO_2 concentrations. However, many species which exhibit enhanced stomatal opening at low concentrations show enhanced closure at higher SO_2 concentrations. At these concentrations, where SO_2-induced reductions in photosynthesis may be considerable, increased concentrations of CO_2 in intercellular spaces may play a significant part in stomatal closure.

To effect changes in stomatal aperture, SO_2 in low concentrations must result in a direct change in the turgor balance of the cells of the stomatal complex. To investigate whether observed increases in aperture of *Vicia faba* resulted either from an increase in turgidity within the guard cells or a reduction within the epidermal cells adjacent to the guard cells, Black and Black (1979a, b) used light microscopy to examine epidermal strips taken from bean plants exposed either to scrubbed or to polluted air. The enhanced opening response induced by low concentrations of SO_2 was associated with extensive destruction of adjacent epidermal cells, whereas guard-cell survival was not reduced significantly. Collapsed epidermal cells were also observed, using scanning electron microscopy, in intact leaf samples taken from *V. faba* polluted with SO_2. This evidence suggests that the enhanced stomatal opening in bean, induced by low concentrations of SO_2, may result from preferential injury to the adjacent epidermal cells. At higher SO_2 concentrations stomatal closure was observed and was found to be associated with cellular disorganization and reduced guard-cell viability.

Kondo and Sugahara (1978) and Kondo, Maruta and Sugahara (1980) observed stomatal closure in many species after exposure to 2 ppm SO_2 and suggested that this resulted from a lowering of the pH on the surface or cytoplasm of guard cells and thus a change in turgor relations of the stomatal complex. They also observed a relationship between leaf abscisic acid (ABA) levels and stomatal responses; the species with the largest amount of ABA exhibited a large, rapid reduction in transpiration on exposure to SO_2, whereas in those species with very low levels of ABA, transpiration rates initially increased and subsequently decreased slightly. In addition, they suggested that the lowered pH of the stomatal complex may also reduce the effective ABA concentration necessary to cause stomatal closure. Thus SO_2 or its products may amplify the ABA-induced stomatal closure or sensitize the guard cells to ABA (*Figure 4.2*). However, this interpretation may be an artefact of the SO_2 concentrations employed. Before ABA and SO_2 can interact within the stomatal complex, ABA must be released from the mesophyll. At the extremely high concentrations of SO_2 used by Kondo and colleagues (Kondo and Sugahara, 1978; Kondo, Maruta and Sugahara, 1980), cell damage, stress, or a decline in leaf-water potential may have occurred and led to a stimulation of ABA release from the mesophyll.

The evidence for an interaction between ABA, stomatal aperture and low pollutant concentrations is only indirect and contradictory. Water stress has been shown to result in enhanced ABA concentrations in cells of the stomatal complex of beans (Davies, 1978). Rich and Turner (1972) found that stomata of water-stressed but nonwilted bean plants closed more quickly when exposed to ozone. In contrast, Unsworth, Biscoe and Pinckney (1972) reported increases in stomatal aperture in response to low concentrations of SO_2 in water-stressed plants. Clearly, there is a need to ascertain whether there is an interaction between ABA, SO_2, and stomatal behaviour at realistic levels of pollutant.

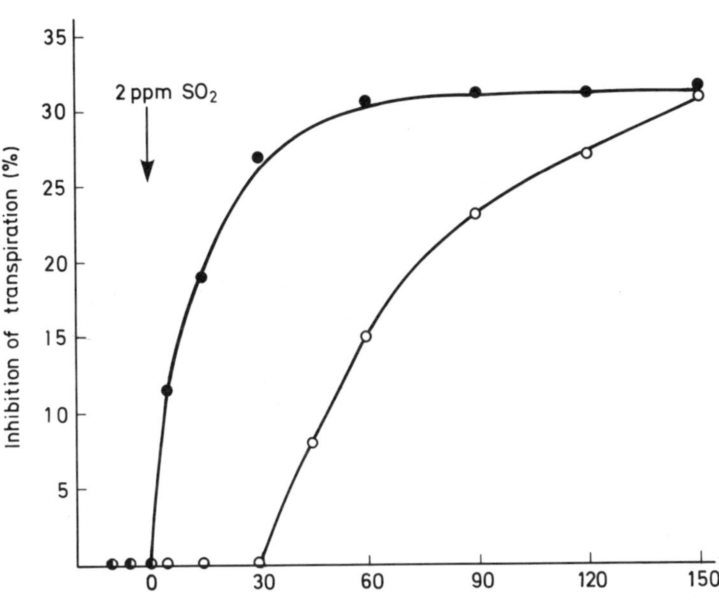

Figure 4.2 Effects of 2 ppm SO_2 on transpiration rates of radish plants sprayed with 10^{-3} M abscisic acid (●) or unsprayed (○) (after Kondo and Sugahara, 1978)

A similar mediating influence on stomatal behaviour could be attributed to ethylene, a hormone known to be released by stressed plants (*see* Lieberman, 1979). Plants which are exposed to pollutants such as ozone and SO_2 also show increased ethylene (ethane) production (Craker, 1971; Bressan *et al.*, 1978; Peiser and Yang, 1979; Yu *et al.*, this volume, page 507). Peiser and Yang, for instance, found that when they exposed alfalfa plants to 0.7 ppm SO_2 for 8 h, ethylene production reached a maximum after 6 h and returned to control levels 18 h following the end of the fumigation period. Could such ethylene production occurring either in the mesophyll or stomatal cells influence stomatal behaviour and act as a protective mechanism against SO_2 uptake and injury, as has been suggested by Craker (1971) for ABA and ozone injury? The evidence for direct ethylene-mediated stomatal responses is not conclusive. Vitagliano and Hoad (1978) reported that ethylene derived from ethephon always decreased stomatal conductance in a number of species, whereas Pallaghy and Raschke (1972) showed that $1-10^5$ ppm ethylene had no effect on stomata of turgid *Zea* and *Pisum* leaves. However, Lieberman (1979) suggested that the response to ethylene may be quite different in water-stressed leaves, especially in association with high concentrations of ABA. Saad (1953–54) reported that concentrations of ethylene greater than 0.1% caused the closure of stomata in *Pelargonium zonale*, and reopening resulted on removal of the gas. However, although exposure to high concentrations of SO_2 may result in sufficient ethylene production to cause stomatal closure, there is as yet no evidence to suggest that this occurs in

(a)

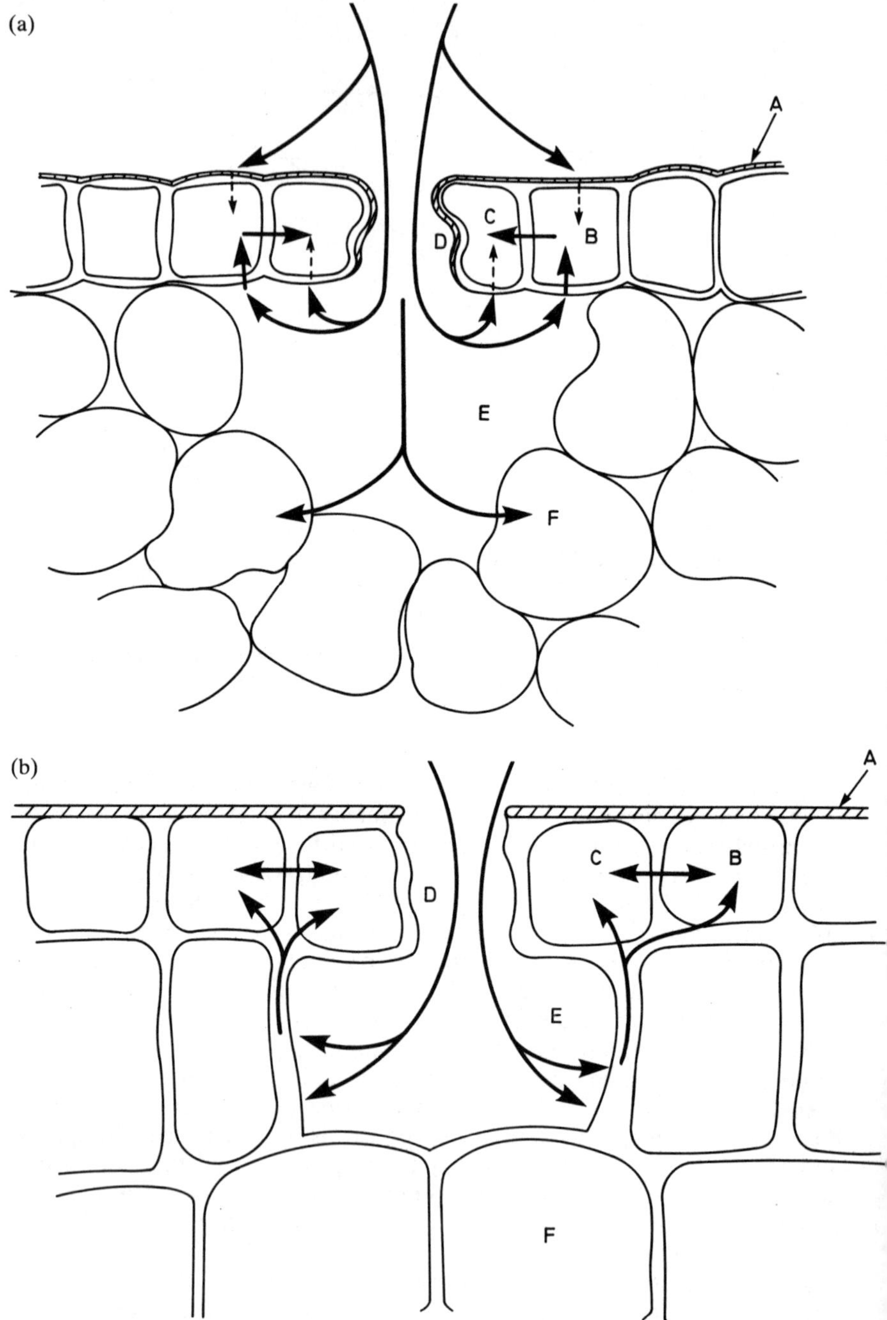

Figure 4.3 Cross-sections of typical substomatal cavity of a leaf, to demonstrate two proposed pathways of pollutant uptake. (a) This review. (b) After Heath (1980)

◄— Principal paths and directions of pollutant flow; ◄--- Possible paths which require investigation; A: cuticle; B: epidermal cell; C: guard cell; D: stomatal pore; E: substomatal cavity; F: mesophyll cell exposures to low concentrations. Clearly, SO_2-induced ethylene production would not explain the enhanced stomatal opening observed in many species. However, Yu *et al.* (this volume, page 507) suggested that ethylene was produced when the free radicals, produced during the oxidation of sulphite to sulphate, induced peroxidation of membrane lipids. Such action of free radicals on membranes may explain the sensitivity of stomatal responses to SO_2.

Although the primary mode of action of SO_2 on stomatal behaviour still remains uncertain, it seems likely that SO_2 initially follows the shortest diffusion pathway and enters the stomatal complex directly (*Figure 4.3(a)*). This is in contrast to the views of Heath (1980) who suggested that SO_2 would pass through the mesophyll before entering the stomatal complex (*Figure 4.3(b)*). However, stomatal responses occur within a few minutes and at low concentrations of SO_2, even in species where the epidermis is partly hydraulically separated from the mesophyll cells. It seems unlikely therefore, that a sufficient concentration of sulphite/bisulphite would pass through the mesophyll without being absorbed or detoxified, at the speed necessary to effect the observed changes in aperture.

There is no conclusive evidence to indicate whether SO_2 enters the cells of the stomatal complex via the adjacent epidermal cells or via guard cells. Direct entry into the guard cells may be impeded by the presence of the cuticle which extends into the substomatal cavity and covers the guard-cell wall in certain species. In addition, Squire and Mansfield (1972) found that the subsidiary cells (specialized epidermal cells) of *Commelina communis* absorbed chemicals applied to epidermal strips faster than did guard cells, and regulated solute entry into the guard cells. Entry of SO_2 via the epidermal cells would be one hypothesis to explain the preferential damage to epidermal cells reported by Black and Black (1979a,b). Alternatively, this differential sensitivity may be explained if the guard cells have a greater capacity to tolerate, detoxify or compartmentalize absorbed sulphite or bisulphite. When fluxes of SO_2 into the plant become large, such mechanisms of resistance may become saturated, or sufficient SO_2 may reach the guard cells to impair guard-cell function, and stomatal closure results. Conversely, the observed stomatal closure in some species could be explained either by direct entry of pollutant into the guard cells or by greater guard-cell sensitivity. Clearly, a study of behavioural and mechanistic differences between species would yield information to increase our understanding of stomatal responses to SO_2.

REVERSIBILITY OF STOMATAL RESPONSES

Whether effects of SO_2 on stomata and stomatal action are temporary or permanent has not been resolved. Several types of response after removal of SO_2 have been reported. These include continuation of responses observed during exposure (Beckerson and Hofstra, 1979; Black and Unsworth, 1980) and a return to unpolluted conductances either as soon as the pollutant is removed (Taniyama *et al.*, 1972; Unsworth, Biscoe and Pinckney, 1972; Sij and Swanson, 1974; Bonte, De Cormis and Louguet, 1977) or several hours or days after removal of the pollutant (Majernik and

Mansfield, 1970; Ashenden, 1979a). Recovery of unpolluted conductances during exposure has been suggested by Bell, Rutter and Relton (1979) and reported by Winner and Mooney (1980c). Accurate assessment of reversibility of stomatal responses is very difficult to achieve. Environmental conditions such as irradiance, temperature, CO_2 concentration and water vapour-pressure deficit, factors which also influence stomatal behaviour, must be maintained constant or monitored before, during and after exposure, and must be comparable between control and treated plants. Interpretation of responses is also difficult because these are usually inferred from measurements of water loss from whole plants, leaves or parts of leaves, rather than from the study of single stomatal pores. At higher pollutant concentrations, changes in aperture which are directly induced by SO_2 may be obscured by changes in nonstomatal water loss or in stomatal responses mediated by changes in bulk leaf-water potential. There is also no definitive evidence to indicate whether reported reversibility can be attributed to complete recovery of both stomatal responses and function. Recovery to pretreatment conductances does not necessarily imply that stomatal cell structure or function has not been permanently impaired. Stomatal responses may solely reflect modifications in behaviour to compensate for changes in water loss induced by the pollutant. It is therefore important to investigate not only whether stomata may recover pretreatment conductances, but also whether basic stomatal function and structure may be permanently impaired. Similarly, it is important to investigate whether SO_2-induced effects on stomata depend on species examined, SO_2 concentration, and length, number and frequency of exposures.

The literature suggests, therefore, that SO_2 induces a number of stomatal responses, which are influenced by pollutant concentration and environmental conditions and differ between species. This complexity of responses emphasizes that it is vital to study stomatal behaviour throughout any investigation. Only when the magnitude and permanence of responses are defined, will the importance of SO_2-induced stomatal effects in determining changes in growth, development and yield be assessed.

Action on respiration

DARK RESPIRATION

The effects of SO_2 on this process are rarely studied, although dark respiration and photorespiration are important components of the carbon budget. From the limited number of investigations, widely contrasting conclusions may be drawn. Sulphur dioxide has been reported (1) to have no effect on respiration (Sij and Swanson, 1974; Furukawa, Natori and Totsuka, 1980), (2) to have a small effect on respiration (Shimazaki and Sugahara, 1980), (3) to decrease respiratory rates (Gilbert, 1968; Taniyama, 1972) or (4) to increase respiratory activity (Keller and Müller, 1958; De Koning and Jäger, 1968; Taniyama *et al.*, 1972; Baddeley, Ferry and Finegan, 1973; Türk, Wirth and Lange, 1974; Black and Unsworth, 1979b). This lack of uniformity of response is not surprising because the experimental procedures for exposure and monitoring respiratory responses to SO_2 were not consistent. Measurements of effects of SO_2 on respiratory rates were made using intact plants in the dark period following exposure in the light (Black and Unsworth, 1979b), with leaves treated with SO_2 for 2 h in the dark (Sij and Swanson, 1974), with *Euglena* exposed to a solution of sulphite (De Koning and Jegier, 1968) and with strips excised following exposure (Shimazaki and Sugahara, 1980). The species selected ranged from higher plants to bryophytes, algae and lichens (Taniyama, 1972; Gilbert, 1968; De Koning and Jäger, 1968; Baddeley, Ferry and Finegan, 1973). Not only did the length of exposure vary but exposure concentrations ranged from 5 ppm (De Koning) to less than 0.04 ppm (Black and Unsworth, 1979b). In addition, plants were studied at different stages of development and under different temperature regimes.

The published reports are so diverse that it is difficult to make generalizations about the effects of SO_2 on respiratory processes. Not only have the magnitude, direction and longevity of responses not been carefully defined, but the mechanisms of response, whether indirectly through impaired photosynthesis or cell damage or as a direct consequence of SO_2 injury on the respiratory process itself, are not understood. Clearly, this important component of CO_2 exchange needs to be investigated fully before it can be assessed whether SO_2 induces a significant alteration in respiratory activity.

PHOTORESPIRATION

The most conveniently measured parameter of carbon dioxide exchange is net photosynthesis, a term which encompasses both gross photosynthetic CO_2 fixation and photorespiratory CO_2 release. Thus, in order to assess whether SO_2 affects either gross photosynthesis or respiration alone, or a combination of both, it is necessary to quantify effects of SO_2 on respiration in the light (photorespiration). This is particularly important in species where photorespiratory rates may be significantly higher than dark respiratory rates, for instance in beans, where rates differ by a factor of at least three (Zelitch, 1971). Because quantitative and unequivocal measurements of photorespiration are difficult to make, there are few published reports where the effects of SO_2 on this process have been examined. Investigations have tended to use high concentrations of pollutant or have studied the effects of SO_2, sulphite or sulphonates on enzymes thought to be involved in photorespiration. The results from these investigations are then extended to postulate effects of SO_2 on photorespiratory CO_2 release in intact plants.

Koziol and Jordan (1978) reported an exponential increase in photorespiration of intact plants of *Phaseolus vulgaris* with increasing SO_2 concentration and suggested that these enhanced rates stemmed from the greater use of energy in repair or replacement processes. However, photorespiration was deduced from the rate of CO_2 release immediately following exposure to light. Ludlow and Jarvis (1971) doubted the validity of this technique, in which results are often subjective and dependent on a large number of environmental variables.

Sij and Swanson (1974) considered that SO_2-induced effects on photorespiration were unlikely because net photosynthesis responses of C_3 and C_4 species tested were similar. In contrast, several workers (Buron and Cornic, 1973; Spedding and Thomas, 1973; Koziol and Cowling, 1978; Furukawa, Natori and Totsuka, 1980) reported that photorespiration was inhibited by SO_2. Such results are in agreement with the indirect evidence that glycollate accumulated in *Hordeum* plants (Bethge, 1958) and in the $^{14}CO_2$-fixation pathway of spinach leaves (Libera, Ziegler and Ziegler, 1974) following fumigations with high concentrations of SO_2. On the basis of these observations, Ziegler (1975) postulated that SO_2 is likely to 'drastically reduce or completely abolish rates of photorespiration'.

If there is such an inhibition of photorespiration at low concentrations of SO_2, enhanced rates of net photosynthesis will result. Any depression in photorespiration therefore would be advantageous to the plant if this process is inessential and wasteful of assimilated CO_2. However, if photorespiration is a protective metabolic process (Tolbert and Ryan, 1976) any SO_2-induced depression of photorespiration may not be completely beneficial. Therefore, before any assessments of this nature can be made, a systematic study of the effects of low concentrations of SO_2 on photorespiration of intact plants needs to be carried out before, during and after exposure to assess the magnitude and longevity of any effects. Only then can we be confident that reported depressions of net photosynthesis are solely attributable to effects on the photosynthetic process itself.

Action on photosynthesis

Numerous reports (*for reviews see* Mudd, 1975; Hällgren, 1978; Heath, 1980) indicate that photosynthesis is a process which is very sensitive to SO_2 in many species. The majority of investigations indicate that SO_2 exposure results in depressed net photosynthesis rates, although a small number of workers report temporary enhanced photosynthesis (Black and Unsworth, 1979b; Winner and Mooney, 1980c). Although these enhanced rates can often be attributed to increased stomatal conductances, or perhaps to depressed photorespiration, many of the reported substantial reductions in net photosynthesis cannot result entirely from changes in these factors. Depressions in photosynthetic rates occur within 30 min to a few hours after the start of exposure, are dependent on SO_2 concentration over a range of concentrations, are readily reversible and are not usually accompanied by major visible injury, at least at low concentrations (Bull and Mansfield, 1974; Black and Unsworth, 1979b). At higher concentrations, responses are less reversible and appear to be associated with the breakdown of biochemical systems, tissues and appearance of visible injury.

Species vary, however, in the magnitude and threshold of photosynthetic responses to SO_2. This variability in apparent sensitivity is illustrated in *Figure 4.4*, which was constructed from the data of several authors. The observed responses of several species contradict the conclusions of Verkroost (1974), that inhibition of photosynthesis occurred only when SO_2 concentrations exceeded a threshold which varied with species but which

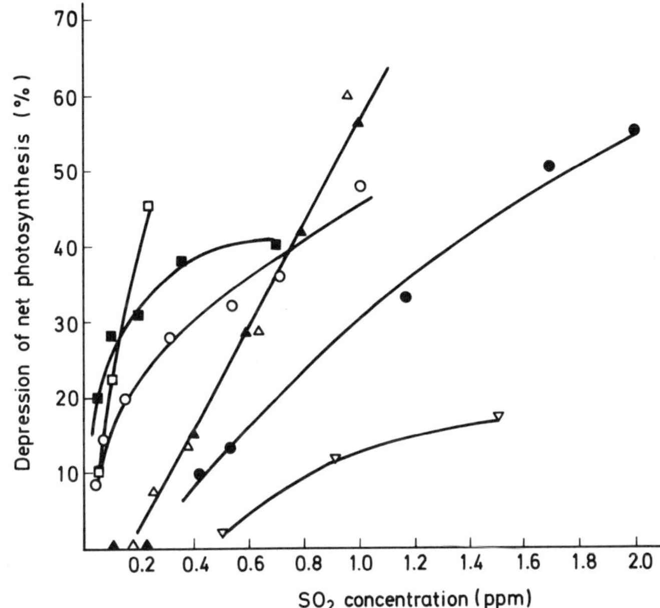

Figure 4.4 Effects of variation in SO_2 exposure concentration on depression of net photosynthesis in several species. □ *Pisum sativum* (after Bull and Mansfield, 1974); ■ *Vicia faba* (after Black and Unsworth, 1979b); ○ Rice (after Taniyama, 1972); ▲ Barley (after Bennett and Hill, 1973); △ *Diplacus aurantiacus* (after Winner and Mooney, 1980a); ● *Heteromeles arbutifolia* (after Winner and Mooney, 1980b); ▽ *Atriplex sabulosa* (after Winner and Mooney, 1980c)

was well above concentrations commonly experienced in the field. However, interpretation and comparison of responses is difficult. Workers commonly relate responses either to the gas concentrations surrounding the plant or to the dose, defined as the product of the gas concentration and duration of exposure. In practice, however, responses to a pollutant are more likely to depend either on the rate at which gas is absorbed by the plant, or on the total quantity absorbed over a period of time. These are factors which strongly depend on the conditions of exposure, stomatal conductance and degree of adsorption of SO_2 on to leaf surfaces, as well as ambient pollutant concentration. For instance, as early as 1921, Loftfield found that lucerne was only susceptible to SO_2 if stomata were open.

INFLUENCE OF ENVIRONMENT

There is considerable evidence also that many environmental factors, such as light, relative humidity, temperature and carbon dioxide concentration, may influence photosynthetic responses to SO_2. These may indirectly modify the influence of SO_2 on net photosynthesis, either by altering the flux of pollutant into the plant via changes in stomatal conductance, or by affecting respiratory and detoxification rates. Alternatively, environmental factors themselves may directly affect the pollutant sensitivity of the photosynthetic process.

Boundary-layer resistance

Uptake of CO_2 and SO_2 is not only restricted by diffusion through stomata, but also by the boundary-layer resistance to gas transfer between leaves and atmosphere. The magnitude of this resistance is modified by wind speed, chamber-mixing characteristics, leaf morphology and leaf surface structure. Wind speed has been shown to influence SO_2-induced reductions in the growth of *Lolium perenne* (Ashenden and Mansfield, 1977) and depression of net photosynthesis of *Vicia faba* (Black and Unsworth, 1979a), the reductions being greatest when wind speed is high and boundary-layer resistance is low. Fluxes of SO_2 into plants, and thus effects of SO_2, will therefore be greater in exposure systems which are well ventilated and mixed, than in badly designed exposure chambers or within canopies.

Carbon dioxide concentration

Because stomatal conductances decrease in response to increasing CO_2 concentrations, Mansfield (1976) suggested that high CO_2 concentrations which are associated with many toxic pollutants generated in fossil-fuel combustion may be sufficiently large to offer protection against the toxic effects of the gas. Indeed, such beneficial interactions between CO_2 concentrations and SO_2 effects have been demonstrated by Hou, Hill and Soleimani (1977). They observed a smaller reduction in the yield of *Medicago sativa* when 645 ppm CO_2 accompanied the SO_2 than when CO_2 concentrations were 315 ppm, typical of ambient. These increased CO_2 concentrations were typical of those found downwind of a power plant in a plume in which SO_2 concentrations were around 1 ppm. Decreased stomatal apertures in response to enhanced CO_2 will result in restricted SO_2 flux into the plant and would thus confer protection against pollutants. However, for such a protective mechanism to be effective, the benefits of restricted SO_2 uptake must outweigh the concomitant decrease in CO_2 uptake.

However, the protective influence of raised CO_2 concentrations may not be solely stomatal in origin. *Figure 4.5* demonstrates the interaction between CO_2, SO_2 and net photosynthesis in *Vicia faba*. To remove the accompanying responses of stomata to changing CO_2 concentrations, percentage inhibitions of net photosynthesis in four CO_2 regimes have been plotted against the SO_2 flux into the plant rather than against ambient SO_2 concentration. *Figure 4.5* shows that SO_2-induced reductions in net .photosynthesis were greater at ambient than at raised CO_2 concentrations; similarly, thresholds for response to SO_2 were lower at ambient than at increased CO_2 concentrations. This demonstrates that the protection conferred by increased CO_2 may be partly nonstomatal in origin and may arise from a reduction in the direct effects on the photosynthetic process itself. Such an interaction may be important, not only for plants growing downwind of point sources, but also in greenhouses with supplemented CO_2. Similarly, if the combination of SO_2 with lower-than-ambient CO_2

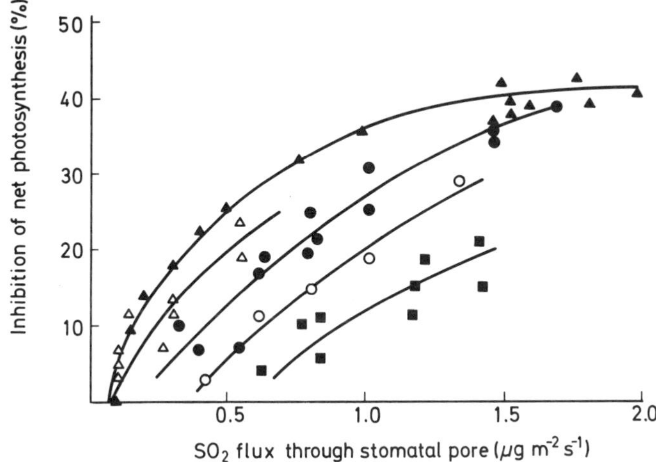

Figure 4.5   Variation in % inhibition of net photosynthesis with SO_2 flux into stomata and CO_2 concentration in *Vicia faba*. ▲ 330 ppm; △ 430 ppm; ● 480 ppm; ○ 580 ppm; ■ 680 ppm CO_2. (Black, previously unpublished)

concentrations is found to result in greater reductions of net photosynthesis than occur with SO_2 and ambient CO_2, responses obtained in poorly designed exposure chambers or in crop canopies where CO_2 concentrations may fall well below ambient, may be misleading.

Irradiance

Irradiance is also important in influencing SO_2-induced changes in net photosynthesis rates, both directly and indirectly. The irradiance usually governs not only whether stomata are open, but also the magnitude of stomatal conductance, and thus the flux of CO_2 and SO_2 into the plant. However, irradiance also seems to influence SO_2-induced changes in the photosynthetic process itself. Taniyama (1972), Matsuoka (1978), Black and Unsworth (1979b) and Winner and Mooney (1980c) have observed greater reductions in net photosynthesis on exposure to SO_2 at high irradiances than at low irradiances. If this is the sole process affected by SO_2, one would expect greater reductions in yield when irradiances are saturating rather than limiting. However, Davies (1980) observed that the effect of 11 pphm SO_2 on the growth of *Phleum pratense* was greatest when irradiance was low and photoperiod short. Davies suggests that 'plants whose growth is severely limited by light will have low reducing power and will not be able rapidly to detoxify SO_2 products', and that 'toxic levels of SO_4^{2-} and SO_3^{2-} are therefore likely to accumulate and growth may be inhibited.' This would support the findings of Bell (this volume, Chapter 11) that SO_2-induced reductions in the yield of *Lolium perenne* were greater in winter than in summer.

Relative humidity and temperature

Interactive effects of relative humidity and SO_2 on yield and physiological processes have been observed frequently. In many species, effects of SO_2 are greater when r.h. is high and stomatal conductances are large than when r.h. is low (Katz and Ledingham, 1939; Barton, McLaughlin and McConathy, 1980). However, there is conflicting evidence regarding the influence of temperature on photosynthetic responses to SO_2. Using concentrations of SO_2 exceeding 1 ppm, Matsuoka (1978) found that, within the range 18–28 °C, the lower the temperature the greater was the SO_2-induced inhibition of photosynthesis in rice. However, Taniyama (1972) found the opposite response. In barley and rape, inhibition of photosynthesis tended to be greater at 20 °C than at 10 °C, and in rice, over the range 25–40 °C, the higher the temperature the greater was the inhibition. This agrees with the evidence of Heck and Dunning (1978) that oats are more sensitive at higher growth temperatures. As temperatures vary diurnally and seasonally, and between laboratory and field investigations, it is vital to ascertain how and in what way temperature influences responses to pollutants.

Temperature and relative humidity may influence the SO_2-induced inhibition of photosynthesis through changes in stomatal conductance, thus changing CO_2 and SO_2 fluxes into the plant. Alternatively, temperature may also interact with SO_2 to modify photosynthesis by changing rates of detoxification or the biochemical processes themselves, as was suggested by Davies (1980) for irradiance.

Clearly, the effect of SO_2 on photosynthetic rates is very dependent on the environmental conditions during exposure, especially in the field or in canopies where climate and microclimate vary spatially and temporally. However, the extent to which environmental factors such as temperatures influence SO_2-induced changes in the photosynthetic process itself, as distinct from indirect changes in stomatal conductance and respiratory rates, has not been well investigated. In particular, whole-plant studies which permit the assessment of the more direct effects on photosynthesis are rare. One exception is the work of Winner and Mooney (1980c) who were able to show that in *Atriplex*, a C_3 species was more sensitive than a C_4 species because of a higher stomatal conductance as well as a higher sensitivity of the photosynthetic processes themselves. Clearly, a mechanistic understanding of the influence of environmental variables such as light, temperature, relative humidity and carbon dioxide on SO_2-induced inhibition of photosynthesis will not be achieved until experiments are conducted which are designed to separate photosynthetic and stomatal responses.

MECHANISMS OF ACTION ON PHOTOSYNTHESIS

The mechanisms of action of SO_2 on the photosynthetic process are poorly understood. Many of the investigations of mechanisms comprise *in vitro* assays, using isolated chloroplasts, membranes or enzyme systems, which are rarely comparable with true physiological conditions. In addition, the use of unrealistically high concentrations of SO_2 may provide results which bear no relationship to the 'subtle' responses that occur within the cell. Reviews by Mudd (1975), Ziegler (1975), Malhotra and Hocking (1976), Hällgren (1978) and Heath (1980) demonstrate the multitude and variety of hypotheses about how SO_2 affects photosynthesis. When SO_2 enters leaves it is metabolized to sulphite, bisulphite and sulphate (Puckett *et al.*, 1973) which have been shown to affect several biochemical and cellular processes and characteristics (Horsman and Wellburn, 1976), thereby causing an inhibition of photosynthesis. For example, high concentrations of SO_2 cause the degradation of chlorophylls to phaeophytins, which leads to earlier senescence (LeBlanc and Rao, 1975). However, although prolonged exposures to low concentrations of SO_2 have been shown to result in injury at the molecular level by affecting enzymes such as chlorophyllase (Malhotra, 1977), effects are too slow and insufficient to account for substantial reductions in photosynthesis.

On the other hand, interference with the structure and permeability of membranes and their associated enzymes will result in alterations in many biochemical processes in the cell. For instance, Guderian and Haut (1970) suggested that sulphite could effect membrane integrity. Similarly, Yu *et al.* (this volume, page 507) postulated that SO_2 interferes with the differential permeability of the plasma membrane by affecting membrane lipids. Puckett *et al.* (1974) suggested that cleavage of disulphide linkages by sulphite was responsible for protein and membrane disruption. Disruption of cellular membranes would explain the increased potassium efflux following exposure to SO_2, observed by many workers (Nieboer *et al.*, 1976).

Sulphur has been shown to be incorporated into chloroplast lamellae during SO_2 fumigation (Ziegler, 1977) and is thought to affect these membranes. Murray and Bradbeer (1971) suggested that bisulphite compounds inhibited CO_2 fixation by interfering with inner and outer chloroplast membranes and their associated transport systems. Alscher-Herman (this volume, page 502) has suggested also, that a possible explanation for the effect of sulphite may be its competitive action for stromal binding sites in chloroplast thylakoid membranes.

Ultrastructural evidence for SO_2-induced effects on membranes was provided by Fischer, Kramer and Ziegler (1973) and Wellburn, Majernik and Wellburn (1972) who reported reversible swelling of thylakoid membranes of chloroplasts in leaves exposed to low concentrations of SO_2. Malhotra (1976) found that ultrastructural changes induced by SO_2 were associated with a depression in the Hill Reaction activity in chloroplasts. At high concentrations, however, irreversible damage to thylakoid membranes has been observed (Wellburn, Majernik and Wellburn, 1972).

Other SO_2-induced structural changes have been reported. Black and Black (1979a) have observed chloroplast swelling in the guard cells of stomata in plants exposed to SO_2, and Malhotra (1976) has reported mitochondrial changes in pine needles treated with aqueous SO_2. At higher concentrations of SO_2, and prolonged exposures, cellular plasmolysis and mesophyll collapse have been observed (Solberg and Adams, 1956; Brandt and Heck, 1968; Stewart, Treshow and Harner, 1973). Effects of these high concentrations are usually irreversible and lead to acute leaf injury and to the appearance of visible injury. Any disruption of these structures and disorganization of enzymes associated with them is likely to affect CO_2 assimilation.

Ribulose bisphosphate (RUBP) and phosphoenol pyruvate (PEP) carboxylase, which effect the first step in CO_2 fixation in the C_3 and C_4 pathways of photosynthesis, are enzymes associated with membranes and may be affected by SO_2. Ziegler (1972, 1973) first reported that the inhibition of photosynthetic CO_2 fixation by SO_2 was due to competition between CO_2 and sulphur products for active binding sites on these enzymes. At higher SO_2 concentrations this inhibition was noncompetitive. Similar reports have been made by Mukerji and Yang (1974) and Martinovič and Plesničar (1977). Such a mechanism would explain the rapid inhibition of photosynthesis on SO_2 exposure, the concentration-dependence of the magnitude of the response and the rapid recovery of pretreatment photosynthetic rates on removal of the pollutant. However, in contrast, Gezeilus and Hällgren (1980) found no indication that the inhibition of photosynthesis by SO_2 was purely competitive, but could not explain the discrepancy between their data and that of Ziegler. They suggested that the concentrations of pollutant used by Ziegler were high and not likely to be found inside chloroplasts *in vivo*. Hällgren (1978) argued that studies of SO_2 on photosynthesis in isolated chloroplasts yield contradictory results through the use of different experimental methods and conditions, and different chloroplast populations. Indeed, Malhotra (1976) showed that the degree of inhibition of oxygen evolution by SO_2 products in pine-needle chloroplasts depended on the developmental stage of needles used for preparation. Thus, the question of whether the action of SO_2 on photosynthesis *in vivo* is by competitive inhibition, has not been resolved.

A similar controversy exists over the action of SO_2 on photosynthetic electron transport and photophosphorylation. Shimazaki and Sugahara (1979) showed that SO_2 fumigation in the light inhibited the activity of photosystem II but not I, i.e. both noncyclic electron flow and photophosphorylation were inhibited, but cyclic photophosphorylation driven by photosystem I was not. In contrast, in aqueous exposures, Asada *et al.* (1965) observed inhibition of both cyclic and noncyclic photophosphorylation by sulphite but no effect on electron flow. Similar results observed under acidic conditions by Silvius, Ingle and Baer (1975), led Shimazaki and Sugahara (1979) to conclude that the nature of inhibitory effects of sulphite will depend on exposure conditions. If photophosphorylation is inhibited, the Calvin cycle will also be inhibited, and electron transport and oxygen evolution will be reduced. The action of SO_2 on this process *in vivo* has yet to be elucidated.

Several other enzymes are affected by SO_2. For instance, Ballantyne (1973) showed that sulphur anions inhibited ATP formation in mitochondria. Similarly, Harvey and Legge (1979) found that ATP was reversibly diminished as the SO_2 concentrations increased, thus restricting the energy available for photosynthesis. Anderson and Avron (1976) and Ziegler, Marewa and Schoepe (1976) reported that SO_2 could interfere with the regulatory processes in the Calvin cycle. Enzyme activity may also be stimulated by SO_2. Wellburn *et al.* (1976) and Jäger and Klein (1977) showed that exposure to SO_2 resulted in enhanced activity of glutamate dehydrogenase and glutamate pyruvate transaminase, enzymes involved in amino acid and nitrogen metabolism. Stimulation of peroxidase and polyphenol oxidase, enzymes associated with the respiratory oxidations which characterize injury responses, also occurs in response to SO_2 (Keller, 1974; Rabe and Kreeb, 1976).

From this consideration of only a few of the reported effects of SO_2 on cellular processes and characteristics it appears that not only may SO_2 act on processes and enzymes directly involved with photosynthesis, but also on many other metabolic pathways, enzymes, membrane integrity and organelle ultrastructure. An effect on even one process at any level of organization within the cell may trigger a variety of subsequent effects and interactions, ultimately to influence carbon fixation and growth and yield. For example, effects on certain biochemical pathways will lead to alterations in carbohydrate and amino acid pools (Koziol and Jordan, 1978) which will alter the balance of source/sink relationships between organs, and lead to a shortage of assimilates in the growing regions. The action of oxidized sulphur radicals has been shown to promote and inhibit cell division (Hanamett, 1930). Indeed, Bleasdale (1973) suggested that SO_2 reduces growth by increasing the level of partly oxidised radicals in plants, resulting in reduced cell division. Nitrogen fixation is also strongly influenced, although reversibly, by $NaHSO_3$ (Hällgren and Huss, 1975), and the levels and action of hormones which regulate vegetative and reproductive growth are modified by SO_2 (Yang and Saleh, 1973).

It is not surprising, therefore, that no single mechanism to explain the action of SO_2 on photosynthesis has been identified. Many processes which will influence photosynthesis, both directly or indirectly, are likely to be affected by SO_2 to varying degrees. The overall effect of SO_2 on these factors will depend also on the number, concentration, duration and frequency of exposures to pollutant. The reduction in photosynthesis will result partly from the inability of cells to sustain photosynthetic rates during pollutant uptake, and will be influenced by the sensitivity of the affected systems themselves and by the effectiveness of any compartmentalization or detoxification process. However, the action of any detoxification mechanism may indirectly impair photosynthesis by competing for energy supplies used in photosynthesis. Similarly, energy may be channelled preferentially into repair mechanisms rather than to photosynthesis, growth and development (Wellburn *et al.*, 1976) and to other energy-requiring processes such as nitrogen fixation.

Thus, *in vivo*, the magnitudes of reduction in photosynthesis are likely to be the combined result of the action of SO_2 on stomata and on a number of respiratory and biochemical processes, and of the cells' ability to tolerate, compartmentalize or detoxify the pollutant. As the concentration of SO_2 or flux into the plant increases, SO_2 may influence a greater number of characteristics and larger reductions in carbon fixation will result. At high concentrations or fluxes, saturation of the capacity of enzyme systems and of mechanisms for compartmentalization or detoxification is likely to occur, which will culminate in irreversible damage and visible injury.

Mechanisms for tolerance or recovery

The sensitivity of plants to SO_2 exposure will not depend solely on physiological responses to the pollutant but rather on the ultimate effect of these responses on growth, development and yield. There are various strategies that could ensure a minimizing of pollutant damage. For instance, stomatal closure in response to SO_2 would be one protective mechanism. This was demonstrated for ozone by Butler and Tibbits (1979), who found that the most resistant cultivars of *Phaseolus vulgaris* were those which exhibited rapid stomatal closure in response to the pollutant, and reopening following exposure. Secondly, the majority of the absorbed SO_2 could be detoxified or compartmentalized, resulting in high thresholds for responses and protection of the photosynthetic process. For instance Miller and Xerikos (1979) have shown that resistant cultivars of soybeans convert sulphite more rapidly than sensitive cultivars. However, the high energy requirement for detoxification may also result in depressed photosynthetic rates. Alternatively, a strategy for consuming less energy would avoid detoxification and leave the components of the photosynthetic process exposed to sulphite or bisulphite. If these components are able to function satisfactorily in the presence of sulphite, little reduction in photosynthesis will occur. Nevertheless, it is unlikely that metabolic activity remains completely unaltered, especially if SO_2 fluxes are large. The plant may then be able to endure reductions in photosynthetic rates and recover unpolluted metabolic rates or compensate by increased metabolic rates, immediately on removal of the pollutant.

Clearly, the relationships between pollutant effects on physiological processes observed in the laboratory, and the ultimate effect on growth, development, yield and reproductive capacity, are very complex. They will depend not only on the number of pollutants present, concentration, the time, duration and frequency of exposure to the pollutant, but also on environmental conditions, plant age, stage of development, pretreatment, growth rates and plant physiological and biochemical status. A considerable amount of well-designed work needs to be carried out, not only in carefully defined and controlled conditions in the laboratory, but also in the field, where physiological processes can be measured in conjunction with growth and yield. Only then will an understanding of the relationship and mechanisms of responses to SO_2 be achieved.

Acknowledgement

The support of the Science Research Council during the preparation of this review is gratefully acknowledged.

References

ANDERSON, L.E. and AVRON, M. (1976). *Plant Physiology*, **57**, 209–213
ASADA, K., KITOH, S., DEURA, R. and KASAI, Z. (1965). *Plant and Cell Physiology*, **6**, 625–629

ASHENDEN, T.W. (1978). *Environmental Pollution,* **15,** 161–166

ASHENDEN, T.W. (1979a). *Environmental Pollution,* **18,** 45–50

ASHENDEN, T.W. (1979b). *Environmental Pollution,* **18,** 249–258

ASHENDEN, T.W. and MANSFIELD, T.A. (1977). *Journal of Experimental Botany,* **28,** 729–735

ASTON, M.M. (1976). *Australian Journal of Botany,* **3,** 489–502

AYAZLOO, M., BELL, J.N.B. and GARSED, S.G. (1980). *Environmental Pollution,* **22,** 295–307

BADDELEY, M.S., FERRY, B.W. and FINEGAN, E.J. (1973). In *Air Pollution and Lichens,* pp. 299–313 (Ferry, B.W. and Baddeley, M.S. Eds), Athlone Press, London

BALLANTYNE, D.J. (1973). *Phytochemistry,* **12,** 1207–1209

BARTON, J.R., McLAUGHLIN, S.B. and McCONATHY, R.K. (1980). *Environmental Pollution,* **21,** 255–265

BECKERSON, D.W. and HOFSTRA, G. (1979). *Atmospheric Environment,* **13,** 1263–1268

BELL, J.N.B. and CLOUGH, W.S. (1973). *Nature,* **241,** 47–49

BELL, J.N.B., RUTTER, A.J. and RELTON, J. (1979). *New Phytologist,* **83,** 627–643

BENNETT, J.H. and HILL, A.C. (1973). *Journal of the Air Pollution Control Association,* **23,** 203–206

BETHGE, H. (1958). *Schriftenreihe des Vereins für Wasser-, Boden- and Luft-hygiene,* **13,** 3–10

BIGGS, A.R. and DAVIS, D.D. (1980). *Journal of the American Society of Horticultural Science,* **105,** 514–516

BLACK, C.R. and BLACK, V.J. (1979a). *Journal of Experimental Botany,* **30,** 291–298

BLACK, C.R. and BLACK, V.J. (1979b). *Plant, Cell and Environment,* **2,** 329–333

BLACK, V.J. and UNSWORTH, M.H. (1979a). *Journal of Experimental Botany,* **30,** 81–88

BLACK, V.J. and UNSWORTH, M.H. (1979b). *Journal of Experimental Botany,* **30,** 473–483

BLACK, V.J. and UNSWORTH, M.H. (1979c). *Nature,* **282,** 68–69

BLACK, V.J. and UNSWORTH, M.H. (1980). *Journal of Experimental Botany,* **31,** 667–677

BLEASDALE, J.K.A. (1973). *Environmental Pollution,* **5,** 275–285

BONTE, J. (1975). PhD thesis, University of Pierre and Marie Curie, Paris

BONTE, J., De CORMIS, L. and LOUGUET, P. (1977). *Environmental Pollution,* **12,** 125–133

BRANDT, C.S. and HECK, W. (1968). In *Air Pollution,* Vol. 1, p.401 (Stern, A.C., Ed.), Academic Press, New York

BRESSAN, R.A., WILSON, L.G., LECUREUX, L. and FILNER, P. (1978). *Plant Physiology Supplement,* **61,** 93

BULL, J.N. and MANSFIELD, T.A. (1974). *Nature,* **250,** 443–444

BURON, A. and CORNIC, G. (1973). *Bulletin de la Société Vaudoise des Sciences Naturelles,* **71,** 451–461

BUTLER, L.K. and TIBBITS, T.W. (1979). *Journal of the American Society for Horticultural Science,* **104,** 213–216

CAPUT, C., BELOT, Y., AUCLAIR, D. and DECOURT, N. (1978). *Environmental Pollution,* **16,** 3–15

COWLING, D.W. and KOZIOL, M.J. (1978). *Journal of Experimental Botany,* **29,** 1029–1036

COWLING, D.W. and LOCKYER, D.R. (1978). *Journal of Experimental Botany,* **29,** 257–265

CRAKER, L.E. (1971). *Environmental Pollution,* **1,** 299–304

CRITTENDEN, P.D. and READ, D.J. (1979). *New Phytologist,* **83,** 645–651

DAVIES, T. (1980). *Nature,* **284,** 483–485

DAVIES, W.J. (1978). *Journal of Experimental Botany,* **29,** 175–182

DAVIS, C.R. (1972). *Journal of the Air Pollution Control Association,* **22,** 964–966

DE KONING, H.W. and JEGIER, Z. (1968). *Atmospheric Environment,* **2,** 321–326

ELKIEY, T. and ORMROD, D.P. (1979). *Zeitschrift für Pflanzenphysiologie,* **91,** 177–181

FARRAR, J.F., RELTON, J. and RUTTER, A.J. (1977). *Journal of Applied Ecology,* **14,** 861–875

FISCHER, K., KRAMER, D. and ZIEGLER, H. (1973). *Protoplasma,* **76,** 83–96

FURUKAWA, A., NATORI, T. and TOTSUKA, T. (1980). *Research Report from the National Institute for Environmental Studies, Yatabe, Japan,* **11,** 1–8

FURUKAWA, A., ISODA, O., IWAKI, H. and TOTSUKA, T. (1980). *Research Report from the National Institute for Environmental Studies, Yatabe, Japan,* **11,** 113–126

GARSED, S.G. and MOCHRIE, A. (1980). *New Phytologist,* **84,** 421–428

GEZELIUS, K. and HÄLLGREN, J.E. (1980). *Physiologia Plantarum,* **49,** 354–358

GILBERT, O.L. (1968). PhD thesis, University of Newcastle upon Tyne

GUDERIAN, R. and HAUT, H.U. (1970). *Staub,* **30,** 22–35

HÄLLGREN, J.-E. (1978). In *Sulphur in the Environment: Part II. Ecological Impacts,* pp.163–209 (Nriagu, J.O., Ed.), Wiley, New York

HÄLLGREN, J.-E. and HUSS, K. (1975). *Physiologia Plantarum,* **34,** 171–176

HANAMETT, F.S. (1930). *Proceedings of the American Philosophical Society,* **69,** 217–223

HARVEY, G.W. and LEGGE, A.A. (1979). *Canadian Journal of Botany,* **57,** 759–764

HEATH, R.L. (1980). *Annual Review of Plant Physiology,* **31,** 395–431

HECK, W.W. and DUNNING, J.A. (1978). *Journal of the Air Pollution Control Association,* **28,** 241–246

HECK, W.W., DUNNING, J.A. and HINDAWI, I.J. (1965). *Journal of the Air Pollution Control Association,* **15,** 511–515

HORSMAN, D.C. and WELLBURN, A.R. (1976). In *Effects of Air Pollutants on Plants,* pp.185–199 (Mansfield, T.A., Ed.), University Press, Cambridge

HOU, L-Y, HILL, A.C. and SOLEIMANI, A. (1977). *Environmental Pollution,* **12,** 7–15

JÄGER, H.-J. and KLEIN, K. (1977). *Journal of the Air Pollution Control Association,* **27,** 464–466

KATZ, M. (1949). *Industrial and Engineering Chemistry, International Edition,* **41,** 2450–2465

KATZ, M. and LEDINGHAM, G.A. (1939). In *Effect of Sulphur Dioxide on Vegetation,* ch.11, pp.262–287. National Research Council Bulletin 815, Ottawa, Canada

KELLER, T.H. (1974). *European Journal of Forest Pathology*, **4**, 11–19

KELLER, T.H. and MÜLLER, J. (1958). *Forstwissenschaftliche Forschungen*, **10**, 5–63

KLEIN, H., JÄGER, H.J., DOMES, W. and WONG, C.H. (1978). *Oecologia*, **33**, 203–208

KONDO, N. and SUGAHARA, K. (1978). *Plant and Cell Physiology*, **19**, 365–373

KONDO, N., MARUTA, I. and SUGAHARA, K. (1980). *Research Report from the National Institute for Environmental Studies, Yatabe, Japan*, **11**, 127–136

KOZIOL, M.J. and COWLING, D.W. (1978). *Journal of Experimental Botany*, **29**, 1431–1439

KOZIOL, M.J. and JORDAN, C.F. (1978). *Journal of Experimental Botany*, **29**, 1037–1043

LEBLANC, F. and RAO, D.N. (1975). In *Responses of Plants to Air Pollution*, pp.237–272. (Mudd, J.B. and Kozlowski, T.T. Eds), Academic Press, New York

LIBERA, W., ZIEGLER, I. and ZIEGLER, H. (1974). *Zeitschrift für Pflanzenphysiologie*, **74**, 420–433

LIEBERMAN, M. (1979). *Annual Review of Plant Physiology*, **30**, 533–591

LOCKYER, D.R., COWLING, D.W. and JONES, L.H.P. (1976). *Journal of Experimental Botany*, **27**, 397–409

LOFTFIELD, J.V.G. (1921). *Carnegie Institution Publication*, **314**, p.1

LUDLOW, M.M. nd JARVIS, P.G. (1971). In *Plant Photosynthetic Production – A Manual of Methods*, pp. 294–315 (Sěsták, Z., Čatský, J. and Jarvis, P.G., Eds), Junk, NV, The Hague

MAJERNIK, O. and MANSFIELD, T.A. (1970). *Nature*, **227**, 3778

MAJERNIK, O. and MANSFIELD, T.A. (1972). *Environmental Pollution*, **3**, 1–7

MALHOTRA, S.S. (1976). *New Phytologist*, **76**, 239–245

MALHOTRA, S.S. (1977). *New Phytologist*, **78**, 101–109

MALHOTRA, S.S. and HOCKING, D. (1976). *New Phytologist*, **76**, 227–237

MANSFIELD, T.A. (1976). In *Commentaries in Plant Science*, pp.13–22 (Smith, H, Ed), Pergamon Press, Oxford

MANSFIELD, T.A. and MAJERNIK, O. (1970). *Environmental Pollution*, **1**, 149–154

MARTINOVIČ, B. and PLESNIČAR, M. (1977). In *4th International Congress on Photosynthesis – Abstracts*, p.240 (Compiled by J. Coombs). London

MATSUOKA, Y. (1978). *Special Bulletin of the Chiba-Ken Agricultural Experiment Station*, **7**, 1–63

MILLER, J.E. and XERIKOS, P.B. (1979). *Environmental Pollution*, **18**, 259–264

MUDD, J.B. (1975). In *Responses of Plants to Air Pollution*, pp.9–22 (Mudd, J.B. and Kozlowski, T.T. Eds), Academic Press, New York

MUKERJI, S.K. and YANG, S.F. (1974). *Plant Physiology*, **53**, 829–834

MURRAY, D.R. and BRADBEER, J.W. (1971). *Phytochemistry*, **10**, 1999

NIEBOER, E., RICHARDSON, D.H.S., PUCKETT, K.J. and TOMASINI, F.O. (1976). In *Effects of Air Pollutants on Plants*, pp.61–85 (Mansfield, T.A., Ed.),University Press, Cambridge

PALLAGHY, C.K. and RASCHKE, K. (1972). *Plant Physiology*, **49**, 275–276

PEISER, G.D. and YANG, S.F. (1979). *Plant Physiology*, **63**, 142–145

PUCKETT, K.J., NIEBOER E., FLORA, W.P. and RICHARDSON, D.H.S. (1973). *New Phytologist*, **72**, 141–154

PUCKETT, K.J. RICHARDSON, D.H.S., FLORA, W.P. and NIEBOER, E. (1974). *New Phytologist*, **73**, 1183–1192

RABE, R. and KREEB, K. (1976). *Angewandte Botanik*, **50**, 70–78

RICH, S. and TURNER, N.C. (1972). *Journal of the Air Pollution Control Association*, **22**, 718–721

RIST, D.L. and DAVIS, D.D. (1979). *Phytopathology*, **69**, 231–235

SAAD, S.I. (1953–4). *Bulletin of the Institute of Egypt*, **36**, 269

SHERTZ, R.D., KENDER, W.J. and MUSSELMAN, R.C. (1980). *Journal of the American Society of Horticultural Science*, **105**, 594–598

SHIMAZAKI, K. and SUGAHARA, K. (1980). *Plant and Cell Physiology*, **20**, 26–35

SIJ, J.W. and SWANSON, C.A. (1974). *Journal of Environmental Quality*, **3**, 103–107

SILVIUS, J.E., INGLE, M. and BAER, C.H. (1975). *Plant Physiology*, **56**, 434–437

SMITH, W.H. (1974). *Environmental Pollution*, **6**, 111–129

SOLBERG, R.A. and ADAMS, D.F. (1956). *American Journal of Botany*, **43**, 755–760

SPEDDING, D.J. (1969). *Nature*, **224**, 1229–1231

SPEDDING, D.J. and THOMAS, W.J. (1973). *Australian Journal of Biological Sciences*, **26**, 281–286

SQUIRE, G.R. and MANSFIELD, T.A. (1972). *New Phytologist*, **71**, 1033–1043

STEWART, D., TRESHOW, M. and HARNER, F.M. (1973). *Canadian Journal of Botany*, **51**, 983

STOKLASA, J. (1923). *Die Beschadigungen der Vegetation durch Rauchgause und Fabriksexhalationen.* Urban und Schwarzenberg, Berlin

SWAIN, R.E. (1923). *Industrial and Engineering Chemistry*, **15**, 296–301

TANIYAMA, T. (1972). *Bulletin of the Faculty of Agriculture, Mie University, Tsu, Japan*, **44**, 11–130

TANIYAMA, T., ARIKADO, H., IWATA, T. and SAWANAKA, K. (1972). *Proceedings of the Crop Science Society of Japan*, **41**, 120–125

TOLBERT, N.E. and RYAN, F.J. (1976). In *CO_2 Metabolism and Plant Productivity*, pp.141–159 (Burris R.A. and Black, C.C. Eds), University Park Press, Baltimore

TÜRK, R., WIRTH, V. and LANGE, O.L. (1974). *Oecologia*, **15**, 33–64

UNSWORTH, M.H., BISCOE, P.V. and PINCKNEY, H.R. (1972). *Nature*, **239**, 458–459

VERKROOST, M. (1974). *Mededelingen Landbouwhogeschool Te Wageningen*, **74**, 1–78

VITAGLIANO, C. and HOAD, G.V. (1978). *Scientia Horticulturae*, **8**, 101–106

WELLBURN, A.R., MAJERNIK, O. and WELLBURN, F.A.M. (1972). *Environmental Pollution*, **3**, 37–49

WELLBURN, A.R., CAPRON, T.M., CHAN, H.-S. and HORSMAN, D.C. (1976). In *Effects of Air Pollutants on Plants*, pp.105–114 (Mansfield, T.A. Ed.) University Press, Cambridge

WINNER, W.E. and MOONEY, H.A. (1980a). *Oecologia*, **44**, 290–295

WINNER, W.E. and MOONEY, H.A. (1980b). *Oecologia*, **44**, 296–302

WINNER, W.E. and MOONEY, H.A. (1980c). *Oecologia*, **46**, 49–54

WISLICENUS, H. (1901). *Zeitschrift für Angewandte Chemie*, **28**, 689–712

YANG S.F. and SALEH, M.A. (1973). *Phytochemistry*, **12**, 1463–1466

ZELITCH, I. (1971). In *Photosynthesis, Photorespiration and Plant Productivity*, p.347. Academic Press, New York and London
ZIEGLER, I. (1972). *Planta,* **103,** 155–163
ZIEGLER, I. (1973). *Phytochemistry,* **12,** 1027–1030
ZIEGLER, I. (1975). *Residue Reviews,* **56,** 79–105
ZIEGLER, I. (1977). *Planta,* **135,** 25–32
ZIEGLER, I., MAREWA, A. and SCHOEPE, E. (1976). *Phytochemistry,* **15,** 1627–1632

5

OXIDES OF NITROGEN AND THE GREENHOUSE ATMOSPHERE

R.M. LAW
T.A. MANSFIELD
Department of Biological Sciences, University of Lancaster

Introduction

The effects of an air pollutant on the productivity of crops must be discussed in relation to the concentrations that are known to occur where crops are grown. There are particular reasons why it is more difficult to delimit these concentrations for NO_x than for any other type of pollutant. For example, crops which are grown within the normally 'protected' environment of a greenhouse are more likely to be subject to NO_x damage than 'out-of-doors' plants, because the concentrations experienced by the former may be more than a hundred times greater than those encountered by the latter. We shall begin by outlining the main sources of NO_x and the likely concentrations in different situations.

FORMATION OF NO_x

Nitrogen oxides, like sulphur oxides, are formed mainly as a result of the burning of fossil fuels. There is, however, little nitrogen in the fuel, and nitric oxide is formed in the heat of combustion when atmospheric nitrogen and oxygen combine:

$$N_2 + O_2 \xrightarrow{\text{heat}} 2NO \tag{5.1}$$

There is then a spontaneous, but not necessarily rapid, reaction between nitric oxide and oxygen:

$$2NO + O_2 \longrightarrow 2NO_2 \tag{5.2}$$

The rate of reaction (5.1) is determined by the conditions of combustion, the flame temperature being particularly important. The further oxidation (5.2) is not instantaneous as is sometimes believed. Its rate depends on the square of the NO concentration, and it is not rapid at the concentrations normally found in polluted air. The conversion of NO to NO_2 can, however, be assisted by other reactions and may, for example, be accelerated in the presence of O_3 (Eggleton, 1974). The balance between

NO and NO_2 in the atmosphere is therefore variable and often unknown, which is why it is more convenient to refer to this form of pollution as NO_x. It is, however, important to know the relative proportions of NO and NO_2 when the phytotoxicity of NO_x is being considered, for although these two gases may damage cells in a similar way, they enter leaves at different rates and toxicity is related to uptake rather than to concentration.

OCCURRENCE OF NO_x

We can illustrate the wide range of NO_x concentrations to which plants are exposed, by reference to three situations:

Rural sites

There are measurable amounts of NO and NO_2 in the lower atmosphere well away from areas of human activity. Estimates of 'background' concentrations vary, which may reflect the fact that biological sources and sinks differ in their intensity in different places. Two ppb seems a likely background level of NO over land at latitudes between 65 degrees N and 65 degrees S (Colucci and Simmons, 1978). The amount of NO_2 in truly unpolluted air is still uncertain, but may be as low as 0.1 ppb (Noxon, 1978). In the most extensive survey in Britain at a rural site remote from obvious sources of NO_x, Martin and Barber (1981) found 50% more NO_x than SO_2 on a volume/volume basis, the annual means being 6–10 ppb and 9–10 ppb for NO and NO_2 respectively. There were important diurnal variations in concentrations of the two gases, and hourly mean concentrations of total NO_x occasionally rose above 100 ppb.

Industrial and traffic sources

There have been few detailed surveys of the effects of industrial sources of NO_x on atmospheric concentrations in otherwise rural areas. However, two independent sets of measurements have been made at sites in Heysham, Lancashire, where background pollution is low, in the vicinity of a factory manufacturing ammonia, nitric acid and ammonium nitrate. Monitoring close to the works, Docherty (1974) found that a 24 h average concentration of 53 ppb was exceeded on 105 days in 2.5 years, but 4 km away the same concentration was exceeded on only 10 occasions. Continuous monitoring by Harrison and McCartney (1979) in the same area broadly confirmed these findings, but it was shown that the passage of a plume over the monitoring site led to sharp peaks ranging from 100–500 ppb for both NO and NO_2.

The high concentrations of NO_x in that particular example of a rural area intruded upon by industry compare in magnitude, but not in duration, with those reported by roadsides in towns. Hickman, Bevan and Colwill (1976) found hourly averages of around 500 ppb for NO_x for 7–8 consecutive hours on some days beside busy roads in Coventry. In this situation, NO constituted the bulk of the NO_x (*Figure 5.1*).

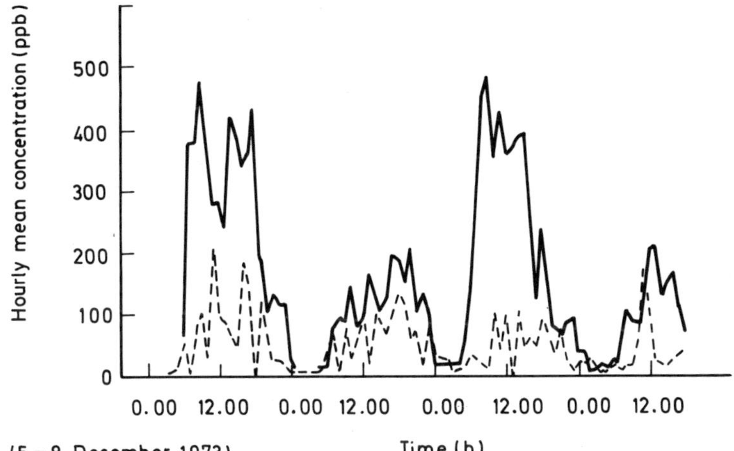

Figure 5.1   Levels of NO_x measured at Foleshill Road, Coventry. ——— NO; ——— NO_2. (From Hickman, Bevan and Colwill, 1976)

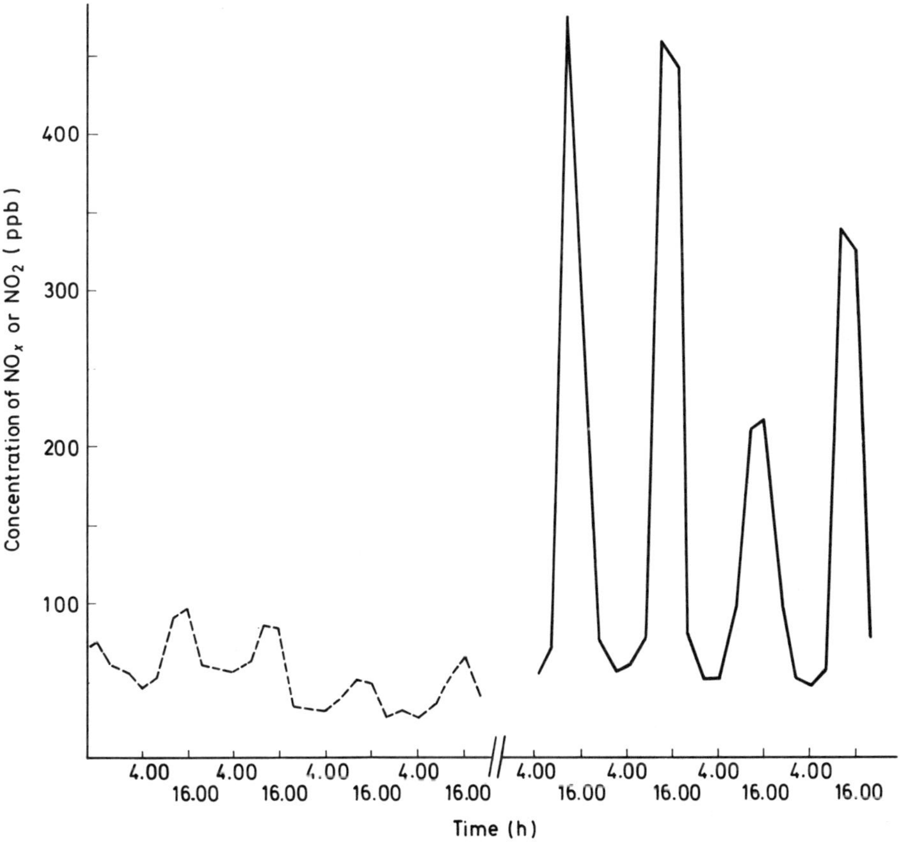

Figure 5.2   Concentrations of NO_2 (———) and NO_x (———) in a greenhouse enriched with CO_2 to about 1100 ppm using a kerosene burner

Greenhouses

The situation in a greenhouse, where propane or kerosene is burnt to enrich the atmosphere with CO_2, is remarkably similar to that by an urban roadside. In *Figures 5.1 and 5.2* we have plotted measurements made beside a busy road (20 000 vehicles each week-day) and in a greenhouse where the atmosphere around a crop of roses was enriched with CO_2 by the burning of propane during the daylight hours. Not only were the peaks of total NO_x of similar magnitude, but also the balance between NO and NO_2 was very similar. This is probably because in both situations the measurements were made near to the pollution source, so there was little time for the conversion of NO to NO_2.

The concentrations of NO_x achieved in *Figure 5.1* are typical of those found in greenhouses where hydrocarbons are burnt to enrich the atmosphere with CO_2 to the normal recommended level of about 1000 ppm. In recent years it has been an increasingly common practice to burn fuels such as kerosene or natural gas as the sole source of heating in a greenhouse. The exhaust gases are ducted directly into the atmosphere around the crop in order to achieve the most economical use of the heat generated, as well as to provide CO_2 enrichment. The NO_x levels resulting from this

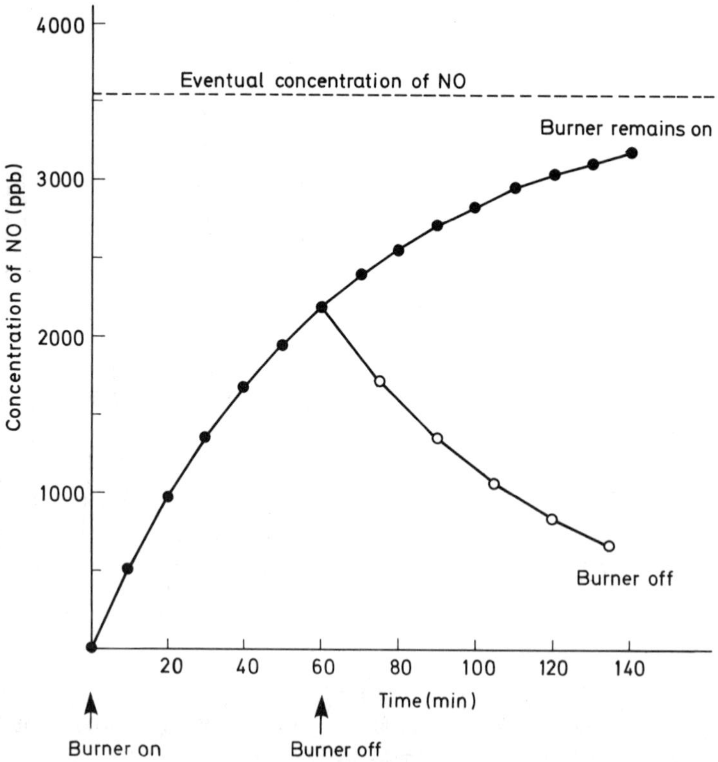

Figure 5.3 Increase of nitric oxide concentrations with time in a greenhouse (volume 2100 m³) heated with a 66 kW kerosene burner. Ventilation rate was 1.25 air changes per hour

procedure can be much higher than those associated with normal CO_2 enrichment. We have recorded up to 5000 ppb total NO_x in extreme cases; more typically, levels are 2000–3000 ppb in a greenhouse with a correctly operated burner. In *Figure 5.3*, data of Law *et al.* (1982) have been plotted to show the accumulation of NO over a period of time in a greenhouse of volume 2100 m³ heated with a 66 kW kerosene burner. The ventilation rate in this case was 1.25 air changes per hour. The greenhouse was a relatively new one and so was better sealed than many of the older ones in commercial use. However, modern polythene tunnel-houses are very well sealed and a substantial accumulation of pollutants would be expected in these, in the absence of ventilation. *Figure 5.4* shows the predicted effect of varying the ventilation rate upon the maximum eventual concentration of NO (*see also* Law *et al.*, 1982). A well-constructed greenhouse which retains the heat produced by the burner can clearly accumulate very high levels of NO. *Figure 5.4* does not take into account any uptake of NO into plants or soil, and the data that follow show that uptake into plants can be rapid. However, NO concentrations up to 7000 ppb are occasionally recorded around crops in well-sealed greenhouses with less than 0.5 air changes per hour. *Figure 5.5* shows that, for the greenhouse monitored by

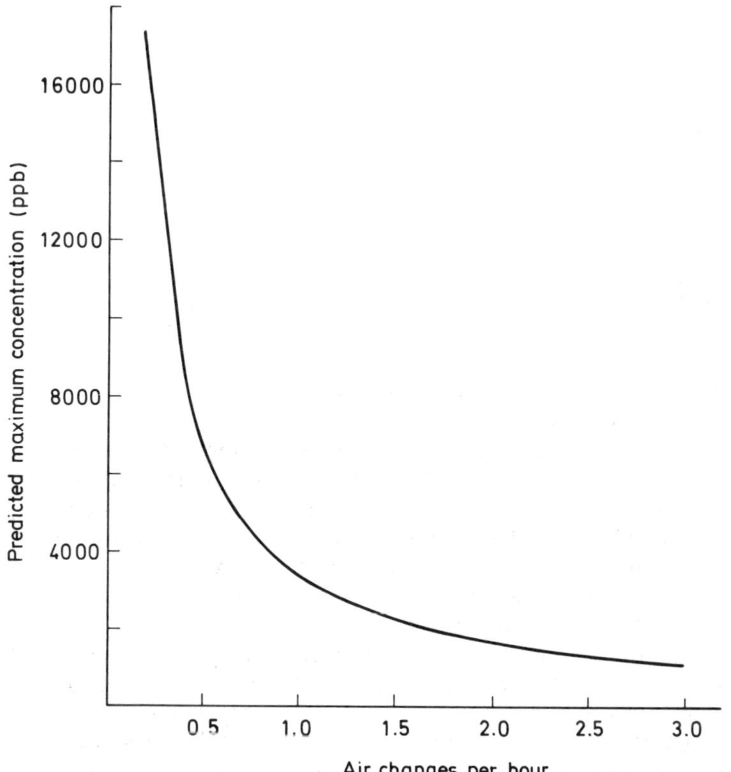

Figure 5.4   Effect of ventilation rate on the maximum eventual concentration of nitric oxide. (Greenhouse volume, 2100 m³; input of NO, 160 mg min⁻¹)

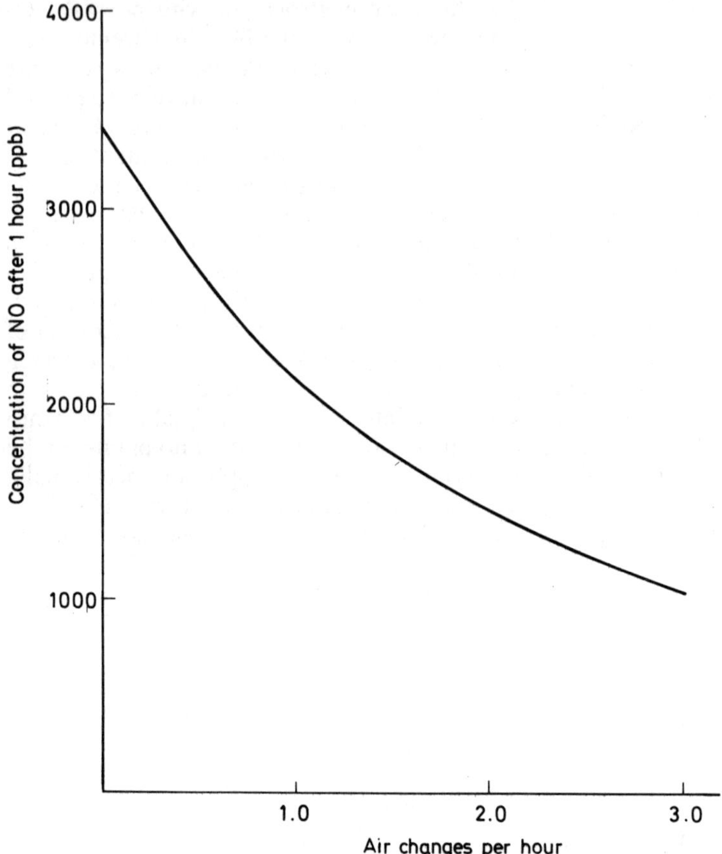

Figure 5.5   Effect of ventilation rate on concentration of NO reached after burner had been switched on for 1 h. (Greenhouse volume, $2100\,m^3$, input of NO, $160\,mg\,min^{-1}$)

Law *et al.* (1982), the concentration of NO would rise to nearly 2000 ppb after 1 h in the absence of uptake by a crop.

It is clear from these data that, in greenhouses equipped with flueless hydrocarbon burners, we are attempting to grow crops in the presence of NO_x levels equivalent to, or greater than, those they would experience in the most heavily polluted situation out of doors. No-one would purposely set out to carry out commercial horticulture within a few metres of a busy urban road, yet in normal recommended practice, greenhouse crops are being grown in comparably high NO_x levels.

Table 5.1 summarizes the levels of NO and NO_2 which might be experienced by plants growing in different situations. The concentrations extend over two orders of magnitude, and it is this disparity which makes simple coverage of this topic impossible. We shall deal primarily with greenhouse crops because these may be the ones exposed to the greatest risk of NO_x damage, although we shall show there are some factors which reduce this risk.

Table 5.1 OCCURRENCE OF NO_x POLLUTION IN DIFFERENT SITUATIONS

Situation	*Concentration* (ppb) *of*		*Duration of exposure to these concentrations,* *etc.*	*Sources of information*
	NO	*NO$_2$*		
Rural site in an industrialized country, but remote from obvious sources of pollution	6–10 >50	9–10 >50	Annual means Hourly means occurring less than 200 times per year, mainly in winter	Martin and Barber (1981)
Rural site near industrial source of NO_x	180	150	Average concentrations during passage of a plume of pollutant; average duration was 20 min	Harrison and McCartney (1979)
Urban roadside	300–500	100	Concentrations which can be experienced throughout the day, but falling at night	Hickman, Bevan and Colwill (1976)
Greenhouse enriched with CO_2 to 1100 ppm by burning kerosene	300	50–100	Daytime concentrations (CO_2 enrichment is unnecessary at night)	Capron and Mansfield (1975)
Greenhouse with flueless kerosene burner as sole source of heat	2700–3500	Approx. 5% of concentration of NO	Concentrations around crop when burner operating normally, with greenhouse ventilation rates of 0.95–1.25 air changes per hour	Law *et al.* (1982)

Entry of NO_x into leaves

The rate of uptake of a gaseous pollutant into a leaf is dependent on several physical factors, such as the aerodynamic resistance and the stomatal resistance (Bennett, Hill and Gates, 1973). However, before the pollutant can cause injury within a plant cell, it must first enter into solution in the extracellular water contained in the cell wall. The simple question of solubility in water is therefore likely to assume importance when we are considering two pollutants which differ markedly in solubility. This is the case with NO and NO_2, for the former is only sparingly soluble in water $(0.047\,m^3\,m^{-3}\,H_2O$ at $20\,°C$ and 1 atmosphere $(\approx101\,kPa)$ pressure). The solubility of NO_2 cannot be quoted in the same terms because of its reactions with water, but it can be regarded as very high.

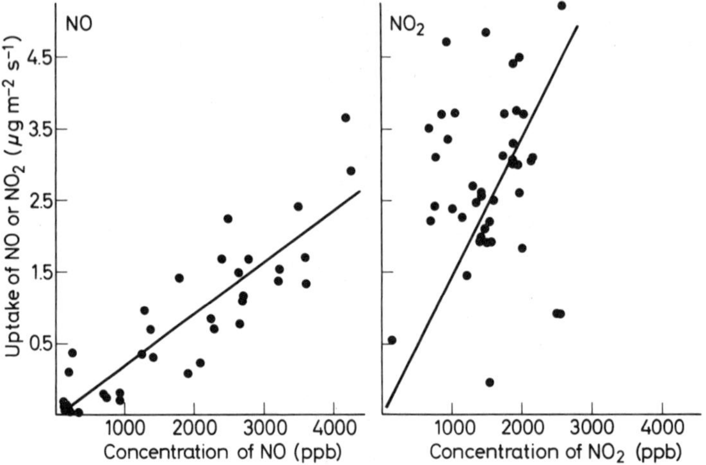

Figure 5.6   Uptake per unit leaf area of NO and NO_2 into leaves of sweet pepper (cv. Bellboy)

In view of this difference in solubility, it would be expected that NO and NO_2 would be taken up at different rates by plants. That this is the case, is illustrated in *Figure 5.6*. The uptake of NO_2 per unit leaf area was almost three times that of NO when the two gases were present at the same concentration. Despite the careful selection of uniform plant material for the determinations in *Figure 5.6*, there was great variation in the data. That this might have been due to variation in stomatal resistance, was shown not to be the case, because when the results were expressed as μg of NO_x taken up per g of H_2O lost, there was no reduction in the scatter of the points (*Figure 5.7*). It therefore seems likely that when the conditions are such that the stomata are not limiting the uptake, then the rate is predominantly determined by another resistance. This resistance must reflect the ability of the cells inside the leaves to take in NO_x, and it is therefore the 'residual internal resistance' (Unsworth, Biscoe and Black, 1976). This resistance would be very low if the cells were able to assimilate the products of NO_x solution so rapidly that their accumulation in the extracellular water was

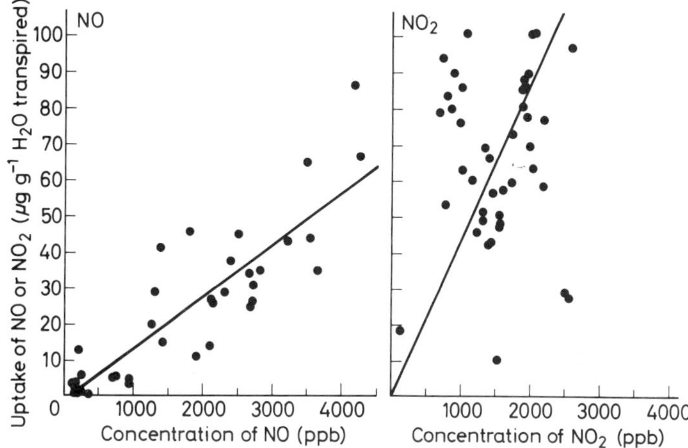

Figure 5.7   Data from *Figure 5.6* expressed as uptake of NO or NO$_2$ per unit of water lost

prevented. The feature of *Figures 5.6 and 5.7* that is of particular interest is the great scatter of the points, which must reflect great differences in internal resistance. If such variation reflects differences in the ability to assimilate NO$_x$, and in the sensitivity of individual plants to damage by NO$_x$ (which might be inversely related to assimilation ability), then it may be possible to select crop varieties which possess some resistance to injury.

Symptoms of NO$_x$ damage

When a plant has been exposed to a pollutant, the amount of damage suffered will vary in its severity according to several factors such as concentration and length of exposure. The symptoms are sometimes regarded as falling into the categories of 'invisible injury' and 'visible injury'. Invisible or hidden injury is usually defined as that where there are no obvious signs of damage on the plant, except an overall reduction in growth. This reduction is frequently apparent only when the polluted plant can be compared with another that has not been exposed to pollution. Several recent investigations have provided illustrations of growth reductions unaccompanied by visible markings (e.g. Bell and Clough, 1973; Ashenden and Mansfield, 1978). In a study by Spierings (1971), tomato plants fumigated with 500 ppb NO$_2$ showed an average reduction in height of 10% after 45 days. A fumigation with 250 ppb NO$_2$ for the entire growing period of 4 months resulted in a decrease in the yield of fruit (fresh weight) of 22%. Yield reductions of this magnitude are obviously important with a commercial crop. They are not, however, always obvious to the grower who will therefore not take any steps to reduce the losses.

Figure 5.8 and 5.9 illustrate the point in question. *Figure 5.8* shows a sweet pepper crop that has been grown in a greenhouse equipped with a flueless kerosene burner to provide heat and CO$_2$ enrichment. The plants are all healthy in appearance and would not give cause for concern to a grower. However, if the appearance of the crop is compared with that in

Figure 5.8   Pepper plants (cv. Bellboy) grown using a kerosene burner for heating and CO_2 enrichment. NO levels were about 2000 ppb for most of the day

Figure 5.9, which shows the same variety of pepper grown in similar conditions, except that there were no pollutants present in the atmosphere, then it is apparent that the NO_x may have caused quite a dramatic reduction in the size of the plants. For the crop in *Figure 5.9*, the heat and CO_2 were controlled so as to duplicate the conditions created by the burner: the heat was from hot-water pipes and the CO_2 came from

Figure 5.9   Pepper plants (cv. Bellboy) grown using a hot-water-pipe system for heating, and pure CO_2 from cylinders for CO_2 enrichment. No pollutants were present

cylinders of the pure gas, so that the major difference between the two sets of plants was that those heated by the burner experienced the pollution, whereas those heated by the pipe-heat system were pollution-free. The concentrations of NO_x in the greenhouse of *Figure 5.8* were monitored, and were approximately 2000 ppb NO with only 50 ppb NO_2 for most of the day, but there were occasions when the NO increased to 5000 ppb. Concentrations of other pollutants, such as sulphur dioxide and ethylene, were also monitored but were found to be negligible. The reduction in growth of the plants was reflected in the weight of the fruit obtained, as shown in *Figure 5.10*. While we cannot entirely rule out the possibility that

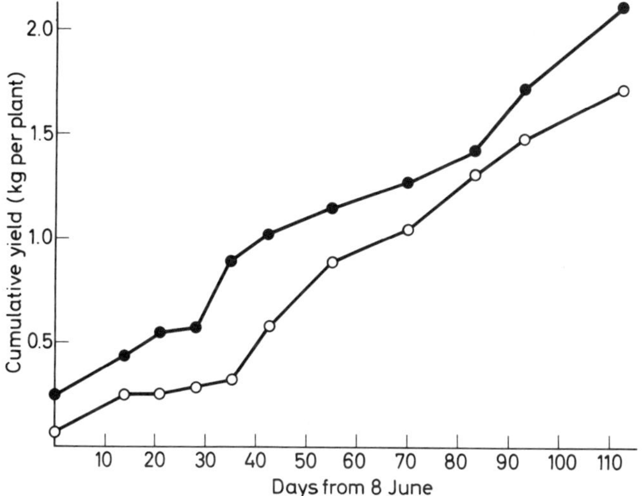

Figure 5.10   Cumulative yields of peppers from plants grown under the two heating systems (*see Figures 5.8 and 5.9*). ●——● hot water pipes and pure CO_2; ○——○ heat and CO_2 from kerosene burner

uncontrolled microclimatic factors may have been partly responsible for this reduction, in separate experiments in controlled-environment chambers it has been found that exposure to NO of up to 5000 ppb for 4 weeks causes a reduction in total dry weight of pepper plants of up to 20% (Law, unpublished data); the plants were growing in an atmosphere containing 2000 ppm CO_2 to which NO was added. Thus, the differences in yield in *Figure 5.10* are most likely to be caused by the action of NO_x, mainly during the early growth of the plants before the fruit was gathered. Apart from differences in yield, which amounted to 17%, the production of fruit was delayed by several days. As it is the early crop that frequently fetches the highest price, the economic losses from this invisible injury may be greater than the simple difference in yield.

Although we have referred to the reduced growth of the plants in *Figure 5.8* as a case of 'invisible injury', it must be emphasized that our use of the term in this context is somewhat unusual. Although the difference in growth of the peppers in *Figures 5.8 and 5.9* is very marked, the growth of the crop in the greenhouse heated by the kerosene burner was *greater* than that in a greenhouse equipped with conventional heating and no CO_2 enrichment. Thus the inhibitory effect of NO_x is more than compensated by the high CO_2 concentration, and the total pollution produced by the burner, including the CO_2, is actually beneficial. This is one reason why the effects of NO_x pollution have been overlooked by horticulturists. The scale of growth of the plants in *Figure 5.9* is something that is rarely seen in commercial horticulture. It could, however, be commonplace if the high CO_2 levels produced by kerosene or gas burners were not accompanied by NO_x pollution.

The mechanism by which NO_x causes injury is still largely a matter of speculation. It has been shown by Capron and Mansfield (1976), using detached tomato leaves, that 500 ppb of NO or NO_2 reduced the net photosynthetic rate by about 30% when compared with unpolluted control leaves. Wellburn, Majernik and Wellburn (1972) have also shown that fumigation with NO_2 causes a reversible swelling of the thylakoids in the chloroplasts of *Vicia faba*. A physical disruption such as this may well be partly responsible for the reduced photosynthetic rates.

A reduction in photosynthesis would, in the long term, inevitably lead to a reduction in growth. It is unlikely, however, that this alone is the cause of invisible injury, for Wellburn et al. (1976) have shown that NO_x fumigation results in changes of several enzyme levels within the cell (*see also* Wellburn, this volume, Chapter 7) These changes must inevitably utilize energy and materials that would otherwise have been available for

Figure 5.11 NO_x damage on pepper leaf. After fumigation with NO_x the leaf on the left shows typical symptoms of damage. Control leaf is on the right

growth and so this diversion of resources may also contribute to the symptoms of injury. If the injury is severe enough to cause either chlorosis or cell death, the injury can be referred to as 'visible'.

Occasional occurrences of visible injury, like invisible injury, have largely been ignored by growers in the past: where they have been sufficient to cause concern, it has not often been possible to attribute them to NO_x (Hand, 1979). Visible or acute injury may take the form of chlorotic areas appearing on the affected leaves, frequently associated with necrotic areas (*Figure 5.11*). These symptoms are often mistaken for the effects of other agents of which the grower is more aware, such as nutrient deficiency or sunscorch. Now that the attention of growers has been drawn to the possible effects of pollution from flueless heaters, it is likely that there may be more reports of acute NO_x damage in future.

Factors influencing the degree of NO_x damage

Although the amounts of NO_x in greenhouses can be very high compared with those out of doors, there are some factors which may operate to reduce the effects on plants.

CARBON DIOXIDE

NO_x pollution occurs in greenhouses when fossil fuels are burnt for heating or for CO_2 enrichment of the atmosphere. The pollution is therefore always associated with increased CO_2 levels, and its effects must be assessed in the context of CO_2 concentrations of about 1000 ppm (as recommended for CO_2 enrichment) or 2000–4000 ppm which occur when flueless burners are a sole source of heat. The physiological effects of 1000 ppm and >2000 ppm CO_2 are probably similar, because in low winter irradiances, photosynthesis is saturated by 1000 ppm CO_2.

It has been shown that the presence of additional CO_2 can reduce the damaging effect of a pollutant. In a study by Hou, Hill and Soleimani (1977) it was found that in alfalfa, twofold CO_2 enrichment of the atmosphere reduced the visible injury caused by SO_2 and NO_2 by over 50%.

It is known that increases in CO_2 concentration lead to stomatal closure (Meidner and Mansfield, 1968) and it has been suggested (Mansfield, 1973) that this reaction could be important in reducing the entry of a gaseous pollutant into a plant. In a study by Majernik and Mansfield (1972) it was shown that the stomatal reaction to CO_2 could still occur in the presence of a pollutant, even, in the case of SO_2, under conditions where SO_2 alone would stimulate stomatal opening. Thus, under the conditions of CO_2 enrichment in a greenhouse it would be expected that some stomatal closure would occur, leading to a greater resistance to pollutant uptake than in normal air.

The additional CO_2 may also act at a metabolic level, for it increases the rate of photosynthesis and so may provide cells with increased capacity for repair processes or for detoxification mechanisms. Little research has been conducted to ascertain the importance of such effects.

Even though CO_2 may help to mitigate the effects of NO_x pollution, it is clear from *Figures 5.8 and 5.9* that it does not eliminate them. The economic effects of NO_x in this situation should perhaps be considered in relation to the beneficial effects of CO_2 enrichment: the full benefits are not realized when there are high levels of NO_x in the atmosphere.

NITRATE METABOLISM

When NO and NO_2 dissolve in the extracellular water within a leaf, they form nitrate and nitrite ions. These are normally present within a plant cell as part of the nitrate assimilation pathway in which nitrate is reduced first to nitrite and then to ammonium ions, which are used in the formation of amino acids and finally proteins. It has been demonstrated that, after fumigation with NO_2, pea plants showed a higher content of nitrate and nitrite ions, with an associated increase in the rate of protein synthesis (Zeevaart, 1976). In addition, in a study by Wellburn, Wilson and Aldridge (1980) it was shown that, after 3 hours' exposure to 400 ppb NO, tomato plants showed significant reductions in the levels of nitrate reductase, and significant increases in the levels of nitrite reductase. However, when the NO level was increased to 1500 or 2500 ppb, the levels of nitrite reductase were much greater in the cultivar Sonato; this is a cultivar that has been shown to be more tolerant of NO pollution (Anderson and Mansfield, 1979). Increased activity of the enzymes glutamate dehydrogenase (involved in reductive amination), glutamate oxaloacetate transaminase and glutamate pyruvate transaminase were also observed. As these enzymes are the ones involved in the pathway by which nitrogen is assimilated into amino acids, then it can be seen that, after fumigation with NO, there is an increased ability to assimilate nitrogen. This could reflect an ability to absorb and incorporate nitrogen from NO to benefit the plant, not only by providing nitrogen if this is limiting, but also by preventing the accumulation of potentially toxic nitrate and nitrite ions.

The output of NO from a 66 kW kerosene burner has been measured as 160 mg min^{-1} (Law *et al.*, 1982). This means that in a growing season of 100 days, with the burner operating for, say, 12 h daily, approximately 11.5 kg of NO will be produced. In a greenhouse with a floor area of 500 m^2, this rate of input is equal to 107 kg N ha^{-1}. *Table 5.2* lists the nitrogen

Table 5.2   QUANTITIES OF NITROGEN REMOVED BY VARIOUS CROPS (ONLY AMOUNTS FOUND IN PARTS NORMALLY HARVESTED ARE GIVEN). (From Mengel and Kirkby, 1978)

Crop	Nitrogen removed (kg ha^{-1})
Tomatoes (fruit)	130
Maize (grain)	150
Barley	40
Wheat	56
Sugar-cane	110
Grapes	110

requirements of several crops: it can be seen that if all the nitrogen in the NO_x were as available to the crop as that in conventional fertilizers, then in theory the burner could provide almost all of the nitrogen requirements of the different crops.

Of course, it is not realistic to suppose that all of the nitrogen could be used in this manner, for not all the NO_x in the atmosphere will enter the crop and, of that which is taken up, not all will be utilized in the nitrogen assimilation pathway. It has, however, been demonstrated that under conditions of low soil-nitrogen content, when plant growth was considered to be reduced by availability of nitrogen, then fumigation with low levels of NO_x can benefit the plant and stimulate growth (Anderson and Mansfield, 1979). *Table 5.3* shows the results of a commercial trial by Talent (1978) in

Table 5.3 COMPARISON OF YIELDS OF TOMATOES (kg m^{-2}) GROWN WITH THREE LEVELS OF NITROGEN NUTRITION (APPLIED AS NUTRIENT SOLUTION) USING TWO HEATING SYSTEMS. VALUES GIVEN ARE TOTAL YIELDS FROM FOUR CULTIVARS: ESTRELLA, ODINE, SARINA AND SONATINE. (From Talent, 1978)

Heating system	*Nitrogen level in nutrient solution* (ppm)					
	170	85	43	170	85	43
	(normal strength)					
	Yield after 1 month's picking			*Yield after 4 months' picking*		
Kerosene burner	2.6	2.8	2.5	13.5	14.2	12.2
Hot water pipes with pure CO_2 for enrichment	3.7	3.9	3.0	15.5	14.6	11.9
% change in yield attributable to NO_x pollution from burner	−30	−29	−17	−13	−3	+3

which tomatoes were grown using a kerosene burner to provide heat and CO_2 enrichment, while others were grown in a greenhouse heated by hot-water pipes, with pure CO_2 added from cylinders; the heat regime and the degree of CO_2 enrichment were carefully matched. The only detectable difference in the aerial environment around the two crops was that those grown in the kerosene-heated house experienced NO_x levels of approximately 2000 ppb for most of the day. The data in *Table 5.3* show that plants grown with a normal supply of nitrogenous fertilizer produced a smaller yield of tomatoes in the presence of NO_x pollution, but nitrogen-deficient plants were either little affected by NO_x or showed a slight improvement in yield.

On the basis of this result we could suggest that a possible way of alleviating the effects of NO_x pollution in greenhouses would be to reduce the amounts of nitrogenous fertilizer applied to the crop. Unfortunately, however, the level of soil nitrogen would have to be reduced to a point where crop yield would be affected, and so any effect of reducing or eliminating NO_x damage would be lost. It is unlikely, therefore, that this could become a recommended practice in commercial horticulture.

AIR MOVEMENT

The resistance to molecular diffusion caused by the thin boundary layer surrounding the leaf can be important in regulating the entry of a gaseous pollutant. The effects of SO_2 on perennial ryegrass were examined by Ashenden and Mansfield (1977) at two wind speeds. In a wind speed of $25\,m\,min^{-1}$, the leaf areas and total weights of the plants after 4 weeks were significantly lowered in the presence of the pollutant, whereas at a lower wind speed of $10\,m\,min^{-1}$ the SO_2 had no significant effect. The thickness of the boundary layer is increased as wind speed falls, and so the resistance to pollutant uptake increases; fewer molecules of the pollutant then enter the leaf to cause damage.

In a greenhouse there is characteristically very little air movement around the crop and so the boundary layer will be large. This diffusion resistance must therefore play some part in reducing the uptake of NO_x and protecting the plants from damage. Thus, although the amounts of NO_x in greenhouses can be much higher than out of doors, the uptake into leaves is unlikely to be as high as the concentrations quoted at the bottom of *Table 5.1* would suggest.

THE FORM OF NO_x

As we read down *Table 5.1* the increasing levels of NO_x encountered in different situations are accompanied by a rise in the $NO:NO_2$ ratio. In a greenhouse we would expect this ratio to be higher than that outdoors, for two reasons. First, the pollutants are monitored close to their source, and the time available for conversion of NO to NO_2 is reduced compared with outdoor situations. Second, other atmospheric components, especially ozone, which would speed the conversion of NO to NO_2, are unlikely to be present in significant amounts in a greenhouse.

Our measurements of the uptake of NO and NO_2 (*Figure 5.6*) show that an increase in the $NO:NO_2$ ratio should decrease the rate of uptake per unit of NO_x. This is, therefore, another factor which can be regarded as offering some modest protection to greenhouse crops. However, some of the smaller greenhouse heaters apparently emit NO_2 rather than NO (Ashenden, Mansfield and Harrison, 1977) and so the converse might apply in some cases.

ABSENCE OF OTHER POLLUTANTS

The fuels recommended for burning in greenhouses are virtually free from sulphur, and therefore emissions of SO_2 concomitant with NO_x are unlikely. Our own monitoring has confirmed that this is the case: we have never detected SO_2 concentrations above 10 ppb in greenhouses in which 'recommended' fuels were burned.

In the outside air, a major hazard to crops seems to be the association of SO_2 and NO_x (Ashenden and Mansfield, 1978). In *Table 5.1*, the concentrations of NO and NO_2 that we have quoted as being typical of an urban roadside and of a greenhouse enriched with CO_2, are very similar. There is, however, the significant difference that there is usually SO_2 by a roadside, both from the traffic emissions and from other sources in a town. Perhaps the greenhouse is the only situation where we can consider the practical consequences of exposure to 'pure NO_x', although in many cases where burner design is deficient there may be appreciable amounts of ethylene (Davison and Wharmby, 1979).

Control measures

DESIGN OF BURNER

With the design of burner currently in use, the hydrocarbon fuel is mixed with air and the fuel/air mixture is then ignited. Carbon dioxide is produced from the oxidation of carbon in the fuel, whereas the NO_x originates only to a minor extent from the oxidation of nitrogen-containing compounds in the fuel, and comes mainly from the oxidation of atmospheric nitrogen. The temperature of the flame is important in determining the amount of NO_x produced: the higher the temperature, the greater the quantity of NO_x formed.

However, it is possible, by recirculating some of the heat from combustion without recycling the products, to use very lean fuel/air mixtures (Lloyd and Weinberg, 1975). This results in much lower flame temperatures, which can be below the critical temperature required for the formation of NO_x. In effect, these burners are pollution-free, for the very lean fuel mixtures ensure that complete combustion occurs: there is therefore less risk of the formation of hydrocarbon pollutants, such as ethylene, which are also known to be detrimental to plants (Davison and Wharmby, 1979). Jones, Lloyd and Weinberg (1978) have reviewed some of the designs of burners in which heat recirculation is employed. Unfortunately, such burners are inevitably more complicated, and therefore more expensive, than the simple type currently used in commercial horticulture. Until this additional cost can be shown to be less than the predicted losses to the crop caused by NO_x, it is unlikely that they will offer a practical means of reducing NO_x in greenhouses.

RESISTANT CULTIVARS

A more promising solution might be the development of cultivars that are less sensitive to NO_x. It has been suggested earlier that differences in residual internal resistances to NO_x (*Figure 5.6*) indicate wide variation in uptake, which may reflect sensitivity/tolerance and provide a basis for selection.

In a trial designed to examine the effects of 400 ppb NO on four varieties of tomato, Anderson and Mansfield (1979) found significant differences in growth between them. After 35 days of fumigation the change in dry weight caused by the NO varied from a reduction of 28% for the variety Adagio to an increase of 89% for Sonato (both statistically significant). It was suggested that the increased growth of Sonato was attributable to its utilization of NO as a source of nitrogen, and thus avoidance of any damaging effects. We can suggest that the scatter of points in *Figure 5.6* represents variation in the ability to utilize NO_x as a source of nitrogen. It will be worth exploring the possibility of screening a population using uptake measurements which can be completed quickly on individual plants.

It can be speculated that some selection may have been performed unknowingly, as trials for the development of new varieties are often performed in CO_2-enriched greenhouses polluted with NO_x. This could explain the tolerance displayed by the variety Sonato, and it encourages the view that, in the foreseeable future, we shall see the development of varieties with greater tolerance of NO_x than those currently grown.

VENTILATION

It has been shown earlier that the ventilation rate is important in determining the level of NO_x that builds up in a greenhouse. The current recommendation to growers in Britain (Hand, 1979) suggests that, in greenhouses which are nearly airtight (with a ventilation rate of <0.3 air changes per hour), accumulation of NO_x can be avoided by purging the atmosphere at least once a day. However, reference to *Figure 5.5* shows that this practice will relieve the situation only temporarily, for the levels will rapidly build up again to over 2500 ppb within an hour. This will occur only if the burner operates continuously for that hour, and, as suggested by Hand (1979), a limitation on the time of operation of the burner might be considered. However, because these burners are generally employed as the only source of heat, their operation can be limited usefully only by a thermostat, unless some supplementary heating of another type is used.

We therefore consider that spasmodic, uncontrolled ventilation is inadequate. The NO_x levels need to be controlled by a ventilation programme which cuts the loss of CO_2 and heat to a minimum. The 'unacceptable' level of NO_x which will determine the degree of ventilation will depend on the tolerance of the crop being grown.

It is possible to predict the ventilation rate of a particular greenhouse on the basis of environmental parameters. Kozai and Sase (1978) have shown that at low wind speeds ($<2\,\mathrm{m\,s^{-1}}$) with the ventilators closed, the ventilation rate is dependent mainly upon the difference in temperature between the inside and outside, whereas at higher wind speeds ($>2\,\mathrm{m\,s^{-1}}$) the rate is dependent more upon the wind speed and the number of spans in the greenhouse. We suggest that this model for predicting the ventilation rate could be applied to a particular greenhouse. If an additional ventilation capacity were added by the provision of extractor fans then, with a knowledge of the NO_x output of the burner being used, it would be possible to predict when the levels of NO_x would be unacceptable, and to ventilate as necessary. A small microprocesser device could be used to control such a system in order to ensure that the loss of heat and CO_2 was kept to a minimum.

Further research is necessary, however, in order to ascertain the concentration of NO_x that can be considered 'tolerable' by any particular crop, for the losses due to NO_x damage must be weighed against the benefits of using these flueless burners. In the long term, the development of cultivars of crop plants that are tolerant to NO_x pollution would seem to be the most promising method of increasing crop yields from polluted greenhouses, by ensuring that the maximum benefit from the heaters is achieved.

Acknowledgment

R.M. Law acknowledges the receipt of a CASE studentship from the Science Research Council, and generous assistance from Mr C.J.W. Talent and others at Fairfield Experimental Horticulture Station.

References

ANDERSON, A.S. and MANSFIELD, T.A. (1979). *Environmental Pollution, 20,* 113–121

ASHENDEN, T.W. and MANSFIELD, T.A. (1977). *Journal of Experimental Botany,* **28,** 729–735

ASHENDEN, T.W. and MANSFIELD, T.A. (1978). *Nature,* **273,** 142–143

ASHENDEN, T.W., MANSFIELD, T.A. and HARRISON, R.M. (1977). *Environmental Pollution,* **14,** 93–99

BELL, J.N.B. and CLOUGH, W.S. (1973). *Nature,* **241,** 47–49

BENNETT J.H., HILL, A.C., and GATES D.M. (1973). *Journal of the Air Pollution Control Association,* **23,** 957–962

CAPRON, T.M. and MANSFIELD, T.A. (1975). *Journal of Horticultural Science,* **50,** 233–238

CAPRON, T.M. and MANSFIELD, T.A. (1976). *Journal of Experimental Botany,* **27,** 1181–1186

COLUCCI, A.V. and SIMMONS, W.S. (1978). *Nitrogen Oxides: Current Status of Knowledge.* Report EA-668. Electrical Power Research Institute, Palo Alto, California

DAVISON, A.W. and WHARMBY, S. (1979). *Journal of the Royal Horticultural Society,* 37–39

DOCHERTY, A.C. (1974). *Analyst,* **99,** 853–858

EGGLETON, A.E.J. (1974). In *Some Gaseous Pollutants in the Environment,* pp.13–15. Natural Environment Research Council, London

HAND, D.W. (1979). *ADAS Quarterly Review,* **33,** 134–143

HARRISON, R.M. and McCARTNEY, H.A. (1979). *Atmospheric Environment,* **13,** 1105–1120

HICKMAN, A.J., BEVAN, M.G. and COLWILL, D.M. (1976). *Atmospheric Pollution from Vehicle Emissions: Measurements at Four Sites in Coventry, 1973.* Report No. LR 695. Transport and Road Research Laboratory, Crowthorne, Berkshire, UK

HOU, L.Y., HILL, A.C. and SOLEIMANI, A. (1977). *Environmental Pollution,* **12,** 7–16

JONES, A.R., LLOYD, S.A. and WEINBERG, F.J. (1978). *Proceedings of the Royal Society*, **360**, 97–115

KOZAI, T. and SASE, S. (1978). *Acta Horticulturae*, **87**, 39–49

LAW, R.M. HARRISON, R.M., McCARTNEY, H.A. and TALENT, C.J.W. (1982). *Journal of Experimental Horticulture*, **32**, 49–55

LLOYD, S.A. and WEINBERG, F.J. (1975). *Nature*, **257**, 367–370

MAJERNIK, A. and MANSFIELD, T.A. (1972). *Environmental Pollution*, **3**, 1–7

MANSFIELD, T.A. (1973). Commentaries in Plant Science No. 2. (In *Current Advances in Plant Science*, **2**, 11–20)

MARTIN, A. and BARBER, F.R. (1981). *Atmospheric Environment*, **15**, 567–578

MEIDNER, AH. and MANSFIELD, T.A. (1968). *Physiology of Stomata*. McGraw-Hill, London

MENGEL, K. and KIRKBY, E.A. (1978) (Eds). *Principles of Plant Nutrition*. International Potash Institute, Worblaufen-Bern

NOXON, J.F. (1978). *Journal of Geophysical Research*, **83**, 3051–3057

SPIERINGS, F.H.F.G. (1971). *Netherlands Journal of Plant Pathology*, **77**, 194–200

TALENT, C.J.W. (1978). In *Annual Report of Fairfield Experimental Horticulture Station*, pp.32–34. MAFF, Kirkham, Lancashire

UNSWORTH, M.H., BISCOE, P.V. and BLACK, V. (1976). In *Effects of Air Pollutants on Plants*, pp.5–16 (Mansfield, T.A., Ed.), University Press, Cambridge

WELLBURN, A.R., MAJERNIK, O. and WELLBURN, F.A.M. (1972). *Environmental Pollution*, **3**, 37–49

WELLBURN, A.R., WILSON, J. and ALDRIDGE, P.H. (1980). *Environmental Pollution*, **22**, 219–228

WELLBURN, A.R., CAPRON, T.M., CHAN, H.S. and HORSMAN, D.C. (1976). In *Effects of Air Pollutants on Plants* (Mansfield, T.A., Ed.), University Press, Cambridge

ZEEVAART, A.J. (1976). *Environmental Pollution*, **11**, 97–108

6

VARIATION IN PLANT RESPONSE TO OZONE: A CONCEPTUAL MODEL OF PHYSIOLOGICAL EVENTS

DAVID T. TINGEY
US Environmental Protection Agency, Corvallis Environmental Research Laboratory, Corvallis, Oregon

GEORGE E. TAYLOR, JR.
Environmental Science Division, Oak Ridge National Laboratory, Oak Ridge, Tennessee

Introduction

Ozone effects on vegetation range from overt injury to subtle modifications of cellular biochemistry and whole-plant physiology. Irrespective of the organizational level at which the response occurs (molecular to whole plant), ozone phytotoxicity involves molecular events that culminate in a perturbed cellular structure and function. Differential plant response to ozone has been related to differences in environmental conditions and genic expression (*Table 6.1*). These factors are the ultimate sources of

Table 6.1 FACTORS THAT INFLUENCE VARIATION IN VEGETATION RESPONSE TO OZONE (From Heck, 1968; Ting and Heath, 1975; Heck, Mudd and Miller, 1977)

Environmental conditions	*Genic expression*
Edaphic	Interspecific
Water stress	Intraspecific
Nutrients	Subspecies
pH	Populations
Temperature	Cultivars
Climatic	Varieties
Light regime	Individuals
Temperature	Clones
Atmospheric	Developmental
Humidity	Plant
Carbon dioxide	Leaf
Other pollutants	Tissue
Biotic	Cell
Pathogens	
Competition	

variation in plant response to ozone, while the immediate causes are disparate physiological states.

The extensive variation among plants in ozone response is often considered to be an enigma but provides an unique opportunity to focus on

"

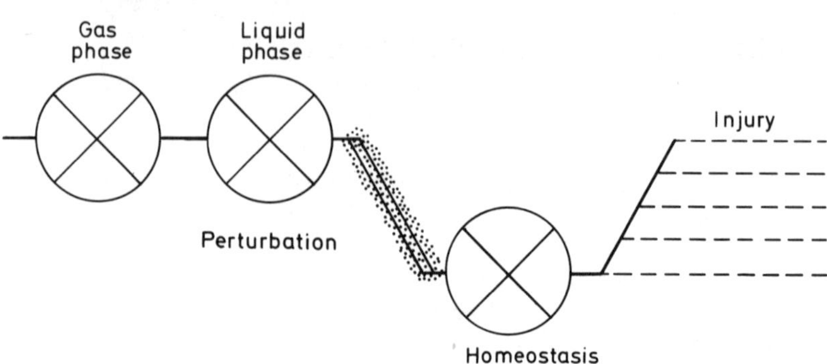

Figure 6.1 Conceptual model of the physiological and biochemical processes controlling plant response to ozone. Major processes underlying plant variation in ozone response are conductance (gas and liquid phases) and homeostasis

the physiological basis of the differences. Plant response may be viewed as the culmination of a sequence of biochemical and physiological events, beginning with ozone uptake and ending with injury. From this perspective, a conceptual model with four sequential components is proposed: leaf conductance (gas and liquid phases), perturbation, homeostasis and injury (*Figure 6.1*). The model is described as follows.

Leaf conductance

Leaf conductance regulates ozone movement from ambient air into cellular perturbation sites. Gas-phase ozone flux depends upon the concentration gradient between the ambient air and deposition sites within the leaf and the resistance to mass transfer along the diffusion path. The flux of ozone following deposition on to wet cell surfaces is governed by liquid-phase processes including diffusion, formation of pollutant derivatives and reactivity with scavenger systems.

Perturbation

This comprises primary ozone-induced changes in cell structure and/or function. Phytotoxicity results from the spontaneous reaction of ozone or its derivatives with macromolecules that are instrumental to the maintenance of cell function. Membranes are the primary sites of ozone attack, resulting in changed permeability.

Homeostasis

Homeostasis is the recovery process following a stress. After perturbation, cells respond with either repair or compensatory processes.

Injury

The preceding sequence of events results in injury, the extent of which is controlled by ozone conductance, perturbation and homeostasis. The susceptibility of membranes and their pervasive importance in cellular metabolism dictates metabolic repercussions throughout the cell. The summation of injury across cells has consequences for whole-plant physiology.

The literature relating to the various model components will be reviewed to illustrate the interplay of environment and genotype as they influence the physiological state of the plant and, therefore, injury expression.

Leaf conductance

Ozone phytotoxicity results from biochemical processes occurring in cells of the leaf interior. The ease with which ozone diffuses from the ambient air to perturbation sites is therefore a determinant of injury. Ozone flux is a function of physical and chemical properties along the gas-to-liquid pathway. Within the gas and liquid phases, physical structures impede flow and chemical reactions may scavenge ozone molecules. Because the conductance properties of the two pathway segments are dissimilar, each phase is discussed separately.

GAS-PHASE CONDUCTANCE

Ozone flux (J) into the leaf results from a chemical potential gradient between the bulk air (C_a) and the leaf interior (C_i). Flux is proportional to the pathway conductance (g) and is inversely related to the resistances to mass transfer as ozone diffuses through the boundary layer (r_a), stomata

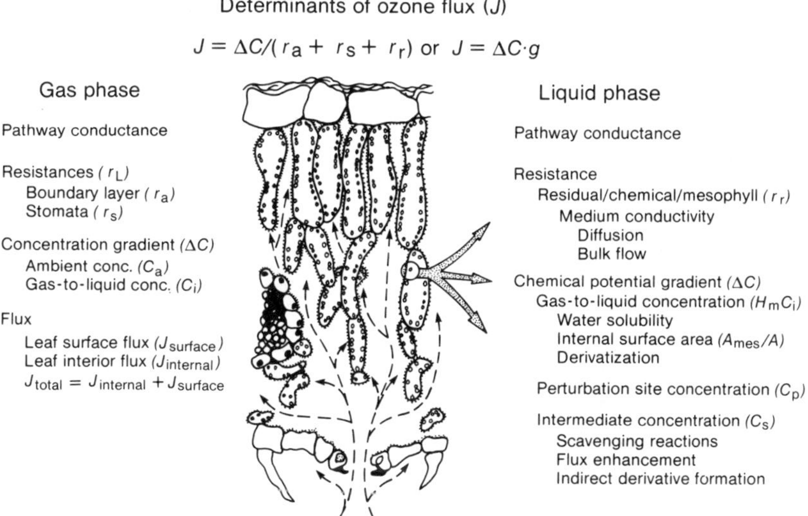

Figure 6.2   The relationship between chemical flux, ozone potential gradient, and leaf conductance or resistances

Table 6.2 ESTIMATES OF OZONE FLUX, CONCENTRATIONS AND RESISTANCES DERIVED FROM ANALOGY OF POLLUTANT FLUX TO OHM'S LAW

Component	Symbol	Units	Solution	Comments
(1) Total O_3 flux to the leaf	J_{total}	$\mu g\,m^{-2}\,s^{-1}$	Experimentally measured	
(2) O_3 flux to leaf surface	$J_{surface}$	$\mu g\,m^{-2}\,s^{-1}$	Experimentally measured	
(3) O_3 flux to leaf interior	$J_{internal}$	$\mu g\,m^{-2}\,s^{-1}$	$J_{total} = J_{surface} + J_{internal}$ $J_{internal} = J_{total} - J_{surface}$	
(4) Ambient O_3 concentration	$C_a^{O_3}$	$\mu g\,m^{-3}$	Experimentally measured	
(5) Boundary-layer resistance to water vapour efflux	r_a^{H2O}	$s\,m^{-1}$	Experimentally measured	
(6) Boundary-layer resistance to O_3 flux	$r_a^{O_3}$	$s\,m^{-1}$	$r_a^{O_3} = r_a^{H2O} \times 1.63$	Assumes resistance is a function of the ratio of the diffusivity of the two gases in air
(7) Stomatal resistance to water-vapour efflux	r_s^{H2O}	$s\,m^{-1}$	Experimentally measured	
(8) Stomatal resistance to O_3 flux	$r_s^{O_3}$	$s\,m^{-1}$	$r_s^{O_3} = r_s^{H2O} \times 1.63$	Same assumption as $r_a^{O_3}$
(9) Leaf surface resistance to adsorption of O_3	$r_e^{O_3}$	$s\,m^{-1}$	$J_{surface} = (C_a^{O_3} - C_e^{O_3})/(r_a^{O_3} + r_e^{O_3})$ $r_e^{O_3} = (C_a^{O_3}/J_{surface}) - r_a^{O_3}$	Assumes O_3 is destroyed on leaf surface after adsorption. C_e is the concentration on the epidermis; $C_e = 0$
(10) Ozone concentration at the leaf surface	$C_c^{O_3}$	$\mu g\,m^{-3}$	$J_{total} = (C_a^{O_3} - C_c^{O_3})/r_a^{O_3}$ $C_c^{O_3} = C_a^{O_3} - (J_{total} \times r_a^{O_3})$	
(11) Residual resistance to O_3 flux	$r_r^{O_3}$	$s\,m^{-1}$	$J_{internal} = (C_c^{O_3} - C_i^{O_3})/r_s^{O_3}$ $J_{internal} = C_c^{O_3}/r_s^{O_3}$ $r_s^{O_3} = C_c^{O_3}/J_{internal}$ $r_s^{O_3} - (r_s^{H2O} \times 1.63) \neq 0$	Assumes, as above, that O_3 is destroyed within the leaf This $r_s^{O_3}$ is derived from the model Compare $r_s^{O_3}$ derived in (11) with measured $r_s^{O_3}$ (8); if (11) is greater than (8), O_3 experiences an additional resistance component other than $r_a^{O_3} + r_s^{O_3}$

and intercellular spaces (r_s) and liquid phase (r_r). The relationships between these parameters and ozone flux are analogous to Ohm's Law (Gaastra, 1959) as shown in *Figure 6.2*. Total ozone flux (J_{total}) consists of leaf-surface($J_{surface}$) and interior ($J_{internal}$) fractions. Both fractions may be large (Elkiey and Ormrod, 1980) although injury results from $J_{internal}$. Variable pathway conductance or a significant $J_{surface}$ means that the gas concentration surrounding the plant or the dose (concentration × duration of exposure) may not accurately reflect the cellular dose to which the plant actually responds (Runeckles, 1974).

The analogy of ozone flux to Ohm's Law can be extended in a manner comparable to the analysis of carbon dioxide exchange (Gaastra, 1959) resulting in a series of equations that provide estimates of ozone flux, concentration gradients and resistances along the pathway (*Table 6.2*). This approach is similar to previous treatments of pollutant flux into leaves (Bennett, Hill and Gates, 1973; Unsworth, Biscoe and Black, 1976; O'Dell, Taheri and Kabel, 1977; Black and Unsworth, 1979), except for the procedure for estimating r_r and the test for equality of pathway conductances for water vapour and ozone.

Many studies have focused on the relationship between gas-phase conductance and foliar injury, and most have concluded that stomatal conductance is the principal or sole factor underlying response differences (*Table 6.3*). This observation is common for both genetically controlled and environmentally induced variation. However, several issues detract from this consensus.

The data in *Table 6.3* imply that changes in stomatal conductance affect $J_{internal}$ with flux being proportional to conductance. However, experimental verification of this relationship is absent from any of the reports

Table 6.3 EXAMPLES OF GENETIC AND ENVIRONMENTAL CONTROL OF OZONE INJURY IN WHICH THE MECHANISM WAS THOUGHT TO BE STOMATAL CONDUCTANCE. OZONE FLUX WAS NOT MEASURED

Genetic variation	*Environmental variation*

Interspecific
  Pinus spp. (Evans and Miller, 1972)
  Multiple spp. (Thorne and Hanson, 1972)
	Edaphic
	  Potassium (Leone, 1976; Noland and
	    Kozlowski, 1979)

Intraspecific
  Acer rubrum (Townsend and Dochinger,
    1974)
  Nicotiana tabacum (Ting and Dugger,
    1971; Dean, 1972; Rich and Turner,
    1972; Turner, Rich and Tomlinson,
    1972)
  Allium cepa (Engle and Gabelman, 1966)
  Phaseolus vulgaris (Thorne and Hanson,
    1976; Butler and Tibbits, 1979)
	Water (Markowski and Grezesiak, 1974)
	  Salinity (Hoffman, Maas and Rawlins,
	    1973; 1975)
	  Nitrogen (MacDowell, 1965)
	
	Climatic
	  Light (Dunning and Heck, 1973)
	  Humidity (Otto and Daines, 1969)

Developmental
  Hypericum sp, (Ledbetter, Zimmerman
    and Hitchcock, 1959)
  Nicotiana tabacum (Glater, Solberg and
    Scott, 1962)
	Atmospheric
	  Ozone dose (MacDowell, 1965; Hill and
	    Littlefield, 1969; Runeckles and Rosen,
	    1977; Domerque, De Cormis and
	    Louguet, 1979)
	  Abscisic acid (Fletcher, Adedipe and
	    Ormrod, 1972)
	
	  Other pollutants (Jacobson and Colavito,
	    1976)

in *Table 6.3*. This omission may explain some of the inconsistencies in the literature. Flux ($J_{internal}$) was less in ozone-resistant than in sensitive cultivars of *Petunia hybrida* (Elkiey and Ormrod, 1980) as was predicted from an earlier study, in the same species, that reported a correlation between ozone sensitivity and leaf conductance (Thorne and Hanson, 1976). However, Elkiey and Ormrod (1979a) found no differences in the stomatal conductance in *Petunia hybrida* cultivars in spite of the dissimilar $J_{internal}$ values. In *Nicotiana tabacum*, pollutant uptake rates accounted for developmental differences in leaf response to ozone, although stomatal aperture did not differ among the leaves (Craker and Starbuck, 1973). However, dissimilar ozone responses in leaves of varying ages of the same species were explained by differences in stomatal size (Glater, Solberg and Scott, 1962).

Table 6.4 EXAMPLES OF VARIABLE PLANT RESPONSES TO OZONE THAT ARE NOT ASSOCIATED WITH DIFFERENTIAL GAS-PHASE CONDUCTANCE

| | Flux measurement | | |
Variation source	*Water vapour*	*Ozone*	*Reference*
Genetic			
Interspecific			
Cucumis sativus	X		Beckerson and Hofstra, 1979a
Raphanus sativus			
Glycine max			
Multiple Species		X	Townsend, 1974
Intraspecific			
Vitis labrusca	X		Rosen, Musselman and Kender, 1978
Glycine max		X	Yingjajaval, 1976
Developmental			
Phaseolus vulgaris	X		Evans and Ting, 1974a
Nicotiana tabacum		X	Leuning *et al.*, 1979
Gossypium hirsutum	X		Ting and Dugger, 1968
Glycine max		X	Yingjajaval, 1976
Environmental			
Biotic			
Nicotiana tabacum virus	X		Brennan, 1975
Glycine max virus	X		Vargo, Pell and Smith, 1978
Atmospheric			
Ethylene diurea chemical	X		Carnahan, Jenner and Wat, 1978
Phaseolus vulgaris			

Several workers have reported that there is no relationship between gas-phase conductance and ozone injury (*Table 6.4*). In *Gossypium hirsutum* and *Phaseolus vulgaris*, ozone susceptibility varied with leaf age, but the differences were not associated with gas-exchange rates (Ting and Dugger, 1968; Evans and Ting, 1974a). Even though stomatal conductance was at a maximum in the early morning, maximum foliar sensitivity to ozone occurred later in the day. In *Vitis labrusca*, an ozone-sensitive cultivar had a lower conductance and fewer stomata than its resistant counterpart (Rosen, Musselman and Kender, 1978). Although *Glycine max* cultivars differed in their ozone susceptibility, their stomatal conductance was similar (Tingey, Fites and Wickliff, 1976). It appears unwarranted to establish an *a priori* relationship between stomatal conductance and foliar injury.

The data relating ozone flux to foliar injury are also conflicting. Ozone flux was three times greater in young than old foliage of *Camellia* (Thorne and Hanson, 1972). Variation in J_{total} was related to leaf-age response differences in *N. tabacum* (Craker and Starbuck, 1973). Townsend (1974) found no consistent pattern between J_{total} and the ozone sensitivity of nine shade-tree species. The lowest ozone flux was $398\,\mu g\,m^{-2}\,s^{-1}$ to *Fraxinus americana*, while the highest was $1058\,\mu g\,m^{-2}\,s^{-1}$ to *Quercus alba*. Both species were ozone-sensitive.

In theory, $J_{internal}$ should be a reliable parameter from which to predict foliar response to ozone: however, the data are contradictory. In *Phaseolus vulgaris*, foliar injury occurred when $J_{internal}$ exceeded $5500\,\mu g\,m^{-2}$ in 1 hour (Bennett, 1979). However, a single 3 h exposure at one-half the concentration (0.27 ppm compared with 0.49 ppm) required a 64% greater $J_{internal}$ to produce leaf injury equivalent to the 1 h exposure. In *P. hybrida*, variable cultivar sensitivity was related to $J_{internal}$ (Elkiey and Ormrod, 1980): the resistant cultivar absorbed approximately 20% less ozone than its sensitive counterpart; $J_{internal}$ values ranged from 2352 to $3040\,\mu g\,m^{-2}$ s^{-1} (leaf area converted to single surface). In contrast, differential sensitivity in *Glycine max* cultivars was not related to $J_{internal}$ (Yingjajaval, 1976). The resistant cultivar consistently absorbed more ozone than the sensitive one.

The absence of a relationship between gas-phase conductance and flux is not uncommon. In two *Zea mays* cultivars, the one with the higher stomatal frequency and leaf conductance had the consistently lower photosynthetic rate, resulting from a lower liquid-phase conductance to carbon dioxide assimilations (Heichel, 1970).

Leaf morphology may affect ozone flux without affecting pathway conductance. Ozone resistance in *P. hybrida* cultivars was associated with trichome density, raised and roughened cuticular features and small epidermal cells (Elkiey, Ormrod and Pelletier, 1979). The positioning of the guard and epidermal cells was distinctly different among cultivars and these workers suggested that the alignment of these cells may significantly affect $J_{internal}$.

If $J_{surface}$ is a significant fraction of J_{total}, $J_{internal}$ may be reduced. Differences in J_{total} among *P. hybrida* cultivars were associated with trichome features that affected $J_{surface}$ (Elkiey, Ormrod and Pelletier, 1979; Elkiey and Ormrod, 1980). Ozone surface flux was estimated for several leaf-surface types (Bennett, Hill and Gates, 1973). The highest flux ($240\,\mu g\,m^{-2}\,s^{-1}$) was to *Atriplex confertifolia* leaf surfaces which had prominent salt glands. Pubescent leaf surfaces absorbed more ozone ($192\,\mu g\,m^{-2}\,s^{-1}$) than glabrous ones ($144\,\mu g\,m^{-2}\,s^{-1}$).

The ozone concentration gradient between the ambient air and cellular perturbation sites may also be reduced by gas-phase reactions that consume ozone. Terpenoids, which occur in leaf tissue, undergo ozonolysis (Grimsrud, Westberg and Rasmussen, 1975) consuming ozone and thereby reducing the concentration gradient. The potential for such reactions has been confirmed: an application of β-pinene, a monoterpene, to the leaf surface of *Malus pumila* reduced ozone injury (Elfving *et al.*, 1976). The significance of monoterpenes as an ozone sink within and around the leaf was investigated using chemical reaction kinetics. The estimated half-life of

Table 6.5   REACTION OF OZONE WITH VOLATILE TERPENOIDS[1]

Terpene	Ozonolysis rate constant	Ozone half-life (min)	Ozone lost s^{-1} (%)
β-Pinene	0.0036	64.18	0.02
Isoprene	0.0030	77.02	0.02
α-pinene	0.0036	6.42	0.18
3-Δ-carene	0.0030	7.70	0.15
β-phellandrene	0.0044	5.25	0.22
γ-terpene	0.0078	3.30	0.35
Carvomethane	0.0138	1.78	0.65
Limonene	0.0160	1.36	0.80
Dihydromyrcene	0.0170	1.36	0.85
Myrcene	0.031	0.75	1.54
Cis-ocimene	0.050	0.46	2.54
Terpinolene	0.25	0.09	11.75
α-phellandrene	0.29	0.08	13.50
α-terpinene	2.20	0.01	66.71

[1]The overall gas-phase reaction of ozone with terpenoids is second-order and assumes only a single reaction between ozone and the terpenoid (Grimsrud, Westberg and Rasmussen, 1975). The reaction rates of ethylene, cymene and methene were much slower than those listed above. The internal ozone concentrations used ranged from 0.05 to 0.5 ppm. A nominal terpenoid concentration of 0.5 ppm within the leaf was assumed, based on flux estimates and stomatal resistances. Ozonolysis rate-constant data are from Grimsrud, Westberg and Rasmussen (1975) and J. Bufalin (personal communication). Data for ozone half-life and ozone lost s^{-1} are unpublished data of Tingey and Taylor

ozone in the presence of selected terpenoids ranged from >1 h to approximately 0.01 min (*Table 6.5*). These half-lives are generally much longer than the estimated ozone turnover rate within the leaves, i.e. 0.1–0.4 s as derived from estimates of the intercellular volume and ozone flux into leaves. The percentage ozone consumed by each terpenoid was linearly dependent on the ozone concentration: the amount of ozone reacting in 1.0 s per terpenoid ranged from 0.02% to > 65% (*Table 6.5*). Only three monoterpenes (terpinolene, α-phellandrene and α-terpinene) consumed significant amounts of ozone, suggesting that terpenoids are insignificant scavengers of ozone in the gas phase.

Patterns of pollutant deposition within the leaf interior should reflect differences in solubility, with deposition increasing with solubility (Hill, 1971). In water (20°C), ozone is 33% and 0.7% as soluble as carbon dioxide and sulphur dioxide, respectively. It is proposed that these differences result in a greater frequency of deposition of soluble gases within the substomatal chamber than for less soluble ones such as ozone. In addition, the rate of pollutant absorption per unit of exposed mesophyll cell surface should be more uniform throughout the leaf for ozone than for more soluble gases.

Gas-phase concentrations within the leaf's intercellular spaces and corresponding fluxes have been studied for carbon dioxide and water vapour. More water evaporates from the region of the stomata, on a per unit area basis, than from elsewhere within the leaf. Between 77–90% of the transpired water is lost from the guard cell subsidiary complex and the substomatal cavity (Cowan, 1977; Rand, 1977). The behaviour of carbon dioxide is postulated to differ from that of water vapour, with only 15% being absorbed in the substomatal cavity (Rand, 1977, 1978). The mosophyll tissue therefore contributes only 10–20% of the transpiration stream, but absorbs approximately 85% of the carbon dioxide. This contrast between carbon dioxide and water-vapour pathways prompted Meidner (1975) to question the use of transpiration data to estimate pathway conductance for carbon dioxide. Because ozone is only 33% as soluble as carbon dioxide, the same warning is appropriate and may, in part, underlie the contradictory data relating pathway conductance and ozone response.

In summary, the gas-phase pathway conductance does affect the rate at which ozone diffuses into the leaf interior. This has obvious consequences for the concentration of toxic ozone derivatives in the liquid phase. However, there are insufficient data to propose an *a priori* relationship between pathway conductance and injury. Therefore, in studies of genetically controlled or environmentally induced variation in ozone response, conductance measurements alone may be misleading when used to investigate mechanisms of ozone resistance.

LIQUID-PHASE CONDUCTANCE

The transfer of ozone from extracellular deposition points to intracellular perturbation sites involves liquid-phase conductance. Along this pathway, ozone may experience diffusion and bulk flow (cyclosis) through water, membranes and cytosol. As each medium is more viscous than air, diffusion alone is several orders of magnitude slower over equivalent distances (Mansfield, 1973). Moreover, transfer may be prevented by scavenger processes that consume ozone. These liquid-phase phenomena are analogous to carbon dioxide assimilation (Gaastra, 1959) and are referred to as a residual resistance (r_r) (i.e. resistance to flux in addition to r_s and r_a). This residual resistance is conceptually similar to the chemical (Leuning, Neumann and Thurtell, 1979; Leuning *et al.*, 1979) and mesophyll (O'Dell, Taheri and Kabel, 1977; Weseley *et al.*, 1978) resistances previously described.

Several reports have alluded to a resistance component located within the leaf interior and not explained by r_a and r_s. Turrell (1942) postulated that variable sulphur dioxide responses among *Medicago sativa* leaves could reflect differences in intercellular volume, mesophyll cell-surface area and cell-wall thickness. The inherent ozone susceptibility of ferns and *N. tabacum* may result from less suberized deposits on meosphyll cell walls facilitating ozone uptake (Glater, 1956; Glater, Solberg and Scott, 1962). Different ozone responses among several *Pinus* species (Evans and Miller, 1972) and *Phaseolus vulgaris* leaf ages (Evans and Ting, 1974a) were linked to r_r which influenced ozone flux. Residual resistances are inversely related to the solubility of a pollutant (O'Dell, Taheri and Kabel, 1977). This would necessitate a greater r_r for ozone than that for either sulphur dioxide or hydrogen fluoride, because of its lower solubility. From field estimates of ozone flux to a *Z. mays* canopy, r_r was estimated at 300–400 s m^{-1} (Weseley *et al.*, 1978), while Leuning *et al.* (1979) assumed r_r to be essentially zero. For alfalfa, a r_r of 60 s m^{-1} was estimated (Heath, 1980).

An analogue modelling technique (*Table 6.2*) was used to calculate r_r for two *G. max* cultivars (cvs Hood and Dare) exposed to a range of ozone doses (Taylor and Tingey, unpublished data). Residual resistance varied as a function of ozone concentration and duration of exposure (*Figure 6.3*). At low ozone concentrations (0.3 ppm or less), r_r was less than 20% of r_s and remained essentially constant over the 4 h exposure period. However, at higher ozone concentrations r_r became the predominant resistant component and increased with both ozone concentration and duration of exposure. The consistent increase in r_r with dose suggested that the liquid-phase conductance to ozone decreased and was responsible for reductions in J_{internal}.

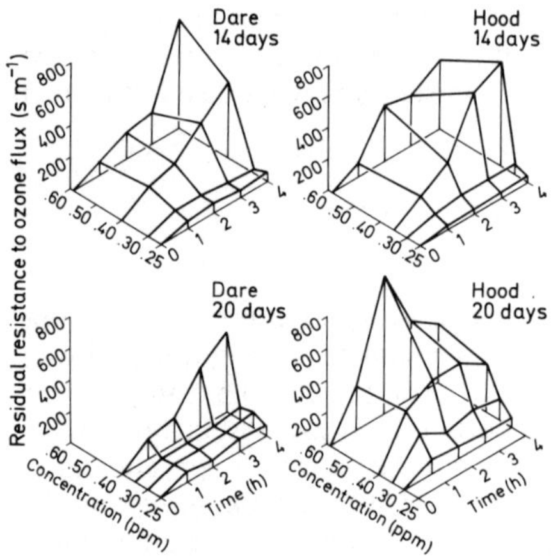

Figure 6.3   Changes in the residual resistance to ozone as influenced by ozone dose in two *Glycine max* cultivars at two leaf ages

The similar pathways of carbon dioxide and ozone flux suggest that the amount of internal surface area would be an important factor in ozone flux and leaf injury (e.g. Turrell, 1942; Glater, 1956). Photosynthetic rates in shade leaves and sun leaves of *Plectranthus parviflorus* were strikingly different, and the fact that the internal-to-external leaf area (A_{mes}/A) was four times greater in sun leaves accounted fully for the higher photosynthetic rates (Nobel, Zaragoza and Smith, 1975). Mechanistically, the larger internal surface area increased liquid-phase conductance, thereby providing more carbon dioxide for carboxylation. The greater susceptibility of palisade versus spongy mesophyll cells was attributed to a higher A_{mes}/A ratio in the palisade tissue (Evans and Ting, 1974a). Similar reasoning was used to explain the greater ozone sensitivity of paraveinal cells in *G. max* (Pell and Weissberger, 1976).

Heath (1980) proposed that bulk water flow within the substomatal chamber redistributes gaseous pollutants or their derivatives to the principal water evaporation sites around the guard-cell complex. This hypothesis requires a steady flow of water, from the mesophyll cells lining the chamber to evaporative sites. Evidence from several tracer studies on bulk water flow in leaves (Crowdy and Tanton, 1970; Tanton and Crowdy,

1972; Meidner, 1975; Sheriff and Meidner, 1975; Byott and Sheriff, 1976) indicates that water moves along preferential conduits emanating from the conducting elements, along sheath extensions to the epidermis and subsequently across the highly conductive apoplasts of the epidermal cells to the stomatal complex. This route has only limited frontage with the substomatal chamber, which would decrease the importance of the redistribution of ozone or its derivatives in bulk water flow around the substomatal chamber.

The chemical reactivity of ozone in the liquid phase may affect its movement. The ease with which any gas moves along a gas-to-liquid pathway reflects the solubility of that gas and the extent to which it can form derivatives in the liquid phase (Liss, 1971). The lower solubility of ozone compared with other pollutants (e.g. sulphur dioxide) would reduce the transfer across the gas–liquid interface. However, the tendency of ozone to form derivatives should facilitate liquid-phase diffusion. Ozone is a highly reactive molecule as a consequence of having two unpaired electrons (Mustafa *et al.*, 1978) and a high redox potential. In solution, ozone may react with unsaturated fatty acids, sulphhydryl and ring-containing compounds. The reaction of ozone with olefinic double bonds in aqueous solution produces an aldehyde or ketone and a Criegee zwitterion (Criegee, 1975). The zwitterion subsequently decomposes to form an additional aldehyde or ketone and hydrogen peroxide.

In water, ozone decomposition products include hydroxyl, hydroperoxyl, superoxide anion and other free radicals (Weiss, 1935; Hoigne and Bader, 1975; Peleg, 1976). These products may be more powerful oxidants than the parent compound (Peleg, 1976). The hydroxyl radical may be the most reactive species of the aqueous decomposition products of ozone, based on the reactions of hydroxyl radicals and ozone with organic compounds.

Ozone increased the free radical concentrations in *Phaseolus vulgaris* and *Glycine max* leaves (Rowlands *et al.*, 1970). The increased free radicals may have resulted directly from ozone decomposition in aqueous solutions or they may have been formed indirectly (*see section on* Perturbation, pages 125–126).

Biological systems have evolved protective scavenging or buffering mechanisms for free radicals, such as superoxide dismutase, catalase and peroxidases (McCord, Keele and Fridovich, 1971; Fridovich, 1975, 1978). Superoxide dismutase converts superoxide anion to hydrogen peroxide, which is subsequently decomposed to water and oxygen by catalase (Levitt, 1975; Fridovich, 1978). Catalase also catalyzes the decomposition of hydrogen peroxide formed through the Criegee reaction. Giese and Christensen (1954) suggested that the cell's first line of defence against ozone was reactive organic matter (e.g. proteins) on the surface coat, which in plants would include both the plasmalemma and cell wall. The dissimilar ozone response of cell cultures of *Saccharomyces cerevisiae* was attributed to a larger battery of unspecified intracellular scavenger molecules in the more resistant stationary cultures (Dubeau and Chung, 1979). Sulphydryls (e.g. glutathione), sulphydryl proteins, cysteine and small amounts of methionine apparently serve as free-radical scavengers and peroxide decomposers (Menzel, 1971; Tappel, 1973; Goldstein *et al.*,

1979). Ozone will oxidize glutathione and other sulphydryls to disulphides and sulphonic acids (Tomlinson and Rich, 1970; Mudd and Freeman, 1977; Freeman, Miller and Mudd, 1979). Tocopherol (vitamin E) is postulated as a scavenger of hydroxyl radicals (Tappell, 1972; Heath, 1980).

The importance of oxidant-scavenging mechanisms in aerobic metabolism (Fridovich, 1975, 1978) would suggest that this liquid-phase process is of paramount importance in limiting ozone transport to cellular perturbation sites. If the scavenger pool is subject to control by the environment or genotype this could be a source of variation in ozone response. Foliar or root applications of an ethylene diurea, EDU, prevented ozone injury (Carnahan, Jenner and Wat, 1978) without affecting pollutant uptake (Bennett, Lee and Heggestad, 1978). EDU-induced resistance was associated with higher tissue levels of superoxide dismutase, catalase and peroxidase (Bennett, Lee and Heggestad, 1979; Bennett *et al.*, 1980). In the case of greater sulphur dioxide resistance in young versus old *Populus euramericana* leaves (Tanaka and Sugahara, 1980), the more resistant foliage contained five times more superoxide dismutase. The differential response to ozone as a function of leaf age may be explained by a similar mechanism.

The overall significance of liquid-phase conductance in controlling leaf response to ozone is difficult to assess. This component may exert a profound influence on the type of ozone derivatives formed and the rate with which derivatives reach (or are generated at) intracellular perturbation sites. Preliminary evidence confirms a potential for control of liquid-phase conductance by both the environment and genotype.

Perturbation

Potential sites for ozone-induced perturbation are those in which the molecular configuration is highly ordered and the function of which is essential to the metabolism of a cell (Demopoulos, 1973a). Experimental evidence, principally from electron microscopy and biochemical investigations, suggests membranes as the site of ozone-induced perturbations (Giese and Christensen, 1954; Rich, 1969).

Initial ultrastructural changes induced by ozone are varied, although most indicate membrane dysfunction. When *Phaseolus vulgaris* was exposed to ozone, the first observed changes were increased granulation and electron density of the chloroplast stroma, followed by rupturing of the chloroplast envelope, plasmalemma and tonoplast and the formation of crystalline fibril arrays (Thomson, Dugger and Palmer, 1966). Distinct membrane alterations were not evident until there was moderate to extensive leaf injury. No membrane changes *per se* were detected in *P. vulgaris* leaf tissue exposed to 0.3 ppm ozone for 2 h, although the chloroplast volume decreased and mitochondrial volume increased (Swanson, Thomson and Mudd, 1973). In *Raphanus sativus* plants exhibiting leaf glazing, the only ultrastructural effect was withdrawal of the plasmalemma from the cell wall (Athanassious, Klyne and Phan, 1978). In *Glycine max*, initial ozone injury was detected in the endoplasmic reticulum, the inner and outer mitochondrial membranes and the limiting chloroplastic membrane (Pell and Weissberger, 1976).

The susceptibility of membranes to ozone attack is also apparent in freeze-fracture preparations. In *P. vulgaris* exposed to 0.25 ppm ozone, protein moieties immersed in the nuclear, chloroplastic and plasmalemma membranes showed marked translational movements (Cunningham and Swanson, 1977). After 30 min some parts of the membrane were devoid of protein moieties. At this stage the membrane exhibited characteristics of a phase transition from liquid to gel in the lipid bilayer. The gel-phase transition has also been observed in *P. vulgaris* cotyledons exposed to ozone, suggesting molecular changes comparable to those occurring in natural ageing processes in both plants and animals (Pauls and Thompson, 1980).

Ozone reacts with protein moieties: sulphydryls are oxidized to disulphides and sulphonates (Tomlinson and Rich, 1970; Menzel, 1971). The susceptibility of methionine, cysteine and cystine to ozone (Mudd and Freeman, 1977) can have major consequences for protein constituents of membranes and normal membrane function.

Lipids, *in vitro*, were oxidized through either ozonolysis or lipid peroxidation (Mudd and Freeman, 1977). *In vivo*, lipid peroxidation followed prolonged ozone exposures or high concentrations, as indicated by the presence of malonaldehyde (Mudd *et al.*, 1971). However, at low doses peroxidation was discounted because the highly reactive and polar ozone molecule was unable to penetrate the apolar lipid bilayer (Demopoulos, 1973b; Swanson, Thomson and Mudd, 1973).

Free radicals have been implicated in ozone toxicity in animal (Menzel, Roehm and Lee, 1972; Chow and Kaneko, 1979; Goldstein *et al.*, 1979), microbial (Epstein and Bishop, 1977) and plant systems (Dass and Weaver, 1968; Rowlands *et al.*, 1970; Bennett *et al.*, 1980). Dubeau and Chung (1979) noted a correlation between the susceptibility of *Saccharomyces cerevisiae* mutants to ozone and ionizing radiation and suggested that the common production of free radicals by both stresses was responsible for the parallel susceptibilities.

Free radicals may be formed by ozone decomposition in aqueous solutions, or indirectly by oxygen reduction in the chloroplasts. Water-stressed leaves face the problem of eliminating photoreductants (Hanson and Tully, 1979). When carbon dioxide uptake is reduced, photosynthetic reductants accumulate and produce highly damaging types such as a superoxide anion and hydrogen peroxide. Failure to dispose of these damaging compounds through scavenging reactions or photorespiration could cause some of the pathological changes induced by water stress (Hanson and Tully, 1979). Superoxide anion may be formed under conditions that cause a carbon dioxide deficiency or inhibit the carbon cycle for carbon dioxide fixation (Asada, 1980).

Air pollutants, including ozone, induce stomatal closure (Hill and Littlefield, 1969) and the resulting internal carbon dioxide deficiency could lead to an accumulation of photochemical reductants. The plant would photoreduce oxygen rather than carbon dioxide, forming superoxide anion (Asada, 1980). Inhibitors of electron flow between photosystems II and I have prevented ozone injury (Koiwai and Kisaki, 1976) suggesting that electron flow through photosystem I, and possibly superoxide anion formation, may be involved in injury development. Ozone exposure increased free radicals associated with photosystem I (Rowlands *et al.* 1970), a site of superoxide anion formation.

The inability of the cell to dissipate excess photochemical reductants, allowing free-radical formation, explains the observation that ozone injury is frequently observed first in the chloroplast (Thomson, Dugger and Palmer, 1966). Ozone inhibited nitrite reductase, a chloroplastic enzyme, more than nitrate reductase, a cytoplasmic enzyme (Leffler and Cherry, 1974). Chang (1971a,b; 1972) reported that ozone reduced polysomal populations, ribosomal sulphydryl content and ribosomal RNA in the chloroplasts more than in the cytoplasm. The indirect action of ozone in forming free radicals in the chloroplasts explains the paradox of ozone diffusing through the cytoplasm without reacting, until it reaches the chloroplast, where many of the early effects are observed.

Heath (1975) has suggested that membrane permeability indices are reliable indicators for monitoring ozone perturbation at the cellular level. The array of plant species exhibiting ozone-induced permeability changes supports this hypothesis. Increases in transmembraneous flux rates are reported for sugars and sugar alcohols (Dugger and Palmer, 1969; Nobel and Wang, 1973; Sutton and Ting, 1977a), amino acids, carbohydrates, water (Ting, Perchorowicz and Evans, 1974), and potassium (Evans and Ting, 1974b; Chimiklis and Heath, 1975; Elkiey and Ormrod, 1979c). The increased fluxes result from changes in membrane permeability, not from a change in cellular utilization (Dugger and Palmer, 1969; Nobel and Wang, 1973).

The effect of ozone on potassium flux across membranes has been extensively investigated. Cells accumulate potassium in an energy-dependent transport process that yields a high concentration difference between the cell interior and exterior. The potassium influx is coupled to the exclusion of sodium. In *Chlorella sorokiniana* cultures, ozone induced a rapid net potassium efflux (Chimiklis and Heath, 1975; Heath and Frederick, 1979). The increased permeability was reversible if the stress was removed (within 1–5 min); however, continued ozonation resulted in irreparable damage to membrane components. The data indicated that a limited number of transport sites were affected by ozone (Chimiklis and Heath, 1975), suggesting that general membrane deterioration did not occur, but rather that ozone selectively inactivated specific membrane sites controlling cellular osmoregulation (Heath, Chimiklis and Frederick, 1974; Heath, 1975).

The distribution of cations across the plasmalemma is in part responsible for the electropotential gradient of a cell, and changes in potassium distribution should be accompanied by altered electrogradients. In *C. sorokiniana*, ozone stimulated both the efflux and influx of potassium (Heath and Frederick, 1979). The efflux rate was 5–8 times larger than the influx rate. The increased influx, 2–3 times greater than in the controls, was attributed to movement along an electrochemical gradient; calculations suggested that ozone decreased the membrane potential for potassium from -90 to $-40\,mV$.

The spatial organization of membrane components is required for electron-transport processes in the chloroplast and mitochondrion. Any agent affecting membrane integrity may spatially disassociate components of the electron-transport system, and consequently impair photosynthesis and respiration. In *Spinacia oleracea*, ozone inhibition of photosystem II activity was attributed to the system's inability to accept electrons from water (Chang and Heggestad, 1974). Similar results in *N. tabacum* suggested an inhibition of photosynthetic electron flow in a membrane-bound system (Koiwai and Kisaki, 1976). Using isolated chloroplasts, Coulson and Heath (1974) found that ozone reduced electron flow in photosystem II more than in photosystem I. However, the coupling of ATP production to electron flow remained unaltered. The authors suggested that'... ozone disrupts the normal pathway of energy flow from light-excited chlorophyll into the photoacts by the disruption of the components of the membrane, but not a general disintegration of the membrane.'

Membranes appear to be primary sites of ozone attack. At low ozone doses, proteinaceous membrane components are oxidized, while at higher doses lipids undergo peroxidation. The maintenance of selective permeability creates an intracellular or intraorganelle environment conducive to metabolism, and ozone-induced changes in permeability will have repercussions throughout the cell. We conclude that membrane permeability is the perturbation event and it results from the susceptibility of membrane macromolecules to ozone attack.

Homeostasis

Living systems attempt to re-establish a normal metabolic state following a perturbation. Two physiological means of recovery are used: repair and/or compensation of the perturbation and its consequences (Levitt, 1972). Even though the end state (i.e. recovery) resulting from each process is the same, the underlying mechanism(s) differ(s). By definition, repair processes remedy the perturbation site, thereby eliminating the physiological dysfunction. In compensation, the perturbation and its physiological consequences remain and the cell responds with processes that counter the detrimental physiological effects of the perturbation. Because recovery may range from partial to complete, injury may vary as well.

On broad-leaved plants, one of the first symptoms of ozone injury is the appearance of water-soaked (shiny or oily) areas on the leaf surface. This results from an increased permeability, with cell contents leaking into the intercellular spaces. If the tissue is not injured irreversibly, water-soaking is temporary: the tissue regains its selective permeability (i.e. repair) and visual necrosis does not develop.

At the cellular level there is some evidence of recovery following ozone stress. Leaf age differences in ozone susceptibility of *P. vulgaris* were attributed to variation in the energy pools which influenced the repair rate of ozone perturbation (Dugger *et al.*, 1962). Ozone increased potassium efflux in *Chlorella*; however, within 2–3 min after ozonation stopped, the normal efflux rate was re-established (Heath, Chimiklis and Frederick, 1974) suggesting an active repair process. The rate at which normal potassium flux was re-established increased with temperature (Chimiklis and Heath, 1975).

In *P. vulgaris* leaves, ozone increased membrane permeability to glucose and 2-deoxy-D-glucose (Perchorowicz and Ting, 1974). The uptake rate of 2-deoxy-D-glucose has been used as an indicator of membrane permeability and repair (Sutton and Ting, 1977a,b). The membrane repair rate was enhanced by continuous light or glucose supplements and retarded by low temperatures or darkness following exposure. These observations support the hypothesis that the repair of altered membrane permeability is an energy-dependent process.

The hexose monophosphate shunt generates reducing power (NADPH) which is used, in part, to reduce disulphides and peroxides (Mudd and Freeman, 1977; Freeman, Miller and Mudd, 1979) and to provide biosynthetic reductant. An increase in the hexose monophosphate shunt activity is a typical response to a wide array of stresses including ozone (Tingey, Fites and Wickliff, 1975, 1976). In *G. max* cultivars that differed in ozone susceptibility, glucose-6-phosphate dehydrogenase activity increased faster in the ozone-sensitive cultivar than in the tolerant one following exposure (Tingey, Fites and Wickliff, 1976). Sutton and Ting (1977a, b) concluded that an endogenous energy source was '... needed for the re-establishment of important membrane sulphydryls and unsaturated lipids oxidized by ozone so that normal transport and permeability properties are restored to the cell membrane.' The magnitude of injury reflects the degree of homeostasis as depicted in *Figure 6.1*. They suggested further that differences in endogenous repair rates may help to explain differences in sensitivity between plants.

Injury

Injury is the consequence of an inability to repair or compensate for altered membrane permeability. Biochemically, injury is expressed as an altered metabolism, including enzyme activities and metabolite pools. Because ozone-induced membrane injury tends to be selective within cells (e.g. chloroplasts or region of plasmalemma), initial metabolic dysfunction may be restricted to specific organelles or cellular regions. The summation across many cells of the altered metabolism affects tissue and organ activities. These, in turn, have consequences that are expressed as foliar pathologies, altered carbon allocation and reduced growth and reproduction. All of these injury responses are consequences of ozone perturbation rather than the primary ozone effect. The magnitude of injury reflects the degree of homeostasis as depicted in *Figure 6.1*.

Effects of numerous metabolic poisons on rat hepatocytes indicated a disruption of the plasmalemma by diverse mechanisms. As a consequence, the membranes lost their selective permeability, which set off a rapid change in the concentration of cellular materials, leading to cell death (Schanne *et al.*, 1979). Analogously, we propose that ozone alters membrane permeability, thus causing functional changes leading to cellular dysfunction and death, in a manner similar to water stress and pathogenesis (Heath, 1975; Thomson, 1975; Tingey, Fites and Wickliff, 1976; Bennett, Heggestad and McNulty, 1977).

Ozone decreased leaf-water potential in a sensitive *Petunia hybrida* cultivar (Elkiey and Ormrod, 1979b) and the relative water content of *Phaseolus vulgaris* leaves by 10% (Evans and Ting, 1974a). However, only a portion of the leaf cells was injured, as indicated by pressure–volume relationships (Turner, 1976).

Water stress readily and reversibly altered protein synthesis, induced polysome dissociation and the accumulation of amino acids, including proline (Hsiao, 1973). Similarly, ozone induced the dissociation of chloroplastic polysomes (Chang, 1971a,b) and an accumulation of amino acids, including proline (Tomlinson and Rich, 1967; Ting and Mukerji, 1971; Craker and Starbuck, 1972; Tingey, Fites and Wickliff, 1973; Bennett, Heggestad and McNulty, 1977). In contrast to the consistent trend with amino acids, the effects on soluble protein are less clear, with decreases (Ting and Mukerji, 1971; Craker and Starbuck, 1972), increases (Beckerson and Hofstra, 1979b) and no effects (Tingey, Fites and Wickliff, 1973) being reported.

Water stress decreased photosynthesis by reducing stomatal conductance and the enzymatic activities of carbon dioxide fixation (Hsiao, 1973). In addition, photosynthate translocation and growth were reduced (Hsiao, 1973). Similarly, ozone inhibited photosynthesis with the magnitude of inhibition depending upon plant species and ozone dose (*Table 6.6*).

Table 6.6 EFFECT OF OZONE ON PHOTOSYNTHESIS

Species	Ozone conc. (ppm)	Exposure duration (h)	Inhibition (%)	Reference
Avena sativa	0.4	0.5	33	Hill and Littlefield, 1969
Nicotiana tabacum	0.4	1.5	78	Hill and Littlefield, 1969
Lycopersicon esculentum	0.6	1.0	43	Hill and Littlefield, 1969
Phaseolus vulgaris	0.6	1.0	29	Hill and Littlefield, 1969
Medicago sativa	0.1	1.0	4	Bennett and Hill, 1974
	0.2	1.0	10	
Quercus velutina	0.5	4.0/2 days*	30	Carlson, 1979
Acer saccharum	0.5	4.0/2 days*	21	Carlson, 1979
Fraxinus americana	0.5	4.0/2 days*	0	Carlson, 1979
Phaseolus vulgaris	0.3	3.0	22	Pell and Brennan, 1973
Glycine max	0.4	4.0	37	Yingjajaval, 1976
	0.6	2.0	19	
Populus euramericana	0.9	1.5	50	Furukawa and Kadota, 1975

*i.e. 4 hours per day for 2 days

Photosynthesis was depressed shortly after the addition of ozone, but it increased when ozone was removed, and returned to the control level within 24 h (Hill and Littlefield, 1969). These workers concluded that the lower photosynthesis resulted from stomatal closure because of the correlation between transpiration and photosynthesis. In contrast, ozone reduced ribulose-1,5-bisphosphate carboxylase activity in both young and old *Oryza sativa* leaves (Nakamura and Saka, 1978). Following exposure, the enzyme activity returned to normal in young leaves but not in old. Carbon dioxide and light compensation points of *Populus euramericana* leaves were increased by ozone (Furukawa and Kadota, 1975), suggesting an inhibition of the carbon-fixation enzymes. In *G. max* cultivars, ozone decreased both the stomatal and mesophyll conductances to carbon dioxide, but affected the stomatal conductance at a lower ozone concentration (Yingajaval, 1976). These studies suggest that ozone affects photosynthesis through several mechanisms.

Immediately following ozone exposure, the reducing-sugar pools in *G. max* leaves were depleted (Tingey, Fites and Wickliff, 1973). Similarly, ozone decreased soluble sugars and increased amino acids and sugar phosphates in *Pinus strobus* (Wilkinson and Barnes, 1973). Reducing sugars and sucrose accumulated in ozone-injured leaf tissue (Bennett, Heggestad and McNulty, 1977), suggesting that translocation was impaired. In addition, low concentrations of photochemical oxidants inhibited starch hydrolysis and translocation in leaves, without substantial foliar injury (Hanson and Stewart, 1970). Soluble carbohydrates and starch accumulated in *Pinus ponderosa* needles with a concurrent reduction in the roots, indicating altered translocation (Tingey, Wilhour and Standley, 1976).

Ozone induced leaf proliferation in *Capsicum annuum* and a retention of photosynthate in leaves at the expense of the fruit (Bennett, Oshima and Lippert, 1979). Similarly, ozone reduced photosynthate partitioning from leaves into roots in several crops (Oshima, Bennett, and Braegelmann, 1978; Oshima *et al.*, 1979). The altered photosynthate partitioning was reflected in the greater ozone inhibition of root versus top growth (Tingey, 1977; Blum and Heck, 1980). In summary, ozone depresses photosynthesis and alters photosynthate pools and partitioning among plant organs, resulting in reduced growth.

Ethylene production was enhanced by increased membrane permeability without the complete loss of cellular compartmentalization (Wright, 1974; Elstner and Konze, 1976). Ozone-induced stress ethylene production was proportional to ozone dose (Craker, 1971; Tingey, Standley and Field, 1976) and ozone-sensitive cultivars produced more ethylene than tolerant ones. Stress ethylene production and either visual injury or plant-growth reduction were correlated (Tingey, 1980). Stress ethylene production preceded the appearance of other pathological manifestations associated with visual injury. Its production was stimulated for only a short period following ozone exposure, before returning to the control level, suggesting a repair process (Tingey, 1980). Ozone-induced stress ethylene has also been implicated in enhanced leaf abscission (Inoue, 1974).

Ethylene was implicated in pathogenesis (Pegg, 1976) and it increased the activities of phenylalanine ammonia lyase (PAL), polyphenol oxidase (PPO), and peroxidase (PRO) (Abeles, 1973). This is compatible with studies that showed increased ethylene production and increased activities of PAL, PPO and PRO in plants exposed to ozone (Tingey, Fites and Wickliff, 1975, 1976; Curtis, Howell and Kremer, 1976). Ozone increased the activities of PAL, PPO and PRO in an ozone-sensitive *G. max* cultivar several hours sooner than in a tolerant one (Tingey, Fites and Wickliff, 1976). Similarly, the activities of only a few PRO isozymes increased in an ozone-tolerant *G. max* cultivar, whereas numerous isozymes showed increased activity in the sensitive cultivar (Curtis, Howell and Kremer, 1976).

In plants subjected to biotic pathogens, lesion formation is associated with increased activites of PAL, PPO and PRO, resulting in the oxidation of phenols (Goodman, Kiraly and Zaitlin, 1967). The polymerized products are responsible for lesion formation. Phenols accumulated in ozone-injured leaf tissue (Menser and Chaplin, 1969; Howell and Kremer, 1973; Howell, 1974; Keen and Taylor, 1975). The phenol accumulations were reflected in the ozone enhancement of PAL activity, the rate-limiting step in phenol biosynthesis (Tingey, Fites and Wickliff, 1975, 1976). Following ozone exposure, sensitive *Arachis hypogaea* cultivars accumulated higher phenol levels than tolerant ones (Howell, 1974). The pigmented lesions that formed in many plants after ozone exposure contained polymerized amino acids, metals, sugars and phenols (Howell and Kremer, 1973).

Visual pigmentation appears to result from increased membrane permeability which induces stress ethylene formation. Ethylene may increase the activities of PAL, PPO and PRO resulting in phenol oxidation and lesion formation. The response of these enzymes to ozone is similar to the responses caused by biotic pathogens, indicating a general wound response and suggesting that similar processes are involved in lesion formation from ozone and biotic pathogens.

Conclusion

The environment and genotype exert profound influences on a plant's physiological state and thereby on its ozone response. A common progression of events is proposed, based on the multiplicity of physiological mechanisms responsible for variable response: ozone conductance (gas and liquid phase), perturbation, homeostasis and finally injury. Injury is the culmination of the preceding events. Gas-phase and liquid-phase conductances are significant sources of variable ozone response. The importance of homeostasis is not well documented.

The association between gas-phase conductance and injury is ambivalent. On the basis of our conceptual model, gas-phase conductance would be the decisive factor controlling injury only if it were the rate-limiting step. Any physiologically significant intervening event between gas-phase conductance and injury would reduce the association between the two. Either liquid-phase conductance or homeostasis could reduce that association. Because injury is not always associated with gas-phase conductance, this is *prima facie* evidence that liquid-phase conductance and homeostasis are important factors in controlling ozone response.

Acknowledgment

The effort of G.E. Taylor, Jr, was made possible under contract W-7405-eng-26 between the US Department of Energy and Union Carbide Corporation. Publication No. 1590, Environmental Sciences Division, ORNL

References

ABELES, F.B. (1973). *Ethylene in Plant Biology*. Academic Press, New York

ASADA, K. (1980). In *Studies on the Effects of Air Pollutants on Plants and Mechanisms of Phytotoxicity*, pp. 165–179. Research Report No. 11. National Institute for Environmental Studies, Iburaki, Japan

ATHANASSIOUS, R., KLYNE, M.A. and PHAN, C.T. (1978). *Zeitschrift Pflanzenphysiologie*, **90**, 183–187

BECKERSON, D.W. and HOFSTRA, G. (1979a). *Atmospheric Environment*, **13**, 1263–1268

BECKERSON, D.W. and HOFSTRA, G. (1979b). *Canadian Journal of Botany*, **57**, 1940–1945

BENNETT, J.H. (1979). In *Methodology for the Assessment of Air Pollution Effects on Vegetation*, pp. 10–1/10–29 (Heck, W.W., Krupa, S.K. and Linzon, S.N., Eds), Air Pollution Control Association, Pittsburgh

BENNETT, J.H. and HILL, A.C. (1974). In *Air Pollution Effects on Plant Growth*, pp. 115–127 (Dugger, M., Ed.), ACS Symposium Series 3. American Chemical Society, Washington D.C.

BENNETT, J.H., HEGGESTAD, H.E. and McNULTY, I.B. (1977). In *Proceedings of the Fourth Annual Plant Growth Regulator Working Group*, pp. 323–330. Agway, Syracuse, NY.

BENNETT, J.H., HILL, A.C. and GATES, D.M. (1973). *Journal of the Air Pollution Control Association*, **23**, 957–962

BENNETT, J.H., LEE, E.H. and HEGGESTAD, H.E. (1978). *Proceedings of the Fifth Annual Plant Growth Regulator Working Group*, pp. 242–246. Agway, Syracuse, NY.

BENNETT, J.H., LEE, E.H. and HEGGESTAD, H.E. (1979). *Plant Physiology*, (Supplement) **63**, 152

BENNETT, J.P., OSHIMA, R.J. and LIPPERT, L.F. (1979). *Environmental and Experimental Botany*, **19**, 33–39

BENNETT, J.H., LEE, E.H., HEGGESTAD, H.E., OLSEN, R.A. and BROWN, J.C. (1981). In *Oxygen and Oxy-radicals in Chemistry and Biology*, pp. 604–605, (Powers, E.L. and Rodgers, M.A.J., Eds.). Academic Press, New York

BLACK, V.J. and UNSWORTH, M.H. (1979). *Nature*, **282**, 68–69

BLUM, U. and HECK, W.W. (1980). *Environmental and Experimental Botany*, **20**, 73–85

BRENNAN, E. (1975). *Phytopathology*, **65**, 1054–1055

BUTLER, L.K. and TIBBITTS, T.W. (1979). *Journal of the American Society for Horticultural Science*, **104**, 213–216

BYOTT, G.S. and SHERIFF, D.W. (1976). *Journal of Experimental Botany*, **27**, 643–649

CARLSON, R.W. (1979). *Environmental Pollution*, **18**, 159–170

CARNAHAN, J.E., JENNER, E.L. and WAT, E.K.W. (1978). *Phytopathology*, **68**, 1225–1229

CHANG, C.W. (1971a). *Biochemical and Biophysical Research Communications*, **44**, 1429–1435

CHANG, C.W. (1971b). *Phytochemistry*, **10**, 1863–2868

CHANG, C.W. (1972). *Phytochemistry*, **11**, 1347–1350

CHANG, C.W. and HEGGESTAD, H.E. (1974). *Phytochemistry*, **13**, 871–873

CHIMIKLIS, P.E. and HEATH, R.L. (1975). *Plant Physiology*, **56**, 723–727

CHOW, C.K. and KANEKO, J.J. (1979). *Environmental Research*, **19**, 49–55

COULSON, C. and HEATH, R.L. (1974). *Plant Physiology*, **53**, 32–38

COWAN, I.R. (1977). In *Advances in Botanical Research*, pp. 117–228 (Preston, R.D. and Woolhouse, H.W. Eds), Academic Press, New York

CRAKER, L.E. (1971). *Environmental Pollution*, **1**, 299–304

CRAKER, L.E. and STARBUCK, J.S. (1972). *Canadian Journal of Plant Science*, **52**, 589–597

CRAKER, L.E. and STARBUCK, J.S. (1973). *Enivironmental Research*, **6**, 91–94

CRIEGEE, R. (1975). *Angewandte Chemie* (International Edition in English), **14**, 745–752

CROWDY, S.H. and TANTON, T.W. (1970). *Journal of Experimental Botany*, **21**, 102–111

CUNNINGHAM, W.P. and SWANSON, E.S. (1977). *Journal of Cell Biology*, **75**, 207a

CURTIS, C.R., HOWELL, R.K. and KREMER, D.F. (1976). *Environmental Pollution*, **11**, 189–194

DASS, H.C. and WEAVER, G.M. (1968). *Canadian Journal of Plant Science*, **48**, 569–574

DEAN, C.E. (1972). *Crop Science*, **12**, 547–548

DEMOPOULOS, H.B. (1973a). *Federation Proceedings*, **32**, 1859–1861

DEMOPOULOS, H.B. (1973b). *Federation Proceedings*, **32**, 1903–1908

DOMERQUE, F., DE CORMIS, L. and LOUGUET, P. (1979). *Comptes Rendus Hebdomadaires des Séances de L'Académie des Sciences, Serie D, Sciences Naturelles*, **288**, 1541–1544

DUBEAU, H. and CHUNG, Y.S. (1979). *Molecular General Genetics*, **176**, 393–398

DUGGER, W.M., Jr. and PALMER, R.L. (1969). *Proceedings of the First International Citrus Symposium*, **2**, 711–715

DUGGER, W.M., Jr., TAYLOR, O.C., CARDIFF, E. and THOMPSON, C.R. (1962). *Plant Physiology*, **37**, 487–491

DUNNING, J.A. and HECK, W.A. (1973). *Environmental Science Technology*, **7**, 824–826

ELFVING, D.C., GILBERT, M.D., EDGERTON, L.J., WILDE, M.H. and LISK, D.J. (1976). *Bulletin of Environmental Contamination and Toxicology*, **15**, 336–341

ELKIEY, T. and ORMROD, D.P. (1979a). *Journal of the Air Pollution Control Association*, **29**, 622–625

ELKIEY, T. and ORMROD, D.P. (1979b). *Zeitschrift Pflanzenphysiologie*, **91**, 177–181

ELKIEY, T. and ORMROD, D.P. (1979c). *Atmospheric Environment*, **13**, 1165–1168

ELKIEY, T. and ORMROD, D.P. (1980). *Journal of Environmental Quality*, **9**, 93–95

ELKIEY, T., ORMROD, D.P. and PELLETIER, R.L. (1979). *Journal of the American Society for Horticultural Science*, **104**, 510–514

ELSTNER, E.F. and KONZE, J.R. (1976). *Nature*, **263**, 351–352

ENGLE, R.L. and GABELMAN, W.H. (1966). *Proceedings of the American Society for Horticultural Science*, **89**, 423–430

EPSTEIN, S.S. and BISHOP, Y. (1977). *Environmental Research*, **14**, 187–193

EVANS, L.S. and MILLER, P.R. (1972). *Canadian Journal of Botany,* **50**, 1067–1071

EVANS, L.S. and TING, I.P. (1974a). *American Journal of Botany,* **61**, 592–597

EVANS, L.S. and TING, I.P. (1974b). *Atmospheric Environment,* **8**, 855–861

FLETCHER, R.A., ADEDIPE, N.O. and ORMROD, D.P. (1972). *Canadian Journal of Botany,* **50**, 2389–2391

FREEMAN, B.A., MILLER, B.E. and MUDD, J.B. (1979). In *Assessing Toxic Effects of Environmental Pollutants,* ch.9. (Lee, S.D. and Mudd, J.B., Eds), Ann Arbor Science, Ann Arbor, Michigan

FRIDOVICH, I. (1975). *American Scientist,* **63**, 54–59

FRIDOVICH, I. (1978). *Science,* **201**, 875–880

FURUKAWA, A. and KADOTA, M. (1975). *Environment Control in Biology,* **13**, 1–13

GAASTRA, P. (1959). *Mededelingen van de Landbouwhoogeschool te Wageningen,* **59**, 1–68

GIESE, A.C. and CHRISTENSEN, E. (1954). *Physiological Zoology,* **27**, 101–115

GLATER, R.A.B. (1956). *Phytopathology,* **46**, 696–698

GLATER, R.A.B., SOLBERG, R.A. and SCOTT, F.M. (1962). *American Journal of Botany,* **49**, 954–970

GOLDSTEIN, B.D., ROZEN, M.G. QUINTAVALLA, J.C. and AMORUSO, M.A. (1979). *Biochemical Pharmacology,* **28**, 27–30

GOODMAN, R.N., KIRALY, N. and ZAITLIN, M. (1967). *The Biochemistry and Physiology of Infectious Plant Disease.* D. Van Nostrand Company, Inc. Princeton, New Jersey

GRIMSRUD, E.P., WESTBERG, H.H. and RASMUSSEN, R.A. (1975). *International Journal of Chemical Kinetics,* **7**, 183–195

HANSON, G.P. and STEWART, W.S. (1970). *Science,* **168**, 1223–1224

HANSON, A.D. and TULLY, R.E. (1979). *Planta,* **145**, 45–51

HEATH, R.L. (1975). In *Responses of Plants to Air Pollution,* pp. 23–55. (Mudd, J.B. and Kozlowski, T.T., Eds), Academic Press, New York

HEATH, R.L. (1980). *Annual Review of Plant Physiology,* **31**, 394–431

HEATH, R.L. and FREDERICK, P.E. (1979). *Plant Physiology,* **64**, 455–495

HEATH, R.L., CHIMIKLIS, P. and FREDERICK, P. (1974). In *Air Pollution Effects on Plant Growth,* pp. 58–75. (Dugger, M., Ed.), ACS Symposium Series 3. American Chemical Society, Washington, D.C.

HECK, W.W. (1968). *Annual Review of Phytopathology,* **6**, 165–188

HECK, W.W., MUDD, J.B. and MILLER, P.R. (1977). In *Ozone and Other Photochemical Oxidants,* pp. 437–585. National Academy of Sciences, Washington, DC

HEICHEL, G.H. (1970). *Journal of Experimental Botany,* **22**, 644–649

HILL, A.C. (1971). *Journal of the Air Pollution Control Association,* **21**, 341–346

HILL, A.C. and LITTLEFIELD, N. (1969). *Environmental Science Technology,* **3**, 52–56

HOFFMAN, G.L., MAAS, E.V. and RAWLINS, S.L. (1973). *Journal of Environmental Quality,* **2**, 148–152

HOFFMAN, G.L., MAAS, E.V. and RAWLINS, S.L. (1975). *Journal of Environmental Quality,* **4**, 326–331

HOIGNE, J. and BADER, H. (1975). *Science,* **190**, 782–784

HOWELL, R.K. (1974). In *Air Pollution Effects on Plant Growth*, pp. 94–103 (Dugger, M., Ed.), ACS Symposium Series 3. American Chemical Society, Washington, D.C.

HOWELL, R.K. and KREMER, D.F. (1973). *Journal of Environmental Quality*, **2**, 434–438

HSIAO, T.C. (1973). *Annual Review of Plant Physiology*, **24**, 519–570

INOUE, T. (1974). *Journal of Japan Society of Air Pollution*, **9**, 375–376

JACOBSON, J.S. and COLAVITO, L.J. (1976). *Environmental and Experimental Botany*, **16**, 277–285

KEEN, N.T. and TAYLOR, O.C. (1975). *Plant Physiology*, **55**, 731–733

KOIWAI, A. and KISAKI, T. (1976). *Plant Cell Physiology*, **17**, 1199–1207

LEDBETTER, M.C., ZIMMERMAN, P.W. and HITCHCOCK, A.E. (1959). *Contributions. Boyce Thompson Institute for Plant Research*, **20**, 275–282

LEFFLER, H.R. and CHERRY, J.H. (1974). *Canadian Journal of Botany*, **52**, 1233–1238

LEONE, I.A. (1976). *Phytopathology*, **66**, 743–736

LEUNING, R., NEUMANN, H.H. and THURTELL, G.W. (1979). *Agricultural Meteorology*, **20**, 115–135

LEUNING, R., UNSWORTH, M.H., NEUMANN, H.N. and KING, K.M. (1979). *Atmospheric Environment*, **13**, 1155–1163

LEVITT, J. (1972). *Responses of Plants to Environmental Stresses*. Academic Press, New York

LEVITT, J. (1975). *What's New in Plant Physiology*, **7**(11), 1–3

LISS, P.S. (1971). *Nature*, **233**, 327–329

McCORD, J.M., KEELE, B.B., Jr and FRIDOVICH, I. (1971). *Proceedings of the National Academy of Sciences of the United States of America*, **68**, 1024–1027

MacDOWELL, F.D.H. (1965). *Canadian Journal of Plant Science*, **45**, 1–12

MANSFIELD, T.A. (1973). *Current Advances in Plant Science*, **2**, 11–20

MARKOWSKI, A. and GREZESIAK, S. (1974). *Bulletin de l'Académie Polonaise des Sciences. II. Serie des Sciences Biologiques*, **22**, 875–887

MEIDNER, H. (1975). *Journal of Experimental Botany*, **26**, 666–673

MENSER, H.A. and CHAPLIN. J.F. (1969). *Tobacco Science*, **73**, 73–74

MENZEL, D.B. (1971). *Archives of Environmental Health*, **23**, 149–153

MENZEL, D.B., ROEHM, J.N. and LEE, S.D. (1972). *Journal of Agricultural and Food Chemistry*, **20**, 481–486

MUDD, J.B. and FREEMAN, B.A. (1977). In *Biochemical Effects of Environmental Pollutants*, pp. 97–133 (Lee, S.D., Ed.), Ann Arbor Science Publishers, Inc., Ann Arbor, Michigan

MUDD, J.B., McMANUS, T.T., ONGUN, A. and McCULLOUGH, T.E. (1971). *Plant Physiology*, **48**, 335–339

MUSTAFA, M.G., ELSAYED, N., LIM, J.S.T., POSTLETHWAIT, E. and LEE, S.D. (1978). In *Nitrogenous Air Pollutants: Chemical and Biological Implications*, pp. 165–178 (Grosjean, D, Ed.) Ann Arbor Science Publ. Inc., Ann Arbor, Michigan

NAKAMURA, H. and SAKA, H. (1978). *Japanese Journal of Crop Science*, **47**, 707–714

NOBEL, P.S. and WANG, C. (1973). *Archives of Biochemistry and Biophysics*, **157**, 388–394

NOBEL, P.S., ZARAGOZA, L.J. and SMITH, W.K. (1975). *Plant Physiology*, **55**, 1067–1070

NOLAND, T.L. and KOZLOWSKI, T.A. (1979). *Canadian Journal of Forestry Research*, **9**, 501–503

O'DELL, R.A., TAHERI, M. and KABEL, R.L. (1977). *Journal of the Air Pollution Control Association*, **27**, 1104–1109

OSHIMA, R.J., BENNETT, J.P. and BRAEGELMANN, P.K. (1978). *Journal of the American Society for Horticultural Science*, **103**, 348–350

OSHIMA, R.J., BRAEGELMANN, P.K., FLAGLER, R.B. and TESO, R.R. (1979). *Journal of Environmental Quality*, **8**, 474–479

OTTO, H.W. and DAINES, R.H. (1969). *Science*, **163**, 1209–1210

PAULS, K.P. and THOMPSON, J.E. (1980). *Nature*, **283**, 504–506

PEGG, G.F. (1976). In *Physiological Plant Pathology*, pp. 582–591 (Heitefuss, R. and Williams, P.H. Eds), Springer–Verlag, New York

PELEG, M. (1976). *Water Research*, **10**, 361–365

PELL, E.J. and BRENNAN, E. (1973). *Plant Physiology*, **51**, 378–381

PELL, E.J. and WEISSBERGER, W.C. (1976). *Phytopathology*, **66**, 856–861

PERCHOROWICZ, J.T. and TING, I.P. (1974). *American Journal of Botany*, **61**, 787–793

RAND, R.H. (1977). *Transactions of the American Society of Agricultural Engineers*, **20**, 701–704

RAND, R.H. (1978). *Journal of Biomechanical Engineering*, **100**, 20–24

RICH, S. (1969). *Annual Review of Phytopathology*, **2**, 253–266

RICH, S. and TURNER, N.C. (1972). *Journal of the Air Pollution Control Association*, **22**, 718–721

ROSEN, P.M., MUSSELMAN, R.C. and KENDER, W.J. (1978). *Scientia Horticulturae*, **8**, 137–142

ROWLANDS, J.R., GAUSE, E.M., RODRIGUEZ, C.F. and McKEE, H.C. (1970). *Electron Spin Resonance Studies of Vegetation Damage*. Final report SRI Project No. 05–2622–01, Southwest Research Institute, San Antonio, Texas

RUNECKLES, V.C. (1974). *Environmental Conservation*, **1**, 305–308

RUNECKLES, V.C. and ROSEN, P.M. (1977). *Canadian Journal of Botany*, **55**, 193–197

SCHANNE, F.A.X., KANE, A.B., YOUNG, E.E. and FARBER, J.L. (1979). *Science*, **206**, 700–702

SHERIFF, D.W. and MEIDNER, H. (1975). *Journal of Experimental Botany*, **26**, 897–902

SUTTON, R. and TING, I.P. (1977a). *Atmospheric Environment*, **11**, 273–275

SUTTON, R. and TING, I.P. (1977b). *American Journal of Botany*, **64**, 404–411

SWANSON, E.S., THOMSON, W.W. and MUDD, J.B. (1973). *Canadian Journal of Botany*, **51**, 1213–1219

TANAKA, K. and SUGAHARA, K. (1980). In *Studies on the Effects of Air Pollutants on Plants and Mechanisms of Phytotoxicity*, pp. 155–164. Research Report No. 11 from the National Institute for Environmental Studies, Iburaki, Japan

TANTON, T.W. and CROWDY, S.H. (1972). *Journal of Experimental Botany*, **23**, 619–625

TAPPEL, A.L. (1972). *Annals of the New York Academy of Sciences*, **203**, 12–28

TAPPEL, A.L. (1973). *Federation Proceedings*, **32**, 1870–1874

THOMSON, W.W. (1975). In *Responses of Plants to Air Pollution*, pp. 179–194 (Mudd, J.B. and Kozlowski, T.T., Eds), Academic Press, New York

THOMSON, W.W., DUGGER, W.M., Jr., and PALMER, R.L. (1966). *Canadian Journal of Botany*, **41**, 1677–1682

THORNE, L. and HANSON, G.P. (1972). *Environmental Pollution*, **3**, 303–312

THORNE, L. and HANSON, G.P. (1976). *Journal of the American Society for Horticultural Science*, **101**, 60–63

TING, I.P. and DUGGER, W. M., Jr. (1968). *Journal of the Air Pollution Control Association*, **18**, 810–813

TING, I.P. and DUGGER, W. M., Jr. (1971). *Atmospheric Environment*, **5**, 147–150

TING, I.P. and HEATH, R.L. (1975). *Advances in Agronomy*, **27**, 89–121

TING, I.P. and MUKERJI, S.K. (1971). *American Journal of Botany*, **58**, 497–504

TING, I.P., PERCHOROWICZ, J. and EVANS, L. (1974). In *Air Pollution Effects on Plant Growth*, pp. 8–21 (Dugger, M., Ed.), ACS Symposium Series **3**, 8–21

TINGEY, D.T. (1977). In *Proceedings of an International Conference on Photochemical Oxidant Pollution and its Control*, pp. 601–609, EPA/600/3-77-00lb (Dimitriades, B, Ed.), US Environmental Protection Agency, Research Triangle Park, NC

TINGEY, D.T. (1980). *HortScience*, **15**, 630–633

TINGEY, D.T., FITES, R.C. and WICKLIFF, C. (1973). *Environmental Pollution*, **4**, 183–192

TINGEY, D.T., FITES, R.C. and WICKLIFF, C. (1975). *Physiologia Plantarum*, **33**, 316–320

TINGEY, D.T., FITES, R.C. and WICKLIFF, C. (1976). *Physiologia Plantarum*, **37**, 69–72

TINGEY, D.T., STANDLEY, C and FIELD, R.W. (1976). *Atmospheric Environment*, **10**, 969–974

TINGEY, D.T., WILHOUR, R.G. and STANDLEY, C. (1976). *Forest Science*, **22**, 234–241

TOMLINSON, H. and RICH, S. (1967). *Phytopathology*, **57**, 972–974

TOMLINSON, H. and RICH, S. (1970). *Phytopathology*, **60**, 1842–1843

TOWNSEND, A.M. (1974). *Journal of the American Society for Horticultural Science*, **99**, 206–208

TOWNSEND, A.M. and DOCHINGER, L.S. (1974). *Atmospheric Environment*, **8**, 957–964

TURNER, N.C. (1976). *Journal of Experimental Botany*, **27**, 1085–1092

TURNER, N.C., RICH, S. and TOMLINSON, H. (1972). *Phytopathology*, **62**, 63–67

TURRELL, F.M. (1942). *American Journal of Botany*, **29**, 400–415

UNSWORTH, M.H., BISCOE, P.V. and BLACK, V. (1976). In *Effects of Air Pollutants on Plants*, pp. 5–16 (Mansfield, T.A., Ed.) Cambridge University Press, New York

VARGO, R.H., PELL, E.J. and SMITH, S.H. (1978). *Phytopathology*, **68**, 715–719

WEISS, J. (1935). *Transactions of the Faraday Society*, **31**, 668–681

WESELEY, M.L., EASTMAN, J.A., COOK, D.R. and HICKS, B.B. (1978). *Boundary-Layer Meteorology*, **15**, 361–373

WILKINSON, T.G. and BARNES, R.L. (1973). *Canadian Journal of Botany*, **51**, 1573–1578
WRIGHT, M. (1974). *Planta*, **120**, 63–69
YINGJAJAVAL, S. (1976). *Ozone Inhibition of Transpiration and Photosynthesis of Two Soybean (Glycine max (L.) Merr.) Cultivars*. Master's thesis, Oregon State University, Corvallis, Oregon

7

PHYSIOLOGICAL RESPONSES OF PLANTS TO FLUORINE

L.H. WEINSTEIN
RUTH ALSCHER-HERMAN
Boyce Thompson Institute, Cornell University, Ithaca, NY

Introduction

The problems of airborne fluoride (F)* in agriculture and forestry were known in Europe for many years before being recognized in the US. This recognition came about with the expansion of the aluminium industry during and shortly after the Second World War (De Ong, 1946) and with extensive mining and processing of phosphate deposits in Florida and Tennessee (MacIntire *et al.*, 1949). The problem has grown worldwide as a result of expanded use of aluminium, coal and fertilizer.

Although Heck, Taylor and Heggestad (1973) ranked F fifth in importance after ozone, sulphur dioxide, oxidants other than ozone (peroxyacyl nitrates, nitrogen oxides) and pesticides with respect to the amount of plant damage produced in the US, F problems have provided more than their share of controversies and litigation throughout the world. F is also the most phytotoxic of the common pollutants, and susceptible species can be injured at atmospheric concentrations 10 to 1000 times lower than those of the other major pollutants (less than 1 ppb or about $0.8\,\mu g\,F\,m^{-3}$). F has one unique and important characteristic—it accumulates in the plant and ingestion can cause disease in herbivores.

Not only has the plant-damaging potential of F been recognized, but more than 100 years ago it was reported that F supplements could stimulate crop growth (*reviewed by* Weinstein, 1977). In the intervening years, growth stimulation from F added to soils or from HF in the atmosphere has often been reported. However, no convincing evidence has been presented that constitutes proof for an essential role of F in plant development, and reports of stimulation of growth could be ascribed to hormoligosis—the phenomenon in which 'subharmful amounts of many stress agents may be helpful when presented to organisms in suboptimal environments' (Luckey, 1959). In contrast to F, oxides of sulphur and nitrogen supplied at low concentrations can directly supplement the nutrient requirements of the plant under some conditions. Ozone falls somewhere between the pollutants that provide essential or nonessential elements to the plant.

---

*'Fluoride (F)' is used as a generic term that refers to the fluoride ion and to combined forms of the element fluorine.

In recent years, several reviews have been published on the effects of F on plants (Thomas and Alther, 1966; Weinstein and McCune, 1970, 1971; McCune and Weinstein, 1971; National Academy of Sciences, 1971; Chang, 1975; Groth, 1975; McCune, 1976; Weinstein, 1977, 1979; Amundson and Weinstein, 1980). In addition to summarizing many published accounts that describe physiological and metabolic effects of F, we have emphasized the role of calcium (Ca) in F movement, accumulation and plant response.

Uptake and accumulation of fluoride

UPTAKE OF FLUORIDE

Plants absorb and accumulate F from the atmosphere, soil, or water. Root absorption is a significant route of entry into the plant, only when the growth medium, whether soil or water, contains a significant amount of soluble F. This can occur near some hot springs and fumaroles, and it is possible that soils near industrial sources of F emissions can contain a significant amount of soluble F (Sidhu, 1977; Flühler, Keller and Scherrer, 1979; Thompson, Sidhu and Roberts, 1979). The most important mode of entry into the plant, particularly of gaseous F (e.g. HF, SiF_4), is through the stomata of the leaf (with perhaps some uptake through lenticels). Soluble particulate forms of F undoubtedly pass through the cuticle and epidermis as well as through stomata (McCune, Silberman and Weinstein, 1977). The rate of uptake of F by the leaf is much more rapid than that of SO_2, O_3, NO_2 or NO (Bennett and Hill, 1973a).

MOVEMENT AND ACCUMULATION

Once inside the leaf or twig, F dissolves in plant liquids and moves in the transpirational stream to its principal sites of accumulation at the tips and margins of the leaf (Rommel, 1941; Ledbetter, Mavrodineanu and Weiss, 1960; Jacobson *et al.*, 1966). F is not uniformly distributed in the leaf tip or margin, but accumulates in specific cells. Garrec *et al.* (1974b) have shown by microscopy and electron microprobe analysis of needles of silver fir, that cells closest to the pathway of penetration and translocation of F exhibit the greatest cytological response (e.g. lacunose parenchyma near the stomata and transfusion parenchyma between the conducting vessels). These cells also have much higher concentrations of F than unaffected cells, such as palisade parenchyma.

Translocation to areas other than the leaf, or movement out of the leaf once it has accumulated, has not been considered to be an important characteristic of F in plants (Ledbetter, Mavrodineanu and Weiss, 1960; Jacobson *et al.*, 1966). Garber (1962) and Keller (1974) presented evidence of F movement Thent from stems to leaves, and Keller also suggested that F is translocated from leaves to stems with photosynthate. At increased root temperatures, Benedict, Ross and Wade (1964) found translocation of F from leaves to roots. Because F can be removed from some leaves by repeated mild washing, movement of F from the interior to the exterior of the leaf must occur (Ledbetter, Mavrodineanu and Weiss, 1960; Jacobson *et al.*, 1966). Kronberger, Halbwachs and Richter (1978) found that, following exposure of the upper or lower portions of the crown of white spruce seedlings to HF, a small amount of F was translocated from the exposed to the protected part of the crown (up to 3% of the total). Kronberger and Halbwachs (1978) found great differences in the accumulation of F in different organs of maize plants and at different stages of growth when cultivated in an area of F contamination. Changes in the F content of the various organs at different times were interpreted to be caused by exposure conditions, growth dilution and translocation. Because the plants were exposed over an extended period, a greater accumulation of F in mature rather than young leaves is not conclusive evidence of a downward translocation of F in phloem, as they have suggested. Neither is the extremely high accumulation of F in the tassel evidence of upward translocation from other parts of the plant. Other plausible explanations for F accumulation in the tassel might be that there was particulate F deposition on stigmatic surfaces, or that there was direct absorption of gaseous F over the great surface area of the tassel. Garrec and Vavasseur (1978) have provided persuasive evidence for the redistribution of F in poplars exposed for a growing season in the field near a source of atmospheric F. The F content of leaf blades was found to be about three times greater than that of petioles and stems combined. F accumulated in cortical tissues of the trunk: in the upper part of the trunk there was an acropetal gradient, while the opposite was true of the lower part of the trunk. The greatest accumulation occurred in the roots. They discounted the possibility of direct F accumulation in the trunk from the atmosphere, or in the roots from the soil. This putative demonstration of the redistribution of F from leaves to other parts of the plant might have been unequivocal had the experimental plants been exposed under more controlled conditions, had time-course measurements been made, and had the amount of F deposited in the soil and its partition between soluble and insoluble forms been known, as was the case in other studies (Sidhu, 1977; Flühler, Keller and Scherrer, 1979; Thompson, Sidhu and Roberts, 1979).

ACCUMULATION AND SUSCEPTIBILITY TO FLUORIDE

As is the case for all pollutants, different species of plants exhibit a wide range of tolerances to F and many compilations ranked for susceptibility have been offered (e.g. Weinstein, 1977). Highly susceptible species such as some conifers (young needles), gladiolus, Chinese apricot, Oregon grape or goatweed may exhibit foliar lesions from an accumulation of <10–20 ppm F. In contrast, cotton, tea, camellia, hickories and flowering dogwood can accumulate very high concentrations of F without foliar injury (as much as 4000 ppm in cotton), even within a nonpolluted environment (Zimmerman and Hitchcock, 1956; Zimmerman, Hitchcock and Gwirtsman, 1957; Jacobson *et al.*, 1966; McClenahen, 1976). In general, broad-leaved species will accumulate more F than conifers when they occur together (Sidhu, 1977; 1978) and great differences can occur between species, with the most tolerant ones often accumulating the most F (Weinstein, 1977). A likely explanation for this response is that, when a susceptible plant is injured metabolically or physiologically by a given dose of F, further absorption and accumulation are reduced.

INFLUENCE OF ENVIRONMENTAL FACTORS

In experiments on F accumulation and foliar injury in light-exposed and dark-exposed plants, Adams, Hendrix and Applegate (1957) reported that the onset of foliar injury in dark-exposed plants occurred with an HF dose only slightly higher than that required in the light, and that, in general, injury occurred in dark-exposed plants that had accumulated about half as much F as that required in the light. Alfalfa plants exposed to HF in the dark accumulated about 40% as much as in the light, despite the fact that the stomata were closed (Benedict, Ross and Wade, 1965). Jerusalem cherry foliage accumulated less F in the dark than in the light and without the occurrence of foliar injury, but when plants were transferred to the light, severe injury was produced (MacLean, Schneider and Weinstein, 1978). The metabolic changes that make Jerusalem cherry tolerant in the dark and susceptible in the light are not known. These authors pointed out that standards based upon 12-hour periods that come closest to separating day-time and night-time exposures might provide the most realistic averaging times with respect to dose-response. Differences in the response of plants to light and dark exposures could be related to air quality in the field if standards were based upon 12-hour averaging periods and if the 12-hour periods were clearly specified.

Other environmental factors influence the response of plants to F. For example, as air temperature was increased, F accumulation was increased in sunflower and decreased in gladiolus; however, foliar injury was greater in gladiolus (MacLean and Schneider, 1971). When soybean plants were exposed to increased temperatures after HF fumigation, the severity of foliar injury was increased (Wiebe and Poovaiah, 1973). An increase in relative humidity resulted in greater F accumulation and injury in gladiolus (Daines, Leone and Brennan, 1952) and tomato leaves (Daines, Leone and Brennan, 1952; MacLean, Schneider and McCune, 1973). Although plants grown under conditions of water stress are more tolerant to F (Daines, Leone and Brennan, 1952; Benedict and Breen, 1955; Zimmerman and Hitchcock, 1956; Applegate and Adams, 1960a), this does not appear to be the case when the stress is imposed after fumigation (Wiebe and Poovaiah, 1973).

Localization and form of fluoride in plants

CELLULAR AND SUBCELLULAR LOCALIZATION

As mentioned earlier, F accumulates at the tips of conifer needles and at the tips and margins of broad-leaved and many narrow-leaved plants. In silver-fir needles it is found nearest the cells where entry (lacunose parenchyma) and translocation (transfusion parenchyma) occur (Garrec *et al.*, 1974b). Before the techniques employed by Garrec and his co-workers were generally available, other methods had been used to determine the sites of F accumulation in the leaf. Differential fractionation, in a sucrose medium, of tomato leaves exposed to HF, suggested that accumulation was greatest in the cytosol, followed in decreasing order by the cell walls, chloroplasts, water-soluble proteins and mitochondria (Ledbetter, Mavrodineanu and Weiss, 1960). Fractionation of orange leaves in various hexane/carbon tetrachloride mixtures showed that the fraction containing cell walls, nuclei, broken chloroplasts and broken cells contained about 75%, and chloroplasts 20% of the total F. Allowing for F in broken chloroplasts and for cross-contamination, the authors concluded that the chloroplast was the site of highest F accumulation (Chang and Thompson, 1966). This is compatible with the reported sensitivity of the photosynthetic apparatus to F (Thomas and Hendricks, 1956; Thomas, 1958; Thomas and Alther, 1966; Bennett and Hill, 1973a,b). Brewer, Sutherland and Guillemet (1969b) reported that 60% of the total F of lemon leaves was in the epidermis and palisade cells. Using polar and nonpolar solvents to extract F from tomato and gladiolus leaves, it was postulated that F remains in a soluble form and maintains the chemical properties of free, inorganic F (Jacobson *et al.*, 1966).

Garrec *et al.* (1973, 1974a) examined the zones of F accumulation in needles of silver fir by electron microprobe analysis. The F was determined from the upper to the lower epidermis, and from the tip toward the base. In the green portion of the needles, three zones of accumulation were identified: the epidermis, palisade parenchyma and conducting tissue. Sclerified and lacunose parenchyma were not major accumulators. In the epidermal layer, some F was detected on the needle surfaces and some just below the surface. In palisade parenchyma, the greatest accumulation was about 30 µm below the upper needle surface where chloroplasts are present. F accumulation decreased from the upper to the lower epidermis (Garrec *et al.*, 1973). When examined along the axis of the needle, the principal area of F accumulation was at the tip (which was necrotic), with a second and lesser area of accumulation in the narrow transitional band immediately below the necrotic area. The latter area of accumulation was due to F-induced tyloses of the conducting vessels (Garrec *et al.*, 1974a).

FLUORO-ORGANIC COMPOUNDS IN PLANTS

A unique feature of certain plants found in the southern hemisphere (mostly Leguminoseae) is their capacity to synthesize fluoro-organic compounds. More than two dozen species of *Gastrolobium, Oxylobium* and *Acacia* have been found in Australia, two species of *Dichapetalum* in Africa, and *Palicourea marcgravii* in South America (*reviewed by* Weinstein, 1977). Most of the species synthesize monofluoroacetic acid, but *Dichapetalum toxicarium* also produces ϖ-fluorofatty acids. The toxicity of these plants to livestock, and even man, led to the identification of their toxic principles, and the elegant research of Sir Rudolph Peters led to the elucidation of the pathway of 'lethal synthesis' (Peters, 1952). The synthesis of fluoroacetic acid has been demonstrated in intact seedlings, tissue cultures and plant macerates of *Acacia georginae.*

If the synthesis of fluoro-organic acids is a prominent feature of plants in general, there should be concern for the potential hazards to livestock and man in relation to environmental contamination by F. In fact, there have been several reports of the presence of fluoroacetate and monofluoroci- trate in forage grass, crested wheat grass and soybeans exposed to atmospheric F in the laboratory or in the field (Cheng *et al.*, 1968; Lovelace, Miller and Welkie, 1968; Yu and Miller, 1970); but in subsequent work, the presence of a significant contamination by inorganic F helped to explain the high concentrations of fluoro-organic acids reported earlier (Yu, Miller and Lovelace, 1971) and the presence of these compounds was not confirmed in later work by the same group (Miller, 1972). Weinstein *et al.* (1972) were not able to detect fluoro-organic acids in grass, hay, crested wheat grass, soybean, corn, alfalfa or tomato when these were exposed to HF in the field, in the laboratory or supplied with NaF in solution. They concluded that, if present at all, the concentration of fluoro-organic acids in these species was below the level of detection of the analytical methods. They were able to identify fluoroacetate in seedlings of *Acacia georginae* provided with NaF in axenic culture, which established the suitability of the extraction and analytical methods.

Plants in general appear to have enzymes that cleave the carbon–fluorine bond (Preuss, Lemmens and Weinstein, 1968; Preuss and Weinstein, 1969; Ward and Huskisson, 1972; Hall, 1974; Vickery and Vickery, 1975), but some species are much more efficient than others (Preuss, Lemmens and Weinstein, 1968). The conversion of fluoroacetate to fluorocitrate has been demonstrated in lettuce (Ward and Huskisson, 1972). Mead and Segal (1972) have proposed a mechanism for biosynthesis of fluoroacetate by fluorination of a C_3 entity linked to pyridoxal phosphate. This would be followed by transamination and release of fluoropyruvate with subsequent oxidative decarboxylation, or by a series of reactions involving decarboxy- lation, transamination and hydrolytic release of fluoroacetaldehyde, which is oxidized to fluoroacetate.

FLUORIDE–CALCIUM RELATIONSHIPS: WHOLE-PLANT STUDIES

Pack (1966) observed that tomato plants with low calcium (Ca) levels were more subject to F injury than were plants maintained at a higher Ca level. Ramagopal, Welkie and Miller (1969) found that the Ca status of wheat roots was crucial to the degree of injury sustained by the roots when exposed to F solutions. Thus, it appears that Ca interacts with F in some vital fashion and can protect the plant tissues from F injury. This forms the basis for the application of Ca salts to protect crops from airborne F (*see later discussions*).

Bligny, Garrec and Fourcy (1973) examined the effect of Ca on the transport of F compounds in cut maize leaves. Using ^{18}F as a radioactive tracer, they showed that the amount of F accumulated at the leaf tip was inversely related to the level of Ca present in the leaves, with higher Ca levels being associated with lower levels of F accumulation at the leaf tips. Thus, Ca appears to increase the retention of F and lessen its rate of transport in the transpiration stream.

When supplied with soluble F salts, *Chara fragilis* thalli were shown by X-ray crystallography to contain crystals of CaF_2. These appeared in increasing amounts with increasing F concentration at the expense of $CaCO_3$ crystals (Lévy and Strauss, 1973).

In normal silver-fir needles there is a decreasing Ca gradient from the basal to the distal end; however, in F-fumigated needles, the highest concentrations of Ca and F occur together and especially at the distal end of the needle (Garrec *et al.*, 1974c). This is particularly marked in needles where necrosis is evident. The simultaneous occurrence of both elements suggests the formation of CaF_2 which may represent a means of detoxification of F. However, this beneficial effect may be offset by the disruption of the normal metabolism of Ca and perhaps other elements. Accelerated ageing may be an expression of such a disruption (Poovaiah, 1979).

The form of Ca in normal and fumigated needles is not always clear, but there is a notable increase in Ca oxalate in fumigated needles that have accumulated more than 500 ppm F (Garrec *et al.*, 1978), and the authors suggest that the increase in total Ca and of Ca oxalate are indications that F may induce premature senescence (Garrec *et al.*, 1974c, 1978).

As normal needles age, the endogenous levels of Ca and F increase, and this has been shown for a number of species of conifers and of broad-leaved plants (Garrec and Vavasseur, 1978). If the increase in the F concentration in needles is less than 300 ppm there is no effect on Ca concentration, but if the concentration exceeds 1000 ppm there is a significant increase in Ca. When the F content of needles reaches 300 ppm, they are generally considered to be from a heavily polluted area, and when the F content is 1000 ppm and higher, severe damage should occur. The authors have computed the following regression equation to describe the Ca–F relationship in fir trees:

$$Ca = 0.011\,F + 19.8 \tag{7.1}$$

in which the Ca concentration (ppm) in polluted needles doubles when the F concentration reaches 1800 ppm (Garrec *et al.*, 1978).

A strong correlation also exists between the concentrations of F and Ca (ppm) in different tissues of black poplar. The regression equations for control and fumigated trees (Garrec and Vavasseur, 1978) are (mean for two trees each)

$$\log\,(F) = 0.0225\,Ca + 0.703 \tag{7.2}$$

$$\log\,(F) = 0.0711\,Ca + 1.325 \tag{7.3}$$

An accumulation of 600 ppm F in silver-fir needles induces disturbances in Ca metabolism equivalent to one year's ageing (Garrec *et al.*, 1978). Although the relationship between F, Ca and senescence is extremely interesting and important, most investigators who have conducted field surveys would probably conclude that premature senescence can be induced at far lower concentrations of F than 300 ppm in many species of plants.

The application of lime to protect plants from F injury has been a practical control method for many years (e.g. Allmendinger, Miller and Johnson, 1950; Benson, 1959; Brewer, Sutherland and Guillemet, 1969a). Soluble Ca salts, such as $CaCl_2$, are as effective as lime sprays (Benson, 1959). It had been generally assumed that the application of lime promoted reaction with F on the plant surfaces, forming relatively insoluble CaF_2 (Allmendinger, Miller and Johnson, 1950). However, because $CaCl_2$ sprays were effective, Benson (1959) stated that Ca was absorbed, 'thus increasing the tolerance of the tissue for fluoride', and it is now known that the protective effect of Ca is due to a Ca–F interaction on and within plant tissues (Lévy and Strauss, 1973; Garrec *et al.*, 1974c; 1978).

FLUORIDE AND OTHER DIVALENT CATIONS

It seems likely that F reacts to form insoluble salts or complexes, not only with Ca, but with other cations in the plant that are essential for normal nutrition. The similarity of chronic F symptoms to Mn deficiency in citrus has been cited as a possible example (Brewer *et al.*, 1960). In addition, some evidence has been presented to suggest that the symptoms may actually be associated with a lower foliar content of Mn and Mg (Brewer *et al.*, 1967). The best evidence of a relationship between F and essential cations (other than Ca) is provided by Garrec, Plebin and Lhoste (1977) who have shown pronounced decreases in the Mg and Mn contents of field-fumigated silver-fir needles as the F content increases. The concentrations of the elements decrease until F accumulates to about 400 ppm, after which they remain constant. No changes were detected in nitrogen, phosphorus or potassium contents. An expression of the decrease in Mg may be the breakdown in chloroplast structure associated with the presence of F (Garrec, Plebin and Lhoste, 1977). The striking resemblance of chronic F symptoms to nutrient deficiencies of Mn, Mg, and Zn, with the evidence cited above, suggests that one action of F may be to induce deficiencies by interfering with metal-ion transport to the leaf and/or inactivating metal ions at their sites of physiological activity.

Role of fluoride in transport

FLUORIDE AND PHOSPHATE UPTAKE

Extensive studies on transport by Penot and his co-workers (Penot, 1967, 1968; Penot and Diouris, 1969; Diouris and Penot, 1972, 1973, 1975, 1977) have yielded interesting data which were ultimately found to be related to Ca–F interactions.

Pretreatment with NaF for 5 min stimulated a light-dependent uptake of ^{32}P in leaf discs. This fluoride-stimulated uptake was subsequently found to occur also in achlorophyllous tissue, where it was light-independent

(Penot, 1968). In both cases, pretreatment with NaF resulted in stimulation of uptake of ^{32}P. However, if F was present at the same time as the ^{32}P, phosphate uptake was inhibited (Penot and Diouris, 1969), with a 63% inhibition of uptake occurring in the presence of 1 mM NaF. Thus, F may act to alter the cellular transport machinery in some manner. Furthermore, as light was not required for the F-induced stimulation to occur, it is unlikely that photosynthesis-related functions are necessarily involved, despite the finding of Chang and Thompson (1966) that F accumulates in the chloroplasts of leaves. Another possibility is that the phosphate economy of the cell is affected by F which, in the case of photosynthesizing cells, might be reflected in a change in chloroplast-related functions (*see* Wei and Miller, 1972; Horvath, Klasova and Navara, 1978, on changes in chloroplast ultrastructure in leaves after fumigation with HF).

Penot and Diouris (1969) compared the effects of pretreatment with 5 mM F on the uptake of some anions (P, S, Cl, NO_3), cations (Rb, Ca, K, Na, NH_4), and glucose into leaf discs, potato tubers and carrot roots. Radio-isotopes were used to monitor uptake in each case. Their data show that F stimulates uptake of phosphate and of other anions in mature, but not in young leaves of tobacco. An analogous effect occurs in potato tuber discs, but only after prewashing in tap water for 24 h. (It is unfortunate that Diouris and Penot chose to carry out all their subsequent investigations of this phenomenon with the washed potato-tuber disc system, as extrapolation of results obtained with one tissue to another, i.e. tubers to leaves, are always subject to question).

The stimulatory effect of NaF on uptake was found to be confined to anion uptake, with cation uptake being inhibited by pretreatment with F. Respiratory activity, as measured by oxygen uptake, was also stimulated. However, since the results were obtained exclusively with whole tissue, it is not possible to differentiate between an indirect effect on transport via, perhaps, an increased rate of oxidative phosphorylation (witness the increased respiratory rate) and an effect on an active transport system *per se*.

FLUORIDE–CALCIUM INTERACTIONS: METABOLIC STUDIES

Diouris and Penot (1973) investigated the effect of Ca on F-stimulated phosphate uptake in washed potato-tuber discs. They found that, in the presence of Ca, F-stimulated phosphate uptake was increased. Furthermore, the Ca-treated tissues contained twice as much F as those incubated with F alone. Diouris and Penot (1975) showed that this F-stimulated phosphate uptake in potato-tuber discs is entirely dependent on the presence of Ca ions and is not purely a function of their washing procedure, since treatment of washed discs with EDTA abolished the effect and re-addition of Ca ions restored it. The effect appeared to be temperature-independent since it occurred between 0°C and 20°C. They conclude that the F-affected process must be nonmetabolic and suggest that Ca adsorption sites at the cell wall or the plasmalemma are involved. This is in contradiction to their earlier work (Penot and Diouris, 1969) where they demonstrated the involvement of an active-transport component in F-stimulated phosphate uptake. Phosphate uptake itself has been demonstrated to involve an electrogenic pump (Dunlop and Bowling, 1978; Bowling and Dunlop, 1978). It should be noted that F inhibited phosphate uptake when present in the absorption medium itself, but stimulated uptake when administered as a pretreatment.

This stimulation was strictly a function of the presence of Ca ions which have been known for some time to affect membrane transport selectivity (Jacobson, Moore and Hannapal, 1960; Jacobson *et al.*, 1961; Moore, Jacobson and Overstreet, 1961; Sze and Hodges, 1977; Lüttge and Higinbotham, 1979). Numerous reports also attest to the stabilizing influence of Ca on cell membranes (*summarized by* Lüttge and Higin botham, 1979). Although Ca has been shown to be essential for membrane function, its concentration in the free form within the cytoplasm is quite low (0.1–10 µM) with most of the cellular Ca (approx. 1–3 mM) existing in a chelated form in the vacuole or as insoluble salts (Raven, 1977). Some of this bound Ca is associated with the plasmalemma and may be involved in the regulation of the levels of free Ca. Ca ions are, in fact, actively transported out of the cell across the plasmalemma (Macklon, 1975), and Ca is an inhibitor of several enzymatic reactions for which Mg is required (Evans and Sorger, 1966). On the other hand, Ca is required for the functioning of some membrane-bound ATPases (*summarized by* Hodges, 1976) and may have a regulatory function in chloroplast protein synthesis (Bouthyette and Jagendorf, 1982).

The data summarized above suggest that a transport system which operates in the absence of Ca is inhibited by F. In the presence of Ca, transport selectivity is altered and F-susceptibility disappears. Furthermore, membrane permeability appears to be altered so that more F penetrates into the tissue itself. To accord with the observations made with whole plants, this additional F which is taken up in the presence of Ca must somehow be sequestered by the Ca and rendered less toxic to the plant cell. Hence, it is not possible to accept an enhanced level of accumulated F as an indicator of increased F injury.

The effect of F on transport and its interaction with Ca can be separated into two categories: effects on uptake itself and the nature of the Ca–F interaction. If an active transport process is involved, it is possible that a membrane ATPase constitutes a part of the site of action of F at the plasmalemma. Using a histochemical technique to localize ATPase activity, Gilder and Cronshaw (1974) showed that 10 mM F inhibited a plasmalemma-associated ATPase in the sieve elements of the phloem of tobacco. Hodges (1976) tested the effect of K salts of various halides, including F, on membrane (plasmalemma) ATPases in wheat roots. He found that KF inhibited the high pH ATPase by 50% . Unfortunately, the concentration used by Hodges in this work was 50mM and his results may have little bearing on the events which take place when F is accumulated in leaves exposed to the levels of HF that commonly occur in polluted areas. Diouris and Penot (1977) fractionated F-treated potato-tuber disc tissue after incubation with ^{32}P in an attempt to determine the fate of the accumulated phosphate, the uptake of which was stimulated by F. They found that inorganic, but not organic, phosphate in the TCA-soluble fraction was increased, and also reported an increase in phosphate levels in TCA-insoluble material (membranes, nucleic acids?). They concluded that, because only inorganic phosphate was affected, the effect of F was being exerted on the plasmalemma. In contrast, an analogous fractionation of tissue which had been exposed to F during the absorption phase showed decreased levels of phosphate in inorganic and organic acid-soluble fractions.

A possible explanation for the Ca-F interaction and its stimulatory effect on phosphate uptake can be proposed, that is related to the documented inhibitory effect of F on metal-requiring enzymes such as phosphatases (Massart and Dufait, 1942; Naganna *et al.*, 1955). Ca alters plasmalemma permeability so that more F enters the cell, where it reaches the site of action of ATPases and other phosphatases, the functioning of which is crucial for the maintenance of adequate intracellular phosphate levels. A feedback system may exist, so that decreases in inorganic phosphate levels lead to increased phosphate uptake.

Existing information on Ca–F interactions and phosphate uptake could be utilized to learn more about the physiological mode of action of F and the mechanisms by which the cell responds to that stress. The intracellular localization (plasmalemma, vacuole, mitochondrion, chloroplast?) of a possible Ca–F complex, and the mechanism by which such a complex is formed, should be determined. It is possible that the basis for differences in F resistance between species or cultivars might reside in the capability of any particular species or cultivar to inactivate F rapidly and efficiently. The susceptibility to F of the various membrane ATPases (plasmalemma, mitochondrial, chloroplast) that are essential for cellular maintenance and growth, might be a basis for differences in F susceptibility.

Effects of fluoride on respiration

Oxygen uptake has been variously reported to be stimulated (Applegate and Adams, 1960b) or inhibited (Hill *et al.*, 1959; Béjaoui and Pilet, 1975) as a result of exposure of intact plants or plant tissues to HF. Many variables are involved, including age, nutrient status and nature of the HF dose administered (Weinstein, 1977). Exposure to F appeared to increase the relative proportion of the pentose phosphate pathway over glycolysis in leaves (Ross, Wiebe and Miller, 1962; McCune *et al.*, 1964; McCune, De Hertogh and Weinstein, 1967; McCune and Weinstein, 1971). As enolase has been demonstrated to be F-sensitive (Miller, 1958) (perhaps because of the formation of a magnesium–fluorophosphate complex), it is possible that an inhibition of glycolysis leads to an increased flow of carbon through the pentose phosphate pathway. Lee, Miller and Welkie (1966) found that fumigation of soybean leaves with HF (approx. $80\,\mu g\,m^{-3}$) resulted in a substantial increase in the activity of glucose-6-phosphate dehydrogenase, the first enzyme of the pentose phosphate pathway. They determined the effect of fumigation on the activities of five cellular oxidases, some mitochondrial and some cytoplasmic in origin. The activities of three of these enzymes (cytochrome oxidase, catalase and peroxidase) were increased as a result of fumigation. However, neither cytochrome oxidase nor catalase activities were affected by F ions *in vitro*. Mechanical injury of leaves produced increases in the activities of those same enzymes and thus the changes observed as a result of HF fumigation are a nonspecific expression of host response to injury. The increase in glucose-6-phosphate dehydrogenase activity after fumigation was much greater than that produced by mechanical injury. This additional increase suggests that F stimulated glucose-6-phosphate dehydrogenase activity in some other, more specific fashion. Lee and his co-workers did not assay for enolase activity, which is unfortunate because they would then have been better able to assess the possibility that the glycolytic pathway was being blocked by F at that step.

If F can be shown to have penetrated into the mitochondria of any particular tissue, it is possible that it is exerting an effect on enzymatic events which take place in that organelle. Pilet and Roland (1972) demonstrated changes in the conformation of mitochondria in NaF-treated (10^{-5}–10^{-3} M) secondary explants of *Rubus hispidus*.

Béjaoui and Pilet (1975) demonstrated that oxygen uptake was more susceptible to inhibition by F in younger than in older tissues. This can be explained by the observed shift from glycolysis to the pentose phosphate pathway which occurs with ageing of the leaf (Gibbs and Beevers, 1955). A similar shift in sensitivity to F with ageing in potato-tuber discs was described by Coutrez-Geernick (1973). Additional evidence which supports this hypothesis can be found in the germinating seed system, where it has been reported that F exerts a greater inhibitory effect in maize seeds as germination proceeds (Belozerova and Fedorova, 1976). It is thought that the pentose phosphate pathway is the major oxidative pathway operating in the seed during the early phases of germination (Roberts, 1969) with a shift to glycolysis occurring later in the course of that process. Some interesting experiments were carried out by Pilet and Béjaoui (1975) to determine whether Ca or Mg influence the effect of F on respiratory activity. As Ca does have a protective function with regard to F susceptibility (*discussed above*), it is of great interest to determine whether it lessens individual physiological expressions of F toxicity. Since Mg is required for the operation of glycolysis, it might act to block the inhibitory effect of F. Pilet and Béjaoui found that the presence of 0.5×10^{-3} M Ca and 1×10^{-3} M Mg did lessen the inhibitory effect of 10^{-3}–10^{-2} M NaF on oxygen uptake in *Rubus hispidus* tissue cultures. Although these workers do not make a point of this, it seems that Ca and Mg can be substituted for each other to some extent in counteracting the inhibitory effect of F. The protective effect of Mg could result from a replenishing of cellular Mg levels to a point where Mg-requiring enzymes such as enolase could again become functional. However, essential enzymatic steps in protein synthesis also require Mg. Pilet and Béjaoui exposed tissues to F for up to 105 min, a period sufficient for an overt expression of protein-synthetic events to become apparent (Bewley and Black, 1978). Assays of enolase activity and the activities of other Mg-requiring respiratory enzymes under the various conditions employed would have given an indication as to whether the mode of action of F was primarily to block glycolysis or whether the *de novo* synthesis of new enzymes(s) was being inhibited by removal of Mg by F.

Glycolysis is known to take place in the cytoplasm. However, the results obtained by Pilet and Roland (1972) in their ultrastructural study of the effect of F on mitochondrial morphology indicated that F must penetrate into that organelle, because massive alterations in mitochondrial structure occur as a result of F treatment. Miller and Miller (1974) demonstrated an inhibition of oxidative phosphorylation in etiolated leaf discs of soybean and corn after 48 h of exposure to 9–12 µg HF m^{-3}. This effect may be related to inhibition by F of ATPase activity (Kielley and Meyerhof, 1948) since oxidative phosphorylation takes place through the operation of cristae-bound, Mg-requiring ATPases. Again, protein turnover might be involved. If fumigation with HF could be shown to result in the formation of mitochondrial F–ATPase complexes, then the action of F at that site would be clearly established.

Effects of fluoride on photosynthesis

APPARENT PHOTOSYNTHESIS

Table 7.1, taken from Amundson and Weinstein (1980), summarizes much of what is known to date concerning effects of HF and NaF on apparent photosynthesis (AP). Long-term exposure of sorghum to relatively low levels of HF (0.7, 1.7 and 3.5 µg m^{-3}) resulted in reversible reductions in the rate of AP (McCune, MacLean and Schneider, 1976). Others have reported no effect of exposure to HF on AP in the absence of foliar injury (Hill *et al.*, 1958; Hill, 1969; Thompson, Sidhu and Roberts, 1979). Keller (1977) determined rates of AP of the foliage of 11 different tree species which had been exposed to varying amounts of atmospheric F over extended periods. In this instance, foliar injury and abscission were brought about by F exposure, with the foliage remaining on the trees maintaining the same rate of AP as that of control trees.

Thomas (1958) and Woltz and Leonard (1964), on the other hand, reported a F-induced decrease in AP which was greater than the amount of injured foliage. Exposures to higher levels of F ($>$10 µg m^{-3}) have been reported to reduce AP in every species tested, with the exception of cotton (Thomas and Hendricks, 1956; Thomas, 1958; Hill, 1969; Bennett and Hill, 1973b).

RECOVERY OF APPARENT PHOTOSYNTHESIS AFTER EXPOSURE TO HYDROGEN FLUORIDE

Horvath, Klasova and Navara (1978) measured the effect of long-term (14 days) and short-term (24 h) exposures to approx. 40 µg HF m^{-3} on AP of *Vicia faba*. After 24 h of fumigation, AP was decreased, with recovery occurring within 24 h of removal of the plants from the HF atmosphere. Recovery of AP was also reported by McCune, MacLean and Schneider (1976) after exposure of field-grown sorghum to HF at up to 1.7 µg m^{-3}. The recovery process appears to be quite rapid in many instances, but is frequently incomplete (Thomas and Alther, 1966).

Table 7.1 REPORTED EFFECTS OF FLUORIDE (HF AND NaF) ON APPARENT PHOTOSYNTHESIS (AP) (AS MEASURED BY CHANGES IN CO_2 UPTAKE) OF HIGHER PLANTS. (After Amundson and Weinstein, 1980)

Genus or species	Concentration ($\mu g\,HF\,m^{-3}$)	Duration	Response	Reference
Gladiolus	0.8–8.0	7 d	% reduction in AP = % leaf injury	Thomas and Hendricks, 1956
Hordeum	32	4–8 h	Total interruption in AP, with recovery	
Medicago	200	4–8 h	in few hours to days	
Fruit trees	16–40	4–8 h		
Hordeum	32	2 h	AP reduced during exposure, with	Bennett and Hill, 1973b
Medicago	32	2 h	recovery after exposure	
Vicia	40	1 and 14 d	AP reduced during exposure, with recovery after exposure	Horvath, Klasova and Navara, 1978
Lycopersicon	1.4–5.2	4 weeks	No effect	Hill et al., 1959
	0.9–11.2	3 weeks	No effect	
Fruit trees	2.1 (av.)	183 h	14% reduction in AP, 10% injury	Thomas, 1958
Gladiolus	3.1–5.2	30–205 d	% reduction in AP = % leaf injury	
Gossypium	13.6	138 h	No effect	
Citrus	0.32–0.77	Growing season	No effect	Thompson et al., 1967
Gladiolus	0.8	39 d	No effect	Hill, 1969
	1.2	27 d	3% reduction in AP over injury	
Fragaria	2.3	63 d	No effect	
	38	1 d	50% reduction in AP	
Lycopersicon	5.1/12	17/21 d	No effect	
Prunus	1.6	42 d	No effect	
Zea	7.7	16 d	No effect	

Species	Treatment	Duration	Effect	Reference
Sorghum	0.7 2.2 then 1.7 3.5 then 5+	14 d 12/2 d 7/7d	No effect Reduced, with recovery after exposure Reduced during 3.5 μg m^{-3} exposure, then severely injured. Little recovery	McCune, MacLean and Schneider, 1976
Pinus sylvestris *P. nigra* *P. strobus* *Larix leptolepsis* *Quercus borealis* *Pseudotsuga menziesii* *Picea excelsa* *Alnus incana* *Sorbus aria* *Acer pseudoplatanus* *Larix decidua*	Ambient near source	Nov–April	Reduced AP of whole plant due to loss of foliage with visible injury on remaining foliage	Keller, 1977
Cornus florida *Liquidambar styraciflua* *Plantanus occidentalis* *Acer rubrum* *Liriodendron tulipifera* *Oxydendrum arboreum* *Pinus strobus* *P. taeda* *P. echinata*	1.9, 19 or 190 ppm NaF	24 h	AP reduced in older needles of *P. taeda* and *P. echinata* at 1.9 ppm NaF AP reduced in all species at 19 and 190 ppm NaF	McLaughlin and Barnes, 1975
Picea excelsa	100 ppm	Winter–spring	AP reduced in old foliage/new injured	Keller, 1980

METABOLIC AND ULTRASTRUCTURAL EFFECTS OF FLUORIDE ON CHLOROPLASTS

Ballantyne (1972) demonstrated an inhibitory effect of KF on the Hill reaction in bean chloroplasts. The addition of stoichiometric amounts of Mg was found to neutralize the inhibitory effect of KF, suggesting that, in this instance, F may be acting to render Mg (which is essential for thylakoid function) physiologically inactive. Venesland and Turkington (1966) reported an inhibitory effect of KF on oxygen evolution at relatively low concentrations, with higher concentrations affecting Hill-reaction activity. An inhibitory effect of NaF on pigment synthesis was suggested by the data of McNulty and Newman (1961) with bush bean, and Bianchi and Laudi (1977) with *Larix* seedlings. The results of Chang and Thompson (1966) seem to indicate that F accumulates in the chloroplast, and the ultrastructural studies of Bligny *et al.* (1973) demonstrated deformations in chloroplast morphology in polluted (350 µg F/g dry wt.) needles of silver fir. Horvath, Klasova and Navara (1978) also reported HF-induced ultrastructural changes in chloroplast morphology. In both cases the internal membranes of the chloroplasts were observed to be dilated and the total amount of green membranous material per cell was reported to be lower in cells from polluted specimens. It may be that this effect is an expression of an inhibitory effect of F on pigment biosynthesis which, in turn, may control thylakoid development directly (Hoober and Stegeman, 1973).

However, Wei and Miller (1972), in an ultrastructural study of the effect of HF on mesophyll cells of soybean, reported changes in the tonoplast, in mitochondria and in the endoplasmic reticulum, with little effect on chloroplast morphology. This may reflect differences in F susceptibility between species or perhaps differences in environmental conditions which obtained during HF exposures.

Effects of fluoride on growth and development

FLUORIDE AND THE SYNTHESIS OF RNA AND PROTEIN

A decrease in the level of Mg and Mn ions in HF-fumigated leaves was demonstrated by Garrec, Plebin and Lhoste (1977) (*see* page 146). As transcription and translation cannot proceed in the absence of Mg, it is probable that exposure to F may affect these processes. The data of Pilet (1968, 1969, 1970) and of Chang (1968, 1970) show that this is indeed the case. Chang (1970) determined the effect of NaF on the ribosomal content of maize roots. A substantial (25–30%) decrease occurred in fluoride-treated roots. Sucrose density-gradient profiles of ribosomal samples from control and fluoride-treated roots showed a decrease in polysomal and an increase in monosomal (single ribosomes) material in the F-exposed sample. This indicates a decrease in the extent of protein synthesis (translation) in fluoride-treated roots. Because total ribosomal RNA decreased too, this suggests that F affects RNA synthesis (transcription) also. Determinations of ribonuclease activity were also carried out by

Chang. Substantial increases in ribonuclease were detected in all subcellular fractions obtained from F-treated roots, with the exception of the plastid fraction. Because ribonuclease activity *in vivo* is thought to be controlled by Ca, Mg, and K ratios (Hanson, 1960) it is possible that fluoride is exerting its action through a disruption of the Ca:Mg balance.

RNA and protein synthesis are essential growth processes. It is therefore not surprising that Chang (1968) demonstrated a direct relationship between F-inhibited growth rate and RNA levels in maize roots. A F-induced increased rate of RNA breakdown was demonstrated by Pilet (1968) in senescing lentil roots. Pilet (1969) demonstrated an increased rate of ribonuclease activity in the same system, and in 1970 he showed a correlation in time between decrease in RNA content and increase in ribonuclease activity in fluoride-treated roots. However, a regression analysis revealed that the F-induced changes in RNA were not solely due to ribonuclease activity. Thus RNA synthesis is also affected by fluoride. It is possible, although by no means proved, that fluoride exerts its inhibitory action on RNA and protein synthesis through a sequestration of Mg. As in so many other instances, the crucial experiments in this area remain to be performed.

FLUORIDE AND GERMINATION

Navara and his co-workers in Bratislava have carried out several studies of the effects of F on germination (Navara, 1964; Holub and Navara, 1966; Navara, Holub and Bedatsova, 1966). They found that F tolerance seemed to be related to high Ca content and to high endogenous F levels. Their data show that, if pea seeds were soaked in increasing concentrations of $CaCl_2$ (up to 0.4 M) in the presence of 1.7 mM NaF, F accumulated in increasing amounts. Pea seeds which were initially soaked in 0.4 M $CaCl_2$ showed increased resistance with regard to germinability in a subsequent exposure to F (19 mM). Mg appeared to exert a protective effect also, but to a lesser degree than did Ca. Again, in the germination system as with the washed potato-tuber discs, the presence of Ca has a protective effect while stimulating greater F uptake.

Hauskrecht (1972) and Chang (1967) have also studied the effects of exposure to F on germination. Chang showed that NaF acts to inhibit phytase activity in germinating maize seeds. Phytin as phytic acid or a salt of Mg, K or Ca is a major phosphate reserve in many seeds, containing up to 50% of total stored phosphate. It has been shown that the activity of phytase—a Mg-requiring enzyme (McCance and Widdowson, 1944)—increases with germination in both cereals and dicots (Hall and Hodges, 1966, *summarized by* Bewley and Black, 1978), and it is thought that the free inorganic phosphate which is released is probably transported to growing regions of the seedlings. This increase in phytase activity may be due, in part, to *de novo* synthesis of the enzyme, as opposed to its activation. Hauskrecht found that 250 μM NaF inhibited phytase activity *in vitro* by 34% with 1 mM NaF inhibiting enzyme activity *in vivo* by 80%. In addition, the phytase activity of germinating seeds was inhibited 24% by germination in a medium containing 1 mM F, and 65% by germination in the presence of 4 mM NaF.

It is not possible to determine whether F is exerting its inhibitory influence *in vivo* on the *de novo* synthesis of phytase or on its activity. Possible effects of F on nucleic acid and protein synthesis are discussed above. As phytase is a Mg-requiring enzyme the formation of a Mg–F complex might result in an impairment of either or both functions.

FLUORIDE AND FRUITING

The work of Pack and his colleague (Pack, 1966, 1971, 1972; Sulzbach and Pack, 1972; Pack and Sulzbach, 1976) suggests that F has a specific effect on fertilization. Pack (1966) demonstrated that tomato plants that had been subjected to fumigation with HF produced smaller fruit which were partially or completely seedless. Plants subjected to a low nutrient level of Ca showed similar symptoms, and Ca deprivation and exposure to HF produced an additive effect when administered together. An analogous effect on seed production was induced in HF-fumigated bean plants (Pack, 1971). In order to determine whether pollen-tube growth was being affected by exposure to F, Sulzbach and Pack (1972) determined the effects of HF fumigation on *in vivo* pollen germination from plants growing in media containing sufficient (5 mM) and low (1 mM) Ca, using fluorescence microscopy. They also tested effects of F (and its interaction with Ca) on pollen germination *in vitro*. The germination of cucumber pollen grains was significantly inhibited by 2.6 mM F, while tomato pollen grains were more resistant, with germination remaining unaffected up to 5.3 mM F. Ca exerted a protective effect on pollen-tube germination (2.5 mM Ca for both species).

The *in vitro* germination of pollen collected from HF-fumigated plants did not differ significantly from the germination of pollen collected from control plants. In contrast, *in vivo* pollen germination and growth were markedly influenced by HF fumigation. It was determined that the HF susceptibility lay in the maternal parent, because reductions in pollen retention on the stigma, germination and percentage of tubes growing as far as the ovules, were all functions of the origin of the pistil, regardless of the origin of the pollen. HF treatment of the pollen resulted in a similar effect, only when that pollen was applied to control pistils. Bonte (this volume, Chapter 10) discusses this and similar work of his own. Pack and Sulzbach (1976) extended their study of the effect of F on pollen-tube germination and growth to 13 different agronomically important species. Qualitatively similar results were obtained in each case. Seed production was totally inhibited in some crops (sweet corn, soybean) and partially inhibited in others. In cucumber, the number of fruits produced by fumigated plants (48.8 μg F m^{-3} days) was greater than that produced by control plants, although the seed number per plant was markedly reduced. In some cases, the weights of the vegetative parts actually increased as a result of fumigation. Two interesting inferences can be drawn from these findings. First, these data constitute further confirmation of a specific effect of F on fertilization. Second, they point to an as yet ill-defined but very important response of growth and morphogenesis to the presence of fluoride (*see also* MacLean, Schneider and McCune, 1977), a response also elicited in carrot and pepper plants by exposure to O_3 (Bennett and Oshima, 1976; Bennett, Oshima and Lippert, 1979). In both cases, i.e. fumigation with HF or with O_3, the development of vegetative organs is maintained, or even stimulated, at the 'expense' of the reproductive tissues. As this phenomenon does not seem to constitute a specific response to F, it will not be discussed further here. (However, it should be pointed out that a lack of specificity was also demonstrated by the results of Lee, Miller and Welkie (1966) in the case of fumigation-induced changes in the activities of respiratory enzymes obtained from soybean leaves. Thus, the specificity of any given response to a pollutant, be it at the subcellular, cellular, tissue or organ level, must be rigorously tested, using the appropriate controls.)

Pack and Sulzbach (1976) determined tissue F concentrations in their investigations. As expected, the leaf was the primary site of accumulation of F, with fruit and seeds being relatively low in F content. Interestingly, plants grown on Ca-sufficient media accumulated more F than did those grown on low Ca. This recalls the results of Diouris and Penot (1975) obtained with potato-tuber discs, and those of Holub and Navara (1966) with pea seeds. Once again, external F appears to be more toxic than F which has entered the tissue in the presence of Ca. Pack and Sulzbach did not determine F concentrations in floral parts; therefore, data are not available from their work which would shed light on Ca–F interactions at the pistil itself.

A possible role for Ca in pollen-tube germination has been suggested by Jaffe and his co-workers (Jaffe, Weinsenseel and Jaffe, 1975; Weisenseel, Nuccitelli and Jaffe, 1975) who implicated Ca in the generation of the large electrical current which traverses germinating pollen tubes. Using ^{45}Ca and an autoradiographic technique, they demonstrated an accumulation of Ca in the growing tips of pollen tubes isolated from *Lilium* anthers. This accumulation appeared to be in the cytoplasm and may amount to as much as 100 mM Ca at the tip of the growing tube.

Facteau, Wang and Rowe (1973) and Facteau and Rowe (1977) have investigated the effect of fumigation with HF on pollen-tube growth in cherry and apricot. Fumigation during anthesis inhibited *in vivo* pollen-tube growth but the same inhibitory effect was produced whether pollination preceded fumigation or vice versa (Facteau, Wang and Rowe, 1973). In fact, it was possible to fumigate 24 h after pollination and still obtain inhibition of pollen-tube growth. Thus, the F-sensitive step cannot be related to stigma receptivity or the initiation of pollen germination, but is probably related to pollen-tube growth itself (*see* Bonte, this volume, Chapter 10).

SUTURE RED SPOT OF PEACH: A POSSIBLE ROLE OF CALCIUM

One of the most interesting F-induced disorders of plants is 'suture red spot' (SRS) or 'soft suture' of peach fruits, because (1) the effect is produced in the fruit which is spatially separated from the leaf, the normal organ of F accumulation; (2) there appears to be an association of SRS with Ca nutrition; and (3) SRS can be induced by extremely low concentrations of atmospheric F. The disorder is characterized by a reddening of the skin and flesh on one or both sides of the suture toward the stylar end of the fruit. The affected areas ripen considerably earlier than other parts of the fruit, so that at harvest they are soft and often decomposing. Splitting of the flesh along the suture is often an ancillary symptom (Benson, 1959). Evidence has been presented that the softening of the flesh along the suture is accompanied by an increase in the firmness of the flesh on the dorsal side of the fruit. The suture area was found to accumulate more F than other parts of the flesh and it was suggested that this 'triggers the ripening sequence' (Facteau and Wang, 1972).

Ca salts are effective in reducing the incidence and severity of SRS: spray applications to trees at the pit-hardening stage of fruit development were reported to be most effective (Benson, 1959).

In a series of experiments conducted over several years with 'Elberta' peaches, MacLean and his colleagues (personal communication) have elucidated some aspects of SRS. They confirmed the following points:

(1)   The peach fruit is extremely susceptible to gaseous HF. For long-term continuous exposures of more than 80 days, the threshold for SRS is below $0.26\,\mu g\,F\,m^{-3}$. This concentration produced SRS on more than 90% of the fruits. For intermittent exposures of about 70 days, the threshold is below (*a*) $1\,\mu g\,F\,m^{-3}$ for 6 h, 3 days a week; (*b*) $0.6\,\mu g\,F$ m^{-3} for 6 h, 5 days a week; or (*c*) $0.3\,\mu g\,F\,m^{-3}$ for 12 h, 5 days a week. Each exposure regime provided about $180\,\mu g\,F\,m^{-3}\,h$ ($7.5\,\mu g\,F\,m^{-3}\,d$) and induced SRS in more than 50% of all fruits.

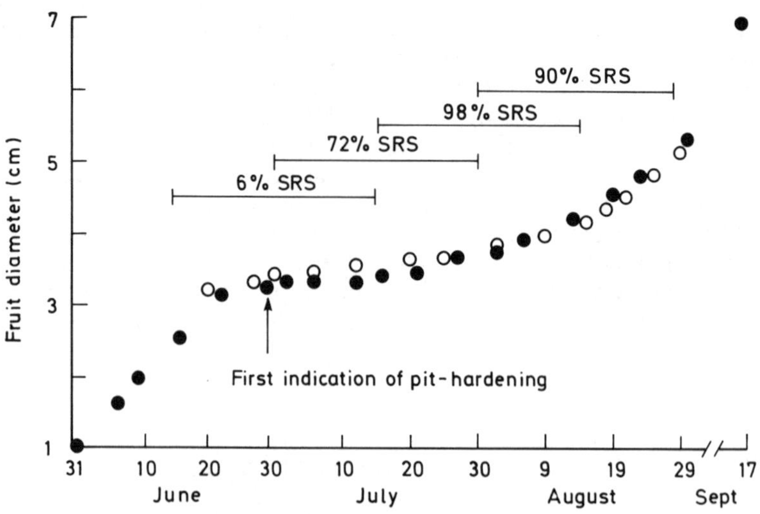

Figure 7.1   Incidence of suture red spot (SRS) in 'Elberta' peach fruits as influenced by time of fumigation (i.e. stage of fruit development). Horizontal lines indicate the dates of the four overlapping 30-day exposures. The percentage of SRS is indicated. Open and closed circles show the course of fruit development for two consecutive years

(2)   Susceptibility to SRS was found to increase with time after pit hardening, reaching a maximum at the period of rapid cell enlargement when there is an increased translocation of photosynthate from leaves to fruit. This was demonstrated by exposing peach trees to four sequential but overlapping 30-day fumigations that were provided at various times during fruit development. The first exposure began 15 days before pit hardening, the second 15 days later, and so on. Only 6% of the fruit exposed during the period that bracketed pit hardening had SRS. Exposures after pit hardening induced high percentages of SRS (*Figure 7.1*).

(3)   In some experiments, applications of boron during early stages of fruit development appeared to reduce the incidence and severity of SRS.

(4)   The incidence and severity of SRS may not be related directly to foliar F accumulation or dose. For example, leaves of trees exposed to $1 \mu g$ F m^{-3} for 6 h, 3 times a week for 7 weeks ($126 \mu g$ m^{-3} h, or $5.25 \mu g$ m^{-3} d) accumulated 36 ppm F, while those exposed to $1 \mu g$ F m^{-3} continously for 72 days ($1728 \mu g$ m^{-3} h, or $72 \mu g$ m^{-3} days) accumulated 170 ppm; but in both treatments 55% of the fruit had SRS.

(5)   In experiments with $^{45}CaCl_2$ applied to leaves, only about 1% was translocated to the fruits. Of the total ^{45}Ca in the fruit, more was found in the suture tissues than in the rest of the fruit, and the proportion found in the suture area was greater in fruits from control than from HF-fumigated trees. Thus, F may block Ca transport to the suture.

From the data available, it is clear that the production of SRS is one of the most sensitive responses of a plant to F. Intermittent exposures providing the equivalent of $1 \mu g$ F m^{-3} for 5.25 days induced SRS in more than 50% of the fruit. Because incidence and severity of SRS are not consistently related to the F content of the foliage or to the dose of F provided, other factors must have an important role in inciting the condition of premature senescence of the suture tissues. Other than biological factors, such as timing of F exposure with respect to stage of development of the fruit, or varietal differences in response, there is strong circumstantial evidence that SRS is related to Ca metabolism of the peach tree, as was suggested by Thomas and Alther (1966). We assume that Ca transport in the peach follows the same pattern as in other fruits, i.e. direct movement to the fruit in the transpirational stream that reaches its maximum early in the fruit development and virtually stops when phloem transport of photosynthate from the leaf becomes rapid (Hangar, 1979). Although the period of maximum accumulation of Ca corresponds to the least sensitive stage of the fruit to F, the accumulation of Ca in the fruit is not great. After pit hardening, the uptake of F may induce a deficiency of physiologically active Ca along the suture, perhaps by formation of a Ca–F complex. Even if circumstances exist under which export of Ca from the leaf to the fruit with photosynthate were a significant factor in Ca distribution, Ca could be immobilized by F in the leaf, and the suture area of the fruit could be deficient in Ca.

Under normal conditions, the flesh of the peach fruit is a reservoir for Ca and there is a gradient from the flesh to the embryo. The major vascular tissues pass through the distal end of the suture where SRS develops (Biddulph, unpublished work, *discussed by* Thomas and Alther, 1966). With arrested Ca transport through these tissues, this region of the fruit presumably would be particularly vulnerable to Ca deficiency. Poovaiah (1979) has reviewed the literature demonstrating the relationship between Ca deficiency and premature senescence and ripening in plants in general.

The prevention of SRS by application of Ca sprays is only one fragment of evidence that SRS results from a localized Ca deficiency. A second is that spray applications of boron have also been shown to be effective in reducing the incidence of SRS. In apples, boron sprays have been reported to increase the Ca content of the fruit (Mason, 1979) and are effective in the control of bitter pit, a Ca-deficiency disease of apples (Delong, 1936). A third fragment of evidence is that SRS has been observed by one of us (L.H. Weinstein) in peach orchards remote from a source of airborne F, where foliar wilting is a common occurrence because of a lack of irrigation. Because the Ca status of the fruit is regulated by the transpirational stream, low soil moisture and an irregular water supply can increase the incidence of Ca deficiency (Millaway and Wiersholm, 1979).

In summary, the evidence from the data of MacLean and his colleagues, already cited, and that of others, although not conclusive, points to a relationship between Ca metabolism and the induction of SRS.

Conclusion

Although not all of the consequences of F pollution are understood, it seems reasonable to conclude that Ca and Mg play a central part in the physiological responses of plants to F. *Figure 7.2* summarizes these various effects. Ca and/or Mg are implicated in responses at the cellular level, such as alterations in transport phenomena and in respiratory and photosynthetic processes. A detoxification mechanism appears to consist of the sequestration and insolubilization of F by virtue of its reaction with Ca. Responses to F at the organ level, such as inhibition of pollen-tube growth

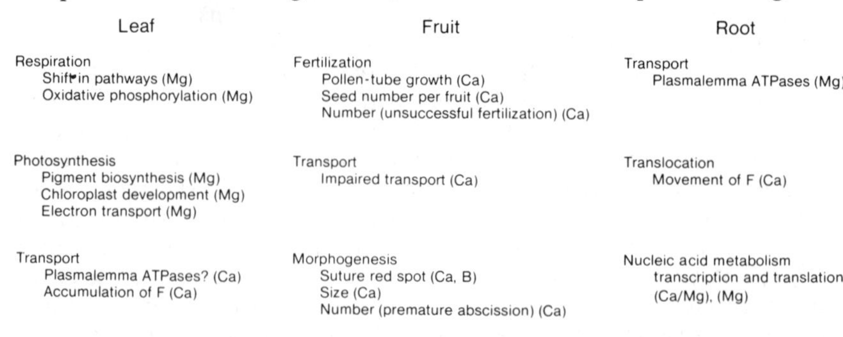

Figure 7.2   Physiological effects of fluoride: involvement of calcium, magnesium and other elements

and suture red spot, also appear to involve Ca. The basis for F susceptibility will not be understood until the plant's responses to it at these various organizational levels are integrated into a coherent overall concept.

In the 100 years or so since injury to plants by airborne F was first recognized, many hypotheses and controversies have attempted to explain how and why F is damaging to plants and plant communities. Such disputes merely fortify our rapport with Hilaire Belloc (Wells, 1936) who wrote:

> '...But Scientists, who ought to know,
> Assure us that they must be so...
> Oh! let us never, never doubt
> What nobody is sure about!'

References

ADAMS, D.F., HENDRIX, J.W. and APPLEGATE, H.C. (1957). *Agricultural and Food Chemistry*, **5**, 108–116

ALLMENDINGER, D.F., MILLER, V.L. and JOHNSON, F. (1950). *Journal of the American Society for Horticultural Science*, **56**, 427–432

AMUNDSON, R.G. and WEINSTEIN, L.H. (1980). Effects of airborne F on forest ecosystems. In *Proceedings of the International Symposium on Effects of Air Pollutants on Mediterranean and Temperate Forest Ecosystems, Riverside, California*, pp.63–78. USDA Gen. Tech. Rept. PSW–3. Pacific Southwest Forest and Range Experiment Station, Berkeley, CA

APPLEGATE, H.C. and ADAMS, D.F. (1960a). *Phyton*, **14**, 111–120

APPLEGATE, H.C. and ADAMS, D.F. (1960b). *Botanical Gazette*, **121**, 223–227

BALLANTYNE, D. (1972). *Atmospheric Environment*, **6**, 267–273

BÉJAOUI, M. and PILET, P.E., (1975). *Comptes Rendus Hebdomadaires des Séances de l'Académie des Sciences, Serie D*, **280**, 1457–1460

BELOZEROVA, L.S. and FEDOROVA, L.K. (1976). *Soviet Plant Physiology*, **23**, 135–138; 163–166

BENEDICT, H.M. and BREEN, W.H. (1955). In *Proceedings of the Third National Air Pollution Symposium, Pasadena, California*, pp. 177–190

BENEDICT, H.M., ROSS, J.M. and WADE, R.W. (1964). *International Journal of Air and Water Pollution*, **8**, 279–289

BENEDICT, H.M., ROSS, J.M. and WADE, R.W. (1965). *Journal of the Air Pollution Control Association*, **15**, 253–255

BENNETT, J.H. and HILL, A.C. (1973a). *Journal of the Air Pollution Control Association*, **231**, 203–206

BENNETT, J.H. and HILL, A.C. (1973b). *Journal of Environmental Quality*, **2**, 526–530

BENNETT, J.P. and OSHIMA, R.J. (1976). *Journal of the American Society for Horticultural Science*, **101**, 638–639

BENNETT, J.P., OSHIMA, R.J. and LIPPERT, L.F. (1979). *Environmental and Experimental Botany*, **19**, 33–39

BENSON, N.R. (1959). *Proceedings of the American Society for Horticultural Science*, **74**, 184–198

BEWLEY, J.D. and BLACK, M. (1978). *Physiology and Biochemistry of Seeds in Relation to Germination. I. Development, Germination, and Growth*. Springer-Verlag, New York

BIANCHI, A. and LAUDI, G. (1977). *Giornale Botanico Italiano*, **111**, 219–226

BLIGNY, R., GARREC, J.-P. and FOURCY, A. (1973). *Comptes Rendus Hebdomadaires des Séances de l'Académie des Sciences, Serie D*, **276**, 961–964

BOUTHYETTE, P.-Y. and JAGENDORF, A.T. (1982). Calcium inhibition of amino acid incorporation by pea chloroplasts and the question of loss of activity with age. In *Proceedings of the 5th International Congress on Photosynthesis*. International Science Services, Jerusalem

BOWLING, D.J.F. and DUNLOP, J (1978). *Journal of Experimental Botany*, **29**, 1139–1146

BREWER, R.F., SUTHERLAND, F.H. and GUILLEMET, F.B. (1969a). *Journal of the American Society for Horticultural Science*, **94**, 302–304

BREWER, R.F., SUTHERLAND, F.H. and GUILLEMET, F.B. (1969b). *Environmental Science and Technology*, **3**, 378–381

BREWER, R.F., GARBER, M.J., GUILLEMET, F.B. and SUTHERLAND, F.H. (1967). *Proceedings of the American Society for Horticultural Science*, **91**, 150–156

BREWER, R.F., SUTHERLAND, F.H., GUILLEMET, F.B. and CREVELING, R.K. (1960). *Proceedings of the American Society for Horticultural Science*, **76**, 208–214

CHANG, C.W. (1967). *Cereal Chemistry*, **44**, 129–142

CHANG, C.W. (1968). *Plant Physiology*, **43**, 669–674

CHANG, C.W. (1970). *Canadian Journal of Biochemistry*, **48**, 450–454

CHANG, C.W. (1975). In *Response of Plants to Air Pollution*, pp.57–95 (Mudd, J.B. and Kozlowski, T.T., Eds), Academic Press, New York

CHANG, C.W. and THOMPSON, C.R. (1966). *Plant Physiology*, **41**, 211–213

CHENG, J.Y.-O., YU, M.-H, MILLER, G.W. and WELKIE, G.W. (1968). *Environmental Science and Technology*, **2**, 367–370

COUTREZ-GEERNICK, D. (1973). *Bulletin de la Société Royale de Botanique de Belgique*, **106**, 237–256

DAINES, R.H., LEONE, I. and BRENNAN, E. (1952). In *Air Pollution*, pp. 97–105 (McCabe, L.C. Ed.), McGraw-Hill, New York

DELONG, W.A. and DELONG, W.A. (1936). *Plant Physiology*, **11**, 453–456

DE ONG, E.R. (1946). *Phytopathology*, **36**, 469–471

DIOURIS, M. and PENOT, M. (1972). *Comptes Rendus Hebdomadaires des Séances de l'Académie des Sciences, Serie D*, **275**, 2647–2650

DIOURIS, M. and PENOT, M. (1973). *Comptes Rendus Hebdomadaires des Séances de l'Académie des Sciences, Serie D*, **276**, 1557–1560

DIOURIS, M. and PENOT, M. (1975). *Fluoride*, **8**, 208–223

DIOURIS, M. and PENOT, M. (1977).*Zeitschrift für Pflanzenphysiologie*, **85**, 215–226

DUNLOP, J. and BOWLING, D.J.F. (1978). *Journal of Experimental Botany*, **29**, 1147–1153

EVANS, H.J. and SORGER, G. (1966). *Annual Review of Plant Physiology*, **17**, 47–76

FACTEAU, T.J. and ROWE, K.E. (1977). *Journal of the American Society for Horticultural Science*, **102**, 95–96

FACTEAU, T.J. and WANG, S.-Y. (1972).*HortScience*, **7**, 505–506

FACTEAU, T.J., WANG. S.-Y. and ROWE, K.E. (1973). *Journal of the American Society for Horticultural Science*, **98**, 234–236

FLÜHLER, H., KELLER, T. and SCHERRER, H.V. (1979). *Bulletin Murithienne*, **96**, 4–22

GARBER, K. (1962). *Wissenschaftliche Zeitschrift der Technischen Universität, Dresden,* **11,** 549–552

GARREC, J.-P. and VAVASSEUR, A. (1978). *European Journal of Forest Pathology,* **8,** 37–43

GARREC, J.-P., PLEBIN, R. and LHOSTE, A.-M. (1977). *Environmental Pollution,* **13,** 159–167

GARREC, J.-P., ABDULAZIZ, P., LAVIELLE, E., VANDEVELDE, L. and PLEBIN, R. (1978). *Fluoride,* **11,** 186–197

GARREC, J.-P., BLANCHARD, B., BRUN, J.-C., BISCH, A.-M., BLIGNY, R. and FOURCY, A. (1973). *Comptes Rendus Hebdomadaires des Séances de l'Académie des Sciences, Serie D,* **277,** 855–808

GARREC, J.-P. LIGEON, E., BONTEMPS, A., BLIGNY, R. and FOURCY, A. (1974a). *Journal of Radioanalytical Chemistry,* **19,** 359–365

GARREC, J.-P, MORLEVAT, J.-P, JOLIVET, J., BISCH-LHOSTE, A.-M and FOURCY, A. (1974b). *Comptes Rendus Hebdomadaires des Séances de l'Académie des Sciences, Serie D,* **279,** 61–64

GARREC, J.-P, OBERLIN, J.-C., LIGEON, E., BISCH, A.-M. and FOURCY, A. (1974c).*Fluoride,* **7,** 78–84

GIBBS, M. and BEEVERS, H. (1955). *Plant Physiology,* **30,** 343–346

GILDER, J. and CRONSHAW, J. (1974). *Journal of Cell Biology,* **68,** 221–235

GROTH, E., III. (1975). *Environment,* **17,** 29–38

HALL, R.J. (1974). *Environmental Pollution,* **6,** 267–280

HANGAR, B.C. (1979). *Communications in Soil Science and Plant Analysis,* **10,** 171–193

HANSON, J.B. (1960). *Plant Physiology,* **35,** 372–379

HAUSKRECHT, I. (1972). *Biologia (Bratislava),* **27,** 961–968

HECK, W.W., TAYLOR, O.C. and HEGGESTAD, H.E. (1973). *Journal of the Air Pollution Control Association,* **23,** 257–266

HILL, A.C. (1969). *Journal of the Air Pollution Control Association,* **19,** 331–336

HILL, A.C. PACK, M.R., TRANSTRUM, L.G. and WINTERS, W.S. (1959). *Plant Physiology,* **34,** 11–16

HILL, A.C., TRANSTRUM, L.G., PACK, M.R. and WINTERS, W.S. (1958). *Agronomy Journal,* **50,** 562–565

HODGES, T.K. (1976). In *Encyclopedia of Plant Physiology, New Series, Vol. 2, Part A,* pp. 260–281 (Lüttge, U. and Pittman, M., Eds). Springer-Verlag, New York

HOLUB, Z. and NAVARA, J. (1966). *Biologia (Bratislava),* **21,** 177–182

HOOBER, J.Y. and STEGEMAN, W.J. (1973). *Journal of Cell Biology,* **56,** 1–12

HORVATH, I., KLASOVA, A. and NAVARA, J. (1978). *Fluoride,* **11,** 89–99

JACOBSON, J.S., WEINSTEIN, L.H., McCUNE, D.C. and HITCHCOCK, A.E. (1966). *Journal of the Air Pollution Control Association,* **16,** 412–417

JACOBSON, L., MOORE, D.P. and HANNAPAL, R.J. (1960). *Plant Physiology,* **35,** 352–358

JACOBSON, L., HANNAPAL, R.J., MOORE, D.P. and SCHAEDLE, M. (1961). *Plant Physiology,* **36,** 58–61

JAFFE, L., WEISENSEEL, M. and JAFFE, T. (1975). *Journal of Cell Biology,* **67,** 488–492

KELLER, T. (1974). *Fluoride,* **7,** 31–35

KELLER, T. (1977). *Eidgenossiche Anstalt für das Forstliche Versuchwesen Mitteilungen,* **53,** 161–195

KELLER, T. (1980). *Oecologia,* **44,** 283–285

KIELLEY, W.W. and MEYERHOF, O. (1948). *Journal of Biological Chemistry,* **176,** 591–601

KRONBERGER, W. and HALBWACHS, G. (1978). *Fluoride,* **11,** 129–135

KRONBERGER, W., HALBWACHS, G. and RICHTER, H. (1978). *Angewandte Botanik,* **52,** 149–154

LEDBETTER, M.C., MAVRODINEANU, R. and WEISS, A.J. (1960). *Contributions. Boyce Thompson Institute for Plant Research,* **20,** 331–348

LEE, C.-J., MILLER, G.W. and WELKIE, G.W. (1966). *International Journal of Air and Water Pollution,* **10,** 169–181

LÉVY, L. and STRAUSS, R. (1973). *Comptes Rendus Hebdomadaires des Séances de l'Académie des Sciences, Serie D,* **277,** 181–184

LOVELACE, J., MILLER, G.W. and WELKIE, G.W. (1968). *Atmospheric Environment,* **2,** 187–190

LUCKEY, T.D. (1959). In *Antibiotics, Their Chemistry and Non-medical Uses,* pp. 173–323 (Goldberg, H.S., Ed.), Van Nostrand, Princeton, New Jersey

LÜTTGE, U. and HIGINBOTHAM, N. (1979). *Transport in Plants.* Springer-Verlag, New York

McCANCE, R.A. and WIDDOWSON, E.M. (1944). *Nature,* **153,** 650

McCLENAHEN, J.R. (1976). *Journal of Environmental Quality,* **5,** 472–475

McCUNE, D.C. (1976). In *Diagnosing Vegetation Injury Caused by Air Pollution,* ch.5 (Lacasse, N.L. and Treshow, M., Eds), US Environmental Protection Agency

McCUNE, D.C. and WEINSTEIN, L.H. (1971). *Environmental Pollution,* **1,** 169–174

McCUNE, D.C., DE HERTOGH, A.A. and WEINSTEIN, L.H. (1967). In *Abstracts of Papers: 153rd Meeting of the American Chemical Society, Miami Beach, Florida, April 9–14,* Abstract 15, p. T–15

McCUNE, D.C., MacLEAN, D.C. and SCHNEIDER, R.E. (1976). In *Effects of Air Pollutants on Plants* pp. 31–46 (Mansfield, T.A., Ed.), University Press, Cambridge

McCUNE, D.C., SILBERMAN, D.H. and WEINSTEIN, L.H. (1977). *Proceedings of the International Clean Air Congress,* **4,** 116–119

McCUNE, D.C., WEINSTEIN, L.H., JACOBSON, J.S. and HITCHCOCK, A.E. (1964). *Journal of the Air Pollution Control Association,* **14,** 465–468

MacINTIRE, W.H. and Associates. (1949). *Industrial and Engineering Chemistry,* **41,** 2466–2475

MACKLON, A.E.S. (1975). *Planta,* **122,** 131–144

MacLEAN, D.C. and SCHNEIDER, R.E. (1971). *Proceedings of the International Clean Air Congress,* **2,** 292–295

MacLEAN, D.C., SCHNEIDER, R.E. and McCUNE, D.C. (1973). *Proceedings of the International Clean Air Congress,* **3,** A143–A145

MacLEAN, D.C., SCHNEIDER, R.E. and McCUNE, D.C. (1977). *Journal of the American Society for Horticultural Science,* **102,** 297–299

MacLEAN, D.C., SCHNEIDER, R.E. and WEINSTEIN, L.H. (1978). *Paper 78–44.3, 71st Annual Meeting of the Air Pollution Control Association, Houston, Texas*

McNULTY, I.B. and NEWMAN, D.W. (1961). *Plant Physiology,* **36,** 385–388

MASON, J.L. (1979). *Communications in Soil Science and Plant Analysis,* **10,** 349–371

MASSART, L. and DUFAIT, R. (1942). *Hoppe-Seyler's Zeitschrift für Physiologische Chemie*, **272**, 157–170

MEAD, R.J. and SEGAL, W. (1972). *Australian Journal of Biological Sciences*, **25**, 327–333

MILLAWAY, R.M. and WIERSHOLM, L. (1979). *Communications in Soil Science and Plant Analysis*, **10**, 1–28

MILLER, G.W. (1958). *Plant Physiology*, **33**, 199–306

MILLER, G.W. (1972). In *Carbon-Fluorine Compounds*, p.117. Elsevier, New York

MILLER, J.E. and MILLER, G.W. (1974). *Physiologia Plantarum*, **32**, 115–121

MOORE, D.P., JACOBSON, L. and OVERSTREET, R. (1961). *Plant Physiology*, **35**, 53–57

NAGANNA, B., RAMAN, A., VENUGOPAL, B. and SRIPATHI, C.E. (1955). *Biochemical Journal*, **60**, 215–223

NATIONAL ACADEMY OF SCIENCES (1971). *Fluorides*. Committee on Biologic Effects of Atmospheric Pollutants, National Research Council, Washington, DC

NAVARA, J. (1964). *Biologia (Bratislava)*, **19**, 589–596, (English translation)

NAVARA, J., HOLUB, Z. and BEDATSOVA, L. (1966). *Biologia (Bratislava)*, **21**, 87–97

PACK, M.R. (1966). *Journal of the Air Pollution Control Association*, **16**, 541–544

PACK, M.R. (1971). *Journal of the Air Pollution Control Association*, **21**, 133–137

PACK, M.R. (1972). *Journal of the Air Pollution Control Association*, **22**, 714–717

PACK, M.R. and SULZBACH, C.W. (1976). *Atmospheric Environment*, **10**, 73–81

PENOT, M. (1967). *Comptes Rendus Hebdomadaires des Séances de l'Académie des Sciences, Serie D*, **264**, 1169–1171

PENOT, M. (1968). *Comptes Rendus Hebdomadaires des Séances de l'Académie des Sciences, Serie D*, **266**, 885–888

PENOT, M. and DIOURIS, M. (1969). *Bulletin de la Societé Francaise de Physiologie Végétale*, **15**, 173–191

PETERS, R.A. (1952). *Proceedings of the Royal Society of London, Series B*, **139**, 143–170

PILET, P.E. (1968). *Comptes Rendus Hebdomadaires des Séances de l'Académie des Sciences, Serie D*, **266**, 1574–1576

PILET, P.E. (1969). *Comptes Rendus Hebdomadaires des Séances de l'Académie des Sciences, Serie D*, **269**, 954–957

PILET, P.E. (1970). *Fluoride*, **3**, 153–159

PILET, P.E. and BÉJAOUI, M. (1975). *Biochemie und Physiologie der Pflanzen*, **168**, 483–491

PILET, P.E. and ROLAND, J.-C. (1972). *Berichte der Schweizerischen Botanischen Gesellschaft*, **82**, 269–283

POOVAIAH, B.W. (1979). *Communications in Soil Science and Plant Analysis*, **10**, 83–88

PREUSS, P.W. and WEINSTEIN, L.H. (1969). *Contributions. Boyce Thompson Institute for Plant Research*, **24**, 151–155

PREUSS, P.W., LEMMENS, A.G. and WEINSTEIN, L.H. (1968). *Contributions. Boyce Thompson Institute for Plant Research*, **24**, 25–31

RAMAGOPAL, S., WELKIE, G.W. and MILLER, G.W. (1969). *Plant and Cell Physiology*, **10**, 675–685

RAVEN, J.A. (1977). *New Phytologist*, **79**, 465–480

ROBERTS, E.H. (1969). *Symposium of the Society for Experimental Biology*, **23**,161–192

ROMMEL, L.G. (1941). *Svensk Botanisk Tidskrift*, **35**, 271–286

ROSS, C.W., WIEBE, H.H. and MILLER, G.W. (1962). *Plant Physiology*, **37**, 305–309

SIDHU, S.S. (1977). *70th Annual Meeting of the Air Pollution Control Association, Toronto, Canada*. Paper 77–30.2

SIDHU, S.S. (1978). *71st Annual Meeting of the Air Pollution Control Association, Houston, Texas*. Paper 78–24.7

SULZBACH, C.W. and PACK, M.R. (1972). *Phytopathology*, **62**, 1247–1253

SZE, H. and HODGES, T.K. (1977). *Plant Physiology*, **59**, 641–646

THOMAS, M.D. (1958). *Agronomy Journal*, **50**, 545–550

THOMAS, M.D. and ALTHER, E.W. (1966). The effects of fluoride on plants. In *Handbook of Experimental Pharmacology. Part I. Pharmacology of Fluorides*, pp. 231–306. Springer-Verlag. New York

THOMAS, M.D. and HENDRICKS, R.H. (1956). In *Air Pollution Handbook*, ch.9 (Magill, P.L., Holden, F.R. and Ackley, C., Eds), McGraw-Hill Co., New York

THOMPSON, C.R., TAYLOR, O.C., THOMAS, M.D. and IVIE, J.O. (1967). *Environmental Science and Technology*, **7**, 644–650

THOMPSON, L.K., SIDHU, S.S. and ROBERTS, B.A. (1979). *Environmental Pollution*, **18**, 221–234

VENESLAND, B. and TURKINGTON, E.O. (1966). *Archives of Biochemistry and Biophysics*, **116**, 153–161

VICKERY, B. and VICKERY, M.L. (1975). *Phytochemistry*, **14**, 423–427

WARD, P.I.V. and HUSKISSON, N. (1972). *Biochemical Journal*, **130**, 575–587

WEI, L. and MILLER, G.W. (1972). *Fluoride*, **5**,67–73

WEINSTEIN, L.H. (1977). *Journal of Occupational Medicine*, **19**, 49–78

WEINSTEIN, L.H. (1979). In *The Proceedings of the Ninth Conference on Environmental Toxicology, AMRL-TR-79-68* pp. 252–282. Aerospace Medical Research Laboratory, Ohio 45433

WEINSTEIN, L.H. and McCUNE, D.C. (1970). In *Impact of Air Pollution on Vegetation Conference*, pp. 81–106 (Linzon, S.N., Ed.), Air Pollution Control Association, Pittsburgh

WEINSTEIN, L.H. and McCUNE, D.C. (1971). *Journal of the Air Pollution Control Association*, **21**, 410–413

WEINSTEIN, L.H., McCUNE, D.C., MANCINI, J.F., COLAVITO, L.J., SILBERMAN, D.H. and van LEUKEN, P. (1972). *Environmental Research*, **5**, 393–408

WEISENSEEL, M.H., NUCCITELLI, R. and JAFFE, L.F. (1975). *Journal of Cell Biology*, **66**, 556–557

WELLS, C. (1936). *The Book of Humorous Verse*. Books for Libraries Press, Freeport, New York

WIEBE, H.H. and POOVAIAH, B.W. (1973). *Plant Physiology*, **52**, 542–545

WOLTZ, S.S. and LEONARD, C.D. (1964). *Proceedings of the Florida State Horticultural Society*, **77**, 9–15

YU, M.-H and MILLER, G.W. (1970). *Environmental Science and Technology,* **4,** 492–495

YU, M.-H., MILLER, G.W. and LOVELACE, C.J. (1971). *Proceedings of the International Clean Air Congress,* **2,** 156–158

ZIMMERMAN, P.W. and HITCHCOCK, A.E. (1956). *Contributions. Boyce Thompson Institute for Plant Research,* **18,** 263–279

ZIMMERMAN, P.W., HITCHCOCK, A.E. and GWIRTSMAN, J. (1957). *Contributions. Boyce Thompson Institute for Plant Research,* **19,** 49–53

8

EFFECTS OF SO$_2$ AND NO$_2$ ON METABOLIC FUNCTION

A.R. WELLBURN
Department of Biological Sciences, University of Lancaster

The problem(s)

It is not the intention of this contribution to consider in detail the metabolic effects of individual pollutants, particularly of SO$_2$. Reviews provided by Ziegler (1973, 1975), Malhotra and Hocking (1976), Wallace and Spedding (1976), Hallgren (1978) and Heath (1980) more adequately provide a fuller guide to the available literature. Nevertheless, one or two salient features of the effects of SO$_2$ may usefully be outlined at this stage. In solution, SO$_2$ is active in the form of either HSO$_3^-$ or SO$_3^{2-}$. Metabolic changes induced in the cell may be reflected by changes in enzyme conformation, which in turn may or may not lead to inhibition of enzymic activity. The more normal mechanisms of substrate inhibition are noncompetitive but in some instances, such as with bicarbonate, competitive inhibition has been detected (*see* Ziegler, 1975). Attention in the past has also been devoted to the possible intermediate creation of inhibitory α-hydroxysulphonates. However, rather large concentrations of atmospheric SO$_2$ are required to cause their formation in plant tissues. Present evidence indicates that such bisulphite addition compounds act in a different manner to sulphite and that the latter is a better model for eliciting metabolic effects similar to those shown by SO$_2$ fumigation.

In recent years few biochemical studies have considered effects at low but realistic atmospheric levels (< 250 ppb) of SO$_2$. This is despite clear indications from physiological, growth-analysis and monitoring studies that such a concentration would be regarded as the upper limit for results to be meaningful to the field situation (Bell and Clough, 1973; Biscoe, Unsworth and Pinckney, 1973; Bleasdale, 1973; Bull and Mansfield, 1974; Ashenden, 1978; Ashenden and Mansfield, 1978; Mansfield and Ashenden, 1978). Notable exceptions are the studies of Jäger and Klein (1977) and of Harvey and Legge (1979). The former demonstrated that in pea plants exposed to 100 or 150 ppb for 18 days there was a greater accumulation of inorganic sulphur, a reduced buffer capacity and a stimulation of the levels of glutamate dehydrogenase activity in comparison with unpolluted controls. The latter workers, using shorter fumigations to mimic their particular environmental situation (< 250 ppb for 0.5 h upon acclimatized pine hybrids) showed that ATP levels were inversely related to SO$_2$ concentration, with little effect upon photosynthetic CO$_2$ assimilation.

Both NO (nitric oxide) and NO_2 (nitrogen dioxide) are produced from atmospheric O_2 and N_2 by combustion processes and are phytotoxic. Various factors affect the ratio of NO to NO_2 in emissions (e.g. flame temperature) and surrounding atmospheres (e.g. radiant light flux at certain wavelengths). The term NO_x is often used collectively to describe the mixture in studies of their effects. Usually most researchers have preferred to employ NO_2 alone to study the phytoxicity of the nitrogen oxides, and ignore the effect of NO. In view of the widely differing properties of the two components, especially with respect to solubility (Bennett and Hill, 1973) the use of the term NO_x may perhaps be considered unwise until a full and long-overdue comparison between the two with respect to pollution sensitivity of various plants has been undertaken. Present evidence would suggest that, of the two, NO_2 is perhaps the more harmful (Capron and Mansfield, 1976, 1977).

At high concentrations (e.g. 1 ppm and above) foliar lesions can be caused by NO_2 which are difficult to distinguish from those produced by SO_2 (Van Haut and Stratmann, 1969). At lower concentrations NO_2 and NO may interfere with physiological processes and reduce growth rate without producing easily identifiable physical damage (Capron and Mansfield, 1976, 1977; Ashenden and Mansfield, 1978).

As NO_2 and SO_2 usually occur together in polluted atmospheres in concentrations which are of the same order of magnitude, and as NO_2 is nearly as toxic to some plants as SO_2, there is a danger that their effects will be confused in the field. There is a strong case for always considering the combined effects of these pollutants, rather than treating them as quite independent phytotoxic agents. There has been little research into this question despite its obvious practical significance. The first important paper on the subject was presented by Tingey *et al.* (1971). They found that mixtures of SO_2 and NO_2 at concentrations of less than 250 ppb caused substantially more damage than the same amount of either of the pollutants alone, and more than the sum of their separate effects. Some would call this a synergistic interaction, whilst others more cautiously would call it a 'more than additive' effect. We use either term without prejudice to each viewpoint.

Hill *et al.* (1974) performed some experiments on native vegetation of the Arizona desert and failed to find an interaction between SO_2 and NO_2, but the species they studied were all xerophytes, the sensitivity of which to pollution may be lower than that of mesophytes. The same Utah-based group went on to study the effect of combinations of SO_2 and NO_2 upon plants of economic importance. Relatively high (> 500 ppb for 1 h) doses of SO_2 alone, or SO_2 combined with NO_2, were required to cause visible injury (Bennett *et al.*, 1975). At much lower mixed concentrations ($<$ 500 ppb for 1–2 h) $SO_2 + NO_2$ caused a synergistic inhibition of apparent photosynthesis in alfalfa (White, Hill and Bennett, 1974). High CO_2 concentrations increased the rate of CO_2 uptake by the plants and protected them from the adverse effects of the $SO_2 + NO_2$ mixtures (Hou, Hill and Soleimani, 1977).

Research at Lancaster has also confirmed that $SO_2 + NO_2$ are more damaging when present in the atmosphere together. Depending on the characteristic studied and the species employed, their effects range from additive to more than additive. An example of simple summation was found in the rates of photosynthesis of relatively resistant peas (Bull and Mansfield, 1974), but even then there was a pronounced synergistic interaction when the levels of activities of peroxidase, ribulose-1,5-bisphosphate carboxylase, glutamate dehydrogenase and various transaminases were determined in the same pea cultivar under similar conditions (Horsman and Wellburn, 1975; Wellburn *et al.*, 1976). With increasing sophistication of exposure-chamber design, subsequent studies at very low levels of each pollutant (68 ppb for 20 weeks) singly and in combination, have been reported. Synergistic effects of SO_2 and NO_2 in combination, recorded as percentage reductions in total dry weight and leaf area, were found for the grasses *Dactylis glomerata, Lolium multiflorum* and *Phleum pratense* although *Poa pratensis* showed only additive reductions in dry weight but more than additive reductions in leaf area (Ashenden and Mansfield, 1978; Mansfield and Ashenden, 1978; Ashenden, 1979; Ashenden and Williams, 1980).

Biochemical investigations of effects of SO_2 and NO_2

While the above studies were in progress at Lancaster, parallel biochemical investigations were undertaken, designed to elucidate some of the important features of the 'invisible' injury caused by $SO_2 + NO_2$ which gives rise to depressions of growth and reductions of leaf area. As, at present, this represents virtually all the biochemical work on the combined effects of SO_2 and NO_2, an account of the experimental techniques will be given in some detail, before describing and discussing the metabolic effects of SO_2 and NO_2 when they occur together.

MATERIALS AND METHODS

Plant material

With the exception of the early experiments using peas (for growth conditions etc. *see* Horsman and Wellburn, 1975) all the remaining experiments employed grasses commonly growing in pasture in the United Kingdom. Seeds of *Dactylis glomerata* L. (cv. Aberystwyth S37), *Phleum pratense* L. (cv. Eskimo) and *Poa pratensis* L. (cv. Monopoly) were germinated and 8-week-old plants were used for the species comparisons. Cloned samples of *Lolium perenne* L. were obtained from Dr J.N.B. Bell of Imperial College, London. One was the 'Helmshore' clone previously described (Bell and Mudd, 1976) and another was originally from 'S23' which had previously shown a high resistance to SO_2 damage during screening tests. This clone is referred to as S23 'Bell resistant'. Seeds of *Lolium perenne* L. (cv. S23 and S24) were germinated and, thereafter, sensitive individual seedlings were cloned in a similar manner to the 'Helmshore' and 'Bell resistant' clones. Particular care was taken over the timing of splitting and ensuring that plants of similar age and stature were employed at the start of each experiment so that comparisons between experiments could be made. The response of plants to pollutants is known to differ with respect to season (Davies, 1980). All the results quoted here refer to experiments harvested in the autumn and winter when the combined effects of SO_2 and NO_2 are maximal.

Fumigation systems

For the long-term fumigations the conditions of growth and fumigation were exactly the same as those described by Ashenden and Mansfield (1978), Ashenden (1979) and Ashenden and Williams (1980). Shorter-term experiments were undertaken according to the system described by Horsman and Wellburn (1975) with the addition of fans to ensure that leaf boundary-layer resistances were small (*see* Ashenden and Mansfield, 1977). Pollutant concentrations were periodically monitored using appropriate Meloy detectors for SO_2 or NO_2. Concentrations of ammonia were measured using the method of Weatherburn (1967).

Biochemical assays

Methods of plastid isolation, enzyme extraction and assay have been described elsewhere (Wellburn, Wilson and Aldridge, 1980). Rates of ascorbate/diaminodurene (DAD)-dependent photophosphorylation and photosystem II and I activities were assessed on shocked plastid preparations using the methods described by Wellburn and Hampp (1979). Amounts of ATP and energy charge ratios were calculated according to procedures described by Hampp and Wellburn (1980). Chlorophyll was estimated by the method of Arnon (1949) and total protein by the method of Lowry *et al.* (1951).

Replication and statistics

All experiments were carried out at least in triplicate and the longer-term ones were blocked and randomized. At least three extracts were made from each experimental treatment and four to six biochemical assays were

Table 8.1   RATIO OF GLUTAMATE DEHYDROGENASE TO GLUTAMINE SYNTHETASE ACTIVITIES IN EXTRACTS FROM *PISUM SATIVUM* L. CV. FELTHAM FIRST (14 DAYS OLD) EXPOSED TO CLEAN AIR OR VARIOUS COMBINATIONS OF SO_2, NO_2 AND/OR NH_3 IN AIR FOR 6 DAYS (S.D. OF THE RATIO WAS $< \pm 0.006$). CONTROLS WERE GROWN IN CHARCOAL-FILTERED AIR

SO_2	*Controls*	NO_2		NH_3		$NO_2 + NH_3$
(ppm)		(1.0 ppm)	(2.0 ppm)	(1.0 ppm)	(2.0 ppm)	(1.0 ppm each)
0	0.128	0.132	0.137	0.125	0.128	0.128
1.0	0.133	0.126	0.126	0.129	0.134	0.143
2.0	0.146	0.145	0.170	0.142	0.174	0.185

The significance of the treatments upon the ratios above were as follows: SO_2, $P < 0.001$; NO_2, $P < 0.05$; NH_3, $P < 0.01$; $SO_2 + NO_2$, $P < 0.001$; $SO_2 + NH_3$, $P < 0.01$; $SO_2 + NO_2 + NH_3$, $P < 0.01$. There were no significant effects on the levels of glutamine synthetase alone but significant effects on the levels of glutamate dehydrogenase activity on a total-protein basis were as follows: SO_2, $P < 0.01$; NO_2, not significant; NH_3, $P < 0.05$; $SO_2 + NO_2$, $P < 0.01$; $SO_2 + NH_3$, $P < 0.01$; $SO_2 + NO_2 + NH_3$, $P < 0.05$

carried out on each extract. Two-way analysis of variance was performed on the data from peas used to compile *Table 8.1*, and distribution-free tests (Colquhoun, 1971), more suitable for biochemical data (Nimmo and Atkins, 1979), were applied to all the data obtained for grasses.

RESULTS AND DISCUSSION

Ratios of glutamate dehydrogenase to glutamine synthetase

Earlier we reported confirmation of the original observation of Pahlich, Jäger and Steubing (1972) that the levels of glutamate dehydrogenase (GDH) activity in the reduction amination mode rise noticeably in pea tissue polluted with SO_2 (Wellburn *et al.*, 1976). In studies of different enzymic changes during SO_2 fumigation, this function has consistently responded in a more sensitive manner than any other enzymic indicator (Horsman and Wellburn, 1976; Jäger and Klein, 1977). Normally in higher plants the accepted mode of ammonia assimilation into amino acids is by the glutamine synthetase (GS)/glutamate synthase (GOGAT) pathway, rather than by using GDH (Lea and Miflin, 1974; Miflin and Lea, 1976). Why SO_2 should cause an increase in the levels of GDH activity is not clear at present. Fumigation with high levels of SO_2 brings about changes in the subunit structure of GDH (Pahlich, Jäger and Steubing, 1972) but this does not explain the elevated levels of GDH provided that the normal GS/GOGAT pathway is unaffected and functional.

We have long been aware that the basis of expression is critically important in studies of changes of enzymic activity in response to fumigation, because any factor selected as a basis can also change in the presence of pollutant (*see* Wellburn *et al.*, 1976). Of all the possible bases (i.e. total protein, chlorophyll, fresh weight, dry weight, per seedling or leaf, or even per whole plant), protein was found to give the least bias to the expression of enzyme activities. A much better method would be to internalize the basis of expression, quoting one enzymic activity in terms of another. With the GDH and GS/GOGAT alternative pathways a perfect opportunity presents itself. GOGAT is not easy to assay routinely but measurement of GDH/GS ratios is quite possible provided that an efficient extraction method is employed which selectively harms neither enzyme. The extraction method of Stewart and Rhodes (1977) has been found to be suitable for either peas or grasses.

Peas are relatively resistant to SO_2 pollution damage (Bull and Mansfield, 1974) but were used in our earlier work (Horsman and Wellburn, 1975; Wellburn *et al.*, 1976) and also initially to check the feasibility of using GDH/GS ratios to monitor the effects of SO_2, NO_2 and NH_3. The summarized results are shown in *Table 8.1* as ratios, with the significances of all the treatments listed below, either as ratios or as individual enzymes expressed on a protein basis. Glutamine synthetase levels appeared to be unaffected by each pollutant separately or in their various combinations. Jäger and Pahlich (1972) found a similar lack of response by GS to SO_2 fumigation. With the exception of NO_2 effects alone, there is little difference if the results are expressed as GDH activities separately on a protein basis or as GDH/GS ratios. The internalizing procedure sharpens the statistical analysis as the protein basis is removed. More than additive effects of $SO_2 + NO_2$, $SO_2 + NH_3$ and $SO_2 + NO_2 + NH_3$ may be identified.

Ammonia is often found in industrial atmospheres, especially in the vicinity of synthetic fertilizer factories, at levels similar to those of SO_2 and NO_2 (Harrison and McCartney, 1979). We have previously also noted significant effects of NH_3 and $SO_2 + NH_3$ upon GDH activity (Wellburn *et al.*, 1976). It is possible in this case that, as NH_3 is a substrate for both GDH and GS, one reason why SO_2 fumigation causes GDH to increase may be due to an inhibitory effect of SO_2 upon the mechanisms involved in the provision of energy (ATP) to assist the GS/GOGAT pathway to operate. It is possible that GDH activity (which is not dependent upon ATP) may be consequently increased to compensate for this deficiency.

Table 8.2   RATIO OF GLUTAMATE DEHYDROGENASE TO GLUTAMINE SYNTHETASE ACTIVITIES IN EXTRACTS FROM *LOLIUM PERENNE* L. VAR ABERYSTWYTH S23 (2 WEEKS OLD AFTER SELFING) EXPOSED TO CLEAN AIR OR SO₂ AND/OR NO₂ FOR DIFFERENT PERIODS OF TIME (S.D. OF THE RATIO WAS <±0.003). CONTROLS WERE GROWN IN CHARCOAL-FILTERED AIR

Days	Controls	SO₂ (250 ppb)	NO₂ (250 ppb)	SO₂ + NO₂ (250 ppb each)
7	0.018	0.023	0.022	0.023
21	0.015	0.025*	0.019	0.028*
35	0.017	0.030*	0.021	0.042**
49	0.020	0.047**	0.023	0.065***
63	0.019	0.069***	0.028*	0.082***

*$P < 0.1$, **$P < 0.05$, ***$P < 0.01$

Applying the internalized assay of both GDH and GS to grasses, which are more sensitive to SO_2 than is *P.sativum*, gave similar results. *Table 8.2* shows the time course of the effects of SO_2, NO_2 or $SO_2 + NO_2$ upon the GDH/GS ratios in cloned *Lolium perenne* L. (cv. S23) laminae, which are known to be sensitive to SO_2 pollution (Bell and Mudd, 1976). Although there was about 50 times as much GS as GDH, highly significant effects of SO_2 and of $SO_2 + NO_2$ were detected. A slight but positive effect of NO_2 alone was also shown. Nevertheless, a comparatively long time (3 weeks) was required for the SO_2-related effects to be clearly evident and even longer (9 weeks) for those of NO_2 to be detected, implying that this type of pollutant-induced enzymic lesion is secondary to those of a more fundamental nature.

The use of GDH/GS ratios was shown to best advantage in a long-term experiment designed to compare the three *Lolium* clones (*Figures 8.1a–c*). The insensitivity of the S23 'Bell resistant' clone in comparison to the normal SO_2-sensitive S23 clone with respect to SO_2 alone was clearly demonstrated (compare *Figure 8.1a* with *Figure 8.1b*) but both were significantly affected by $SO_2 + NO_2$. In contrast, the 'Helmshore' clone of *Lolium perenne* (*Figure 8.1c*) appeared to be unaffected by either SO_2 or $SO_2 + NO_2$. Thus the 'Helmshore' and the S23 'Bell resistant' clones must differ in their respective SO_2-resistance mechanisms. This is to be expected

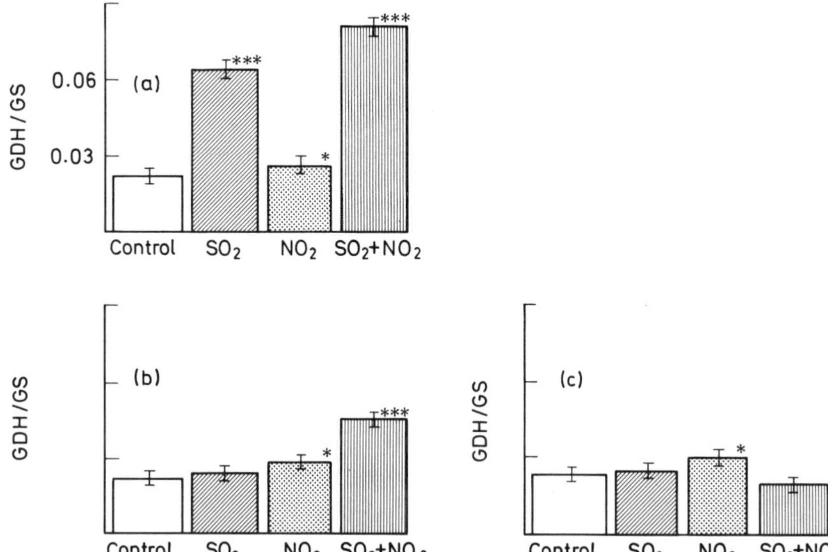

Figure 8.1 Ratio of glutamate dehydrogenase to glutamine synthetase activities in extracts from clones of *Lolium perenne* L. (2 weeks old after selfing) exposed to atmospheres containing SO_2 or NO_2 alone (680 ppb) or in combination for 11 weeks (error bars indicate S.D. $\pm$ 0.002, $*P < 0.1$, $***P < 0.01$). (a) *Lolium S23*; (b) *Lolium S23* 'Bell resistant'; (c) *Lolium* 'Helmshore'

as the 'Helmshore' clone was isolated from Pennine grassland above industrial Lancashire which, presumably, had wide genetic variation, whereas the S23 'Bell resistant' clone came originally from S23 stock of high SO_2 sensitivity. All three clones responded slightly to NO_2 in a positive manner. This phenomenon may well be linked to the possibility of NO_2 uptake and utilization, which is described and discussed later.

Although these results concerning the GDH/GS ratios are useful and allow some conclusions to be drawn concerning both the nature of the specific metabolic effects of SO_2 and the more than additive effects of SO_2 + NO_2, they are unlikely to be involved with the primary lesions but could be a consequence of them. They shed little, if any, light on why NO_2 has an effect upon plant growth although aspects of nitrogen metabolism were being examined. In view of this, it was decided to concentrate upon the effect of SO_2, NO_2, and SO_2 plus NO_2, upon the biochemical mechanisms of ammonia formation in grasses.

Effects on nitrite reductase activities

There appears to be general agreement that, while nitrate reductase is located in the cytosol, nitrite reductase activity is confined to plastids (Miflin, 1974; Hewitt, 1975). The induction of nitrate reductase by nitrate is well established (Beevers and Hageman, 1969) and fumigation with

NO$_2$, admittedly at a high concentration (12 ppm), has the same stimulatory effects upon nitrate reductase (Zeevaart, 1974). This may be followed by an intensified reduction of nitrite to ammonia and amino acids (Zeevaart, 1976), presumably involving nitrite reductase and the GS/GOGAT pathway. Recent studies with ^{15}N-labelled NO$_2$ have confirmed that ^{15}NO$_2$ may be converted into nitrate and nitrite (Yoneyama and Sasakawa, 1979) and that the reduced ^{15}N may be assimilated into amino acids.

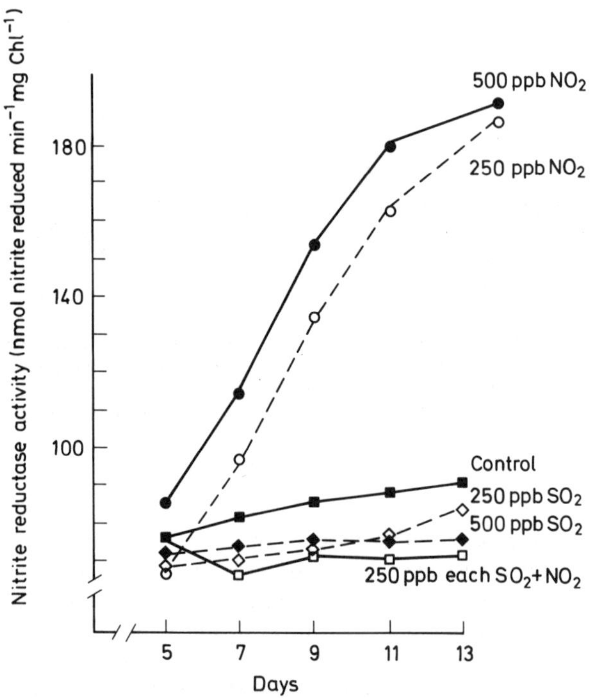

Figure 8.2 Levels of nitrite reductase activity in extracts from *Lolium perenne* L. laminae (6 weeks old from selfing, before experiment started) which have been exposed to clean air or various combinations of SO$_2$-polluted and/or NO$_2$-polluted air for 5–15 days

Using plastid preparations from control and polluted grasses, the possibility of changes in the levels of nitrite reductase activity due to SO$_2$, NO$_2$ or SO$_2$ + NO$_2$ in the atmosphere were investigated. The results are shown in *Figures 8.2–8.4*. The time course of responses to different pollutant concentrations using the SO$_2$-sensitive S23 *Lolium* clone (*Figure 8.2*) revealed many interesting features. Sulphur dioxide had no direct effect upon the levels of nitrite reductase (NiR) activity even at a relatively high concentration (1 ppm) but NO$_2$ induced a significant increase in NiR activity (P < 0.01 after 9 days at 250 ppb or after 7 days at 500 ppb). This feature was also shown by the 'Helmshore' clone after 13 days of fumigation. Most important of all were the effects of SO$_2$ + NO$_2$. In these instances the presence of SO$_2$ completely prevented the induction of increased activity by the NO$_2$. This phenomenon was shown in all of the

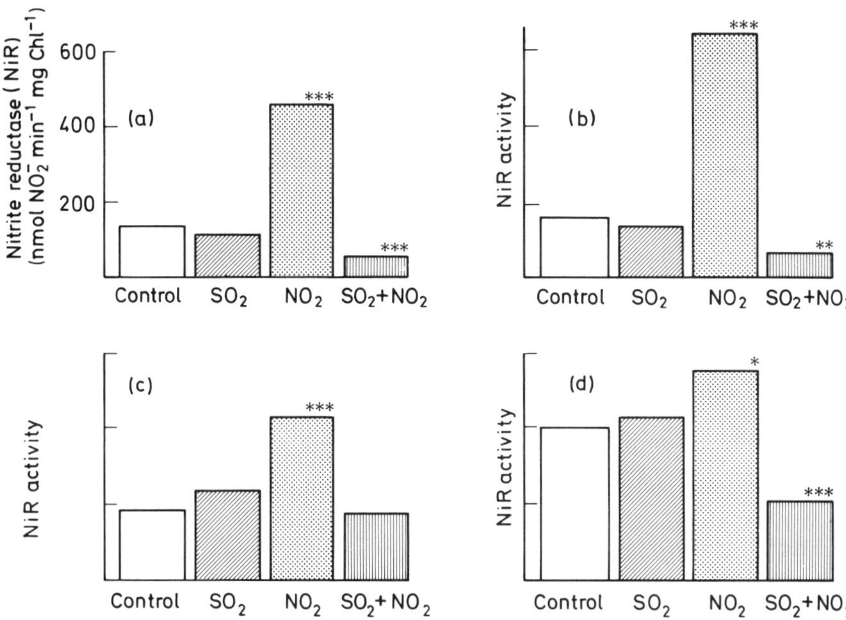

Figure 8.3   Levels of nitrite reductase activity in extracts from *Lolium perenne* L. laminae (6 weeks old after selfing) which have been exposed to clean air or to air containing SO_2 and/or NO_2 (68ppb) over a period of 20 weeks (The asterisks indicate the significance of the difference between treatment and control: $*P<0.1$, $**P < 0.05$; $***P < 0.01$). (a) *Lolium* S23; (b) *Lolium* S24; (c) *Lolium* S23 'Bell resistant'; (d) *Lolium* 'Helmshore'

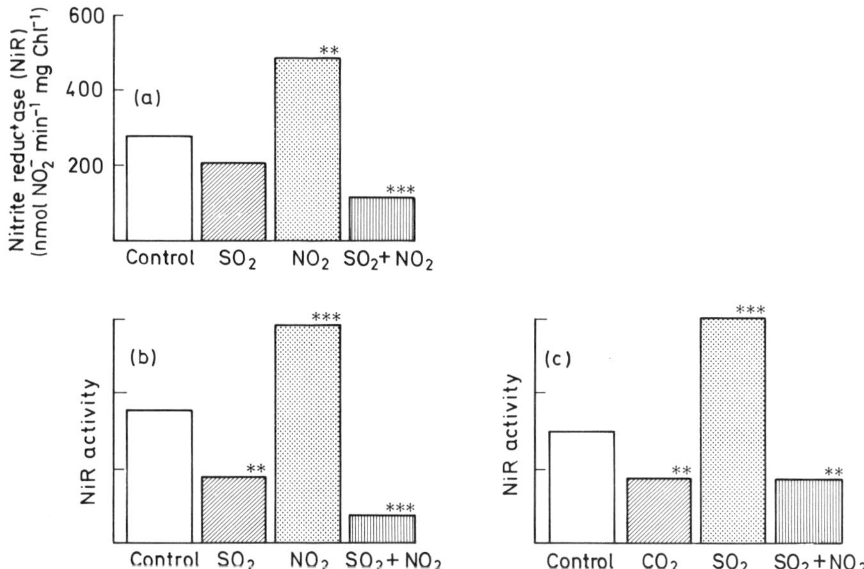

Figure 8.4   Levels of nitrite reductase activity in different grasses (8 weeks old after germination, before experiment started) exposed to atmospheres containing SO_2 and NO_2 alone (68ppb) or in combination for 20 weeks ($**P < 0.05$, $***P < 0.01$). (a) *Dactylis glomerata*; (b) *Phleum pratense*; (c) *Poa pratensis*

clones of *Lolium* available (*Figures 8.3 a–d*) and also in the other grasses at low levels of pollutants (*Figures 8.4 a–c*). After 20 weeks, levels of NiR activity in plants grown in NO$_2$-polluted air (68 ppb) were approximately double those in plants grown in clean air. *Lolium perenne* (any clone) or *Dactylis glomerata* samples, after SO$_2$ fumigation (68 ppb) for the same length of time, showed no significant changes in the levels of NiR activities, although both *Phleum pratense* and *Poa pratensis* exhibited significant depressions of activity due to the SO$_2$ treatment. In all grasses the SO$_2$ + NO$_2$ treatment failed to increase the levels of NiR in ways similar to those shown by the NO$_2$ treatment. Indeed, with the exception of the S23 'Bell resistant' (*Lolium*) clone (*Figure 8.3c*) all levels of NiR were significantly depressed below normal clean-air control levels.

In these studies and those of others at Lancaster, when very low levels of NO$_2$ have been employed, visually there is little difference between control plants and those which have experienced NO$_2$ fumigation. Frequently they even appear to be greener and healthier than those plants grown in clean air, and tend to senesce at the same time or slightly later. In these experiments no convenient internal basis of expression, such as the GDH/GS ratios previously used, was available. As a consequence the amounts of chlorophyll have been used: these would tend to reduce or eliminate the differences reported above, rather than amplify them. Despite this, the induction of NiR by NO$_2$ fumigation, and the inhibition of the induction by SO$_2$ + NO$_2$, were clearly demonstrated.

The inescapable conclusion to be drawn is that the presence of SO$_2$ prevents the induction of additional NiR activity normally associated with NO$_2$ fumigation. As a consequence the plants, harmed by the SO$_2$ and unable properly to utilize or detoxify the NO$_2$, are exposed to damage by both pollutants in one or a number of ways at the same time.

Nitrite reductase activity is intimately linked to the photosynthetic electron-transport chain to receive the necessary reductant. At the same time photophosphorylation is carried out. Sulphur dioxide is known to have deleterious effects upon photochemical and bioenergetic processes (*see* Ziegler, 1973; Malhotra and Hocking, 1976). Consequently, attention was diverted to examination of various fundamental bioenergetic processes in grasses under these particular fumigation regimes.

Effects on bioenergetic functions

Many methods for the measurement of rates of cyclic photophosphorylation have been described, most of which employed reduced phenazine methosulphate (PMS). This particular compound is now known to react directly with P$^+$-700 (the oxidized side of the active centre of photosystem I), supporting cyclic electron flow at high rates with almost no light saturation (Witt, Rumberg and Junge, 1968). A much better source of electrons is the couple, ascorbate/diaminodurene (DAD), which uses some of the natural photosynthetic electron-transport intermediates to generate a proton gradient which may be harnessed to drive ATP formation by coupling factors. Rates of photophosphorylation measured in this way (Wellburn and Hampp, 1979) have been shown to reflect more accurately the *in vivo* situation. Similar techniques were used to examine the ATP-forming capability of various grass clones after fumigation with low levels of SO_2 and/or NO_2 over 20 weeks.

The results of these studies are shown in *Figures 8.5 a–d*. Treatment with SO_2 or $SO_2 + NO_2$ significantly reduced the rates of photophosphorylation in all of the *Lolium* clones and also those of *Phleum*. In contrast, fumigation with NO_2 significantly enhanced the ability to form ATP, even when chlorophyll was used as the basis of expression. This means that not only are rates of reduction of nitrite increased by very low levels of NO_2 but, at the same time, increased cycling of electrons is possible to allow

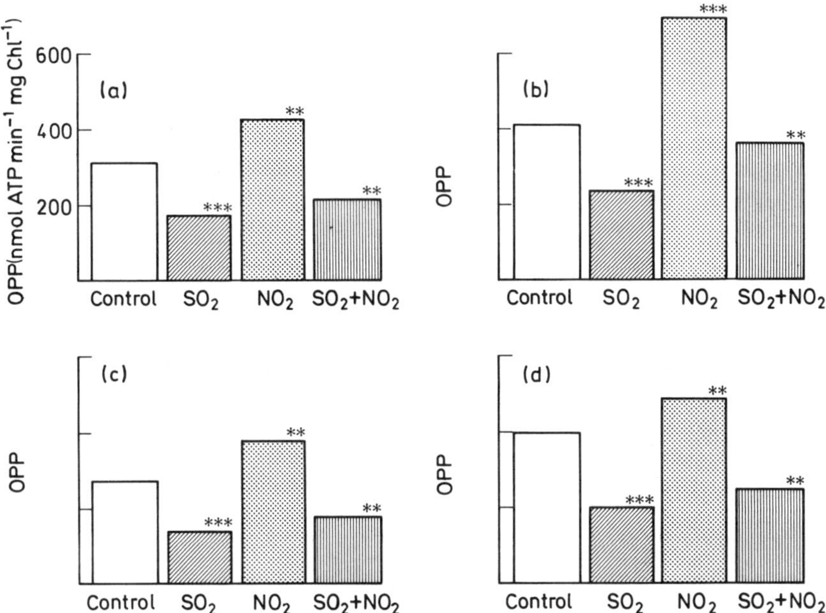

Figure 8.5   Rates of ascorbate/diaminodurene-dependent photophosphorylation (OPP) in extracts from grasses (for ages before experiment started, see *Figures 8.3, 8.4*) exposed to clean air or to air containing SO_2 and/or NO_2 (68 ppb) over a period of 20 weeks (S.D. $\pm <$ 10%, **$P < 0.05$, ***$P < 0.01$). (a) *Lolium* S23; (b) *Phleum pratense*; (c) *Lolium* S23 'Bell resistant'; (d) *Lolium* 'Helmshore'

additional photophosphorylation. The advantages of measuring cyclic photophosphorylation in a more authentic manner are clearly demonstrated when contrasted with the use of PMS-dependent assays. Shimazaki and Sugahara (1979, 1980) were unable to detect any changes in PMS-catalyzed ATP formation in spinach, even at the high SO_2 level of 1 ppm. This has the implication that coupling factors themselves are not the site of SO_2 attack, despite the well-known inhibitory effects of sulphate on Ca^{2+}-dependent ATPase activities (Ryrie and Jagendorf, 1971).

More detailed series of experiments were undertaken in the shorter term with the S23 clone of *Lolium*. The length and level of fumigation were chosen to allow measurements of the changes in various bioenergetic parameters which coincide with the exponential phase of the increase in

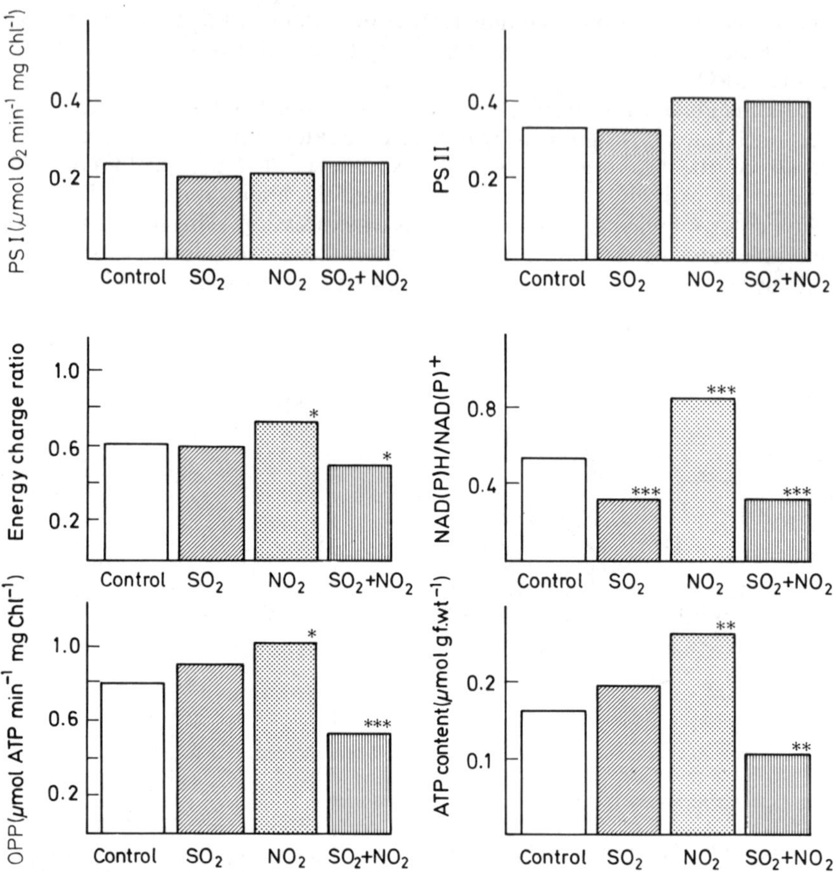

Figure 8.6 Effect upon various cellular parameters of treating *Lolium perenne* L. (S23 clone) with clean air or with air containing SO$_2$ and/or NO$_2$ (250 ppb) for 11 days (***$P <$ 0.01, **$P <$ 0.05, *$P <$ 0.10)

levels of nitrite reductase elicited by NO$_2$, as shown earlier in *Figure 8.2*. The summarized data are shown in *Figure 8.6a–f*. No significant changes of the rates of either photosystem I or photosystem II were detected under any fumigation regime. At least maximum electron-flow capabilities appear to be unaffected by pollutants at these low levels. Differences in pollutant concentration may explain the disagreement with recent reports by Shimazaki and Sugahara (1979, 1980), which have suggested that SO$_2$ inhibits electron transfer at a site close to the reaction centre of photosystem II. Their studies on lettuces employed SO$_2$ at levels between 1 and 2 ppm.

Examination of the rates of ascorbate/DAD-dependent ATP formation was more revealing. Significant reductions of cyclic photophosphorylation below those in unpolluted *Lolium* (S23 clone) were given by SO$_2$ + NO$_2$, while NO$_2$-polluted air alone gave enhanced rates of ATP formation. Estimations of the total ATP levels and the energy charge ratios are also in accordance with the rates of cyclic photophosphorylation. Significantly higher total ATP amounts and energy charge ratios for NO_2 fumigation were noted. The opposite situation was found for $SO_2 + NO_2$ fumigation. These laminae exhibited significant reductions in both ATP levels and energy charge ratios.

General levels of reductant which are available are indicated by the assays of the $NAD(P)H/NAD(P)^+$ ratios. Both SO_2 and $SO_2 + NO_2$ treated tissues had lower levels of reduced cofactors than unpolluted tissues, while those treated with NO_2 appeared to have an excess of available reductant even in the presence of enhanced nitrite-reducing capability.

Taken together, these indications of changes in the capacity for ATP formation and general levels of reductant support our previous contentions that low levels of NO_2 alone may be beneficial but, in the case of $SO_2 + NO_2$ pollution, exactly the reverse holds. These plants appear to be incapacitated bioenergetically even in comparison with those which have experienced SO_2 fumigation only.

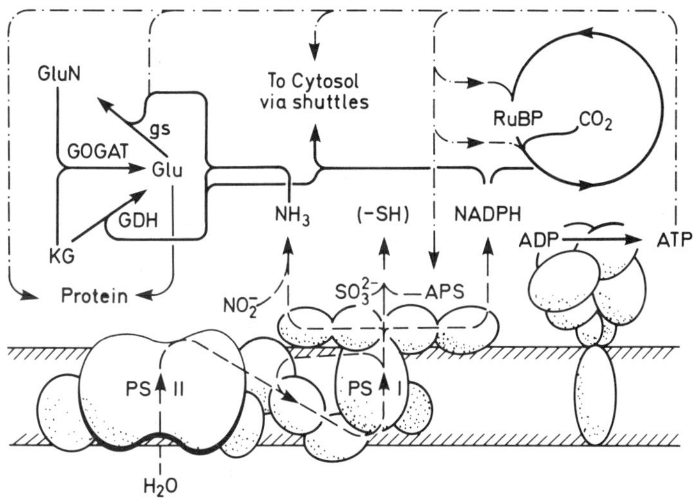

Figure 8.7   Scheme to illustrate the various destinations for electrons from the photosynthetic membrane (dashed lines), some of the uses of the reduced products and the involvement of ATP in various processes which are dependent upon photosynthesis (alternate dots and dashes). As discussed in the text, gaseous pollutants are thought to exert their effects simultaneously upon several of the processes outlined in this scheme

Conclusions may be drawn from these studies with the aid of the scheme shown in *Figure 8.7*. Electrons, derived from water, may be either recycled around photosystem I, providing a proton gradient, which in turn may be harnessed by coupling factor particles to make ATP, or they may pass to a variety of electron acceptors. These would normally be $NADP^+$ or the nitrite reductase complex. However, as sulphur metabolism is involved (especially in the presence of SO_2), the electrons either could also pass to the thiosulphonate reductase associated with APS-sulphotransferase, or could enable the direct reduction by sulphite reductase (where it exists), to yield thiol groups of various compounds such as cysteine in both cases. The mechanisms and hierarchical relationships by which electrons are apportioned between these various acceptors are not known. NADPH is required mainly for CO_2 fixation and export to the cytosol by means of the triose phosphate and dicarboxylate shuttles, while NH_3, the product of nitrite reduction, is incorporated into amino acids mainly by means of the GS/GOGAT pathway. Where (or if) GDH operates, reductive amination of α-ketoglutarate would be carried out with assistance from NAD(P)H. As indicated in the scheme shown in *Figure 8.7*, ATP is required for nearly all these processes and many others, particularly so for protein biosynthesis. It is also required for carbon fixation and the provision of phosphorylated sugars for export, as well as for the formation of APS by ATP sulphurylase and glutamine by GS. Harvey and Legge (1979) have demonstrated reductions of ATP levels in plants fumigated with SO_2. The results described here strongly support this finding, suggesting that the overall ATP level may be a critical limiting factor.

Sulphite ions and SO_2 are known to have an inhibitory effect upon the rate of CO_2 fixation (Tanaka, Takanashi and Yatazawa, 1974; Silvius, Ingle and Baer, 1975) but at low levels of sulphite ($< 1\,mM$) there is an enhancement of CO_2 fixation which is initiated by the superoxide radical (Ziegler and Libera, 1975). Under the conditions of our experiments the levels of SO_2 employed are likely to bring about the latter conditions. Furthermore, the levels of superoxide dismutase activity may have a direct effect upon the rates of CO_2 fixation in the different grass clones. That superoxide dismutase may participate in the defence against SO_2 toxicity was originally suggested by Ziegler and Libera (1975). This variability of superoxide dismutase levels in populations has recently been used to explain different degrees of SO_2 tolerance (Tanaka and Sugahara, 1980).

At no stage in all these experiments has the development of a tolerance towards NO_2 been indicated. However there was evidence from the GS/GDH studies (*Figure 8.1d*) of the 'Helmshore' clone that tolerance to combinations of air pollutants could evolve. This suggestion was recently made in a growth study of 36 individual *Lolium* clones obtained from different regions of Liverpool (Horsman, Roberts and Bradshaw, 1978). Nevertheless, the SO_2 inhibition of the induction of nitrite reductase shown even by the 'Helmshore' clone indicates that, even here, there is no resistance to this particular lesion and that multiple biochemical factors are involved in pollutant injury.

As may be seen in *Figure 8.7*, the availability of ATP affects many processes such as CO_2 fixation, glutamine and thiol formation and many other aspects of growth, especially in the areas of protein biosynthesis. Even at the relatively low levels ($< 250\,ppb$) of this study, deprivation of ATP was indicated in the $SO_2 + NO_2$ fumigation. As there are indications that coupling-factor activity is not affected by SO_2, and cycling of electrons is at the expense of reductant for NO_2^- or $NADP^+$, then other SO_2-induced or NO_2-induced lesions must occur within the photosynthetic membranes themselves, possibly affecting the ability to generate effective proton gradients. The importance of pH in modulating the toxicity of SO_2 has recently been recognized (Spedding *et al.*, 1980). Ionic disturbances of this kind are the most likely basis for the ultrastructural disturbances of the thylakoids by SO_2 (and also NO_2), first observed by ourselves (Wellburn, Majernik and Wellburn, 1972) and subsequently confirmed by many other ultrastructural studies of air-pollution effects (Fischer, Kramer and Ziegler, 1973; Godzik and Sassen, 1974; Wong, Klein and Jäger, 1977).

We believe that the more-than-additive inhibitory effects of low atmospheric concentrations of SO_2 and NO_2 may be explained partly by failure to induce additional nitrite reductase activity and partly by the combined effects of sulphite and nitrite. These may act by the intermediate formation of free radicals damaging the photosynthetic membrane and preventing sufficient proton gradients, which would normally have allowed extra ATP to be formed. This additional ATP is required to counteract the stromal and cytoplasmic consequences of the two pollutants: deprivation results in the depression of growth and reduction of leaf area reported by the Lancaster and Utah groups (e.g. Bennett *et al.*, 1975; Ashenden, 1979) while studying the combined effects of SO_2 and NO_2.

Where next?

Many problems in this area still remain. A comprehensive biochemical study of the differences between the effects of both NO and NO_2 is urgently required. The study of NO effects has been started (Wellburn, Wilson and Aldridge, 1980) and further studies using ^{15}N-labelled NO and NO_2 may well prove rewarding.

Bennett and Hill (1975) found that 6.5 ppm NO for 1 h rapidly reduced photosynthesis by 35% in alfalfa, recovery being equally rapid when fumigation ceased. Responses to NO were much faster than those to NO_2, an observation that does not quite correlate with the low solubility and uptake of NO compared with the ready solubility and uptake of NO_2, reported by Bennett and Hill (1975) in the same paper. They attribute the slow recovery rates from NO_2 to the time taken for reducing the amount of the NO_2-derived solutes built up during fumigation. Uptake mechanisms for NO not dependent on solubility alone could equally well be in operation. Bennett and Hill (1975) also speculate on the possibility of inhibition of ferredoxin by the formation of a concentration-dependent iron–NO complex, although this does not explain the ingress of NO from the peristomatal cavity, across the extracellular fluid layer, cellulose wall, plasmalemma, cytoplasm and plastid envelopes to the site of ferredoxin. Nitric oxide does not directly react with either ferredoxin or nitrite reductase under strictly anaerobic conditions (Pickard and Hewitt, 1972). Nevertheless, under aerobic conditions, reaction of NO with some form of oxygen (produced perhaps by photosynthesis) cannot be ruled out. The products of such reactions could well be both nitrite and nitrosonium ions (NO^+), which may be responsible for the toxicity of NO. Plant injury by nitrite is well known (Mevius, 1958), but Zeevaart (1976) concluded, from his experiments with NO_2, that acidification rather than nitrite accumulation was the damaging factor. The pK_a of 3.4 for the reaction $HONO \rightleftharpoons NO^+ + OH^-$ would favour the acidification theory. Noncyclic electron flow in photosynthesis is competitively required for both $NADP^+$

and nitrite reduction. Acidification changes may explain the inverse relationship observed (Magalhaes, Neyra and Hageman, 1974) between CO_2 fixation and NO_2 reduction.

Until more progress has been in elucidating the differences, and possible interactions, between NO and NO_2, it is not realistic to speculate further or to introduce experiments involving a three-component mixture. Some appreciation of just how difficult this type of experiment may be was gained from the work on peas using SO_2 + NO_2 + NH_3 (*Table 8.1*). Undoubtedly this is the ultimate direction which interaction studies should take, but the dearth of basic information on single pollutants, such as NH_3 and NO, relegates this prospect to the distant future despite the fact that field environments are invariably a complex mixed-pollutant situation for some, if not most, of the time.

Two promising avenues of pollution research present themselves as capable of yielding useful information in the short term. Nondestructive methods available for measurement of leaf chlorophyll fluorescence are capable of providing detailed information on the efficiency of the photosynthetic membranes. Schreiber *et al.* (1978) were able to predict ozone injury 20 h before visible signs of bean-leaf necrosis, using such techniques. The first effect they detected was on the water-splitting systems, followed by inhibition of electron transport between the photosystems. Clearly, such sensitive techniques could equally well be usefully employed for studies on SO_2, NO_2 and their various interactions. The other line to be pursued in the biochemical area is the use of isolated plant-cell protoplasts. Precise control of pollutant dosage over the short term in such systems could remove some of the variation experienced at present between different plants and populations, leading to a greater insight into basic disturbances within the cell due to sulphite, nitrite and ammonia. At Lancaster we are currently in the process of introducing both techniques into our studies of the effects of atmospheric pollutants.

Acknowledgement

I am grateful for a grant from the Natural Environment Research Council to enable the bulk of these studies to be carried out.

References

ARNON, D.I. (1949). *Plant Physiology*, **24**, 1–15

ASHENDEN, T.W. (1978). *Environmental Pollution*, **15**, 161–166

ASHENDEN, T.W. (1979). *Environmental Pollution*, **18**, 249–258

ASHENDEN, T.W. and MANSFIELD, T.A. (1977). *Journal of Experimental Botany*, **28**, 729–735

ASHENDEN, T.W. and MANSFIELD, T.A. (1978). *Nature*, **273**, 142–143

ASHENDEN, T.W. and WILLIAMS, I.A.D. (1980). *Environmental Pollution*, **21**, 131–139

BEEVERS, L. and HAGEMAN, R.H. (1969). *Annual Review of Plant Physiology*, **20**, 495–522

BELL, J.N.B. and CLOUGH, W.S. (1973). *Nature*, **241**, 47–49

BELL, J.N.B. and MUDD, C.H. (1976). In *Effects of Air Pollutants on Plants*, pp 87–103 (Mansfield, T.A. Ed.). S.E.B. Seminar Series. No.1. Cambridge University Press, Cambridge

BENNETT, J.H. and HILL, A. (1973). *Journal of the Air Pollution Control Association*, **23**, 203–206

BENNETT, J.H. and HILL, A.C. (1975). In *Responses of Plants to Air Pollution.* pp. 273–306 (Mudd, J.B. and Kozlowski, T.T., Eds). Academic Press, New York

BENNETT, J.H., HILL, A.C., SOLEIMANI, A. and EDWARDS, W.H. (1975). *Environmental Pollution*, **9**, 127–132

BISCOE, P.V., UNSWORTH, M.H. and PINCKNEY, H.R. (1973). *New Phytologist*, **72,** 1299–1306

BLEASDALE, J.K.A. (1973). *Environmental Pollution*, **5**, 275–285

BULL, J.N. and MANSFIELD, T.A. (1974). *Nature*, **250**, 443–444

CAPRON, T.M. and MANSFIELD, T.A. (1976). *Journal of Experimental Botany*, **27**, 1181–1186

CAPRON, T.M. and MANSFIELD, T.A. (1977). *Journal of Experimental Botany*, **28**, 112–116

COLQUHOUN, D. (1971). *Lectures on Biostatistics.* Clarendon Press, Oxford

DAVIES, T. (1980). *Nature*, **284**, 483–484

FISCHER, K., KRAMER, D. and ZIEGLER, H. (1973). *Protoplasma*, **76**, 83–96

GODZIK, S. and SASSEN, M.M. (1974). *Phytopathologische Zeitschrift*, **79**, 155–159

HÄLLGREN, J.E. (1978). In *Sulfur in the Environment. Part II. Ecological Impacts*, pp. 164–209 (Nriagu, J.A., Ed.), Wiley, New York

HAMPP, R. and WELLBURN, A.R. (1980). *Zeitschrift für Pflanzenphysiologie*, **98**, 289–303

HARRISON, R.M. and McCARTNEY, H.A. (1979). *Atmospheric Environment*, **13**, 1105–1120

HARVEY, G.W. and LEGGE, A.H. (1979). *Canadian Journal of Botany*, **57**, 759–764

HEATH, R.L. (1980). *Annual Review of Plant Physiology*, **31**, 285–431

HEWITT, E.J. (1975). *Annual Review of Plant Physiology*, **26**, 73–100

HILL, A.C., HILL, S., LAMB, C. and BARRETT, T.W. (1974). *Journal of the Air Pollution Control Association*, **24**, 153–157

HORSMAN, D.C. and WELLBURN, A.R. (1975). *Environmental Pollution*, **8**, 123–133

HORSMAN, D.C. and WELLBURN, A.R. (1976). In *Effects of Air Pollutants on Plants*, pp. 185–199 (Mansfield, T.A., Ed.). Cambridge University Press

HORSMAN, D.C., ROBERTS, T.M. and BRADSHAW, A.D. (1978). *Nature*, **276**, 493–494

HOU, L-Y., HILL, A. and SOLEIMANI, A. (1977). *Environmental Pollution*, **12**, 7–16

JÄGER, H-J. and KLEIN, H. (1977). *Journal of the Air Pollution Control Association*, **27**, 464–466

JÄGER, H-J. and PAHLICH, E. (1972). *Oecologia*, **9**, 135–140

LEA, P.J. and MIFLIN, B.J. (1974). *Nature*, **251**, 614–616

LOWRY, O.H., ROSEBROUGH, N.J., FARR, A.L. and RANDALL, R.J. (1951). *Journal of Biological Chemistry*, **193**, 265–275

MAGALHAES, A.C., NEYRA, C.A. and HAGEMAN, R.H. (1974). *Plant Physiology,* **53**, 411–515

MALHOTRA, S.S. and HOCKING, D. (1976). *New Phytologist,* **76**, 227–237

MANSFIELD, T.A. and ASHENDEN, T.W. (1978). *Nature,* **273**, 142–143

MEVIUS, W. (1958). *Handbuch der Pflanzenphysiologie,* **8**, 166–175

MIFLIN, B.J. (1974). *Plant Physiology,* **54**, 550–555

MIFLIN, J. and LEA, P.J. (1976). *Phytochemistry,* **15**, 873–885

NIMMO, I.A. and ATKINS, G.L. (1979). *Trends in Biochemical Science,* **4**, 236–238

PAHLICH, E., JÄGER, H-J and STEUBING, L. (1972). *Angewandte Botanik,* **46**, 183–197

PICKARD, M. and HEWITT, E.J. (1972). *Annual Report, Long Ashton Research Station,* 70–71

RYRIE, I.J. and JAGENDORF, A.T. (1971). *Journal of Biological Chemistry,* **246**, 582–588

SCHREIBER, U., VIDAVER, W., RUNECKLES, V.C. and ROSEN, P. (1978). *Plant Physiology,* **61**, 80–84

SHIMAZAKI, K. and SUGAHARA, K. (1979). *Plant & Cell Physiology,* **20**, 947–955

SHIMAZAKI, K. and SUGAHARA, K. (1980). *Research Report of the National Institute of Environmental Studies,* **11**, 79–89

SPEDDING, D.J., ZIEGLER, I., HAMPP, R. and ZIEGLER, H. (1980). *Zeitschrift für Pflanzenphysiologie,* **96**, 351–364

SILVIUS, J.E., INGLE, M. and BAER, C.H. (1975). *Plant Physiology,* **56**, 434–437

STEWART, G.R. and RHODES, D. (1977). *New Phytologist,* **79**, 257–268

TANAKA, K. and SUGAHARA, K. (1980). *Research Report of the National Institute of Environmental Studies,* **11**, 155–164

TANAKA, H., TAKANASHI, T. and YATAZAWA, M. (1974). *Water, Air & Soil Pollution,* **3**, 11–16

TINGEY, D.T., REINERT, R.A., DUNNING, J.A. and HECK, W.W. (1971). *Phytopathology,* **61**, 1506–1511

VAN HAUT, H. and STRATMANN, H. (1969). *Berichte, Landesanstalt für Immissions und Bodennutzungschutz des Landes Nordreinwestfalen, Essen,* **7**, 50–70

WALLACE, R.G. and SPEDDING, D.J. (1976). *Clean Air (Melbourne),* **10**, 61–64

WEATHERBURN, M.W. (1967). *Analytical Chemistry,* **39**, 971–974

WELLBURN, A.R. and HAMPP, R. (1979). *Biochimica et Biophysica Acta,* **547**, 380–397

WELLBURN, A.R., MAJERNIK, O. and WELLBURN, F.A.M. (1972). *Environmental Pollution,* **3**, 37–49

WELLBURN, A.R., WILSON, J. and ALDRIDGE, P.H. (1980). *Environmental Pollution (Series A),* **22**, 219–228

WELLBURN, A.R., CAPRON, T.M., CHAN, H.S. and HORSMAN, D.C. (1976). Biochemical effects of atmospheric pollutants on plants. In *Effects of Air Pollutants on Plants,* pp. 105–114 (Mansfield, T.A., Ed.), Cambridge University Press, Cambridge

WHITE, K.L., HILL, A.C. and BENNETT, J.H. (1974). *Environmental Science & Technology,* **8**, 574–576

WITT, H.T., RUMBERG, B. and JUNGE, W. (1968). In *Biochemie des Sauerstoffs*, pp. 262–306 (Hess, B. and Standinger, H.J., Eds), Springer Verlag, Heidelberg

WONG, C.H., KLEIN, H. and JÄGER, H-J. (1977). *Angewandte Botanik*, **51**, 311–319

YONEYAMA, T. and SASAKAWA, H. (1979). *Plant & Cell Physiology*, **20**, 263–266

ZEEVAART, A.J. (1974). *Acta Botanica Neerlandica*, **23**, 345–346

ZEEVAART, A.J. (1976). *Environmental Pollution*, **11**, 97–108

ZIEGLER, I. (1973). In *Environmental Quality and Safety*, pp. 182–208 (Coultson, F and Korte, F, Eds), Academic Press, New York and London

ZIEGLER, I. (1975). *Residue Reviews*, **56**, 79–105

ZIEGLER, I. and LIBERA, W. (1975). *Zeitschrift für Naturforschung*, **30c**, 634–637

9

EFFECTS OF OXIDANTS ON METABOLIC FUNCTION

J.B. MUDD
*MSU/DOE Plant Research Laboratory, Michigan State University,
East Lansing*

Introduction

In areas of high motor-traffic density, intense sunlight and stagnant air, the
primary pollutants released from exhausts (unburned hydrocarbons, oxides of nitrogen) are photochemically converted to an array of new
compounds. The principal components of the oxidants in this mixture are
ozone and the peroxyacyl nitrates.

The reactions which generate ozone and peroxyacetyl nitrate are well
described in *Ozone and Other Photochemical Oxidants* (National Academy
of Sciences, 1977). *Figure 9.1* shows the conversion of NO and propylene
to products including ozone and peroxyacetyl nitrate in an irradiation
chamber. In a polluted atmosphere such as that of Los Angeles, the
recorded concentrations of peroxyacetyl nitrate may be 5–10% that of
ozone. Depending on the hydrocarbon irradiated, a family of compounds

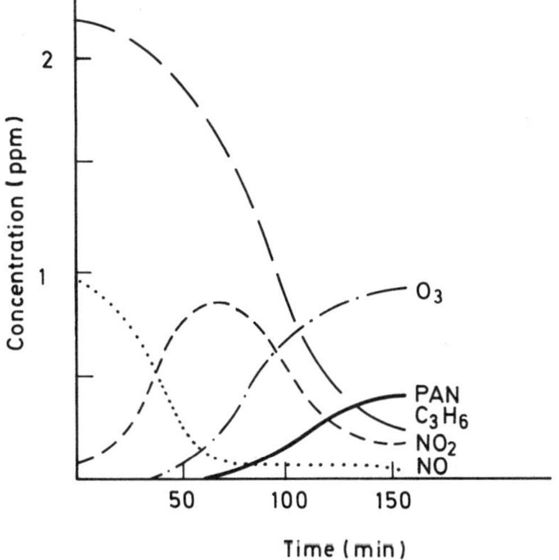

Figure 9.1 Generation of ozone and peroxyacetyl nitrate from nitric oxide and propylene in
an irradiation chamber, simulating photochemical reactions in the atmosphere. The major
organic product in this system was acetaldehyde. (Modified after Niki, Daby and Weinstock,
1972)

may be generated: $RCO\,OONO_2$; $R = CH_3$, peroxyacetyl nitrate; $R = CH_3CH_2$, peroxypropionyl nitrate, and so on. The higher the analogue in the series, the more toxic it is.

Plant damage by these oxidants is very well documented. What is not well known is the mechanism by which these pollutants cause lesions on leaves. In this chapter it is taken for granted that these mechanisms are worth knowing, and it is taken as axiomatic that the lesions are the manifestations of a series of chemical reactions. I have therefore emphasized what is known about the reactions of ozone and peroxyacetyl nitrate with biochemical compounds in the belief that the reactions will eventually be fitted into a complete description of the mechanism of toxicity of these oxidants.

Ozone

REACTIONS OF OZONE WITH LIPIDS

The use of ozone in preparative and analytical organic chemistry has a history of more than 130 years (Criegee, 1975). These uses depend on the reaction of ozone with double bonds, forming first an addition compound, the primary ozonide, which is then cleaved to form a carbonyloxide:

$$-\overset{\backslash}{C}=\overset{/}{C}- \;+\; O_3 \longrightarrow \text{primary ozonide} \longrightarrow \overset{|}{\underset{|}{C}} = O \;+\; \overset{|}{\underset{|}{\underset{\underset{O-}{|}}{C}}}=O^+$$

The carbonyl and carbonyloxide recombine to form the ozonide,

$$\overset{\displaystyle O{-}O}{\underset{\displaystyle C{-}O{-}C}{}}$$

which can be used to generate alcohols, aldehydes or carboxylic acids. In these reactions of organic chemistry, water is rigorously excluded, but in biochemistry the presence of water is inevitable and the reaction sequence is modified at the second step so that the ozonide is not formed

$$\overset{\backslash}{\underset{\underset{O^-}{|}}{C}}{}^+\!=O \;+\; H_2O \longrightarrow -\overset{}{\underset{\underset{OH}{/}}{C}}-\overset{}{\underset{\underset{OH}{|}}{O}} \longrightarrow -\overset{\backslash}{C}=O + H_2O_2$$

but rather a hydroxyhydroperoxide which breaks down to a carbonyl compound and hydrogen peroxide (Criegee, 1975).

The most abundant olefinic materials in biochemistry are the unsaturated fatty acids. These are usually esterified to a glycerol molecule and, frequently in the form of phospholipids and glycolipids, generate the basic structure of the biological membrane. The unsaturated fatty acids in plants usually have a double bond at position 9 and the polyunsaturated fatty acids have double bonds at other positions towards the methyl end of the fatty acid (*see Table 9.1*). If ozone reacted with these fatty acids in the manner described above, in all cases the fragment remaining attached to the glycerol backbone would be only nine carbons long and end in an aldehyde group rather than a methyl group. It is most unlikely that a membrane so modified would be stable. It is known that phospholipids with ten-carbon fatty acid substituents are lytic when added to normal cells. However, when this prediction of instability was tested, cells were not readily lysed by ozone (Teige, McManus and Mudd, 1976). Furthermore, the treatment of pure phospholipid with an equivalent amount of ozone generated a product which did indeed lyse the cells. The conclusion was that ozone reacted with the pure phospholipid as expected, but did not react with the unsaturated fatty acids in the cellular membrane (*Figure 9.2*). Analysis of the ozonized phospholipid demonstrated that one mole of hydrogen peroxide was generated for each double bond broken and that the nine carbon fragments expected from degradation of oleate were generated (Teige, McManus and Mudd, 1976).

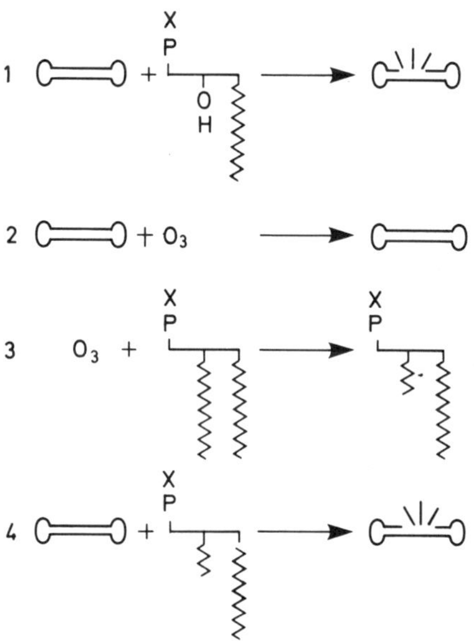

Figure 9.2   Reaction of ozone with red blood cells. 1. Addition of lysophospholipid (e.g. monoacylglyceryl phosphorylcholine) to red blood cells causes lysis. 2. Treatment of red blood cells with ozone does not cause lysis. 3. Treatment of phospholipid with ozone shortens fatty acid substituents by ozonolysis at double bonds. 4. Addition of ozonized phospholipid to red blood cells causes lysis. Conclusion: Treatment of red blood cells with ozone does not generate ozonized phospholipid

The course of ozonolysis outlined above does not involve radical intermediates. The literature on exposure of biological materials to ozone frequently mentions peroxidation as responsible for the damage observed. Peroxidation mechanisms are quite different from the classic mechanisms of ozonolysis and they do involve free radical intermediates. The confusion between peroxidation and ozonolysis in biological systems comes about because one product, malonaldehyde, is formed in both cases and unfortunately this is the most readily measured product. Many authors have concluded that peroxidation was taking place in biological material because they could measure the production of malonaldehyde. There are several ways of distinguishing peroxidation products from ozonolysis products (*Table 9.1*). Mudd, McManus and Ongun (1971) and Teige, McManus and Mudd (1976) concluded that exposure of fatty acids or phospholipids to ozone resulted in ozonolysis and not peroxidation. Heath and Tappel (1976) also concluded that ozonolysis was the oxidation mechanism when fatty acids were exposed to ozone. Malonaldehyde is

Table 9.1   COMPARISON OF OZONOLYSIS AND PEROXIDATION OF UNSATURATED FATTY ACIDS

Fatty acid	Symbol	Ozonolysis	Peroxidation
Oleic	$18:1^9$	$CH_3(CH_2)_7CHO$ $OHC (CH_2)_7COOH$ H_2O_2	none
Linoleic	$18:2^{9,12}$	$CH_3(CH_2)_4CHO$ $OHC\ CH_2CHO$ $OHC(CH_2)_7COOH$ H_2O_2	13-hydroperoxy-9c,11t-octadecadienoic acid 9-hydroperoxy-10t,12c-octadecadienoic acid
Linolenic	$18:3^{9,12,15}$	CH_3CH_2CHO $OHCCH_2CHO$ $OHC(CH_2)_7COOH$ H_2O_2	$OHCCH_2CHO$ 13-hydroperoxy-11t-pentadecenoic acid

generated when biological materials are exposed to ozone. Tomlinson and Rich (1970) concluded that this effect on bean leaves was caused by degradation well after the initial damaging event. Frederick and Heath (1970) measured malonaldehyde production in *Chlorella* after ozone exposure and concluded that malonaldehyde production and viability loss were simultaneous. Many methods of damaging plant tissue will give rise to malonaldehyde by either enzymic or nonenzymic peroxidation. The crucial questions are these: (1) If malonaldehyde is detected, is it formed by ozonolysis or peroxidation? (2) If malonaldehyde is detected, is it a primary effect of ozone exposure or a delayed response to tissue disruption? At the moment, one cannot give definitive answers to these questions.

Pryor *et al.* (1981) have concluded that ozone does induce lipid peroxidation. They have exposed pure linoleic acid and pure methyl linoleate to ozone and measured the production of free radicals by electron-spin resonance. The reactions involved are

$$LH \longrightarrow L\cdot \xrightarrow{O_2} LOO\cdot$$

The peroxy radical LOO· could not be detected, but spin-trapping experiments indicated that a compound formed from the fatty acid and ozone, that was stable at $-78°C$, released radicals when warmed to $-45°C$. The authors state that the previous report of radicals produced when ozone was introduced to linoleic acid (Goldstein *et al.*, 1968) was probably incorrect. Pryor *et al.* (1981) suggest that, when linoleic acid at room temperature is exposed to ozone, transient radicals are produced which generate lipid peroxides: however, there must be some scepticism that experiments with pure fatty acids at $-74\,°C$ are related to what happens in biological material.

There is less debate about the initiation of lipid peroxidation by nitrogen dioxide (Roehm, Hadley and Menzel, 1971). Pryor *et al.* (1980) have made the striking observations that the reaction of nitrogen dioxide with unsaturated compounds is different at high and low concentrations of the gas:

High [NO$_2$], NO$_2$ + $-\overset{\diagdown}{C}=\overset{\diagup}{C}- \longrightarrow O_2N-\overset{|}{\underset{|}{C}}-\overset{|}{\underset{|}{C}}\cdot \longrightarrow O_2N-\overset{|}{\underset{|}{C}}-\overset{|}{\underset{|}{C}}\,OO\cdot$

Low [NO$_2$], NO$_2$ + $-\overset{\diagdown}{C}=\overset{\diagup}{C}-CH_2- \longrightarrow \overset{\diagdown}{C}=\overset{\diagup}{\underset{\diagup}{C}}-\overset{\cdot}{CH}- + HONO$

The second reaction is analogous to the classical initiation of lipid peroxidation.

It should be emphasized that, if initiation of lipid peroxidation is the basis for toxicity of both ozone and nitrogen dioxide, one might expect them to be equally toxic. This is certainly not the case in plants, ozone being much more toxic.

There are mechanisms by which ozone may generate radicals, other than those given above. Ozone decomposes more readily in water at alkaline pH. Hoigné and Bader (1975) have concluded that hydroxyl radicals are the main oxidizing species produced in the decomposition of ozone in water. It is difficult to reconcile this conclusion with the production of classic ozonolysis fragments when phospholipid vesicles, suspended in water, are exposed to ozone. It is also well known that hydrogen peroxide and organic peroxides react with ozone to produce radicals

$$ROOH + O_3 \longrightarrow RO_2\cdot + HO\cdot + O_2$$

(*see* Barnard, McSweeney and Smith, 1960). But hydrogen peroxide could be measured as an ozonolysis product in the predicted stoichiometric amounts when phospholipid was exposed to ozone (Teige, McManus and Mudd, 1976), suggesting that the concentrations of hydrogen peroxide or ozone were too low to produce significant reaction. In biological material, hydrogen peroxide could not be measured after ozone exposure (Mudd *et al.*, 1971), but this can be attributed to enzymic degradation of hydrogen peroxide.

Although reaction of ozone with unsaturated lipids as a basis for ozone toxicity is intellectually appealing, the body of evidence does not clearly support such a hypothesis.

REACTIONS OF OZONE WITH AMINO ACIDS AND PROTEINS

The reaction of ozone with amino acids dissolved in anhydrous formic acid was studied by Previero and Scoffone (1963). The reactions in aqueous buffered solutions were quite similar, revealing that the most susceptible amino acids were cysteine, tryptophan and methionine, and there was less reactivity with histidine and tyrosine (Mudd *et al.*, 1969) (*Table 9.2*). The extent of reaction is dependent on the amino acid concentration, for example, oxidation of the thiol group of glutathione (γ-glutamylcysteinylglycine) is stoichiometric only above 1 mM.

Table 9.2 REACTION OF OZONE WITH AMINO ACIDS

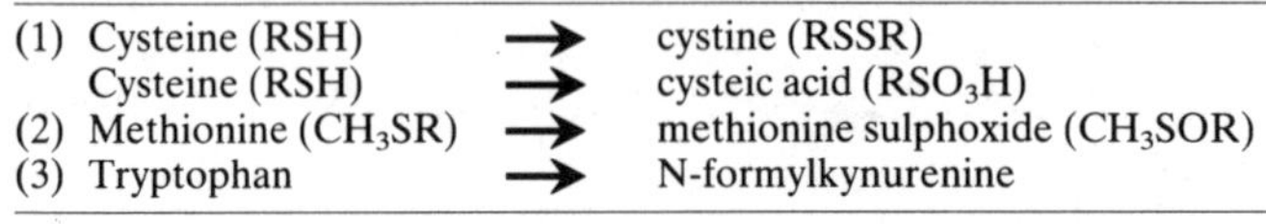

(1) Cysteine (RSH)	$\longrightarrow$	cystine (RSSR)
Cysteine (RSH)	$\longrightarrow$	cysteic acid (RSO$_3$H)
(2) Methionine (CH$_3$SR)	$\longrightarrow$	methionine sulphoxide (CH$_3$SOR)
(3) Tryptophan	$\longrightarrow$	N-formylkynurenine

Although the oxidation of free amino acids in the cell would have harmful consequences, the reaction of ozone with amino acid residues in proteins, especially enzymes, would have an amplified effect depending on the turnover number of the enzyme. It was demonstrated by Todd (1958) that papain, a sulphydryl-requiring enzyme, is much more susceptible to ozone than are urease, catalase and peroxidase. Further characteristics of ozone treatment of proteins are shown in *Table 9.3*. As predicted from the

Table 9.3 REACTION OF OZONE WITH PROTEINS

Protein	Suscceptibility	Residue affected
G3PD	High	Cysteine
Papain	High	Cysteine
Lysozyme	High	Tryptophan
Avidin	High	Tryptophan
Ribonuclease	Moderate	Histidine
α-1-antiprotease	Moderate	Methionine
Urease	Low	–
Peroxidase	Low	–
Catalase	Low	–

reactivity of pure amino acids, proteins requiring cysteine or tryptophan for their biological activity are very susceptible. The requirement for methionine as a functional group in proteins is not well documented and this has prevented accumulation of evidence that oxidation of methionine by ozone has a deleterious effect. The best example so far is the decrease in the anti-elastase activity of α-1-antiprotease after exposure to ozone, since this is also observed after treatment of the protein with a methionine-specific reagent, N-chlorosuccinimide.

One of the most susceptible enzymes to ozone is glyceraldehyde-3-phosphate dehydrogenase (G3PD), an enzyme well known for its suscepti-bility to reagents which react with thiols. The decrease in the catalytic activity of this enzyme after exposure correlates well with the loss of thiol groups, but analysis of the amino acids of the protein shows that tryp-tophan is also oxidized, in the later stages of oxidation, to a greater extent than cysteine (K.L. Knight and J.B. Mudd, unpublished data). This result emphasized that ozone reacts nonspecifically with amino acid residues in proteins, provided that they are equally exposed.

The most-studied protein with respect to ozone oxidation is lysozyme from egg white (Previero, Colletti-Previero and Jolles, 1967; Leh and Mudd, 1974; Kuroda, Sakiyama and Narita, 1975). Both Previero, Colletti-Previero and Jolles (1967) and Leh and Mudd (1974) reported oxidation of tryptophans 108 and 111, but the studies of Kuroda, Sakiyama and Narita (1975) identify the oxidation of tryptophan 62. Differences could be caused by different conditions of exposure (pH, ozone concentration, solvent). The product of ozone treatment of lysozyme is an N-formylkynurenine lysozyme. A preparation containing 0.9 mol N-formylkynurenine per mol protein had 15% of the enzymic activity of the native protein (Yamasaki *et al.*, 1976). This product was deformylated with dilute acid at −10°C, and the N-kynurenine lysozyme now had recovered some of the enzymic activity. Tryptophan 62 is involved in substrate binding, and it was demonstrated that the N-formylkynurenine lysozyme did not bind a substrate analogue, whereas the kynurenine lysozyme *did* bind the substrate analogue. It should be emphasized that the inactivation of an enzyme by ozone could be caused by an effect on substrate binding (e.g. lysozyme), or by affecting the residue which forms a covalent bond with the substrate (e.g. G3PD), or by affecting the residues participating in the catalysis. Ribonuclease is an example of the latter, because the inactivation by ozone correlates with oxidation of histidine residues (Mudd *et al.*, 1969).

Just as the oxidation of free amino acids by ozone is proportional to amino acid concentration, so also is the oxidation of amino acid residues in proteins. There are fast- and slow-reacting thiol groups for reagents such as mercurials, and so are there thiol groups susceptible and resistant to ozone oxidation. The active-site thiol of G3PD is more susceptible to ozone than the thiol of glutathione.

Is the inactivation of a protein by ozone reversible? Oxidation of glutathione by ozone generates mainly disulphide, and biological systems could readily re-reduce such a bond, e.g. with excess thiol. However, the inactivation of pure G3PD is not significantly reversible in the presence of excess thiol, implying that the active-site thiol is converted to a higher oxidation state, such as the sulphonic acid. Although some activity of ozonized lysozyme can be regained by de-formylation, this is not a reversal of the oxidation.

Can proteins be protected from oxidation by ozone? It might be expected that substrate could protect an enzyme from ozone oxidation, as it would cover the amino acid residues, the oxidation of which results in loss of activity. Some evidence for substrate protection is now being amassed.

REACTION OF OZONE WITH MEMBRANES AND INTACT CELLS

Are the reactions of ozone with purified lipids and purified proteins applicable when intact tissues are exposed? The assay of tissues after ozone exposure has the inherent difficulty that changes measured are secondary to the initial effect of ozone. There may have been changes in protein synthesis and degradation. In some cases there may have been cell proliferation.

In order to avoid these complications it seems that mammalian red blood cells make a good choice as a model system. In our experience, exposure to ozone gave no detectable changes in lipid composition (Freeman, Miller and Mudd, 1979; Freeman, Sharman and Mudd, 1979). These lipid analyses included cholesterol, all the phospholipids of the membranes, and the fatty acids. Assays for lipid ozonization or peroxidation included those for malonaldehyde and for conjugation of double bonds. In contrast, the ozone markedly inhibited enzyme activity, both in the case of acetylcholinesterase, considered to be a marker for the outside face of the membrane, and for G3PD, a marker for the inner face of the membrane. The simplest explanation of these results is that the proteins of the membrane are more susceptible than the lipids. The treatment of the erythrocytes with ozone did not change the activities of several intracellular enzymes such as glucose-6-phosphate dehydrogenase glutathione reductase and 6-phosphogluconate dehydrogenase. However, the activity of the pentose phosphate pathway was stimulated several-fold. Assay of glutathione showed that it was oxidized by the ozone exposure, and the increase in pentose phosphate pathway activity would be a consequence of this oxidation.

$$2GSH + O_3 \longrightarrow GSSG + H_2O + O_2$$

$$GSSG + NADPH + H^+ \longrightarrow 2GSH + NADP^+$$

$$NADP^+ + G\text{-}6\text{-}P \longrightarrow 6\text{-}P\text{-gluconate} + NADPH + H^+ \text{ (1st step of the pentose pathway)}$$

We are now wondering whether the oxidation of intracellular GSH is a general phenomenon or restricted to the red blood cell. An important conclusion from these data is that ozone can pass through the plasma membrane and thus affect the contents of the cell.

Tingey and co-workers (Tingey, Fites and Wickliff, 1975, 1976) have measured enzyme activities in soybean leaves after ozone exposure. They reported that the activity of G3PD decreased to less than 80% of the control value, and mentioned that the presence of cysteine in the assay mixtures decreased the apparent ozone effect. This implies that the oxidation of G3PD was reversed by excess thiol. [It has been reported that G3PD inactivation in intact red blood cells can be reversed by thiols, but in fragments from these cells the inactivation is not reversible (Koontz, 1979)]. Tingey, Fites and Wickliff (1975) also reported a stimulation in glucose-6-phosphate dehydrogenase activity, and this might be consistent with an increase in the activity of the pentose phosphate pathway.

There has been much discussion about the depth of penetration of ozone. Some authors conclude that ozone gets no further than the plasma membrane. Swanson, Thomson and Mudd (1973) could observe no changes in fatty acid composition of tobacco leaves exposed to ozone, but did report shrinkage of chloroplasts and swelling of mitochondria. Chang (1971a, b) reported that, before symptoms developed on tobacco leaves after exposure to 0.3 ppm O_3, there was a decrease in the polysome content of chloroplasts but not of the cytoplasm. A very sensitive prediction of ozone damage to leaves is the induction of chlorophyll fluorescence (Schreiber *et al.*, 1978). The simplest explanation of these results is that

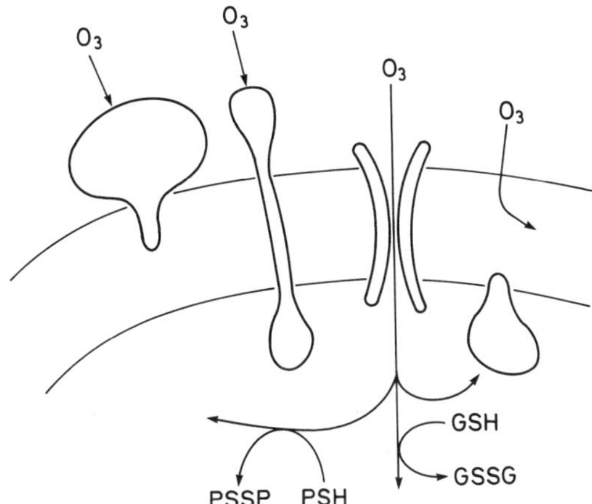

Figure 9.3   Potential points of attack by ozone on the intact cell. First, proteins on the exterior surface of the membrane (most likely). Second, lipids of the membrane (less likely). Third, proteins of the inner surface of the membrane (likely). Fourth, susceptible components of the cytoplasm (likely)

ozone can pass the plasma membrane and affect subcellular organelles such as the chloroplast (*Figure 9.3*).

OZONE-INDUCED CHANGES IN METABOLIC PATHWAYS

The increase in pigmentation observed in some plants (Koukol and Dugger, 1967; Howell and Kramer, 1973) after ozone exposure, implies the induction of enzymes for their synthesis or blockage of pathways utilizing common intermediates. Tomlinson and Rich (1971) reported an increase in steryl glucoside and acylated steryl glucoside in leaves exposed to ozone. Hodgson and Hoffer (1977) have found that the metabolism of the herbicide diphenamide (N,N-dimethyl-1,1-diphenylacetamide) is markedly changed after ozone exposure. Normal metabolites include hydroxymethyl and hydroxyphenyl derivatives and their glucosides, but after ozone exposure there is a marked increase in all of these glucosides. These results imply an increase in the capacity of the plant to transfer glucose from UDP-glucose to hydroxyls, but there is no indication of how close these events are to initiation of the process of toxicity. Is the increase in glucosylation a response to increase in hydroxylated compounds?

Increase in pigmentation is consistent with increases in enzymes such as phenylalanine ammonia lyase and polyphenyl oxidase, measured by Tingey, Fites and Wickliff (1976).

ANTIOXIDANTS

Antioxidants such as glutathione or ascorbic acid have been considered as naturally occurring compounds which might prevent or reverse the effects of ozone (Freebairn, 1957; Barnes, 1972). Hanson, Thorne and Jativa (1971) reported that the resistance of petunia varieties correlated with ascorbic acid content, but Menser (1964) found this was not the case with tobacco varieties ranging in resistance, although leaves of a single variety conditioned to contain more ascorbate were resistant.

Chang (1971b) has shown the analogy between ozone-induced breakdown of chloroplast polysomes and that caused by *p*-mercuribenzoate, implying that the effect of ozone is mediated by oxidation of sulphydryl groups. According to Foyer and Halliwell (1976), the concentration of glutathione in the chloroplast may be as high as 3.5 mM. If protein thiols are oxidized in the plastid after exposure of the leaf to ozone, one would certainly expect that some of the glutathione also would be oxidized. It would be of interest to determine the changes in glutathione reductase as well as glutathione after ozone exposure. Esterbauer and Grill (1978) have reported large changes in glutathione and glutathione reductase in needles of spruce (*Picea abies*) during the season of growth. If the glutathione concentration can be manipulated in leaves, it may be possible to manipulate resistance to ozone. The inactivation and activation of enzymes of carbon fixation in the chloroplast are dependent on the oxidation state of cysteine residues (Wolosink and Buchanan, 1977). Oxidants such as ozone could upset the balance of regulation of these enzymes.

Exogenous chemicals can be used to prevent ozone damage to vegetation. Striking protection has been reported by Carnahan, Jenner and Wat (1978) for the compound N-[2-2(2-oxo-1-imidazolidinyl)ethyl]-N-phenylurea. Unfortunately we have no information about the mechanism of action of this compound.

Koiwa and Kisaki (1973) reported that SKF 525-A (2-diethyl-aminoethyl-2,2-diphenylvalerate) and DPDA (2,4-dichloro-6-phenyl phenoxyethyldiethylamine) protected plants against ozone damage. They have worked more extensively on the antiozone properties of piperonyl butoxide (Koiwa *et al.*, 1974). All three of these compounds are inhibitors of 'mixed-function oxidases' in animal tissue. These oxidases use molecular oxygen to form a hydroxyl group at methyl and phenyl positions. The inhibitors of mixed-function oxidases may intercept the activated oxygen responsible for forming these hydroxyls. However, one cannot say with certainty that the mixed-function oxidase inhibitors react with ozone in a similar way. Koiwa, Fukuda and Kisaki (1977) have found that piperonyl butoxide prevents the formation of malonaldehyde when chloroplasts are exposed to ozone. It would be premature to conclude that piperonyl butoxide prevents ozone damage to leaves by preventing lipid oxidation (ozonolysis or peroxidation).

Peroxyacetyl nitrate (PAN)

TOXICITY

PAN is at least as toxic to plants as ozone. The lesions produced are characteristically different, PAN typically producing a bronzing appearance on the lower surface of the leaf.

Peroxyacyl nitrates were reasonably well studied after their discovery, but research recently has greatly diminished. This may be partly because these compounds are not as important quantitatively as ozone, but also because they are much more difficult to produce in the laboratory and much more dangerous to handle.

Because an earlier review has dealt with this subject (Mudd, 1975), an abbreviated account only will be given here.

REACTIONS WITH THIOLS

Peroxyacetyl nitrate reacts with glutathione forming two products:

$$CH_3 CO\cdot OONO_2 + 3GSH \longrightarrow GSSG + GSCO\cdot CH_3$$

Thus the anhydride character is exemplified in the production of the thioester, and the peroxide character in production of glutathione disulphide. In the thiols so far tested, only glutathione produces the thioester (Mudd, 1966). Reaction of PAN with cysteine and lipoic acid generated only the disulphides (Leh and Mudd, 1974), and reaction of PAN with coenzyme A formed no acetyl-CoA (Mudd and McManus, 1969). Methionine is fairly reactive to PAN, the product being methionine sulphoxide (Leh and Mudd, 1974).

REACTIONS WITH PROTEINS

PAN readily reacts with protein thiols. Of the six thiols in human haemoglobin, two are exposed to solvent and so accessible to mercurials such as *p*-mercuribenzoate. Exactly the same accessibility is shown to PAN (Mudd, Leavitt and Kersey, 1966). This suggests either that PAN cannot penetrate the tertiary structure of the protein, or that the half-life of PAN in water (7 min at pH 7) does not permit the intact molecule to penetrate that far.

The thiols of ovalbumin show different rates of reactivity with mercurials, but none react with PAN (Mudd, Leavitt and Kersey, 1966). This case exemplified the lesser degree of reactivity of PAN with amino acid residues than of ozone. Pancreatic ribonuclease, which contains no thiols, is inactivated by ozone (presumably because of oxidation of histidine residues), but is totally resistant to PAN.

Enzymes which are susceptible to PAN are also inhibited by reagents which react with thiol groups. For example, glucose-6-phosphate dehydrogenase can be protected from PAN by allowing the enzyme to bind one of the substrates, NADP, before exposure to PAN. The second substrate, G6P, does not protect the enzyme from inactivation by PAN. Malate dehydrogenase was protected from PAN by neither malate nor NAD. In both cases, protection and susceptibility correlated exactly with the behaviour of reagents known to react with thiol groups (Mudd, 1963). The implication is that PAN is a relatively specific and relatively mild reagent for thiol groups.

OTHER REACTIONS OF PAN

PAN quantitatively oxidizes reduced nicotinamide nucleotides, NADH and NADPH. The oxidation is relatively mild, for the biologically active forms, NAD and NADP, are produced (Mudd and Dugger, 1963). This contrasts with ozone which reacts with the reduced (but not the oxidized) nicotinamides to cleave the nicotinamide ring. Thus the reaction with PAN is reversible, but that with ozone is irreversible (Mudd, Leh and McManus, 1974). If the reaction of glucose-6-phosphate dehydrogenase is followed as the increase in absorbance at 340 nm, characteristic of NADPH, the introduction of PAN lowers the absorbance because of oxidation to NADP, but after cessation of PAN the enzymic reaction continues unabated. This demonstrates that the enzyme is unaffected and the nucleotide is still the biologically active form.

PAN has been shown to generate epoxides by reaction with olefins, but the biological significance of this reaction has not yet been tested.

PHYSIOLOGICAL OBSERVATIONS

The most striking and puzzling observation on conditions required for plant damage by PAN, is the requirement for illumination before, during and after fumigation (Taylor *et al.*, 1961). It has been suggested previously that proteins which are regulated by interchange between disulphide and thiol forms in dark and light would be more susceptible to PAN in the thiol (illuminated) state (Mudd, 1975). This may be a factor but does not satisfactorily account for the requirement for illumination after the PAN exposure. It may be that PAN interferes with the plants' mechanisms to prevent photoxidation. In darkness these mechanisms can be repaired, but illumination after the PAN exposure comes at the time most susceptible to photoxidation.

Summary and conclusions

There are two extremes of approaches which biochemists may take towards understanding the toxicity of air pollutants. One approach is to expose the biological material to the pollutant and subsequently to analyse the chemical composition, enzymic activity and physiological properties of the material. The second approach is to study reactions of the pollutants in simple well-defined systems and, having understood these, to work up the scale of biological organization to determine the relevance of the chemical reactions to the biological phenomena. Professor Last (this volume, Chapter 20) has referred to these approaches as Analytical and Synthetic. The Analytical approach may measure effects at the end of a long sequence of events and therefore may not perceive the initial chemical event. The Synthetic approach may describe many chemical reactions which have no correspondence to the biological lesions.

Ideally the two approaches should proceed together. At least the investigators using one approach should be aware of the results of the other approach.

In this chapter I have emphasized the Synthetic approach. In some cases I feel the results point out a primary point of attack by the pollutant: ozone is likely to have its first effect on the proteins of the biological material. In other cases it appears that the well-defined chemical reactions are misleading: oxides of nitrogen initiate free-radical peroxidation reactions in well-defined systems, but the evidence from exposures of plants suggests that the effects should be ascribed to solvation products, nitrite and nitrate (*see* Wellburn, this volume, Chapter 8).

I hope that either the success or failure of one of these approaches will not lead to either the exclusive use or the atrophy of either.

Acknowledgement

The preparation of this manuscript was supported by the US Department of Energy under contract EY-76-C-02-1338. The research in the author's laboratory was supported in part by Research Grant ES-00917 from the National Institute of Environmental Health Sciences.

References

BARNARD, D., McSWEENEY, G.P. and SMITH, J.F. (1960). *Tetrahedron Letters,* **14**, 1–4

BARNES, R.L. (1972). *Canadian Journal of Botany,* **50**, 215–219

CARNAHAN, J.E., JENNER, E.L. and WAT, E.K.W. (1978). *Phytopathology,* **68**, 1225–1229

CHANG, C.W. (1971a). *Phytochemistry,* **10**, 2863–2868

CHANG, C.W. (1971b). *Biochemical and Biophysical Research Communications,* **44**, 1429–1435

CRIEGEE, R. (1975). *Angewandte Chemie,* **87**, 765–771

ESTERBAUER, H. and GRILL, D. (1978). *Plant Physiology,* **61**, 119–121

FOYER, C.H. and HALLIWELL, B. (1976). *Planta,* **133**, 21–25

FREDERICK, P.E. and HEATH, R.L. (1970). *Plant Physiology,* **55**, 15–19

FREEBAIRN, H.T. (1957). *Science,* **126**, 303–304

FREEMAN, B.A., MILLER, B.E. and MUDD, J.B. (1979). In *Assessing Toxic Effects of Environmental Pollutants,* pp. 151–171 (Lee, S.D. and Mudd, J.B., Eds), Ann Arbor Science, Ann Arbor, Michigan

FREEMAN, B.A., SHARMAN, M.C. and MUDD, J.B. (1979). *Archives of Biochemistry and Biophysics,* **297**, 264–272

GOLDSTEIN, B.D., BALCHUM, O.J., DEMOPPOULOS, H.B. and DUKE, P.S. (1968). *Archives of Environmental Health,* **17**, 46

HANSON, G.P., THORNE, L. and JATIVA, C.D. (1971). In *Proceedings of the 2nd International Clean Air Congress,* pp. 261–266, (Englund, H.M. and Beery, W.T., Eds). Academic Press, New York

HEATH, R.L. and TAPPEL, A.L. (1976). *Analytical Biochemistry,* **76**, 184–191

HODGSON, R.M. and HOFFER, B.L. (1977). *Weed Science,* **25**, 331–337

HOIGNÉ, J. and BADER, H. (1975). *Science,* **190**, 782–784

HOWELL, R.K. and KRAMER, P.F. (1973). *Journal of Environmental Quality,* **2**, 434–438

KOIWA, A. and KISAKI, T. (1973). *Agricultural and Biological Chemistry,* **37**, 2449–2450

KOIWA, A., FUKUDA, M. and KISAKI, T. (1977). *Plant Cell Physiology,* **18**, 127–139

KOIWA, A., KITANO, H., FUKUDA, M. and KISAKI, T. (1974). *Agricultural and Biological Chemistry,* **38**, 301–307

KOONTZ, A.E. (1979). PhD thesis, University of California

KOUKOL, J. and DUGGER, W.M., Jr. (1967). *Plant Physiology,* **42**, 1023–1024

KURODA, M., SAKIYAMA, F. and NARITA, K. (1975). *Journal of Biochemistry,* **78**, 641–651

LEH, F. and MUDD, J.B. (1974). *American Chemical Society, Symposium Series,* **3**, 22–3a

MENSER, H.A. (1964). *Plant Physiology,* **39**, 564–566

MUDD, J.B. (1963). *Archives of Biochemistry and Biophysics,* **102**, 59–65

MUDD, J.B. (1966). *Journal of Biological Chemistry,* **241**, 4077–4090

MUDD, J.B. (1975). In *Responses of Plants to Air Pollution,* pp. 97–119 (Mudd, J.B. and Kozlowski, T.T., Eds) Academic Press, New York

MUDD, J.B. and DUGGER, W.M., Jr. (1963). *Archives of Biochemistry and Biophysics,* **102**, 52–58

MUDD, J.B. and McMANUS, T.T. (1969). *Archives of Biochemistry and Biophysics,* **132**, 237–241

MUDD, J.B., LEAVITT, R. and KERSEY, W.H. (1966). *Journal of Biological Chemistry,* **241**, 4081–4085

MUDD, J.B., LEH, F. and McMANUS, T.T. (1974). *Archives of Biochemistry and Biophysics,* **161**, 408–419

MUDD, J.B., McMANUS, T.T. and ONGUN, A. (1971). In *Proceedings of the 2nd International Clean Air Congress,* pp. 256–260, (Englund, H.M. and Beery, W.T., Eds). Academic Press, New York

MUDD, J.B., LEAVITT, R., ONGUN, A. and McMANUS, T.T. (1969). *Atmospheric Environment,* **3**, 669–682

MUDD, J.B., McMANUS, T.T., ONGUN, A. and McCULLOUGH, T.E. (1971). *Plant Physiology,* **48**, 335–339

NATIONAL ACADEMY OF SCIENCES (1977). *Ozone and Other Photochemical Oxidants,* pp. 13–44. NAS, Washington, DC

NIKI, H., DABY, E.E. and WEINSTOCK, B. (1972). *Advances in Chemistry Series,* **113**, 16–57

PREVIERO, A. and SCOFFONE, E. (1963). *Gazzetta Chimica Italiana,* **93**, 841–866

PREVIERO, A., COLLETTI-PREVIERO, M.A. and JOLLES, P. (1967). *Journal of Molecular Biology,* **24**, 261–268

PRYOR, W.A., PRIER, D.G., LIGHTSEY, J.W. and CHURCH, D.F. (1981). In *Autoxidation in Food and Biological Systems,* pp.1–16 (Simic, M.G. and Korel, M., Eds). Plenum Publ., New York

ROEHM, J.N., HADLEY, J.C. and MENZEL, D.B. (1971). *Archives of Environmental Health,* **23**, 132

SAKIYAMA, F. and NATSAKI, R. (1976). *Journal of Biochemistry,* **79**, 225–228

SCHREIBER, M., VIDAVER, W., RUNECKLES, V.C. and ROSEN, P. (1978). *Plant Physiology,* **61**, 80–84

SWANSON, E.S., THOMSON, W.W. and MUDD, J.B. (1973). *Canadian Journal of Botany*, **51**, 1213–1219

TAYLOR, O.C., DUGGER, W.M., Jr., CARDIFF, E.A. and DARLEY, E.F. (1961). *Nature*, **192**, 814–816

TEIGE, B., McMANUS, T.T. and MUDD, J.B. (1976). *Chemistry & Physics of Lipids*, **12**, 153–171

TINGEY, D.T., FITES, R.C. and WICKLIFF, C. (1975). *Physiologia Plantarum*, **33**, 316–320

TINGEY, D.T., FITES, R.C. and WICKLIFF, C. (1976). *Physiologia Plantarum*, **37**, 69–72

TODD, G.W. (1958). *Physiologia Plantarum*, **11**, 457–463

TOMLINSON, H. and RICH, S. (1970). *Phytopathology*, **60**, 1531–1532

TOMLINSON, H. and RICH, S. (1971). *Phytopathology*, **61**, 1404–1405

WOLOSINK, R.A. and BUCHANAN, B.B. (1977). *Nature*, **266**, 565–567

YAMASAKI, N., TSUJITA, T., SAKIYAMA, F. and NARITA, K. (1976). *Journal of Biochemistry*, **80**, 409–412

III

Air Pollutants and the Growth and Quality of Crops

10

EFFECTS OF AIR POLLUTANTS ON FLOWERING AND FRUITING

J. BONTE
Institut National de la Recherche Agronomique, Montardon, Morlaas, France

Introduction

When describing the effects of gaseous air pollution on vegetation, particular note is taken of phenomena which originate at the leaf–atmosphere interface. Contamination of the epidermis, disturbance of metabolism, partial destruction of the foliar surface; whatever the importance of these phenomena, they directly or indirectly modify the trophic relationships between leaf and fruit and generally result in a loss of yield.

Apart from these well-known responses, some effects on yield result from the direct action of pollutants on the mechanisms of flowering and fruiting. These effects, independent of those observed on foliage, are sometimes difficult to distinguish from the consequences of leaf injury.

In agricultural regions, loss of yield may have serious economic repercussions. In forests and woodlands, inhibition of seed production diminishes the capacity for natural regeneration and causes, in time, the disappearance of some species unable to reproduce vegetatively.

Studies on the topic of this paper have been comparatively few and have been principally concerned with the effects of fluorine compounds. Studies of the effects of sulphur dioxide and ozone on flowering and fruiting began later, and have been mainly descriptive. Some surveys deal with the effects of global pollution without specifying any particular pollutant (Dobrovolsky, 1974). We will see that, as with effects on foliage, combinations of pollutants can have additive or greater effects.

Fluorides

Independent of effects on foliage, which have been dealt with in many surveys (e.g. Weinstein, 1977), decreases in productivity and defects in fruit formation have been noted on plants grown in atmospheres polluted by fluoride compounds.

Responses differ considerably from one species to the next, and in many cases it has been necessary to resort to experiments to reveal them.

INJURY SYMPTOMS ON FLESHY FRUIT

As apricots mature, a superficial brownish-black injury develops at the fruit apex; very often a reddish halo appears around the injured part (Bolay *et al.*, 1971).

On peaches, the symptoms are similar; the lesions generally appear at the tip of the fruit, spreading along the suture line and are surrounded with a reddish halo due to the presence of anthocyanins (suture red spot, SRS or soft suture) (Griffin and Bayles, 1952; Benson, 1959; Bolay *et al.*, 1971; Mezzetti and Sansavini, 1977; Weinstein, 1977).

On pears, deformation of the fruit and sometimes blackish collapsed tissue around the calyx are noted; varieties such as Bon Chrétien William's are particularly sensitive (Dässler and Grumbach, 1967; de Cormis, 1968).

On apples, some reddening appears round the calyx, and, in that part, inhibition of growth results in a ring-shaped malformation (Rosenbaum, 1939; Bredemann, 1956; Hölte, 1960; Bolay *et al.*, 1971; Seeley, 1979).

In plums (Kotte, 1929; Hölte, 1963) and cherries (Dässler and Grumbach, 1967; de Cormis, 1968), injury appears at the tip of the fruit and collapse of the cellular tissue round the stone can be noted.

On strawberries, malformations at the apex of the receptacle seem to be closely linked with the number of non-fertilized achenes (Pack, 1972).

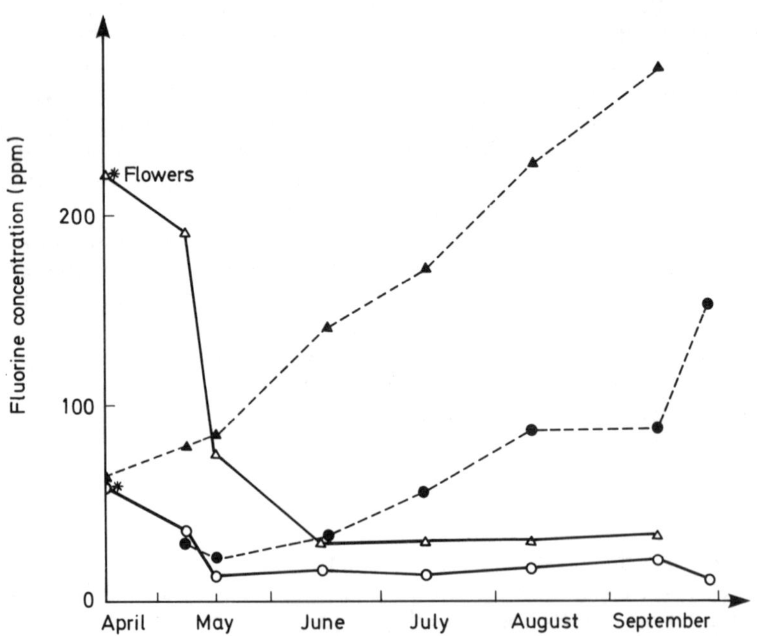

Figure 10.1   Fluorine content of pear fruit (△—△, ○—○) and leaves (▲—▲, ●—●). Symbols: △ ▲, cv. Passe Crassane from highly polluted area $(20\,\mu g\,F\,dm^{-2}\,d^{-1})$; ○ ●, cv. Doyenné du Comice from moderately polluted area $(1.65\,\mu g\,F\,dm^{-2}\,d^{-1})$. After de Cormis (1972a)

INJURY SYMPTOMS ON VEGETABLES AND FIELD CROPS

In some vegetables and field crops, pollution by fluorinated compounds influences seed production. Continuous exposure at $0.64\,\mu g\,F\,m^{-3}$ totally inhibited seed production in soybean, whereas $8\,\mu g\,F\,m^{-3}$ had no effect on cotton. Bell peppers, sweet corn, cucumbers, peas, sorghum, oats, wheat and barley were intermediate between these two species, in decreasing order of sensitivity (Pack and Sulzbach, 1976).

Damage appears at very low atmospheric concentrations, some tenths of $\mu g\,F\,m^{-3}$ in the most sensitive species (Pack and Sulzbach, 1976), often below the threshold for leaf damage (Pack, 1972; MacLean, Schneider and McCune, 1977).

Analyses for fluorine in different parts of the fruit have not revealed accumulations of that element such as are seen in leaves (*Figure 10.1*).

MECHANISMS OF INJURY

It has previously been noted that symptoms of injury have been described quite often, but strangely enough there have been few detailed studies to explain the mechanisms of injury. Should we look for a single mechanism or, taking into account the multiple aspects of the symptoms, for many different mechanisms according to the species?

Effectively, fluorine brings about alterations of organs which, botanically, are as basically different from one another as the pericarp of the pear, the receptacle of the strawberry or the kernel of corn.

In pear, growth of the ovary is inhibited; in strawberry the receptacle (which corresponds to the floral stalk) is deformed; and in corn and grasses, seed growth is absent. Nevertheless among these species, as among many others, there exists a common factor: the absence of seed; in botanical terms, the final stage of ovular growth.

In fact, in polluted strawberry plants, the deformation of the receptacle is proportional to the number of undeveloped achenes (Pack, 1972) and it is known that the growth of receptacle cells occurs only when achenes have been fertilized and produce normal development of the embryos (Nitsch, 1950). The hormones induced by pollination and fertilization initially establish a metabolic gradient between the fruit and sites of synthesis. Once established, that gradient maintains itself independently and induces movement of metabolite to the young fruit (Poli, 1977). Thus the final shape of the receptacle depends on the number of living achenes and on their location on that receptacle.

In pear, counting the seeds of fruit injured by pollution reveals a close relationship between the percentage deformation and the number of undeveloped seeds. Continuous pollution at $3.8\,\mu g\,F\,m^{-3}$ leads to total suppression of development in varieties such as Bon Chrétien William's and Doyenné du Comice (Bonte *et al.*, 1980) (*Table 10.1*).

In contrast to strawberry where in the absence of fertilization the receptacle cannot develop, pear trees are able to produce fruit parthenocarpically. Nevertheless, in certain varieties, William's in particular, parthenocarpic fruit production is unsatisfactory, being accompanied by

Table 10.1 INFLUENCE OF HF POLLUTION ON YIELD OF BON CHRÉTIEN WILLIAM'S AND DOYENNÉ DU COMICE PEAR TREES ($3.80 \pm 0.20\,\mu g\,F\,m^{-3}$) (After Bonte *et al.*, 1980)

Variety	Number of seeds per fruit		Average weight per fruit (g)		Deformed fruit (%)	
	Polluted	Unpolluted	Polluted	Unpolluted	Polluted	Unpolluted
William's	0	6	122 ± 4	153 ± 3	98	0
Doyenné du Comice	0.06	8.8	159 ± 5	163 ± 4	0	0

characteristic developmental and storage problems (deformation and injury around the calyx).

In tomatoes polluted with HF ($2.9\,\mu$g F m^{-3}) the number of seeds per fruit is reduced. For this species, parthenocarpic development leads to the production of smaller fruit (Pack, 1966).

With bean, HF treatments ($2\,\mu$g F m^{-3}) result in a decrease in fresh weight. This decrease is due to the mechanisms already mentioned for other species, i.e. a reduction in the number of seeds per pod, or even their total absence (Pack, 1971).

DURATION OF SUSCEPTIBILITY

The presence of fluorides in the atmosphere inhibits ovular development. Does this inhibition have some effect on one of the processes of fertilization or subsequently on the development of the embryo? Sulzbach and Pack (1972) and Facteau, Wang and Rowe (1973) concluded that the plant is particularly sensitive to fluoride during the flowering stage. Discontinuous fumigations (*Table 10.2*) have shown conclusively that in strawberry, fluoride action is limited to flowering time (Bonte, Bonte and de Cormis, 1980).

Table 10.2 RELATIONSHIP BETWEEN RECEPTACLE MALFORMATION OF STRAWBERRY AND DEVELOPMENTAL STAGE DURING TREATMENT. PLANTS WERE EXPOSED TO F ($5.4 \pm 0.4\,\mu$g F m^{-3}) IN A GREENHOUSE DURING PERIODS MARKED ▨, AND WERE KEPT IN A CLEAN AIR 'CONTROL' GREENHOUSE AT OTHER TIMES ☐. (After Bonte, Bonte and de Cormis, 1980)

Combination	Treatments			Results	
	Before anthesis	*Flowering: fertilization*	*Maturation*	*Malformations (%)*	*Average weight per fruit* (g)
1	▨	▨	▨	57	3.47 a
2	▨	▨	☐	58	4.49 b
3	▨	☐	☐	1.33	5.58 c
4	☐	☐	☐	2.7	5.71 c
5	☐	☐	▨	5.4	5.56 c
6	☐	▨	▨	41.9	5.45 c

Values followed by the same letter are not significantly different for $P = 0.05$

Only those pollution episodes during the period between anthesis and fertilization (fall of the first petal) can damage the strawberry fruit. At other periods, pollution affects only the foliar system. The interval during which the plant is vulnerable is very short, from two to four days according to species and climatic conditions. Pack (1972) showed that continuous exposure of strawberry plants to $0.55\,\mu$g F m^{-3} resulted in deformation of the receptacle. In fact, only a few days' exposure would have been enough to bring about similar effects. This observation explains the apparently random occurrence of damage. At a polluted site the atmospheric concentration of fluoride is seldom constant, but varies with natural fluctuations of climatic conditions. Therefore, damage can appear on plants at a site where pollution is generally low, provided that it has been higher during the period when plants are susceptible.

SITES OF ACTION

Which is the floral organ concerned: the pistil or the pollen?

The germination potential of pollen does not seem to be reduced by fluoride treatment *in vivo*. On the contrary, Sulzbach and Pack (1972) showed that HF treatment of cucumber and tomato flowers tended to stimulate subsequent germination of pollen on the nutritive medium. This phenomenon also occurs in corn (Cochelin, 1974). When the germination medium was polluted with small concentrations the same stimulant effect was noted (Portyanko and Kudrya, 1966). Conversely, at HF concentrations above 10 ppm the opposite effects occurred: reductions of pollen germination and pollen-tube growth (Lai Dinh, 1972). This inhibition is proportional to the concentration of F in the medium, becoming complete in cucumber and tomato when the concentration reaches about 16 mM NaF (Sulzbach and Pack, 1972).

When the pistil was polluted *in vivo*, Facteau, Wang and Rowe (1973) observed that, in emasculated flowers of *Prunus avium*, pollen-tube growth was the same whether pollination took place before or after a pollution episode. This result corroborates the insensitivity of pollen to pollution and implies an effect on the pistil. In fact, cross fertilizations with separately polluted organs revealed that pollution of only the pistil resulted in significant inhibition of pollen germination (Sulzbach and Pack, 1972). In strawberry (*Table 10.3*), malformation of the receptacle observed during maturation is essentially concerned with pollution of the pistil (Bonte, Bonte and de Cormis, 1980).

Table 10.3　EFFECT OF HF POLLUTION ON REPRODUCTIVE ORGANS OF STRAWBERRY PLANTS, CV. SEQUOIA. (After Bonte, Bonte and de Cormis, 1980)

HF concentration (μg F m^{-3})	*Nature of the fertilization*	*Yield*	
		Average weight (g)	*Malformations* (%)
4.10	♀ polluted ♂ unpolluted	3.58 a	74
4.28	♀ unpolluted ♂ polluted	4.53 b	11

a and b are significantly different for $P = 0.05$

Are the defects in fertilization caused by inhibition of pollen germination or of pollen-tube growth in the stigma (as *Figure 10.2*, from Facteau and Rowe (1977) seems to indicate), or by a direct action of HF on the ovule or the embryo?

For the strawberry plant, application of a growth-promoting substance (naphthyl acetic acid) immediately after fertilization in a polluted atmosphere, suppressed growth inhibition and allowed parthenocarpic development of the receptacle tissues and achene pericarps (Bonte, Bonte and de Cormis, 1980) (*Table 10.4*). These results illustrate that ovaries and adjacent tissues kept their growth potential. It seems that the fertilization stimulus was not transmitted to the ovule. Fluoride would then have some localized effect on the style or stigma. In fact, analysis with an electron microprobe of the style and the stigma of polluted plants detected a significant accumulation of fluorine at the apex of the stigma (*Figure 10.3*) (Bonte and Garrec, 1980). Thus, just as pollen does not develop on an

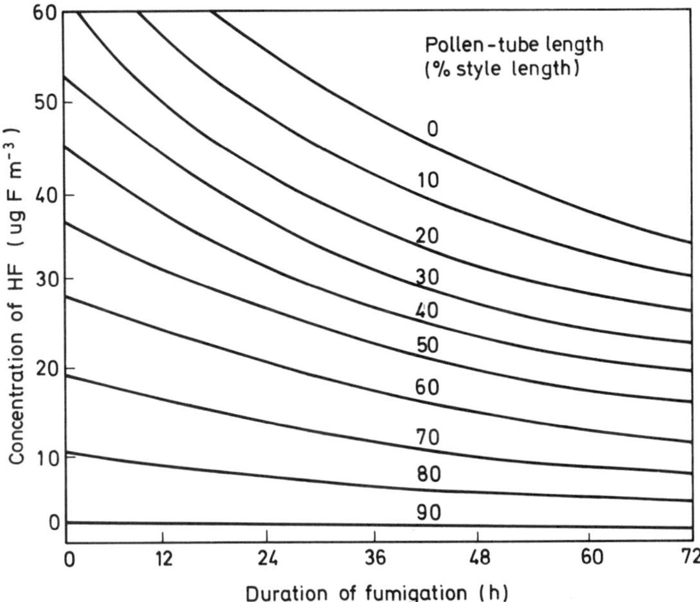

Figure 10.2   Response plot of pollen-tube lengths as a percentage of style length, as a function of concentration of HF and duration of exposure (After Facteau and Rowe, 1977)

artificially polluted medium (agar + sucrose + NaF or HF) it also will not develop on a stigma on which the superficial concentration of F can reach several thousand ppm dry weight.

Fluorine is also noxious to pollinating insects. At polluted sites a high death-rate has often been recorded among bee colonies. Analysis of dead bees revealed an abnormal increase in fluorine concentration (Maurizio and Staub, 1956; Guilhon, Truhaut and Bernuchon, 1962). This increase, which could not be attributed to fluorine in the atmosphere, was no doubt due to honey-gathering on highly contaminated materials. In flowers, de Cormis (1972a) and Börtitz and Reuter (1977) noted significant concentrations (*Figure 10.1* and *Table 10.5*), but analyses of flowers or parts of flowers did not indicate that much higher concentrations might occur in particular parts of the flower.

Table 10.4   EFFECT OF AUXIN APPLICATION ON RECEPTACLE DEVELOPMENT OF STRAWBERRY PLANTS SUBJECTED TO HF (3.9 ± 0.3 µg F m⁻³) DURING FLOWERING. α-NAPHTHYL ACETIC ACID (ANA) (2000 ppm IN AGAR) WAS APPLIED TWO DAYS AFTER FERTILIZATION. (After Bonte, Bonte and de Cormis, 1980)

Treatment	Fruit average weight (g)	Fruit malformation (%)	Number of seeds (achenes) per fruit	Fertilized achenes (%)
HF + ANA	6.25 a	2 a	190 a	16 b
HF − ANA	3.20 b	80 b	55 b	75 a
Control	6.40 a	2 a	233 a	79 a

In the same column the values followed by the same letter are not significantly different. (For $P = 0.05$)

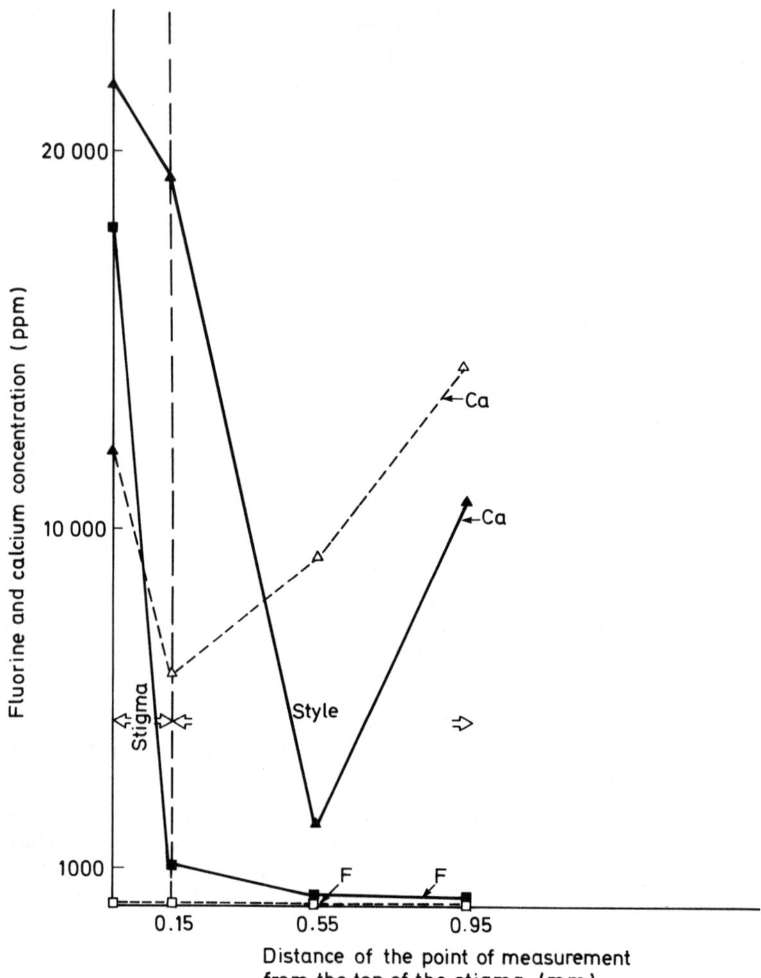

Figure 10.3 Fluorine and calcium concentrations on stigmas and along the style of strawberry plants grown in air containing $14\,\mu g\,HF\,m^{-3}$, and in clean air. $\triangle$---$\triangle$ calcium in unpolluted plants; $\blacktriangle$——$\blacktriangle$ calcium in polluted plants; $\square$---$\square$ fluorine in unpolluted plants; $\blacksquare$——$\blacksquare$ fluorine in polluted plants. (After Bonte and Garrec, 1980)

Table 10.5 FLUORIDE CONCENTRATION IN FLOWERS OF POLLUTED AND UNPOLLUTED APPLE TREES. 0 = CONTROL; 1 = FUMIGATION FOR 63 HOURS BY $140\,\mu g\,HF\,m^{-3}$ (OVER 11 DAYS); 2 = FUMIGATION FOR 86 HOURS BY $140\,\mu g\,HF\,m^{-3}$ (OVER 14 DAYS). (After Börtitz and Reuter, 1977)

| *Variety* | *Fluorine concentration* (ppm dry weight) | | | | |
| | In petals | | | In reproductive organs | |
	0	*1*	*2*	*0*	*1*
Auralia	6.3	245	558	7.4	670
Gelber	3.8	141	200	5.8	467
James Grieve	5.0	136	137	5.0	439

MECHANISMS OF ACTION

Several hypotheses can be put forward:

(1)   The degree of contamination injures the tissues of the stigma, causing changes which prevent development of the pollen tube.
(2)   Fluorine acts directly on pollen germination. Studies *in vitro* have shown that the addition of fluorine to an agar medium inhibits pollen germination and delays pollen-tube growth.
(3)   Fluorine acts indirectly on calcium metabolism. This element is essential for pollen germination and pollen-tube growth (Brewbaker and Kwack, 1963; Pack, 1966) and it attracts fluorine.

In support of this last hypothesis, *Figure 10.3* shows that a spectacular increase of calcium occurred, associated with the accumulation of fluorine. This phenomenon may be related to the antitoxic role attributed to calcium: either by direct action (the plant reacts to fluorine by increasing its calcium concentration, in order to bind the former as CaF_2 (Garrec *et al.*, 1974)) or by indirect action (fluorine brings about metabolic alterations similar to those which occur in senescent plants). Among the changes which mark senescence, development of organic acids is particularly observed. These acids are mainly neutralized by calcium as oxalate precipitates (Garrec *et al.*, 1978).

In both cases, whatever the increase of calcium concentration, the greater part of it is insoluble. We can thus assume that in the stigma in spite of the high calcium concentrations measured, there is in fact a deficiency of physiologically active calcium.

These patterns are common to a great number of cultivated plants, but do not explain the phenomena which appear in stone fruit, such as peach, apricot and cherry. On the one hand these fruit trees only occasionally produce parthenocarpic fruit and this fruit does not resemble that injured by fluorinated compounds: on the other hand, it seems possible to produce the same injury on fruit by treatment with fluorine away from the sensitive flowering time. Benson (1959) and Facteau and Wang (1972) obtained characteristic damage by repeated spraying with fluoride after setting. Seeley (1979) obtained injury similar to that produced by atmospheric fluoride by dipping apples in solutions of NaF.

It seems that apart from well-known effects on fertilization, other processes exist which imply a direct action of fluorides on fruit growth: an action on the internal trophic relations of the fruit?; an inhibition of the role played by auxins? (it should be noted that malformed or injured fruit contain only small amounts of fluorine at maturity); an inhibition of calcium metabolism?

Further studies will be necessary to complete our knowledge of this topic.

Sulphur dioxide

The effects of SO_2 on vegetation are well known and have been the subject of numerous surveys (Mudd, 1975; Linzon, 1978). However, its effects on reproductive organs have received little attention. Pelz (1963) noted a significant reduction in the size of cones and in the weight of seeds of fir sampled in highly polluted areas. The same relationships have been established for *Pinus sylvestris* by Podzorov (1965), Mrkva (1969), and Mamejev and Shkarlet (1970). If cereals are treated at different physiological stages—emergence, flowering, earing and maturation—treatments during flowering are the only ones which affect production (de Cormis, 1972b). These observations suggest that SO_2 may directly alter the process of fruiting.

Döpp (1931) studied the effects of SO_2 on the pollen germination of a great number of species. For *Pinus montana* and *Pinus sylvestris* he showed that a treatment of 10 ppm for 6 days had no effect on pollen germination if exposure took place in a dry atmosphere. However, when the pollen was wet, germination decreased and pollen tubes burst after an *in vitro* treatment of 10 ppm for 45 min. Houston and Dochinger (1977) studies production of Eastern White and Red Pine at two sites, one with high, the other with very low pollution (*Tables 10.6, 10.7*). They concluded that reproductive organs of pines were affected by chronic pollution at concentrations lower than those which generally cause apparent leaf damage. The study by Karnosky and Stairs (1974) on pollen germination *in vitro* (*Figures 10.4, 10.5*) corroborated Döpp's results, although it should be

Table 10.6 EFFECT OF AMBIENT AIR POLLUTION ON REPRODUCTIVE CHARACTERISTICS OF EASTERN WHITE PINE (After Houston and Dochinger, 1977)

Characteristics	Blue Rock State Forest (high pollution incidence)	Tappan Lake (low pollution incidence)
Average cone length (mm)	122	124
Average cone width (mm)	20	21
Average no. of seeds per cone	55	67**
Average 100-seed weight (g)	1.532	1.850*
Filled seed (%)	85	84
Seed germination (%)	70	70
Pollen germination (%)	60	83*
Average pollen tube length (μm)	99	113

*Significant difference at 0.05 level
**Significant difference at 0.01 level

Table 10.7 EFFECT OF AMBIENT AIR POLLUTION ON REPRODUCTIVE CHARACTERISTICS OF RED PINE (After Houston and Dochinger, 1977)

Characteristics	Blue Rock State Forest (high pollution incidence)	Mohican State Forest (low pollution incidence)
Average cone length (mm)	44	47**
Average cone width (mm)	23	23
Average 100-seed weight (g)	0.666	0.805**
Filled seed (%)	50	68**
Seed germination (%)	50	66**
Pollen germination (%)	65	84*
Average pollen tube length (μm)	92	113*

*Significant difference at 0.05 level
**Significant difference at 0.01 level

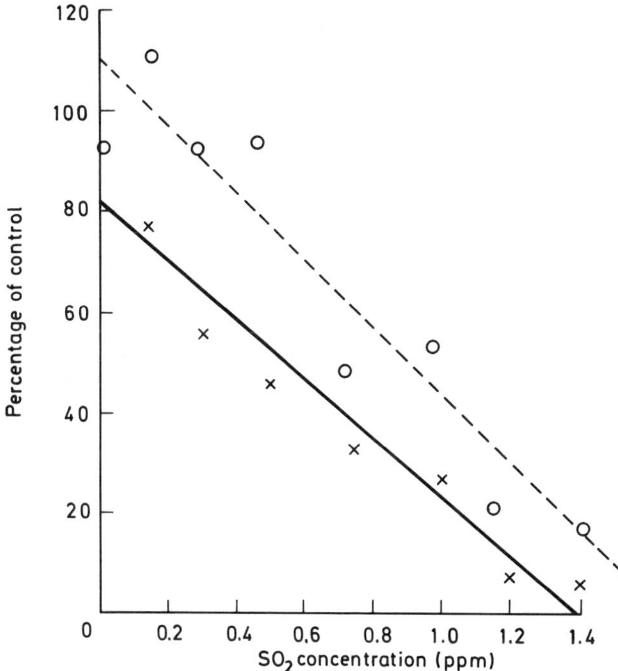

Figure 10.4   Effects of SO_2 on germination (O---O) and tube elongation (×——×) of moist *Populus deltoides* pollen. The linear regressions between % control (y) and ppm SO_2 (x) are: $y = 111.0 - 66.4x$ (germination); $y = 81.8 - 59.8x$ (tube elongation) (After Karnosky and Stairs, 1974)

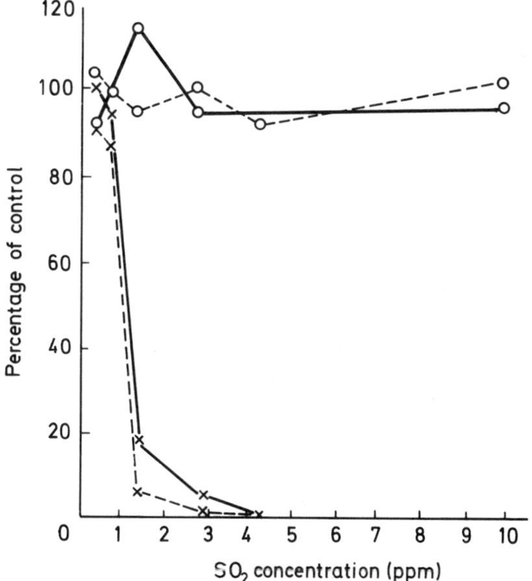

Figure 10.5   Effects of SO_2 and moisture on conifer pollen germination and tube elongation. Pooled data from *Pinus resinosa*, *Pinus nigra* and *Picea pungens*. O---O germination, dry pollen; ×---× germination, moist pollen; O——O elongation, dry pollen; ×——× elongation, moist pollen. (After Karnosky and Stairs, 1974)

noted that the sensitivities of the species being studied were much greater. In addition, a study of the influence of SO_2 on the germination medium revealed that treatment at 1.4 ppm for four hours lowered the pH from 7 to 5. This acidification may explain the reduction of *Populus deltoides* pollen germination but seems to have no effect on conifer pollen germination. Contamination of the germination medium seems to have a greater effect on pollen-tube growth than pollen pollution: the results obtained *in vitro* suggest that *in vivo*, as was shown for fluorinated compounds, stigma pollution could be the cause of the phenomena observed. However, Shkarlet (1972) succeeded in showing a modification of the pollen characteristics. Repeated sampling of pollen from areas treated with different levels of pollution revealed that from polluted trees (*Pinus sylvestris*) the average size of pollen grains was significantly reduced in comparison with those from nonpolluted trees (*Figure 10.6*).

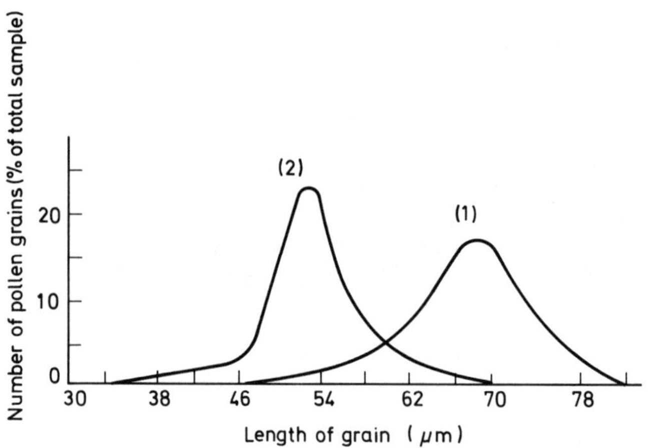

Figure 10.6   Percentage of pollen grains with length L, taken from Scots pine from a clean area (1) and a highly polluted area (2). (After Shkarlet, 1972)

In most experiments carried out in areas exposed to industrial waste, the degree of pollution is expressed in terms of SO_2 concentration because it is the main pollutant, yet in most instances it is accompanied by other pollutants and additive effects must be reckoned with. Nakada, Fukuis and Kanno (1976) showed that, *in vitro*, the addition of SO_2 to NO_2 or O_3, or HCHO, considerably increased the percentage of inhibition compared with the action of each product examined separately.

For the mixture of $SO_2 + F$, Roques, Kerjean and Auclair (1980) found that in areas not polluted by fluorine but moderately polluted by SO_2 (F = 14.5 ppm dry weight, total S = 0.17% dry weight), the characteristics of seeds and cones of *Pinus sylvestris* (cone weight, cone size, growth rate, number of seeds per cone, % empty seeds) were similar to those of controls from unpolluted sites. On the other hand, in areas where the level of SO_2 pollution was similar but the level of fluorine pollution was increased (F = 30 ppm dry weight, S = 0.19% dry weight) the differences were significant. These differences were even greater when the F concentration was the same and the SO_2 concentration was raised (F = 32 ppm, S = 0.34% dry weight).

There is little information available on the mechanisms of action of SO_2 on the pollen tube. Ma and Khan (1976) measured the pollen mitotic index[a] of *Tradescantia paludosa* polluted *in vitro* by SO_2. The mitotic index fell from 38.7% for the control sample to 24.3% for exposure at 0.075 ppm and to 3.8% for exposure at 50 ppm. They assumed that SO_2 broke down the chromosomes of the vegetative and reproductive nuclei of the pollen tube (Ma *et al.*, 1973).

Ozone

There is little evidence available to indicate a direct action of ozone on reproduction.

At concentrations typical of polluted atmospheres, ozone mainly produces leaf damage (Heath, 1975). Loss of yield has been attributed to damage of foliar surface. Yet, according to Heagle, Philbeck and Knott (1979), threshold doses of O_3 affecting kernel yield are higher than those causing foliar injury, demonstrating that field corn can withstand some injury with no loss of yield. The correlation between injury, growth, and yield effects across treatments often was low. Foliar injury and effects on stover weight were a poor indication of the magnitude of the effects on kernel weight. For example, at 0.11 ppm O_3, foliar injury averaged about 7% greater than that at 0.02 ppm, stover weight was decreased by 23% and yield was decreased by only 2%. At 0.15 ppm, foliar injury was 10% greater than at 0.02 ppm, stover weight was decreased by 45% and yield was decreased by 15%.

Ozone treatments delayed and reduced the floral yield of carnation and geranium but those treatments (from 0.05 to 0.10 ppm for 9.5–24 h day^{-1} for 1–3 months) also resulted in a reduction of top growth, side branching and leaf size (Feder and Campbell, 1968; Feder, 1970).

Treatments of 8–10 ppm O_3 for 5 h day^{-1} increased the number of days required for fruit set of tomato. At harvest time this treatment resulted in a significant reduction in the number and weight of fruit (Manning and Feder, 1976). Is this phenomenon due to the action of O_3 on pollination or to an effect on plant growth? (In fact, this treatment decreases plant growth). The studies of Feder (1968) and Harrison and Feder (1974) of the effects of O_3 on pollen, support the first hypothesis but, if leaf-damage is eliminated by treatment with benomyl, yield is significantly increased (Manning and Feder, 1976). It is interesting to observe that the O_3 concentrations in the previous experiments occur over large areas. In fact, Oshima *et al.* (1977) observed a reduction in size of tomatoes cultivated in areas exposed to different levels of pollution.

INFLUENCE OF O_3 ON POLLEN

In tobacco (cv.Bel W_3) both pollen germination and pollen-tube growth are inhibited by exposure to O_3. These effects were similar whether the

---

[a] Mitotic index = number of metaphases/100 nuclei under observation.

pollen was exposed *in vitro* (agar plate), or *in vivo* (intact plant) (Feder, 1968).

There are varietal differences in pollen sensitivity to O_3, similar to those found for leaf damage. For example ozone treatment of 0.1 ppm for 5.5 h reduced pollen germination and tube growth of two susceptible varieties, tobacco cv. Bel W_3 and petunia cv. White Cascade, by about 50%. The resistant varieties, tobacco cv. Bel B and petunia cv. Blue Lagoon, were unaffected by the same treatment (Feder and Sullivan, 1969).

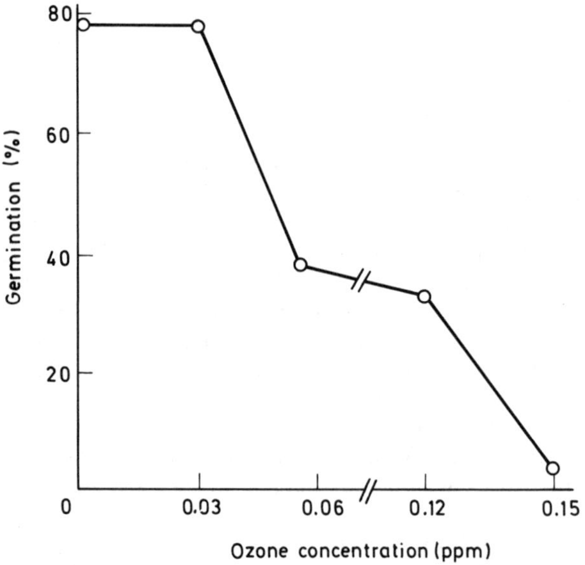

Figure 10.7 Effect of ozone concentration on germination of maize pollen exposed for 5.5 h d^{-1}. (After Mumford *et al.*, 1972)

Free amino acids of maize pollen increased by 50% after an O_3 treatment of 0.03 ppm for 5.5 h day^{-1} for 60 days. Higher concentrations, 0.06 and 0.12 ppm, increased the accumulation of amino acids and of peptides, and inhibited germination by 40–90% (*Figure 10.7*) (Mumford *et al.*, 1972). The results suggest that O_3 induces the autolysis of structural glycoproteins and stimulates amino acid synthesis.

Conclusion

The first problem in this survey has been to distinguish between the various effects of the principal pollutants on the production of seeds and fruit, and to separate those effects that can be attributed to direct action on the sexual organs from those which are brought about by an indirect action on other parts of the plant.

Numerous observations complemented by experiments have enabled us to show that, for fluorides and SO_2, there exists a direct mode of action on the processes of fruiting. In some species, these phenomena are sensitive to concentrations of pollutants lower than those which generally result in apparent leaf damage. They result in a loss of seed yield and in alterations of fruit quality. On the other hand, for O_3, it seems impossible to dissociate leaf damage from production loss.

There are very few studies of the mechanisms of action of pollutants on the different organs of reproduction. With fluorides, it has been demonstrated that for certain species the only sensitive period is flowering time. During that phase, fluoride appears to affect the female part of the flower, more precisely the stigma, on whose surface significant amounts of fluorides have been measured. From these results some hypotheses have been put forward, but they do not explain all the phenomena observed, in particular the fruiting defects of stone fruits.

It seems that the mechanism of action of SO_2 is similar in some ways to that of fluorine. Pollen *in vivo* seems comparatively insensitive to pollution, whereas contamination of the germination medium inhibits pollen-tube growth.

Considering the importance of studies dealing with the effects of O_3 on vegetation, it is surprising how few of them deal with its effects on fruiting. In most cases the loss of yield is correlated with leaf damage and cannot be attributed to more insidious effects. Nevertheless, experimental studies have shown that ozone can inhibit pollen germination.

References

BENSON, N.R. (1959). *Proceedings. American Society for Horticultural Science*, **74**, 184–198

BOLAY, A., BOVAY, E., NEURY, G., QUINCHE, J.P. and ZUBER, R. (1971). *Revue Suisse de Viticulture et d'Arboriculture*, **III,(3)** 82–91

BONTE, J. and GARREC, J.P. (1980). *Comptes Rendus Hebdomadaires des Séances de l'Académie des Sciences, Series D: Natural Sciences*, **290**, 815–818

BONTE, J., BONTE, C. and DE CORMIS, L. (1980). *Compte Rendu Hebdomadaire des Séances de l'Académie d'Agriculture de France*, Discussion, 16 Jan, pp. 80–89

BONTE, J., BONTE, C., DE CORMIS, L. and BAUVILLE, G. (1980). *Pollution Atmospherique*, (89), 31–34

BÖRTITZ, S. and REUTER, F. (1977). *Archiv für Gartenbau*, **25**, 147–255

BREDEMANN, G. (1956). *Biochimie und Physiologie des Fluors*. Akademie-Verlag, Berlin

BREWBAKER, J.L. and KWACK, B.H. (1963). *American Journal of Botany*, **50**, 859–865

COCHELIN, J.C. (1974). *DEA*, Université Paul Sabatier de Toulouse

DÄSSLER, H.G. and GRUMBACH, H. (1967). *Obstbau (DDR)* **7**, 27–29

DE CORMIS, L. (1968). *Compte rendu d'activité du Laboratoire de Phytopharmacie (Pollution Atmosphérique)*. INRA, Montardon, Morlaas

DE CORMIS, L. (1972a). *Rapport d'activité du Laboratoire d'Etude de la Pollution Atmosphérique*, INRA, Montardon, Morlaas

DE CORMIS, L. (1972b). *Sulphur Institute Journal*, **8**, 8

DOBROVOLSKY, I.A. (1974). *Ukryins kÿ Botanichnij Zhurnal*. **31**, 1

DÖPP, W. (1931). *Bericht der Deutschen Botanischen Gesellschaft*, **49**, 172–221

FACTEAU, T.J. and ROWE, K.E. (1977). *Journal of the American Society for Horticultural Science*, **102**, 95–96

FACTEAU, T.J. and WANG, S.Y. (1972). *HortScience*, **7** 505

FACTEAU, T.J., WANG, S.Y. and ROWE, K.E. (1973). *Journal of the American Society for Horticultural Science*, **98**, 234–236

FEDER, W.A. (1968). *Science*, **160**, 1122

FEDER, W.A. (1970). *Environmental Pollution*, **1**, 73–79

FEDER, W.A. and CAMPBELL, F.J. (1968). *Phytopathology*, **58**, 1038–1039

FEDER, W.A. and SULLIVAN, F. (1969). *Phytopathology*, **59**, 399

GARREC, J.P., ABDULAZIZ, P., LAVIELLE, E., VANDEVELDE, L. and PLEBIN, R. (1978). *Fluoride*, **11**, 186–197

GARREC, J.P., OBERLIN, J.C., LIGEON, E., BISCH, A.M. and FOURCY, A. (1974). *Fluoride*, **7**, 78–84

GRIFFIN, S.W. and BAYLES, B.B. (1952). In *Proceedings of the US Technological Conference on Air Pollution*, pp. 106–115. McGraw-Hill, New York

GUILMON, J., TRUHAUT, R. and BERNUCMON, J. (1962). *Comptes Rendus de l'Académie d'Agriculture de France*, **12**, 607–615

HARRISON, B.H. and FEDER, W.A. (1974). *Phytopathology*, **64**, 257–258

HEAGLE, A.S., PHILBECK, R.B. and KNOTT, W.M. (1979). *Phytopathology*, **69**, 21–26

HEATH, R.L. (1975). *Ozone in Responses of Plants to Air Pollution* (Mudd, J.B., Kozlowski, T.T., Eds), Academic Press, New York

HÖLTE, W. (1960). *Bericht Landesanstalt für Bodennutzungsschutz des Landes Nordrhein - Westfalen. Bochum*, pp. 43–62

HÖLTE, W. (1963). In *Raucheinwirkung im Gartenbau*, pp. 29–34. Landwirtschaftsverlag GmBH, Hiltrup, Münster

HOUSTON, D.B. and DOCHINGER, L.S. (1977). *Environmental Pollution*, **12**, 1–5

KARNOSKY, D.F. and STAIRS, G.R. (1974). *Journal of Environmental Quality*, **3**, 406–409

KOTTE, W. (1929). *Nachrichtenblatt für den Deutschen Pflanzenschutzdienst*, **9**, 91–92

LAI DINH, D. (1972). Thesis, Hohenheim University

LINZON, S.N. (1978). In *Sulfur in the Environment, part II* pp. 109–162 (Nriagu, J.O., Ed.), Wiley, New York

MA, T.H. and KHAN, S.H. (1976). *Environmental Research*, **12**, 144–149

MA, T.H., ISBANDI, D., KHAN, S.H. and TSENG, Y.S. (1973). *Mutation Research*, **21**, 93–100

MacLEAN, D.C., SCHNEIDER, R.E. and McCUNE, D.C. (1977). *Journal of the American Society for Horticultural Science*, **102**, 297–299

MAMEJEV, S.A. and SHKARLET, O.D. (1970). In *Proceedings of an International Conference on Air Pollution, VII (Essen, West Germany, Sept.)*, pp. 443–450

MANNING, W.J. and FEDER, W.A. (1976). In *Effects of Air Pollutants on Plants*, pp. 45–60 (Mansfield, T.A., Ed.), Cambridge University Press, Cambridge

MAURIZIO, A. and STAUB, (1956). *Schweizerische Bienenzeitung*, **11**, 476–486

MEZZETTI, A. and SANSAVINI, S. (1977). *Genio rurale,* **5**, 19–31

MRKVA, R. (1969). *Acta Universitatis Agricultura Brno,* **38**, 345–360

MUDD, J.B. (1975). In *Responses of Plants to Air Pollution,* pp. 9–22 (Mudd, J.B. and Kozlowski, T.T., Eds), Academic Press, New York

MUMFORD, R.A., LIPKE, H., LAUFER, D.A. and FEDER, W.A. (1972). *Environmental Science and Technology,* **6**, 427–430

NAKADA, M., FUKUIS, S. and KANNO, S. (1976). *Environmental Pollution,* **11**, 181–187

NITSCH, J.P. (1950). *American Journal of Botany,* **37**, 211–215

OSHIMA, R.J., BRAEGELMANN, P.K., BALDWIN, D.W., VAN WAY, V. and TAYLOR, O.C. (1977). *Journal of the American Society for the Horticultural Science,* **102**, 289–293

PACK, M.R. (1966). *Journal of the Air Pollution Control Association,* **16**, 541–544

PACK, M.R. (1971). *Environmental Science and Technology,* **5**, 1128–1132

PACK, M.R. (1972). *Journal of the Air Pollution Control Association,* **22**, 714–717

PACK, M.R. and SULZBACH, C.W. (1976). *Atmospheric Environment,* **10**, 73–81

PELZ, E. (1963). *Archiv für Forstwesen* **12**, 1066–1077

PODZOROV. N.V. (1965). *Lesneya zhizn khozyajstvo* **18**, 47–49

POLI, M. (1977). *Pomologie francaise,* **19**, 105–113

PORTYANKO, V.F. and KUDRYA, L.M. (1966). *Fiziologiya rastenii,* **13**, 1086–1089

ROQUES, A., KERJEAN, M. and AUCLAIR, D. (1980). *Environmental Pollution (series A),* **21**, 191–201

ROSENBAUM, H. (1939). *Die kranke Pflanze,* **16,** (1) 3–7; (2) 33–36; (3) 50–54

SEELEY, E.J. (1979). *HortScience,* **14**, 162–163

SHKARLET, O.D. (1972). *Soviet Journal of Ecology,* **3**, 38–41

SULZBACH, C.W. and PACK, M.R. (1972). *Phytopathology,* **62**, 1247–1253

WEINSTEIN, L.H. (1977). *Journal of Occupational Medicine,* **19**, 49–78

11

SULPHUR DIOXIDE AND THE GROWTH OF GRASSES

J.N.B. BELL
Department of Botany, Imperial College at Silwood Park, Ascot

Introduction

Grasslands, both sown and natural, represent the largest element of agricultural primary production in many countries, with meadows and pastures covering about 20% of the earth's land surface (Semple, 1970). About 73% of mainland Britain consists of some type of grassland, and grass and forage legumes are currently the most important crop in this country (Spedding and Diekmahns, 1972). In 1969 it was estimated that, in the United Kingdom, grasslands contributed 24% and 67% of the protein diet and 11% and 46% of the energy requirements of humans and animals, respectively (Spedding, 1971).

In view of the agricultural importance of grasses, it is surprising that until relatively recently very little interest has been shown in their response to sulphur dioxide (SO_2), in comparison with other types of crop. Thus Zimmerman and Hitchcock (1956), in scoring 67 species for sensitivity to acute SO_2 injury, omitted grasses completely, despite a wide range of plants being included in their programme. Similarly, Wood (1968), reviewing the literature on SO_2 injury, quoted only two examples of fumigations of grasses, out of a list of 97 species. In the extensive field-exposure trials conducted by Guderian and Stratmann (1968) at Biersdorf, Germany, there were no grasses among the 21 agricultural, horticultural and forestry species used.

In North America the limited work on SO_2 effects on grasses has usually employed acute fumigations, followed by scoring the leaves for visible injury: for example Brennan and Halisky (1970) fumigated 11 cultivars of seven turf grass species with 2145–5148 µg SO_2 m^{-3} for 6 h, and Hill *et al.* (1974) included 16 grass species in a wide range of desert plants which they screened for sensitivity to 1430–28 600 µg SO_2 m^{-3} for 2 h. Only in the last few years has attention been focused in North America on the prolonged effects of lower levels of SO_2 on grass species, with the use of open-air fumigation systems to predict the impact of large coal-fired power stations on the performance of natural grasslands in the northern Great Plains (Heitschmidt, Lauenroth and Dodd, 1978; Dodd *et al.*, 1979).

The effects of urban air pollution on grass growth

Research into the effects of air pollution on plants has developed in the United Kingdom only during the last decade. However, in contrast to work elsewhere, much of the research has been devoted to the growth responses of pasture grasses to long-term exposure to low and moderate levels of SO_2 ($<600\,\mu g\,m^{-3}$), which are generally typical of British conditions. This may partly reflect the importance of grasslands in British agriculture, but such research has also been stimulated by the findings of Bleasdale (1952, 1973), who demonstrated that polluted ambient urban air in the early 1950s produced substantial growth reductions in *Lolium perenne*, without any visible symptoms on the foliage. Bleasdale followed up reports of poor productivity in grasslands re-sown with bred cultivars in polluted hill-farming regions of East Lancashire, by growing *L. perenne* cv. S23 for long periods in a pair of glasshouses, sited in an inner suburb of Manchester, and ventilated with either the ambient air or water-scrubbed clean air. *Table 11.1* shows that significant reductions in shoot dry weight occurred consistently in ambient air compared with clean air, in both summer and winter, over periods when the mean SO_2 concentration was below $200\,\mu g$ m^{-3} (in some experiments SO_2 measurements were not available). At this time it was generally considered that the threshold concentration for SO_2 injury on the most sensitive higher plants was between 300 and $400\,\mu g\,m^{-3}$ (Thomas and Hill, 1937; Katz, 1949).

Since Bleasdale's experiments, SO_2 concentrations in urban areas of the United Kingdom have fallen considerably (Anonymous, 1974), largely in response to reduced emissions from domestic fires, following the Clean Air Act of 1956. The effects of these lower concentrations on the growth of grasses were investigated by Crittenden and Read (1978, 1979). They grew *L. perenne* cv. S23, *L. multiflorum* cv. S22, and *Dactylis glomerata* cv. S143 in the early 1970s in a pair of glasshouses, in an inner suburb of Sheffield and ventilated these with ambient air or charcoal-filtered clean air. *Table 11.2* shows that air containing mean SO_2 concentrations of $45–70\,\mu g\,m^{-3}$ substantially reduced the dry weight of the shoots of all three species compared with the clean-air controls, in a series of experiments carried out in spring and/or summer.

SO_2 fumigation experiments

The ambient-air versus filtered-air experiments of Bleasdale, and of Crittenden and Read, clearly indicate that some adverse factor for grass growth was present in the air in parts of British cities which are not generally classified as highly polluted. In both cases it was believed that SO_2 was a major factor in reducing growth, but the mean levels were much lower than the accepted threshold concentration for injury. In order to elucidate the importance of low-to-moderate SO_2 levels on the growth of grasses in the field, several independent groups have developed facilities for long-term fumigation experiments. Initially, experiments were per-formed only with *L. perenne*, which is the most important agricultural grass in the United Kingdom (Anonymous, 1977) and is also apparently very

Table 11.1 MEAN SHOOT DRY WEIGHTS OF *LOLIUM PERENNE* CV. S23 GROWN IN AMBIENT MANCHESTER AIR AND CLEAN AIR (1950–51)

Mean SO_2 concentration ($\mu g\,m^{-3}$)	Duration (d)	Season	Dry wt in clean air (g)	Dry wt in ambient air (g)	% reduction in ambient air	$P <$	Reference
175	104	Winter	0.053	0.044	17	0.05	Bleasdale (1973)
–	115	Winter	0.099	0.072	27	0.05	Bleasdale (1973)
164[a]	187	Winter	1.38	0.93	33	0.001	Bleasdale (1973)
–	48	Summer	0.341	0.280	18	0.01	Bleasdale (1952)
137	59	Summer	1.064	0.728	32	0.001	Bleasdale (1973)

[a]Mean concentration over last 79 d only.

Table 11.2 MEAN SHOOT DRY WEIGHTS OF *LOLIUM PERENNE* CV. S23, *LOLIUM MULTIFLORUM* CV. S22 AND *DACTYLIS GLOMERATA* CV. S143 GROWN IN AMBIENT SHEFFIELD AIR AND CLEAN AIR (1973–74)

Species	Mean SO_2 concentration	Duration (d)	Season	Dry wt in clean air (g)	Dry wt in ambient air (g)	% reduction in ambient air	$P <$	Reference
Lolium perenne	70	56	Spring	0.793	0.507	36	0.001	Crittenden and Read (1978)
	59	131	Spring/summer	5.843	4.679	20	0.01	
	69	86	Spring	2.721	2.054	25	0.001	
	63	116	Spring/summer	2.500	1.857	26	0.001	
Lolium multiflorum	67	56	Spring	1.080	0.689	36	0.001	Crittenden and Read (1979)
Dactylis glomerata	45	72	Spring/summer	1.848	1.068	42	0.001	Crittenden and Read (1979)

Table 11.3 EXPERIMENTAL SYSTEMS FOR LONG-TERM SO$_2$ FUMIGATION

Location	*Design*	*References*
Imperial College	Outdoor Perspex chambers	Bell and Clough (1973) Bell, Rutter and Relton (1979) Ayazloo, Bell and Garsed (1980) Ayazloo and Bell (1981)
	Glass chambers in glasshouse	Ayazloo and Bell (1981)
Grassland Research Institute (GRI)	Controlled-environment cabinets	Cowling, Jones and Lockyer (1973)
	Perspex chambers in glasshouse	Lockyer, Cowling and Jones (1976) Cowling and Lockyer (1976, 1978) Cowling and Koziol (1978)
University of Lancaster	Controlled-environment wind tunnels	Ashenden and Mansfield (1977) Ashenden (1978) Davies (1980)
	Hemispherical glasshouses	Ashenden and Mansfield (1978) Ashenden (1979) Ashenden and Williams (1980)
University of Liverpool	Controlled-environment wind tunnels	Horsman, Roberts and Bradshaw (1978, 1979) Horsman *et al.* (1979) Roberts *et al.* (1979)
	Outdoor open-top chambers	Roberts *et al.* (1979)

sensitive to SO_2 (Bell and Mudd, 1976), but subsequently they have been extended to several other common species.

Fumigation of grasses has been carried out in four main centres, using different procedures which unfortunately hinder comparisons between experiments (*Table 11.3*). The Imperial College facilities comprise large outdoor chambers with no environmental control and a pair of chambers inside a glasshouse (Bell and Clough, 1973; Bell, Rutter and Relton, 1979; Ayazloo, Bell and Garsed, 1980; Ayazloo and Bell, 1981). At the Grassland Research Institute (GRI), fumigation experiments were initially performed in a controlled-environment cabinet (Cowling, Jones and Lockyer, 1973), but subsequently in small chambers in a glasshouse with only partial environmental control (Lockyer, Cowling and Jones, 1976; Cowling and Lockyer, 1976, 1978; Cowling and Koziol, 1978). The University of Lancaster group employs an environmentally controlled system of wind tunnels (Ashenden and Mansfield, 1977; Ashenden, 1978; Davies, 1980). A modified version of the Lancaster system (Horsman *et al.*, 1979) is used for experiments at the University of Liverpool (Horsman, Roberts and Bradshaw, 1978, 1979; Roberts *et al.*, 1979). Further developments at Lancaster have utilized outdoor hemispherical glasshouses, with no environmental control (Ashenden and Williams, 1980). An outdoor open-top chamber system has been developed at Liverpool, for both ambient-air versus clean-air and SO_2-fumigation experiments (Roberts *et al.*, 1979).

LOLIUM PERENNE

Only in the case of *L. perenne* have sufficient fumigations been performed at different concentrations to permit the investigation of a dose-response relationship. All four groups have worked with this species and it has become increasingly obvious that there are major discrepancies, both between different experiments of any one group and between the results of similar fumigations, using different systems. *Table 11.4* shows all the published changes in shoot dry weight of *L. perenne*, following chronic SO_2 fumigations in the United Kingdom, in ascending order of gas concentration. The greatest range of concentrations has been used at Imperial College and GRI, which include the only examples of fumigation with less than $200\,\mu g\,SO_2\,m^{-3}$. There are particularly marked discrepancies between the results of these two groups: Bell, Rutter and Relton (1979) showed that there were significant reductions in growth following exposure to $43\,\mu g\,SO_2$ m^{-3} for 173 days, $106\,\mu g\,m^{-3}$ for 194 days, $122\,\mu g\,m^{-3}$ for 44 days, and $220\,\mu g\,m^{-3}$ for 133 days, respectively, whereas the GRI group found no adverse effect of SO_2 on growth at concentrations below $400\,\mu g\,m^{-3}$. Even when the same fumigation apparatus is used, there are sometimes major differences between consecutive experiments. Thus Bell, Rutter and Relton (1979) found no significant effect of $66\,\mu g\,SO_2\,m^{-3}$ for 133 days, in an attempt to repeat their experiment in which a 173 d fumigation with $43\,\mu g\,m^{-3}$ caused a substantial growth reduction. Similarly, Cowling and Koziol (1978) were unable to detect any influence on shoot growth of $400\,\mu g\,SO_2\,m^{-3}$ after 72 days, whereas Lockyer, Cowling and Jones (1976)

Table 11.4 CHANGES IN SHOOT DRY WEIGHT OF *LOLIUM PERENNE* AFTER CHRONIC SO_2 FUMIGATION

SO_2 concentration ($\mu g\,m^{-3}$)	Duration (d)	Spaced plants (P) or swards (S)	Mean % change in total shoot dry weight in SO_2	Reference
43	173	P	−68	Bell, Rutter and Relton (1979) (I)
50	72	S	−2 (N.S.)	Cowling and Koziol (1978) (GRI)
50	77	S	−5 (N.S.)	Lockyer, Cowling and Jones (1976) (GRI)
50	87	S	−3 (N.S.)	Cowling and Lockyer (1976) (GRI)
55	85	S	−3 (N.S.)	Cowling and Lockyer (1978) (GRI)
62	42	P	+11 (N.S.)	Bell, Rutter and Relton (1979) (I)
66	144	P	+8 (N.S.)	Bell, Rutter and Relton (1979) (I)
100	77	S	−3 (N.S.)	Lockyer, Cowling and Jones (1976) (GRI)
106	194	P	−24	Bell, Rutter and Relton (1979) (I)
122	44	S	−21	Bell, Rutter and Relton (1979) (I)
122	44	S	−25	Bell, Rutter and Relton (1979) (I)
131	59	S	−16 (N.S.)	Cowling, Jones and Lockyer (1973) (GRI)
136	154	S	−5 (N.S.)	Bell, Rutter and Relton (1979) (I)
183	138	S	+1 (N.S.)	Bell, Rutter and Relton (1979) (I)
191	180	P	−50	Bell and Clough (1973) (I)
200	77	S	+6 (N.S.)	Lockyer, Cowling and Jones (1976) (GRI)
220	133	P	−21	Bell, Rutter and Relton (1979) (I)
312	28	P	−34	Ashenden and Mansfield (1977) (La)
312	28	P	−18	Ashenden and Mansfield (1977) (La)
343	62	P	−40	Bell and Clough (1973) (I)
347	61	P	−21	Ayazloo and Bell (1981) (I)
362	105	P	−19	Ayazloo and Bell (1981) (I)
367	131	P	−8 (N.S.)	Ayazloo and Bell (1981) (I)
380	64	P	−24	Ayazloo, Bell and Garsed (1980) (I)
400	72	S	−4 (N.S.)	Cowling and Koziol (1978) (GRI)
400	77	S	−24	Lockyer, Cowling and Jones (1976) (GRI)
423	108	P	−43	Bell, Rutter and Relton (1979) (I)
600	42	P	−16	Bell, R. (in Horsman *et al.*, 1979) (Li)
650	35	P	−16	Horsman *et al.* (1979) (Li)
650	56	P	−16	Horsman, Roberts and Bradshaw (1978) (Li)
650	56	P	−18	Horsman, Roberts and Bradshaw (1979) (Li)
700	56	P	−30	Horsman *et al.* (1979) (Li)

N.S.: not significant at $P = 0.05$
Fumigation systems: I (Imperial College); GRI (Grassland Research Institute); La (University of Lancaster); Li (University of

found a 24% reduction after exposure to the same concentration for 77 days. It is well known that the degree of SO_2 injury is affected by a range of environmental factors (Tamm and Aronsson, 1972) and so it is not unexpected that discrepancies may occur between the results of experiments at Imperial College and GRI, where there is no full control of climatic conditions. However, even where experiments have been repeated under controlled environments, differences in SO_2 effects are sometimes reported: for example Ashenden and Mansfield (1977) found reductions in shoot dry weight of 18% and 34%, respectively, in two fumigations with 312 $\mu g\,SO_2\,m^{-3}$ for 28 days, which were identical except for a 2°C temperature difference.

In view of these major discrepancies, it is not surprising that no very clear relationship has so far been found between SO_2 concentration and growth effects, either for ambient-air versus filtered-air experiments (Crittenden and Read, 1979) or fumigation experiments (Bell, Rutter and Relton, 1979). One obvious major difference between experiments is that in some cases the plants are grown as spaced individuals, whereas in others dense swards are used (*Table 11.4*). The significance of this will be discussed later, but it is reasonable to consider these two types of experiment separately when attempting to establish a dose-response relationship or minimum concentration for growth reduction.

Figure 11.1 shows all published percentage changes in shoot dry weight of spaced *L. perenne* plants in SO_2 compared with clean-air controls, in relation to the total doses of gas (concentration × time) administered throughout the course of the experiments. Nonsignificant data are included and among these are all the cases of apparent stimulation of growth by SO_2. There is no indication of any correlation between the magnitude of growth reduction and pollutant dose. However, it is possible that a

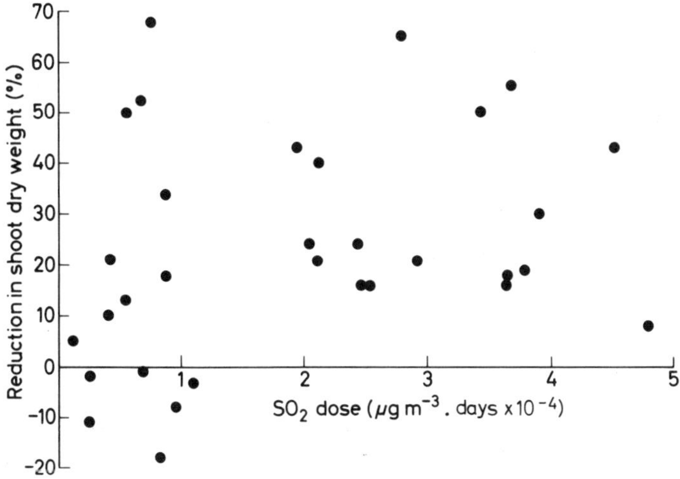

Figure 11.1 Relationship between total SO_2 dose and percentage reduction in shoot dry weight of spaced *Lolium perenne* plants compared with clean-air controls (Data of Bell and Clough, 1973; Ashenden and Mansfield, 1977; Bell, Rutter and Relton, 1979; Horsman, Roberts and Bradshaw, 1978, 1979; Horsman *et al.*, 1979; Ayazloo and Bell, 1981)

relationship between SO_2 concentration and suppression of yield may be obscured when presented in this manner: in experiments of different duration, a constant difference in relative growth rate between fumigated and control plants will result in an increasing difference in dry weights with time. However, the logarithm of the ratio of the dry weights from the two treatments will be linearly related to time and in *Figure 11.2* these values,

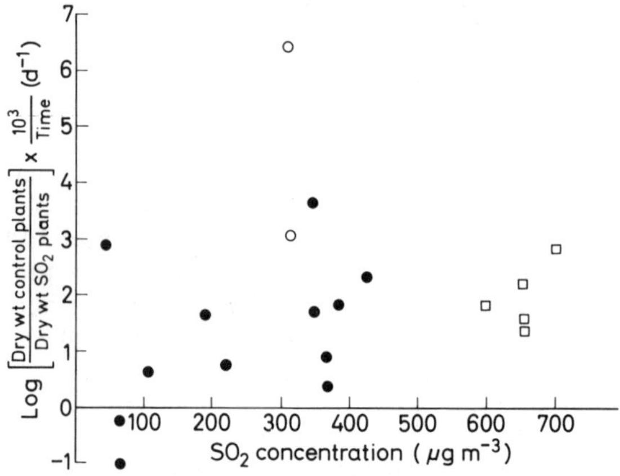

Figure 11.2　Ratios of dry weights of spaced *Lolium perenne* plants from control and fumigation treatments, normalized for duration of exposure, in relation to mean SO_2 concentration. ● Imperial College system (Bell and Clough, 1973; Bell, Rutter and Relton, 1979; Ayazloo and Bell, 1981). □ Liverpool system (Horsman, Roberts and Bradshaw, 1978, 1979; Horsman *et al.*, 1979). ○ Lancaster system (Ashenden and Mansfield, 1977)

for spaced plants in the experiments listed in *Table 11.4*, have been divided by the duration of the experiment and plotted against the SO_2 concentration in order to minimize the effects of time. This still fails to demonstrate a significant correlation and thus it remains impossible at present to establish a dose-response relationship for chronic SO_2 injury and there is an obvious need for further experiments with a range of SO_2 concentrations for each of a range of different durations, under controlled environmental conditons representative of the field.

GRASS SPECIES OTHER THAN *LOLIUM PERENNE*

Both the Imperial College and the Lancaster groups have examined the effects of chronic SO_2 concentrations on the productivity of several other grass species, as well as of *L. perenne*. The results of all such published work are shown in *Table 11.5*. In view of the results of the *L. perenne* fumigations, meaningful comparisons between experiments are not possible. However, in *Table 11.5* there are two examples of several species being fumigated simultaneously in the same experiment: Ashenden (1979) and Ashenden and Williams (1980) subjected *D. glomerata*, *L. multiflorum*, *Phleum pratense* and *Poa pratensis* simultaneously to 194 µg SO_2 m⁻³ for 140 days, while Ayazloo and Bell (1981) similarly fumigated together

Table 11.5 CHANGES IN SHOOT DRY WEIGHT OF GRASS SPECIES (EXCLUDING *LOLIUM PERENNE*) AFTER CHRONIC SULPHUR DIOXIDE FUMIGATIONS (ALL PLANTS SPACED)

Species	SO_2 concentration ($\mu g\,m^{-3}$)	Duration (d)	Mean % change in total shoot dry weight in SO_2	Reference
Dactylis glomerata	194	140	−44	Ashenden (1979) (La)
	312	28	−41	Ashenden (1978) (La)
	312	28	−38	Ashenden (1978) (La)
	367	131	−26	Ayazloo and Bell (1981) (I)
	461	180	−21	Ayazloo and Bell (1981) (I)
Festuca rubra	367	131	−18 (N.S.)	Ayazloo and Bell (1981) (I)
	482	105	−28	Ayazloo and Bell (1981) (I)
Holcus lanatus	250	133	−21	Ayazloo and Bell (1981) (I)
	367	131	−44	Ayazloo and Bell (1981) (I)
Lolium multiflorum	194	140	−14 (N.S.L?)	Ashenden and Williams (1980) (La)
Phleum bertolonii	319	72	−20	Ayazloo and Bell (1981) (I)
Phleum pratense	194	140	−36	Ashenden and Williams (1980) (La)
	343	35	−50[a]	Davies (1980) (La)
	343	35	+ 1 (N.S.)	Davies (1980) (La)
Poa pratensis	194	140	−39	Ashenden (1979) (La)

[a] At low irradiance and short day-length
N.S.: not significant at $P = 0.05$
Fumigation systems: I (Imperial College); La (University of Lancaster)

D. glomerata, Festuca rubra and *Holcus lanatus* (*Table 11.5*), together with
L. perenne (*Table 11.4*), using 367 µg m^{-3} for 131 days. The Table shows
that, although *L. perenne* has generally been classified as highly sensitive
to SO_2, under the conditions of Ayazloo and Bell (1981) there was no
effect of the pollutant on shoot dry weight, in contrast to significant
reductions in *D. glomerata* and *H. lanatus*. Ashenden (1979) and Ashenden and Williams (1980) found that SO_2 caused similar large reductions in
growth of *D. glomerata, Phleum pratense* and *Poa pratensis*. Thus there is
good evidence that several other important pasture-grass species may be
even more seriously affected than *L. perenne* by ambient SO_2 pollution.
On the other hand, *F. rubra* and *L. multiflorum* do not appear to be very
sensitive to chronic levels of SO_2.

The effect of sward formation on SO_2 injury

It has been mentioned already that a major difference between experiments lies in the use of either single-spaced plants or densely sown swards.
All the GRI experiments used swards, which in many cases were placed in
the chambers at a well-developed stage of growth, varying from 21 to 111
days from germination. Although there is evidence that the earlier
experiments at GRI may have used lower gas concentrations than those
published (Unsworth and Mansfield, 1980), it is also possible that some
characteristics of a sward may reduce the impact of SO_2 compared with the
effect on spaced plants.

Morphological differences between spaced plants and swards cause them
to have different values for the boundary-layer resistance (r_a) to gaseous
diffusion between the atmosphere and the leaf surface. Horsman *et al.*
(1979) found that, at a wind speed of 54 m min^{-1}, taken as typical of wind
speeds close to the ground in lowland Britain, r_a for spaced *L. perenne*
plants was 40 s m^{-1}, but that this increased to 115 s m^{-1} in a sward of the
same species, so reducing uptake of SO_2. In many experiments, spaced
plants are sown sufficiently close to each other to develop eventually a leaf
area index similar to that of a densely sown sward. Thus, as the plants grow
larger, the imapct of SO_2 may decline because of a reduction in pollutant
uptake. Support for this theory is provided by the substantial reductions in
dry weight caused by SO_2 in the early stages of most experiments where
harvesting has been carried out at intervals (*Figure 11.3*). In some cases
there is a fall in growth reduction with time, indicating a recovery in the
relative growth rate of the fumigated plants. Crittenden and Read (1979)
similarly found marked effects on *L. multiflorum* and *D. glomerata*
seedlings after only 3 weeks' growth in ambient Sheffield air, and
suggested that this was caused by an increased uptake of SO_2 per unit leaf
area in young spaced plants. Furthermore, Bell, Rutter and Relton (1979)
found the greatest reduction in relative growth rate at an early stage of
long-term fumigations, using 43 and 423 µg SO_2 m^{-3}. They suggested that
this might be caused by a greater susceptibility of young plants. At present
it is not certain whether this susceptibility may also reflect an inherently
lower physiological tolerance to SO_2, in addition to a low value for r_a.

Most agricultural crops are grown as spaced individual plants, but this

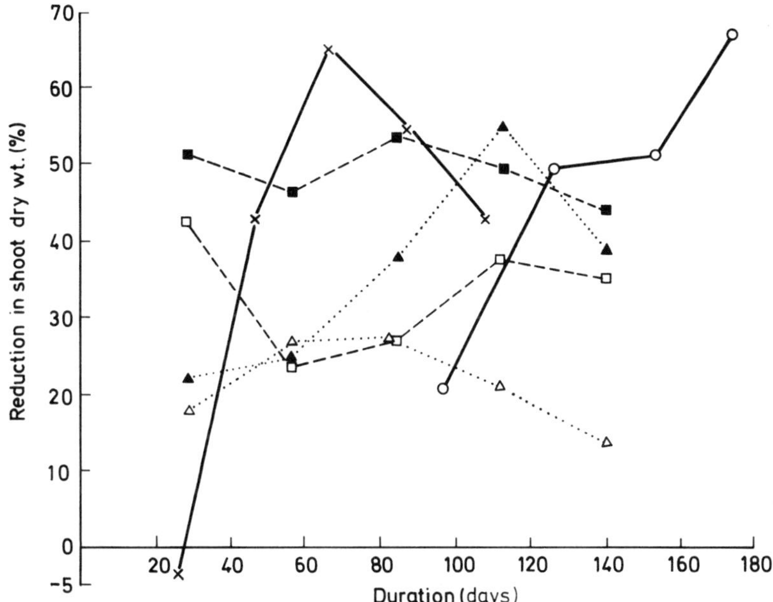

Figure 11.3   Reduction in shoot dry weight of spaced grass plants in SO_2 in relation to duration of fumigation. ×——× 423 µg m^{-3}, O——O 43 µg m^{-3}, *Lolium perenne* (Bell, Rutter and Relton, 1979); ▲······▲ 194 µg m^{-3} *Poa pratensis*, ■----■ 194 µg m^{-3} *Dactylis glomerata* (Ashenden, 1979); △······△ 194 µg m^{-3} *Lolium multiflorum*, □----□ 194 µg m^{-3} *Phleum pratense* (Ashenden and Williams, 1980)

applies to grassland only at a very early stage of development of a newly sown ley. As a closed sward develops, severe intraspecific and interspecific competition takes place, which has important implications for the effects of SO_2 on grass growth in the field. Intraspecific variability in sensitivity to SO_2 in both wild plants and bred cultivars, and the resulting selection for tolerance in polluted areas, has now been demonstrated for several grass species (Bell and Mudd, 1976; Horsman, Roberts and Bradshaw, 1978, 1979; Ayazloo and Bell, 1981). It is possible that, even in a relatively short-term fumigation experiment, more tolerant genotypes may compensate for the reduced growth of sensitive individuals, so that the effect of SO_2 on the sward as a whole is minimized.

Little is known about the rate of selection for SO_2 tolerance within a sward and whether this can be rapid enough to contribute to the generally smaller effect of SO_2 on swards compared with spaced plants. Bell, Ayazloo and Wilson (1982) have recently demonstrated the development of tolerance to acute SO_2 injury in *L. perenne* and *Phleum pratense* populations within 5 years of sowing at Philips Park in a polluted industrial area of Manchester, where mean winter SO_2 concentrations are currently between 150 and 200 µg m^{-3}. Samples of these species were collected at one- to two-year intervals from monoculture plots sown in 1975 by the Sports Turf Research Institute of Bingley, Yorkshire. On each occasion they have been screened, together with similar-aged plants grown from the original seed, for sensitivity to acute SO_2 injury by fumigating with 4000–

$10\,000\,\mu g\,SO_2\,m^{-3}$ for 6 h in a controlled-environment cabinet; injury was assessed on the basis of percentage of total leaf-length destroyed. Collections in 1976 and 1978 showed no difference between populations in response to the fumigations, but by 1979 *L. perenne* plants from the plots were significantly more tolerant than plants grown from the original seed (*Table 11.6*). This differential tolerance was maintained in the 1980

Table 11.6 DEVELOPMENT OF TOLERANCE TO ACUTE INJURY AT PHILIPS PARK, MANCHESTER. VALUES GIVEN ARE THE MEAN PERCENTAGE OF TOTAL LEAF-LENGTH INJURED, AFTER FUMIGATION WITH $4000–10\,000\,\mu g\,SO_2\,m^{-3}$ FOR 6 h, OF *LOLIUM PERENNE* AND *PHLEUM PRATENSE* FROM FIELD PLOTS ESTABLISHED IN 1975, IN COMPARISON WITH PLANTS GROWN FROM THE ORIGINAL SEED. After Bell, Ayazloo and Wilson (1982)

Date of collection	*Lolium perenne cv. S23*			*Phleum pratense cv. S48*		
	Original seed	*Plots*	*P<*	*Original seed*	*Plots*	*P<*
1976	20.5	18.5	N.S.	59.8	61.4	N.S.
1978	23.5	20.1	N.S.	67.0	62.5	N.S.
1979	9.4	4.3	0.01	5.8	9.0	N.S.
1980	42.7	33.5	0.05	67.3	55.0	0.05

N.S: not significant at $P = 0.05$

collection, when a similar effect was detected for the first time in *P. pratense*. Horsman *et al.* (1979) and Ayazloo and Bell (1981) have demonstrated a complete absence of any correlation between the degree of sensitivity of individual grass plants to acute and chronic SO_2 injury. Thus the results of Bell, Ayazloo and Wilson (1982) do not necessarily indicate the evolution of tolerance to chronic injury at Philips Park. However, in view of the moderate SO_2 concentrations prevailing there over the last 5 years, it could be anticipated that tolerance to chronic injury might have evolved even more rapidly than for acute injury.

Bell, Rutter and Relton (1979) described an experiment in which densely sown swards of *L. perenne* were fumigated with $122\,\mu g\,SO_2\,m^{-3}$ for 44 days from emergence. The swards were then harvested by clipping and showed a 25% reduction in shoot dry weight caused by SO_2. However, after allowing the plants to regrow under the same experimental conditions it was found that the adverse effect of SO_2 had disappeared at a second harvest 35 days later. It was suggested that clipping might have reduced the impact of SO_2 by removing stressed leaves. In view of the accumulating evidence of the capacity of grass populations to select for SO_2 tolerance, an alternative hypothesis can be proposed: that, between the harvests, tolerant individuals within the sward increased their growth at the expense of more sensitive genotypes, resulting in an overall compensation in productivity.

There remains a fundamental question as to whether spaced plants or swards should be fumigated in order to predict the effects of SO_2 on grasses in the field. There is evidence that, if SO_2 reduces the growth of seedlings in a newly sown ley, then both reduction in gas uptake and selection for tolerance may result in an amelioration in productivity after the establishment of a closed sward. In this case the frequency of reseeding may be an important factor, with SO_2 having the greatest impact on short-term temporary leys. Many grasslands in Britain are semipermanent, either of a considerable but unknown age for indigenous pastures or else derived from bred cultivars in more recent times. In contrast, most fumigation experiments are of less than one year's duration. The longevity of individual perennial grass plants is uncertain, but is believed to be considerable in many species; for example *Phleum pratense* plants survive for at least 20 years (Spedding and Diekmahns, 1972). Thus it may be possible for SO_2 to have a cumulative adverse impact over a long period, even on genotypes which appear relatively tolerant on the basis of short-term chronic fumigations. A further complicating factor is that SO_2 concentrations at most sites vary considerably on a seasonal basis, with lower concentrations in the summer. Roberts *et al.* (1979) have shown, both in fumigation experiments and in field-exposure trials, that the effects of high levels of SO_2 in the winter may be manifested in slower growth in summer, when pollutant levels have fallen. The importance of this effect for long-term overall productivity of grasslands is at present uncertain, but it is an obvious field for investigation.

Modification of chronic injury by environmental conditions

In this chapter no detailed analysis will be made of the contribution that different environmental conditions have made to the discrepancies between the results of individual fumigations of grass species. However, recent publications by Mansfield and his colleagues at Lancaster have identified two environmental factors which appear to have an important role in this respect.

The influence of r_a on the rate of uptake of SO_2 by single grass plants compared with swards has already been discussed. Ashenden and Mansfield (1977) demonstrated the importance of wind speed in modifying chronic SO_2 injury via effects on r_a. They fumigated *L. perenne* plants with $312\,\mu g\,SO_2\,m^{-3}$ for 28 days, at wind speeds of $25\,m\,min^{-1}$ or $10\,m\,min^{-1}$, which gave measured values of r_a to individual leaves of about $90\,s\,m^{-1}$ and

Table 11.7 MODIFICATION BY WIND SPEED OR LIGHT REGIME OF SO_2-INDUCED REDUCTIONS IN MEAN SHOOT DRY WEIGHT (g) OF *LOLIUM PERENNE* AND *PHLEUM PRATENSE*

Species/Reference	Environmental conditions	Dry wt. in clean air	Dry wt. in $SO_2{}^a$	P <
Lolium perenne (Ashenden and Mansfield, 1977)	Wind speed $25\,m\,min^{-1}$ ($r_a\,89\,s\,m^{-1}$)	0.799	0.529	0.001
	Wind speed $10\,m\,min^{-1}$ ($r_a\,797\,s\,m^{-1}$)	1.251	1.405	N.S.
Phleum pratense (Davies, 1980)	Photosynthetically active radiation $125\,\mu E\,m^{-2}\,s^{-1}$, $12\,h\,d^{-1}$	0.026	0.013	0.001
	Photosynthetically active radiation $480\,\mu E\,m^{-2}\,s^{-1}$, $16\,h\,d^{-1}$	0.504	0.507	N.S.

N.S: not significant at $P = 0.05$.
[a]$312\,\mu g\,m^{-3}$ for 28 d (*L. perenne*); $343\,\mu g\,m^{-3}$ for 35 d (*P. pratense*)

$800 \, \text{s m}^{-1}$, respectively. At the higher wind speed, which is the more representative of normal ambient conditions, SO_2 caused a substantial reduction in shoot dry weight, whereas at the lower wind speed no pollutant effect was detected (*Table 11.7*). This emphasizes the dangers in using fumigation experiments to predict effects in the field, without taking into account the influence of environmental factors. On the basis of their results, Ashenden and Mansfield suggested that earlier fumigation experiments which showed an apparent threshold of $300–400 \, \mu\text{g SO}_2 \, \text{m}^{-3}$ for chronic injury were unrealistic, because the use of a low air-flow rate through the chambers had resulted in an artificially high value for r_a.

The lowest SO_2 concentration so far shown to reduce grass growth ($43 \, \mu\text{g m}^{-3}$) was administered in an experiment carried out under winter conditions (Bell, Rutter and Relton, 1979). Cowling and Lockyer (1978) found that $55 \, \mu\text{g SO}_2 \, \text{m}^{-3}$ increased senescence in *L. perenne* plants which were growing slowly due to limiting nitrogen nutrition. In many cases it is difficult to compare growth rates from published data, but plants grown in the GRI system appear to be very productive compared with the Imperial College experiments (Bell, Rutter and Relton, 1979). Thus there is circumstantial evidence that slow-growing plants have an increased sensitivity to low levels of SO_2. Clear support for this has recently been provided by Davies (1980), who fumigated *Phleum pratense* plants for 5 weeks with $343 \, \mu\text{g SO}_2 \, \text{m}^{-3}$, under two light regimes, representative of winter and summer conditions respectively. No effect of SO_2 was observed at $480 \, \mu\text{E m}^{-2} \text{s}^{-1}$ photosynthetically active radiation (PAR) for $16 \, \text{h d}^{-1}$ (summer conditions), but the pollutant produced a 50% reduction in shoot dry weight in plants growing slowly under $125 \, \mu\text{E m}^{-2} \text{s}^{-1}$ PAR for $12 \, \text{h d}^{-1}$ (winter conditions) (*Table 11.7*). Davies suggested that, because laboratory fumigations are often performed under favourable conditions, losses in the field may be higher than anticipated, due to increased sensitivity when growth rates are reduced by adverse environmental conditions. It is probable that such differences in growing conditions may explain many of the discrepancies found between experiments with different fumigation systems.

Growth reductions in ambient air in relation to fumigation experiments

Growth reductions of *L. perenne* recorded in ambient-air versus filtered-air experiments (Bleasdale, 1973; Crittenden and Read, 1978) are generally larger than those from fumigation experiments, using similar mean concentrations of SO_2. Thus ambient SO_2 appears to have a greater effect on *L. perenne* than the same concentration in fumigation experiments. There are two probable explanations for this phenomenon.

In all fumigations reviewed here, the SO_2 concentration was approximately constant whereas, in the ambient air, fluctuations occur about the mean, with daily mean concentrations of up to $1416 \, \mu\text{g m}^{-3}$ and $259 \, \mu\text{g m}^{-3}$ recorded by Bleasdale (1952, 1973) and Crittenden and Read (1978), respectively. Even the daily means may conceal considerably higher short-term peaks of SO_2, which can be detected only by a continuous monitor. Very little is known about the role of these transient high concentrations of SO_2 in chronic injury to plants. However, the preliminary results of a recently completed experiment at Imperial College suggest that in a two-year fumigation of *Pinus sylvestris* L., exposure to peak SO_2 concentrations within an overall mean of $100\,\mu g\,m^{-3}$ can produce a greater reduction in growth than continuous exposure to the same constant concentration (Garsed and Rutter, personal communication; *see also* Garsed (this volume, page 455). Thus it is probable that fluctuations in SO_2 concentrations may contribute to the discrepancies between ambient-air versus filtered-air and fumigation experiments.

Another reason for the large growth reductions observed at low mean SO_2 concentrations may be the presence of other phytotoxic pollutants in the field. In particular, nitrogen oxides have generally been neglected with regard to effects on vegetation. However, recent monitoring programmes in urban areas have revealed that nitrogen oxide concentrations are very similar to SO_2 concentrations, on a volume/volume basis (Apling, Potter and Williams, 1979). At Lancaster, *Dactylis glomerata, Lolium multiflorum, Phleum pratense,* and *Poa pratensis* have been subjected to prolonged fumigations with moderate levels of SO_2 and NO_2, both singly and in combination (Ashenden and Mansfield, 1978; Ashenden, 1979; Ashenden and Williams, 1980). *Table 11.8* summarizes the reductions in total dry

Table 11.8 PERCENTAGE REDUCTIONS IN TOTAL DRY WEIGHT OF FOUR GRASS SPECIES GROWN IN SO_2, NO_2, OR $NO_2 + SO_2$ FOR 20 WEEKS COMPARED WITH CLEAN-AIR CONTROLS (After Ashenden and Mansfield, 1978)

Species	$194\,\mu g\,SO_2\,m^{-3}$	$139\,\mu g\,NO_2\,m^{-3}$	$194\,\mu g\,SO_2\,m^{-3}$ $+ 139\,\mu g\,NO_2\,m^{-3}$	*Effect*
Dactylis glomerata	40	22	78	Synergistic
Lolium multiflorum	5 (N.S.)	10 (N.S.)	52	Synergistic
Phleum pratense	51	1 (N.S.)	86	Synergistic
Poa pratensis	46	38	84	Additive

N.S: not significant

weight recorded in these experiments. In all species, the pollutant combination produced more injury than the individual gases and this effect was synergistic, except for *Poa pratensis*. Although there are no published data on the impact of a mixture of SO_2 and NO_2 on *L. perenne*, in view of the above results it is likely that at many polluted sites NO_2 may be partly responsible for effects previously ascribed solely to SO_2. Furthermore, there are other phytotoxic pollutants which may also contribute to injury in the field: recently, ozone injury on crops has been identified unequivocally for the first time in the United Kingdom (Ashmore *et al.*, 1980)

Mechanisms of chronic SO_2 effects on growth

So far, the effects of SO_2 on total shoot weight only have been considered, because no other characteristic has been measured in all experiments. However, in many investigations more detailed growth analyses have been performed, which give some indication of the mechanisms by which SO_2

reduces the growth of grasses. These analyses are mainly confined to spaced plants, few measurements (other than shoot dry weight) having been made on swards.

The only complete growth analyses, with estimates of relative growth rate (RGR), net assimilation rate (NAR), and leaf area ratio (LAR), on grass plants fumigated with SO_2 are reported for *L. perenne* by Bell, Rutter and Relton (1979). They found significant reductions compared with controls in the RGR in the early and final stages of a 173 d fumigation with $43\,\mu g\,SO_2\,m^{-3}$ (*Figure 11.4*). The effects of SO_2 on RGR were closely

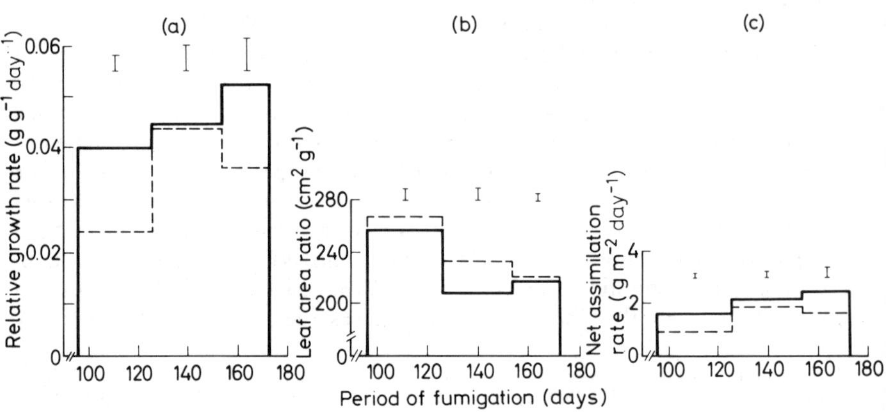

Figure 11.4 Growth analysis on *Lolium perenne* cv. S23 after growth for 173 days in $43\,\mu g\,SO_2\,m^{-3}$ or clean air. —— Control; ––– $43\,\mu g\,SO_2\,m^{-3}$. (a) Relative growth rate; (b) leaf area ratio; (c) net assimilation rate. Bars indicate LSD at $P = 0.05$. (Courtesy of *New Phytologist*: Bell, Rutter and Relton, 1979)

parallelled by proportionally similar fluctuations in NAR, with very little influence of the pollutant on LAR. Thus it appears that SO_2 reduces growth by affecting some physiological process connected with photosynthesis or respiration, rather than a redistribution of assimilate from leaves to non-photosynthetic tissue. In another fumigation, using $423\,\mu g\,SO_2\,m^{-3}$ for 108 d, it was confirmed that the major effect of SO_2 on RGR was via a reduced NAR, but in the later stages of the experiment this was compensated to some extent by an increase in LAR. The specific leaf area was higher in the SO_2-treated plants throughout this experiment, which is in agreement with the results of Ayazloo, Bell and Garsed (1980) after a fumigation of *L. perenne* with $380\,\mu g\,SO_2\,m^{-3}$ for 64 d. Thus, although the efficiency of the foliage in producing assimilate may be reduced by SO_2, there is evidence that in some cases this may be partly offset by changes in leaf area.

Table 11.9 summarizes the effects of SO_2 on numbers of tillers, foliar senescence (dead-leaf:living-leaf dry weights) and root, shoot and root:shoot dry weights for all published fumigation and ambient versus clean-air experiments with spaced grass plants. There are a number of differences between experiments in the nature of the effect of SO_2, even within a single species. In many cases the reduction in shoot dry weight is accompanied by a proportionately similar decrease in the number of tillers (e.g. Bell and Clough, 1973; Bell, Rutter and Relton, 1979; Ashenden and Williams, 1980). On the other hand, substantial reductions in growth have been reported without changes in the rate of tillering (e.g. Ashenden and Mansfield, 1977; Horsman *et al.*, 1979; Davies, 1980). In general, ambient polluted air has reduced tiller number (Crittenden and Read, 1978; 1979), but Bleasdale (1973) reported conflicting results in this respect.

In most cases where dead leaves were weighed separately, reductions in growth were shown to be accompanied by an increase in senescence (*Table 11.9*). However, in a series of fumigation experiments with five grass species, Ayazloo and Bell (1981) demonstrated that, although senescence was consistently accelerated, the weight of dead leaves was small in relation to the living foliage, so that reduced leaf growth appeared to be primarily responsible for the observed depressions in yield. Furthermore, in two replicate fumigations with $215–226\,\mu g\,SO_2\,m^{-3}$ reported by Bell, Rutter and Relton (1979), a reduction in shoot growth was accompanied by increased senescence in one replicate only. Horsman *et al.* (1979) and Crittenden and Read (1978) also showed reductions in growth of *L. perenne*, without changes in senescence, in fumigation and ambient-air versus clean-air experiments, respectively. It must be concluded that, while increased senescence may be one mechanism contributing to growth reduction by SO_2, it is not a prerequisite for this to occur and does not appear to be the principal factor involved.

Relatively little attention has been paid to the effects of SO_2 on root growth. *Table 11.9* indicates that, in many cases, SO_2 exposure reduced root dry weight to a significantly greater extent than shoot dry weight. Out of nineteen estimations of root:shoot ratios, there are only two cases of significant increases in the presence of SO_2, in contrast to eight occasions when reductions occurred. Among the remaining nine cases with no significant effect are all the ambient-air versus clean-air experiments of Crittenden and Read (1979), where SO_2 produced an almost identical reduction in shoot and root dry weights. The fall in root:shoot ratios caused by SO_2 is probably associated with an increase in LAR: in the experiment of Bell, Rutter and Relton (1979), where SO_2 exposure increased LAR and specific leaf area, there was a large reduction in the root:shoot ratio. Plants show a general decline in root:shoot ratio with increasing size (Gales, 1977), which may explain some of the cases in which SO_2 has had no influence on this character: it is possible that the higher root:shoot ratio associated with the smaller plants resulting from a fumigation will counteract the impact of SO_2 in diverting assimilate from the roots to the foliage. The greater effect of SO_2 on root than shoot growth may eventually be manifested under field conditions in a further reduction in shoot yield. When plants are subjected to drought, a smaller root system will increase the degree of water-stress experienced by the plant as a whole. A lower root weight will also reduce the potential for regrowth, following defoliation by grazing or mowing and damage by trampling.

The wide range of types of response of grass plants to SO_2 merits further investigation. It raises the possibility that chronic injury may be caused by a number of different mechanisms, the relative importance of which may be altered by factors such as the age of the plant and prevailing environmental conditions.

Table 11.9 EFFECTS OF SO_2 ON INDIVIDUAL ORGANS OF GRASS SPECIES (SPACED PLANTS ONLY)

Species	SO_2 concentration ($\mu g\,m^{-3}$)	Duration (d)	% change in tiller no. in SO_2	% change in dead/living leaf dry wt in SO_2	% change in root dry wt in SO_2	% change in shoot dry wt in SO_2	% change in root/shoot dry wt in SO_2	Reference
Lolium *perenne* (fumigations)	343	62	−42	+238		−40		Bell and Clough (1973)
	191	180	−41	+256		−50		
	43	173	−68	+948	−56	−68	+23	Bell, Rutter and Relton (1979)
	423	108		+146	−60	−43	−32	
	215	133		+136		−23		
	226	133		N.S.		−20		
	380	64	−11	+ 78	−52	−24	−32	Ayazloo, Bell and Garsed (1980)
	362	105		+ 94	−33	−19	−20	Ayazloo and Bell (1981)
	367	131		+146	N.S.	N.S.	N.S.	
	347	61		+217	−10	−21	+11	
	312	28	N.S.		−48	−34	−35	Ashenden and Mansfield (1977)
	650	35	N.S.	N.S.		−16		Horsman *et al.* (1979)
	700	56	N.S.			−30		
	650	56		+ 95		−18		Horsman, Roberts and Bradshaw (1979)
Lolium *perenne* (ambient/ clean air)	164[a]	187	−30			−33		Bleasdale (1973)
	175	104	N.S.			−17		
	70	56	−20		−36	−36	N.S.	Crittenden and Read (1978)
	59	131	−20	N.S.	−20	−20	N.S.	
	69	86	−25		−24	−25	N.S.	
	63	116	−15	N.S.	−26	−26	N.S.	
Lolium *multiflorum* (fumigation)	194	140	−23		N.S.	−14 (N.S?)	N.S.	Ashenden and Williams (1980)

Lolium multiflorum (ambient/ clean air)	67	56	−24			−36		Crittenden and Read (1979)
Festuca rubra (fumigations)	367	131		+288	−40	N.S.	N.S.	
	482	105		+180	−39	−28	−14	
Holcus lanatus (fumigations)	250	133		+72	−20	−21	N.S.	Ayazloo and Bell (1981)
	367	131		+73	−58	−44	N.S.	
Phleum bertolonii (fumigations)	319	72		+142	−38	−20	−24	
Phleum (fumigations)	343	35	N.S.	+804	−58	−50		Davies (1980)
pratense (fumigations)	194	140	−33		−58	−36		Ashenden and Williams (1980)
Poa pratensis (fumigation)	194	140	−27		−54	−39		Ashenden (1979)
Dactylis glomerata (fumigations)	367	131		+424	−50	−26	N.S.	Ayazloo and Bell (1981)
	461	180		+171	−45	−21	−30	
	312	28	−33		−52	−41	−27	Ashenden (1978)
	194	140	−10		−37	−44		Ashenden (1979)
Dactylis glomerata (ambient/ clean air)	45	72	−28			−42		Crittenden and Read (1979)

[a]Mean concentration over last 79 d only.

N.S: not significant at $P = 0.05$

The absence of data indicates that measurements were not made or that significant effects cannot be determined from the literature

Conclusions

Although considerable effort has been made in the last decade to determine the impact of prolonged, relatively low concentrations of SO_2 on the growth of agricultural grasses, many questions remain. It is still not possible to establish with confidence a threshold concentration for chronic injury, although growth reductions and effects on senescence have been recorded after fumigations with concentrations below $100\,\mu g\,m^{-3}$. There are many discrepancies between the results of different experiments, but some progress has been made towards explaining these in terms of environmental characteristics of the fumigation systems and the culture conditions of the plants. Ultimately the economic impact of SO_2 in the field will be determined only by long-term fumigations of agricultural grasslands, with plants rooted in the ground rather than confined to pots, and using realistically fluctuating concentrations of mixtures of the principal phytotoxic pollutants. It is to be hoped that recently developed open-air fumigation systems can be used for this purpose. However, it must be emphasized that every experiment of this type is unique and will still need to be interpreted on the basis of laboratory fumigations carried out under a range of strictly controlled environmental conditions.

References

ANONYMOUS (1974). *Clean Air Today*. HMSO, London

ANONYMOUS (1977). *Statistics on Grass Usage 1975–1976*. Ministry of Agriculture, Fisheries and Food. HMSO, London

APLING, A.J., POTTER, C.J. and WILLIAMS, M.L. (1979). *Air Pollution from Oxides of Nitrogen, Carbon Monoxide and Hydrocarbons*. Warren Spring Laboratory, Stevenage

ASHENDEN, T.W. (1978). *Environmental Pollution*, **15**, 161–166

ASHENDEN, T.W. (1979). *Environmental Pollution*, **18**, 249–258

ASHENDEN, T.W. and MANSFIELD, T.A. (1977). *Journal of Experimental Botany*, **28**, 729–735

ASHENDEN, T.W. and MANSFIELD, T.A. (1978). *Nature*, **273**, 142–143

ASHENDEN, T.W. and WILLIAMS, I.A.D. (1980). *Environmental Pollution (Series A)*, **21**, 131–139

ASHMORE, M.R., BELL, J.N.B., DALPRA, C. and RUNECKLES, V.C. (1980). *Environmental Pollution (Series A)*, **21**, 209–215

AYAZLOO, M. and BELL, J.N.B. (1981). *New Phytologist*, **88**, 203–222

AYAZLOO, M., BELL, J.N.B. and GARSED, S.G. (1980). *Environmental Pollution (Series A)*, **22**, 295–307

BELL, J.N.B. and CLOUGH, W.S. (1973). *Nature*, **241**, 47–49

BELL, J.N.B. and MUDD, C.H. (1976). In *Effects of Air Pollutants on Plants*, pp. 87–103 (Mansfield, T.A., Ed.), Cambridge University Press, Cambridge

BELL, J.N.B., AYAZLOO, M. and WILSON, G.B. (1982). In *Proceedings of the Second European Ecological Symposium, Berlin, Sept. 1980*, pp. 171–180 (Bornkan, R., Lee, J.A. and Seaward, M.R.D., Eds). Blackwell, Oxford

BELL, J.N.B., RUTTER, A.J. and RELTON, J. (1979). *New Phytologist,* **83**, 627–643

BLEASDALE, J.K.A. (1952). *Atmospheric Pollution and Plant Growth.* PhD thesis, University of Manchester

BLEASDALE, J.K.A. (1973). *Environmental Pollution,* **5**, 273–285

BRENNAN, E. and HALISKY, P.M. (1970). *Phytopathology,* **60**, 1544–1546

COWLING, D.W. and KOZIOL, M.J. (1978). *Journal of Experimental Botany,* **29**, 1029–1036

COWLING, D.W. and LOCKYER, D.R. (1976). *Journal of Experimental Botany,* **27**, 411–417

COWLING, D.W. and LOCKYER, D.R. (1978). *Journal of Experimental Botany,* **29**, 257–265

COWLING, D.W., JONES, L.H.P. and LOCKYER, D.R. (1973). *Nature,* **243**, 479–480

CRITTENDEN, P.D. and READ, D.J. (1978). *New Phytologist,* **80,** 49–62

CRITTENDEN, P.D. and READ, D.J. (1979). *New Phytologist,* **83,** 645–651

DAVIES, T. (1980). *Nature,* **284**, 483–485

DODD, J.L., LAUENROTH, W.K., THOR, G.L. and COUGHENOUR, M.B. (1979). In *The Bioenvironmental Impact of a Coal-Fired Power Plant, 4th Interim Report.* pp. 384–493. (Preston, E.M. and Gullet, T.L., Eds), US Environmental Protection Agency, Corvallis.

GALES, K. (1977). *A study of the Effect of Drought on Root Growth and Root-Shoot Relationships in* Lolium perenne *L.* PhD thesis, University of London

GUDERIAN, R. and STRATMANN, H. (1968). *Forschungberichte des Landes Nordrhein-Westfalen,* Report No. 1920, Westdeutscher Verlag Köln and Oplades

HEITSCHMIDT, R.K., LAUENROTH, W.K. and DODD, J.L. (1978). *Journal of Applied Ecology,* **15**, 859–868

HILL, A.C., HILL, S., LAMB, C. and BARRETT, T.W. (1974). *Journal of the Air Pollution Control Association,* **24**, 153–157

HORSMAN, D.C., ROBERTS, T.M. and BRADSHAW, A.D. (1978). (1978). *Nature,* **276**, 493–494

HORSMAN, D.C., ROBERTS, T.M. and BRADSHAW, A.D. (1979). *Journal of Experimental Botany,* **30**, 495–501

HORSMAN, D.C., ROBERTS, T.M., LAMBERT, M. and BRADSHAW, A.D. (1979). *Journal of Experimental Botany,* **30**, 485–493

KATZ, M. (1949). *Industrial and Engineering Chemistry. Industrial Edition,* **41**, 2450–2465

LOCKYER, D.R., COWLING, D.W. and JONES, L.H.P. (1976). *Journal of Experimental Botany,* **27**, 397–409

ROBERTS, T.M., BELL, R., HORSMAN, D.C. and BRADSHAW, A.D. (1979). In *Sulphur in the Environment.* Proceedings of Society for Chemical Industry Symposium, London, 1979

SEMPLE, A.T. (1970). *Grassland Improvement.* Leonard Hill, London

SPEDDING, C.R.W. (1971). *Grassland Ecology.* The Clarendon Press, Oxford

SPEDDING, C.R.W. and DIEKMAHNS, E.C. (1972). *Grasses and Legumes in British Agriculture.* Commonwealth Agricultural Bureaux, Farnham Royal

TAMM, C.O. and ARONSSON, A. (1972). *Plant Growth as Affected by Sulphur*

Compounds in Polluted Atmosphere. A Literature Survey. Department of Forest Ecology and Forest Soils, College of Forestry, Stockholm

THOMAS, M.D. and HILL, G.R. (1937). *Plant Physiology*, **12**, 309–383

UNSWORTH, M.H. and MANSFIELD, T.A. (1980). *Environmental Pollution (Series A)*, **23**, 115–120

WOOD, F.A. (1968). *Journal of Occupational Medicine*, **10**, 524–534

ZIMMERMAN, P.W. and HITCHCOCK, A.E. (1956). *Contributions. Boyce Thompson Institute for Plant Research*, **18**, 263–279

12

EFFECTS OF SULPHUR DIOXIDE ON THE GROWTH AND YIELD OF AGRICULTURAL AND HORTICULTURAL CROPS

STEFAN GODZIK
Polish Academy of Sciences, Institute of Environmental Engineering, Zabrze, Poland

SAGAR V. KRUPA
Department of Plant Pathology, University of Minnesota, St. Paul

Introduction

Currently there are two major lines of thought concerning the effects of sulphur dioxide (SO_2) on growth and yield of plants. According to some scientists, reduction in growth and yield (damage) occurs only if visible injury is observed and if the injured area exceeds 5% of the total leaf area. Other scientists consider that significant growth and yield reductions can occur even in the absence of visible injury. This latter belief has serious implications in economics, because the magnitude of growth and yield reduction may be far greater than would be estimated according to the first concept.

It has been known for many years that, in addition to soil sulphates, sulphur dioxide can serve as a sulphur source for plants. However, until recently no effort has been made to demonstrate the importance of sulphur dioxide as a sulphur source (fertilizer) for vegetation in geographic areas with sulphur-deficient soils (Noggle, 1980).

In this chapter discussion is restricted to those publications where growth and yield effects were reported in units of height and/or weight or as differences in their percentages in relation to the control. Further, no attempt is made to discuss dose–injury relationships even when damage (in economic terms) occurs, as with necrosis on either vegetable crops such as spinach, lettuce or on ornamental plants, but where growth and biomass are not affected.

Plants may respond to SO_2 by either increased or decreased growth and yield. Either effect can occur with or without visible injury. Distinct differences in the response, both in degree and kind (beneficial or detrimental), have been observed depending on the harvest characteristics or parts of plants used for evaluation. Such disparity does not allow easy generalization of the available data. Results of many experiments are far from being in agreement with each other. For this reason, available information is discussed as follows:

1. Field experiments with ambient SO_2 and no environmental control;
2. Field experiments with ambient SO_2, and standardized soils;

3. Field fumigation experiments eg. chambers, open-air fumigation systems etc.
4. Experiments in controlled environment (glasshouse or growth chambers).

Effects of ambient SO_2 in the field

Surprisingly, very little information is available from field experiments conducted with normal soils, using crop management procedures identical to those used by farmers. Maly (1974) conducted experiments on plots of $100\,m^2$ in the vicinity of different pollutant sources (SO_2 and particulate matter). Both SO_2 concentration and dustfall input varied with locality and with the year of the experiment. During the experiment, at the site of pilot pollutant measurements, SO_2 concentrations varied from $3350\,\mu g\,m^{-3}$ to $3640\,\mu g\,m^{-3}$ (weekly averages). The largest yield decrease was observed with flax: seed, 28.3%; fibre, 23.8% (relative to the control). With other crops the following decreases were observed: potatoes, 16.2%; maize (forage), 16.7%; oats, grain, 12.2%; oats, straw, 8.1%; clover, 15.5%. In other years and at other locations, the following yield losses were also found: cereals (wheat, barley, rye and oats), 20%; beets, 7–23%. For broad bean, values varied from 5% to 20.2% and in one year no yield was obtained at all, apparently because of the occurrence of high pollutant concentrations (especially of SO_2) during the flowering period. All values were presented as differences between the control versus the polluted area and were expressed both in units of weight (quintal ha^{-1}) and as a percentage. Unfortunately no information was given concerning injury ratings. From the data presented, no correlation could be found between crop losses and pollutant concentrations. The mean daily SO_2 concentrations given for the entire study period fluctuated from 80 to $450\,\mu g\,m^{-3}$ with a maximum value of $150–960\,\mu g\,m^{-3}$ at one location, and $330–1300\,\mu g\,m^{-3}$ (maximum of $500–1900\,\mu g\,m^{-3}$) at the second location.

Other studies on this subject were conducted by Jones, Noggle and Weatherford (1978) and Haase, Morgan and Salem (1980). Jones, Noggle and Weatherford (1978) were unable to detect yield loss in soybean fields injured visibly by SO_2. They concluded that the plants had sufficient time to recover and compensate from the initial adverse effects of SO_2. Similar results were reported by Haase, Morgan and Salem (1980), based on a total of 43 years of examination of vegetation in the vicinity of a copper smelter in Arizona (USA). They concluded that:

1. No yield loss occurred in the absence of visible injury;
2. The percentage yield loss was less than the percentage leaf area injured;
3. There was no evidence for the existence of a critical growth stage;
4. Leaf injury induced by SO_2 did not always result in a yield loss;
5. Experimental SO_2 exposures could result in beneficial effects on both growth and yield.

It is appropriate to comment that the situation can be distinctly different when one compares the information on the vegetation effects caused by the emissions from a single source, with a corresponding situation where multiple sources exist in the same geographic area. With a single point source, plants are periodically exposed to short-term ($<$ few hours) peak concentrations above a low background. The frequency distributions of these peaks are not always log-normal. On the contrary, with multiple point sources, plants are exposed continuously to approximately steady SO_2 concentrations (increased above the background). Under these conditions, the exposure also consists of more frequent occurrences of short-term peaks, generally at a higher dose (concentration $\times$ time).

Field experiments with ambient SO_2 and standardized soils

To measure more precisely the direct effect of air pollutants (SO_2) on growth and yield of a number of agricultural (and to a lesser extent horticultural) plant species, methods involving standardized soils have been used.

Significant decreases in growth and yield of potatoes, beet, lettuce and spinach, ranging from 20 to 60% of control crops were observed in areas where sulphation rates on lead candles were 1.05–1.23 mg/(100 cm^2·day) (roughly 100 µg SO_2 m^{-3} in air) (Szalonek and Warteresiewicz, 1966a, b). In these cases, decreases in yield were measured in the absence of visible injury.

Responses of additional plant species have been evaluated by Garber (1967) at two polluted sites (*Table 12.1*) where the mean values of short-term SO_2 measurements (30 min) were: Site 1, 330 µg SO_2 m^{-3} in the first year and 340 µg SO_2 m^{-3} in the second year; Site 2, 470 µg m^{-3} in the first year and 380 µg m^{-3} in the second year. The maximum 30-minute concentration of the second site exceeded 1200 µg m^{-3}. It is not clear whether the decreases in yield reported for radish, lettuce, bush bean and potato were the result of peak concentrations, especially at Site 2. It can be concluded, however, from these experiments that the aerial parts were less affected than the underground parts. For radish, this was confirmed by Tingey, Heck and Reinert (1971), who found that in the absence of visible shoot injury, root weight decreased by 17% (dry weight) or by 30% (fresh weight) and with no change in dry weight, fresh weight decreased by 7%. On the other hand, an increase in the biomass of radish, lettuce, bush bean and potatoes has also been reported (*Table 12.1*, Site 1). In these studies the average SO_2 concentration over 42 h exceeded 500 µg m^{-3}, and peak concentrations were in excess of 800 µg m^{-3}.

Guderian and Stratmann (1968) have conducted one of the most extensive studies to date. Plants were grown in large containers and placed above the soil level. Sulphur dioxide concentrations were measured at six different sites and detailed observations and measurements of effects on plants were performed. Concentrations were expressed in two ways: (1) mean concentrations during 'exposure' periods corresponding to SO_2 fumigations when concentrations exceeded 0.10 ppm; (2) mean concentrations over the whole of the 'monitoring' time. *Table 12.2* shows the concentrations at the six sites, and the fraction of monitoring time when 'exposures' occurred. Using average SO_2 concentrations (*Table 12.2*), and

Table 12.1   EFFECT OF SO$_2$ ON HEIGHT AND YIELD OF VARIOUS PLANT SPECIES GROWN AT TWO POLLUTED SITES: 1, MODERATELY POLLUTED; 2, HEAVILY POLLUTED; C, CONTROL. After Garber (1971)

Plant species	Site	Plant height (cm)	Dry wt (g/pot)		Level of statistical significance
Spinach	C	24	58.9		
	1	18	46.6		***
	2	15	40.5		***
			Aerial	Roots	
Radish A	C	12	13.1	9.5	
	1	11	11.2	8.7	***
	2	8	11.2	8.0	***
Radish B	C	16	7.6	11.8	
	1	20	9.6	11.0	N.S.
	2	11	2.2	1.1	***
Head lettuce A	C	16	34.7		
	1	16	33.6		N.S.
	2	14	29.9		**
Head lettuce B	C	15	45.8		
	1	14	30.0		*
	2	13	23.5		*
Leaf lettuce	C	15	47.1		
	1	16	38.2		N.S.
	2	11	18.6		**
Peas (all aerial parts)	C	38	36.8		
	1	42	44.3		N.S.
	2	35	27.7		***
Peas (pods)	C	9 (length)	52.6		
	1	10	56.3		N.S.
	2	6	36.1		***
Bush beans (all aerial parts)	C	38	52.3		
	1	37	31.9		N.S.
	2	18	3.7		***
Tomato	C	112			
	1	110			N.S.
	2	64			**
			Tubers		
Potato	C	56	199		
	1	56	103		*
	2	46	44		***
			Fresh wt	No. of flowers	
Marigold	C	66	873	20	
	1	68	873	21	N.S.
	2	50	272	12	**
Carrot	C	37	212		
	1	37	164		N.S.
	2	30	91		**

measurements of growth and yield (*Table 12.3*) the threshold limits for damage to potatoes (the most sensitive species studied) fluctuated from 0.21–0.23 ppm during SO$_2$ episodes and from 0.010–0.015 ppm during the total measurement time. The threshold limits for tomato (the least sensitive) were respectively 0.31–0.57 ppm and 0.051–0.124 ppm.

Results of a 3-year experiment at 11 sites in an industrialized area (Warteresiewicz, 1979), showed a good correlation between increase in air pollution and yield decrease (*Table 12.4*). Analysis of the data indicates

Table 12.2 CHARACTERISTICS OF THE EXPERIMENTAL SITES FOR FIELD EXPERIMENTS NEAR AN SO_2 SOURCE. FOR EXPLANATIONS OF TERMS 'EXPOSURE TIME' AND 'MONITORING TIME' SEE TEXT, From Guderian and Stratmann (1968)

Site	Distance from source (m)	Degree of plant injury	Exposure time (t_i) as percentage of the monitoring time (t_m) (%)	Average concentrations (ppm)[a]	
				c_i during exposure time t_i	c_m during monitoring time t_m
1	325	Very severe to total injury	24.1	0.59	0.141
2	600	Very severe injury	18.9	0.44	0.083
3	725	Severe injury	14.7	0.34	0.051
4	1350	Slight injury	7.6	0.26	0.020
5	1900	Slight injury to very sensitive plants	4.3	0.22	0.010
6 (control)	6000	No injury	No measurable SO_2	zero	zero

[a] Average of two monitoring periods, each from 1 April to 31 October

Table 12.3 EFFECTS OF AMBIENT SULPHUR DIOXIDE ON GROWTH AND YIELD OF SOME AGRICULTURAL AND HORTICULTURAL PLANTS. Compiled from Guderian and Stratmann (1968)

Crop and harvest characteristics	Percentage of control value				
	Site 1	2	3	4	5
Apple (*Malus communis* Mill cv. Ellisons Orange)					
Yield of fruits	–	24.5	46.5	90.2	105.1
Shoot growth (elongation)	–	86.6	103.9	91.9	137.1
Trunk radial increment	–	71.2	91.2	103.7	106.2
Trunk area increment	–	65.5	88.8	105.7	108.6
Sweet cherry (*Prunus avium* L. cv. Primavese)					
Shoot growth (elongation)	–	71.5	97.6	89.3	89.2
Trunk radial increment	–	81.5	88.1	92.7	107.4
Trunk area increment	–	76.4	83.4	91.2	114.3
Morello cherry (*Prunus cerasus* L.)					
Shoot growth (elongation)	–	87.4	85.5	87.2	97.2
Trunk radial increment	–	79.7	92.3	98.5	105.3
Trunk area increment	–	72.6	87.4	97.9	106.4
Plum (*Prunus domestica* L.)					
Shoot growth (elongation)	70.3	90.9	94.3	89.2	93.6
Trunk radial increment	60.5	80.1	77.1	94.0	95.0
Trunk area increment	53.2	75.1	71.8	92.0	93.8
Currant (*Ribes rubrum* L.)					
Yield	–	19.4	30.7	60.1	90.7
Shoot growth (elongation)	–	47.2	62.3	85.9	100.1
Gooseberry (*Ribes uva-crispa* L.)					
Yield	–	0	7.8	52.1	79.0
Shoot growth (elongation)	–	39.6	63.1	73.4	70.1
Winter rye (*Secale cereale* L.)					
Yield of grain	47.7	57.7	85.4	96.5	99.2
Yield of straw	62.4	80.5	100	101.2	104.1
Thousand-grain weight	62.4	79.9	92.0	96.5	99.3
Winter wheat (*Triticum sativum* L.)					
Yield of grain	42.7	55.6	85.0	92.0	98.9
Yield of straw	42.6	55.6	100.7	98.4	100.5
Thousand-grain weight	66.8	86.0	94.9	97.9	95.0
Spring wheat (*Triticum sativum* L.)					
Yield of grain	64.0	73.4	88.3	98.6	99.0
Yield of straw	63.7	77.1	92.5	97.1	97.9
Thousand-grain weight	84.7	91.1	99.7	100.2	101.5
Oats (*Avena sativa* L.)					
Yield of grain	62.4	76.1	84.7	93.0	102.4
Yield of straw	69.6	78.1	88.5	93.1	100.0
Thousand-grain weight	81.1	85.7	93.1	99.5	99.5
Spring rape (*Brassica napus* L.)					
Yield of grain	69.3	90.9	91.4	95.7	96.8
Thousand-grain weight	88.2	94.7	92.1	97.4	97.4
Number of pods	87.0	87.0	95.6	100.0	100.0
Potato (*Solanum tuberosum* L.)					
Yield of tubers	44.4	68.0	83.1	90.4	98.3
Number of tubers	84.2	89.5	102.6	110.5	102.5
Beet (*Beta vulgaris* L.)					
Yield of roots	68.3	83.1	90.7	100.0	98.0
Yield of foliage	79.9	101.1	117.0	96.2	108.7
Alfalfa (*Medicago sativa* L.)					
Yield	63.8	81.0	93.1	97.4	95.6
Red clover (*Trifolium pratense* L.)					
Yield	30.2	63.6	90.5	98.8	96.7

Table 12.3 CONTINUED

Crop and harvest characteristics	Percentage of control value				
	Site 1	*2*	*3*	*4*	*5*
Tomato (*Lycopersicon esculentum* L.)					
Yield of fruits	90.2	94.4	104.8	104.6	102.0
Shoot biomass	65.7	92.0	95.3	102.7	107.7
Carrot (*Daucus carota* L.)					
Yield	68.5	84.5	89.8	94.7	92.3
Spinach (*Spinacia oleracea* L.)					
Yield	94.7	92.8	101.4	93.3	103.5

that growth (yield) of aerial parts of plants (barley and alfalfa) was not affected at low SO_2 concentrations. However, yield of grain (barley) was reduced. Results of the analysis of thousand-grain (seed) weight (TGW, TSW) compared with total grain (seed), suggests that the number (of grains or seeds) per unit ground area was more affected than the weight. Of all the species investigated, the yield was most reduced in potatoes. In these experiments, apart from two sites where the SO_2 concentrations were high, leaf injury was observed during the late part of the growing season only (Warteresiewicz and Szalonek, 1972).

We believe that long-term, time-averaged concentrations of SO_2 are of little value in relating pollutant stress and plant response because responses often depend critically on short-term peak concentrations. However, it can be concluded from the results presented that a good correlation exists between time-integrated SO_2 concentrations (expressed as sulphation rates) and crop loss. The lead candle method may therefore be valuable in evaluating long-term effects of SO_2 on plant growth and productivity in areas with multiple point sources and fairly steady ambient SO_2 concentrations. This method, however, seems unsuitable for monitoring single point sources. Results of a 2-year experiment in the vicinity of a single source (Warteresiewicz and Szalonek, 1972) appear to confirm this, as the

Table 12.4 EFFECTS OF SO_2 ON GROWTH AND YIELD OF PLANTS IN THE FIELD (MEANS OF THREE SEASONS: SO_2 CONCENTRATION ($\mu g\,m^{-3}$) IS APPROXIMATELY 100 × SULPHATION RATE QUOTED. After Warteresiewicz (1979)

Sulphation rate (mg sulphate/ (100 cm²·d⁻¹))	Yield (% reference point)						
	Barley			Bean		Potato Tuber yield	Alfalfa Forage yield
	Grain yield	*Thousand-grain weight*	*Straw yield*	*Seed yield*	*Thousand-seed weight*		
0.40	100	100	100	100	100	100	100
0.77	98.0	98.0	98.0	99.3	99.6	94.7	99.2
0.95	94.0	98.0	96.9	90.0	98.2	85.7	98.2
1.00	92.2	96.0	101.0	90.5	99.0	89.7	100.4
1.07	90.2	95.4	98.2	89.4	98.1	86.0	98.6
1.26	85.9	92.0	96.9	84.7	85.9	73.3	88.0
1.54	79.7	89.5	97.3	78.3	91.8	63.0	85.7
1.60	76.3	83.5	83.7	71.5	87.1	56.0	82.0
1.66	71.8	80.9	83.4	65.8	75.4	52.7	76.3
1.80	70.6	88.1	88.9	69.8	77.3	51.7	78.3
2.10	64.7	78.0	79.0	58.5	65.6	41.0	70.0
2.18	65.1	76.1	78.5	60.5	67.1	45.0	69.7

correlation coefficients between yields and sulphation rates were low in this case. However, differences in yield between control plots and the experimental plots were statistically significant when the sulphation rates exceeded $0.80\,mg/(100\,cm^2.d)$. Yields of potato and beet were reduced by 30% to 40%, and barley by about 20%. None of these studies included a detailed statistical correlation analysis between atmospheric SO_2 concentrations and leaf injury. The results of Warteresiewicz (1979) lead to the conclusion that most injury and yield loss occurred where sulphation rates exceeded $1.8\,mg/(100\,cm^2.d)$ in areas with multiple sources and approximately $1.2\,mg/(100\,cm^2.d)$ in areas near single sources.

Field fumigation experiments

Sprugel *et al.* (1978) used an open-air fumigation system in the field and observed statistically significant reductions in yield of soybeans (*Table 12.5*). Sulphur dioxide concentrations in their experiments at low and

Table 12.5 YIELD OF SOYBEANS SUBJECTED TO VARYING CONCENTRATIONS OF SO_2 IN AN OPEN-AIR FUMIGATION SYSTEM, 1977–78. VALUES ARE GIVEN ± STANDARD ERROR. From Sprugel *et al.* (1978)

Plot	Mean SO_2 concentration (ppm)	Dose (ppm h)	Yield (kg ha^{-1}) Fumigated	Control	Yield reduction (%)
1977					
Ambient	0.005–0.015	–	2566 ± 77	2577 ± 73	0.3 ± 4.2
Low	0.117	13.3	3052 ± 92	3478 ± 101	12.3 ± 4.3 (P <0.01)
Medium	0.300	34.2	2482 ± 65	3140 ± 48	20.5 ± 3.1 (P <0.001)
High[a]	0.790	89.6	1636 ± 52	2992 ± 120	45.3 ± 3.7 (P <0.001)
1978					
Low 1	0.095	6.8	2370 ± 27	2531 ± 47	6.4 ± 2.1 (P <0.01)
Low 2	0.108	7.8	2256 ± 39	2379 ± 44	5.2 ± 2.5 (P <0.05)
Medium 1	0.192	13.5	2191 ± 65	2494 ± 41	12.2 ± 3.1 (P <0.001)
Medium 2	0.255	18.9	2008 ± 27	2485 ± 42	19.2 ± 2.0 (P <0.001)
High[a]	0.362	26.1	1859 ± 48	2209 ± 45	15.9 ± 3.0 (P <0.001)

[a]Visible injury observed on fumigated plants: Concentration exceeded US Environmental Protection Agency Secondary Air-Quality Standards for sulphur oxides (0.50 ppm for 3 h)

medium concentrations were below the current US EPA secondary air quality standards for sulphur oxides (0.5 ppm for 3 h) and no visible injury was observed, except at the highest SO_2 concentration. The authors were able to express the relationship between yield reduction y (%) and SO_2 dose x (ppm h):

$$y = 0.803x - 0.003x^2$$

A different relationship (*Figure 12.1*) was observed by the same group for acute SO_2 exposures at concentrations varying from 0.8 to 2.0 ppm for 260 minutes, generated as a gradient from an SO_2 source (Miller *et al.*, 1978). In this study severe chlorosis and occasional necrosis (2–5% of total leaf area) was observed on plants close to the fumigation source. A dose of

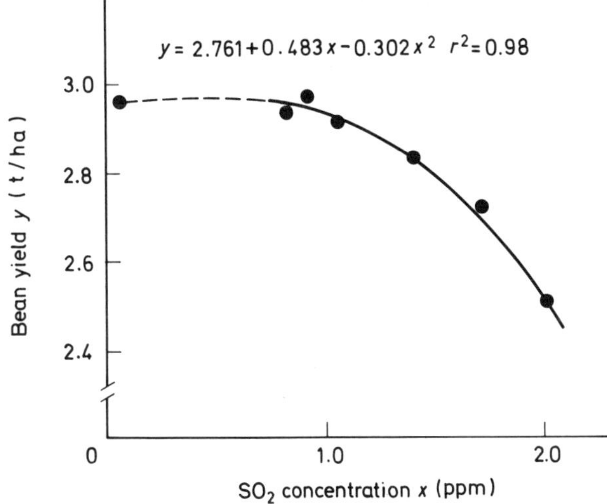

Figure 12.1   Variation of yield of soybeans with SO_2 concentration. Plants were exposed for a single period of 4.3 h. The regression equation is applicable only above a concentration of 0.8 ppm. (From Miller *et al.*, 1978)

8.7 ppm h caused a 15% reduction in yield. In the experiments described previously, at lower concentrations but with a similar dose, a reduction in yield of approximately 6% was observed. This inconsistency between dose and response should be considered when comparing results from different field experiments, even those with partial environmental control. In other studies, SO_2 doses of 65 ppm h^{-1} in field chambers caused thousand-grain weight reductions of 19% (barley), 12% (rye), 10% (wheat) and 9% (oats). Reductions in the straw yield were even greater (*Table 12.6*; Engman, Daessler and Boertitz, 1969).

Some results previously discussed seem to support the view that a yield reduction can occur in the absence of visible injury, suggesting that it should not be accepted as a general rule that a yield loss and/or growth reduction can occur only in the presence of visible leaf injury (Thomas and Hendricks, 1956; Davis, 1972; Heck and Dunning, 1978; Jones, Noggle

Table 12.6   EFFECT OF SO_2 ON STRAW YIELD AND THOUSAND-GRAIN WEIGHT (TGW) OF CEREAL CROPS IN FIELD (1,2) AND FUMIGATION CHAMBER (3,4) EXPERIMENTS (EQUIVALENT SOIL CONDITIONS). EXPERIMENT 1: FIELD CONTROL; 2: POLLUTED FIELD (0.4 km FROM A SULPHURIC ACID FACTORY); 3: FIELD CHAMBER CONTROL; 4: SO_2 FIELD CHAMBER, 400 μg m^{-3} FOR 6–7 h DAILY (TOTAL 65.4 ppm h). From Engman, Daessler and Boertitz (1969)

| Experi- | Summer barley | | Rye | | Wheat | | Oats | |
ment	Straw yield (g)	TGW (g)	Straw yield (g)	TGW (g)	Straw yield (g)	TGW (g)	Straw yield (g)	TGW (g)
1	20.5	51.1	41.7	35.2	43.6	40.1	53.5	42.2
2	19.5	50.5	44.2	36.3	41.6	40.6	51.7	39.7
3	24.2	38.2	31.4	37.6	37.2	39.6	53.3	41.7
4	21.1	31.0	29.6	32.9	35.9	35.7	50.4	38.1

and Weatherford, 1978; Haase, Morgan and Salem, 1980). Some components of yield, mainly aerial parts, can be stimulated in the field when SO_2 concentrations are low to moderate (*see Tables 12.1, 12.3 and 12.4*). This is usually not the case of grain (seed) or storage roots where yield may decrease when yield of aerial parts increases (*see Tables*).

Effects of SO_2 in enclosed environments

The effects of exposure to SO_2 can vary from an increase to a decrease when identical experimental procedures are used, but with different plants and cultivars (Heagle and Johnston, 1978). In this study there was no correlation between visible foliar injury and changes in shoot fresh weight at any of the concentrations used (0.5, 1.0 and 1.5 ppm SO_2). These results indicate the importance of growing conditions on plant response.

In experiments where the final crop yield was not evaluated, a major consideration is the period in the growth cycle when the injury was assessed. For example, in the experiments of Heagle and Johnson (1978), at the time of evaluation the first trifoliate leaves of bean plants were fully expanded and the second trifoliate leaves were 2.5–5.0 cm long (21–26 days after planting). Foliar injury was evaluated 2 days after exposure and fresh weight after 7 days.

More extensive studies of effects of SO_2 exposure on leaf injury and/or weight of aerial parts of several plant species, cultivars and/or breeding forms were carried out by Grzesiak (1979a, b) and Laurence (1979). A lack of correlation between the magnitude of foliar injury and the dry weight of aerial parts was observed under optimum moisture as well as drought conditions (*Table 12.7*). Maize, which had least leaf injury, exhibited a decrease in dry weight comparable to that of sunflower and barley, in which leaf injury was severe. A decrease in both the extent of leaf injury and in dry weight was observed in plants under water stress, except with maize, where leaf injury was not modified. No major differences in response were found when ten maize cultivars or breeding forms were investigated (*Table 12.8*). It seems that the sensitivity of breeding forms resembled the sensitivity of the more tolerant parent e.g. compare the

Table 12.7　EFFECTS OF SO_2 UNDER DIFFERENT SOIL-MOISTURE CONDITIONS ON LEAF INJURY AND DRY WEIGHT OF FOUR PLANT SPECIES. PLANTS WERE FUMIGATED WITH 500 µg SO_2 m^{-3} FOR 5 h d^{-1} AND 6 DAYS PER WEEK FOR 6 WEEKS. After Grzesiak (1979a)

| *Plant species* | *Soil moisture* | | | |
| | *Optimum* | | *Just above wilting point* | |
	Leaf area injured (% control)	*Dry wt of aerial parts (% control)*	*Leaf area injured (% control)*	*Dry wt of aerial parts (% control)*
Sunflower	40.7	56.5	5.4	87.1
Bean	17.7	72.7	2.5	90.7
Barley	40.6	54.4	14.2	80.2
Maize	2.7	57.3	2.8	81.8

Table 12.8 EFFECTS OF SO_2 ON LEAF INJURY AND DRY WEIGHT OF AERIAL PARTS OF TEN MAIZE BREEDING FORMS AND CULTIVARS GROWN WITH OPTIMUM SOIL MOISTURE. After Grzesiak (1979a)

Maize cultivar	*Leaf injury (% control)*		*Dry weight (% control)*	
	14 days	28 days	14 days	28 days
EP-1	3.2	1.9	71.0	68.5
W-79a	1.1	1.1	73.0	63.1
EP 1 × W-79a	1.9	1.2	77.6	68.1
W-459	2.4	1.4	72.1	61.9
W-153r	2.4	1.6	74.5	67.3
W-459 × W-153r	2.6	1.0	78.5	70.3
KB-310	3.0	1.7	79.4	69.6
Golden Bok	3.5	1.6	69.4	67.8
Golden Bok Elita	2.0	1.8	73.5	65.6
Aztek	1.5	1.8	69.2	66.9

sensitivity of EP-1 × W-79a with that of W-79a. Depending on cultivar, after 14 days' exposure to SO_2 both the percentage leaf-area injured and percentage dry-weight loss were generally larger than after 28 days. The data in *Table 12.8* strongly support the view that there is no correlation between leaf injury and dry-weight loss. Again, there was no correlation between the magnitude of leaf injury and dry-weight loss after 14 and 28 days of the experiment.

Both increases and decreases in the dry weight of seven wheat cultivars were observed by Laurence (1979), depending on the concentration and duration of exposure. Statistically significant increases in dry weight were found with six of the seven cultivars investigated after a 100-hour exposure to relatively low SO_2 concentrations (*Table 12.9*) i.e. biomass production was increased on a dry-weight basis. However, when these values are compared as time elapses, the increase (as % of control) becomes smaller (e.g. the increase for cv. Waldron after 78 h is 131%, but after 100 h is 114%, *see Table 12.9*). In two cultivars (Ticonderoga, Yorkstar) in which an increase was found after a 30-hour exposure, the highest relative increase occurred after 100 h. However, these differences were not statistically significant.

A detailed study of the influence of temperature and humidity on the response of two varieties of oats was conducted by Heck and Dunning (1978). A reduction in biomass was observed at SO_2 concentrations roughly equal to the US EPA secondary air quality standard (0.5 ppm for 3 h). Lowering the temperature and the humidity increased plant resistance. With older plants, root growth was affected more than the shoot growth. Exposed plants recovered within a few weeks of the SO_2 exposure if they had been exposed at an early growth stage to relatively low or medium SO_2 concentrations. Foliar injury was highly correlated with the growth reduction and the authors suggested that, at low concentrations, SO_2 acted as a sulphur source for the plants and was capable of stimulating plant growth and, possibly, of increasing yield.

As Tingey, Heck and Reinert (1971) showed in a study on SO_2 exposure of radish, differing conclusions can be drawn depending upon the harvest criteria used for evaluation. Analysis of the effects of SO_2 on sunflower growth (Shimizu, Furukawa and Tolsuka, 1980; *Table 12.10*) can also be

Table 12.9 EFFECT OF FOUR CONCENTRATIONS OF SO₂ AT THREE DURATIONS OF EXPOSURE ON DRY WEIGHT OF CULTIVARS OF HARD RED SPRING WHEAT (HRS) AND SOFT WHITE WINTER WHEAT (SWW). From Laurence (1979)

Cultivar	SO_2 concn (ppm)	Dry weight (g) after exposure for:		
		30 h	78 h	100 h
Era (HRS)	0.0	0.089	0.112 ab*	0.095 ab
	0.2	0.113	0.138 a	0.114 a
	0.4	0.101	0.099 ab	0.098 ab
	0.6	0.105	0.084 b	0.076 b
	MSE	0.00083	0.0013	0.00063
Waldron (HRS)	0.0	0.168	0.164 b	0.152 ab
	0.2	0.178	0.215 a	0.173 a
	0.4	0.142	0.165 b	0.135 ab
	0.6	0.159	0.127 b	0.108 b
	MSE	0.0019	0.0016	0.0022
Thatcher (HRS)	0.0	0.127	0.144	0.132 ab
	0.2	0.144	0.182	0.166 a
	0.4	0.125	0.143	0.127 ab
	0.6	0.130	0.140	0.092 b
	MSE	0.0015	0.0017	0.0015
Prelude (HRS)	0.0	0.165	0.167	0.168
	0.2	0.179	0.207	0.186
	0.4	0.138	0.172	0.164
	0.6	0.171	0.156	0.122
	MSE	0.0023	0.0027	0.0035
Arrow (SWW)	0.0	0.146	0.201 b	0.159 ab
	0.2	0.163	0.268 a	0.178 a
	0.4	0.178	0.197 b	0.138 ab
	0.6	0.167	0.148 b	0.117 b
	MSE	0.0028	0.0029	0.0021
Ticonderoga (SWW)	0.0	0.156	0.154	0.151 ab
	0.2	0.147	0.172	0.174 a
	0.4	0.149	0.158	0.153 ab
	0.6	0.136	0.126	0.103 b
	MSE	0.0015	0.0020	0.0024
Yorkstar (SWW)	0.0	0.162	0.145	0.157 a
	0.2	0.154	0.158	0.177 a
	0.4	0.161	0.145	0.145 ab
	0.6	0.141	0.120	0.095 b
	MSE	0.0025	0.0022	0.0021

*Means followed by the same letter are not significantly different ($P = 0.05$) based on Tukey's test for comparison of means. Absence of letters indicates no significant difference. All comparisons are made within cultivar and time period. Mean of 8 plants.
MSE = mean square for error or experimental estimate of r^2

used as an example. The dry weight of the whole plant was not significantly affected even after 6 weeks of the experiment. The leaf area increased in plants exposed to 0.05 ppm SO₂, but at the same time flower dry-weight and plant height decreased.

Timing, duration of exposure and concentration of SO₂ alter injury (Zahn, 1961), physiological processes and dry weight (McLaughlin *et al.*, 1979). Doses of SO₂ equal to or below the current US secondary air-quality standard (3 h average < 0.5 ppm) were applied to kidney beans to examine the effects of varying peak:mean ratios and of exposure frequency (McLaughlin *et al.*, 1979; *Figure 12.2*). Experiments were conducted using

Table 12.10 THE EFFECTS OF SO_2 EXPOSURE ON SOME CHARACTERISTICS OF SUNFLOWER PLANTS AT HARVEST (6-WEEK EXPERIMENT). From Shimizu, Furukawa and Tolsuke (1980)

Characteristics	*SO_2 concentration* (ppm)		
	0.00 (Control)	0.05	0.10
Total dry weight (g)	43.63 ± 3.10	45.74 ± 7.51	42.79 ± 2.59
Leaf area (dm^2)	24.30 ± 1.68	29.24 ± 4.62^b	25.99 ± 2.51
Withered leaf weight (g)	0.83 ± 0.32	1.08 ± 0.52	1.27 ± 0.55^a
Flower dry-weight (g)	0.89 ± 0.36	0.38 ± 0.28^b	0.43 ± 0.34^b
Plant height (cm)	142.4 ± 4.45	129.5 ± 11.5^b	124.5 ± 10.8^c

Mean of 10 plants and the standard deviation are indicated for each sample. [a,b] and [c] indicate that mean values are significantly different from control at the 0.05, 0.01 and 0.001 levels, respectively

weekly and daily exposures on 4-week-old plants. Reduced shoot weights, compared with controls, were found in experiments where exposures were repeated at 1-week intervals for 4 weeks. Differences between treatment 1–2.0 (*Figure 12.2*) and the control were statistically significant. Other treatments did not show any significant differences, in plants treated weekly or daily (*Table 12.11*).

As mentioned earlier, SO_2 can act as a sulphur source and can stimulate some of the harvest parameters. Faller, Herwig and Kuehn (1970) grew plants in sulphur-depleted nutrient solutions and supplied them with SO_2 from the atmosphere. However, in these studies no comparison was made with a treatment with adequate sulphur in the nutrient solution, to determine whether atmospheric sulphur inputs had an effect equivalent to that of sulphur supplied via the roots (*Tables 12.12, 12.13*). Except in the case of horseradish, an early increase in biomass was observed even at the highest SO_2 concentrations (which caused visible injury in tobacco).

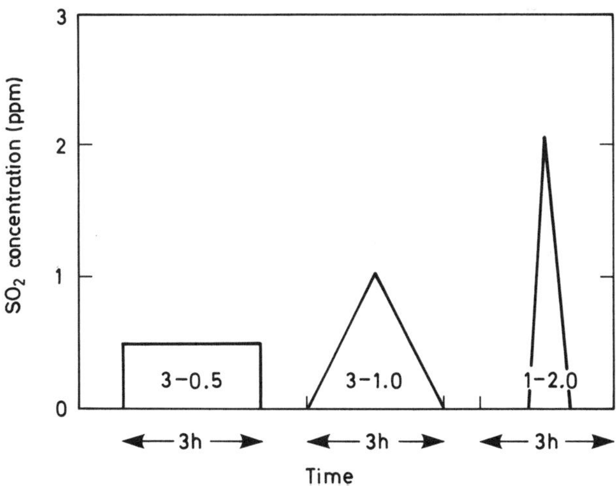

Figure 12.2 Exposure kinetics of three SO_2 treatments. 3–0.5: 3 h exposure with both average and peak concentration of 0.5 ppm; 3–1.0: 3 h exposure, average concentration 0.6 ppm, peak 1.0 ppm; 1–2.0: 1 h exposure, 3 h average concentration 0.37 ppm, peak 2.0 ppm. (From McLaughlin *et al.*, 1979)

260

Table 12.11 COMPARATIVE EFFECTS OF SO_2
TREATMENTS WITH VARYING EXPOSURE KINETICS
ON SHOOT DRY-WEIGHT AT HARVEST. TREATMENTS
WERE APPLIED ONCE WEEKLY FOR 2–4 WEEKS OR
DAILY FOR 4 DAYS. VALUES GIVEN ARE MEANS OF
TWO TRIALS. DETAILS OF THE EXPOSURES ARE
SHOWN IN *FIGURE 12.2*. Adapted from Mclaughlin *et al.*
(1979)

Treatment	Shoot dry-weight (g)	% of control
Weekly		
Control	13.64	
3–0.5	12.05	89
3–1.0	13.40	99
1–2.0	10.71	78
Daily		
Control	8.54	
3–0.5	8.08	96
Control	6.32	
1–2.0	6.94	108

Table 12.12 EFFECT OF SO_2 ON THE HEIGHT OF PLANT SPECIES GROWN IN
SULPHUR-DEFICIENT NUTRIENT SOLUTIONS AND EXPOSED TO SO_2 FOR 12 h
DAILY. (VALUES ARE PERCENTAGES OF THE CONTROL). From Faller, Herwig
and Kuehn (1970)

Plant species	Fumigation time (days)	SO_2 concentration ($\mu g\, m^{-3}$)				
		0	200	500	1000	1500
Sunflower	15	100	136.7	163.7	157.8	163.7
Corn	13	100	108.7	116.8	117.2	116.1
Tobacco	9	100	161.3	146.2	152.2	156.5
Horseradish	20	100	92.2	90.6	89.2	89.0

Table 12.13 EFFECT OF SO_2 ON YIELD EXPRESSED AS PERCENTAGES OF
THE CONTROL VALUES OF PLANTS GROWN IN SULPHUR-DEPLETED
NUTRIENT SOLUTION AND EXPOSED TO SO_2 FOR 12 h DAILY. From Faller,
Herwig and Kuehn (1970)

Plant species	Fumigation time (days)	SO_2 concentration ($\mu g\, m^{-3}$)			
		200	500	1000	1500
Sunflower	15				
Whole plant		130.9	131.3	144.0	127.0
Aerial part		142.2	152.4	177.8	159.8
Roots		93.4	88.4	74.8	59.7
Corn	13				
Whole plant		109.3	118.1	110.5	107.2
Aerial part		111.8	123.7	116.8	112.7
Roots		100.0	97.7	86.9	86.9
Tobacco	9				
Whole plant		132.2	138.2	145.2	168.1
Aerial parts		147.4	165.6	177.7	181.9
Roots		99.0	78.8	74.7	74.7
Horseradish	20				
Whole plant		106.6	104.3	103.9	119.5
Aerial part		109.1	109.5	109.1	131.3
Roots		92.7	75.6	75.6	54.9

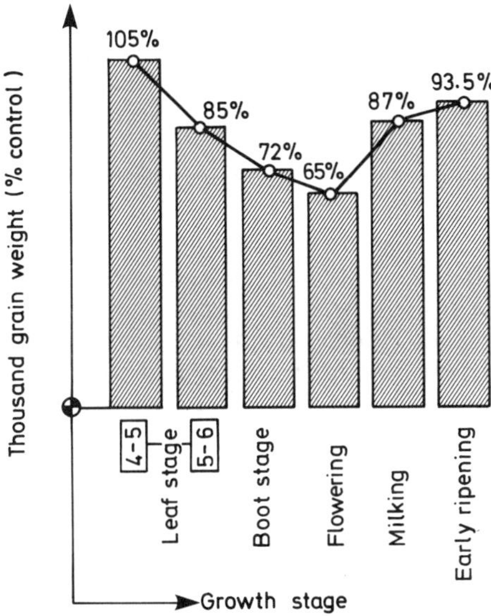

Figure 12.3   Effect of SO_2 exposure at different growth stages, on thousand-grain weight of wheat. From van Haut, 1961

According to Thomas *et al.* (1943), sulphur for atmospheric SO_2 is not equivalent to that supplied via the roots (*see also* Cowling and Koziol, this volume, Chapter 17). One reason for this could be the low mobility of sulphur within the plant if supplied as SO_2. Godzik (1976) found that large proportions of sulphur supplied to bean leaves as $^{35}SO_2$ remained fixed in the leaves, which were thus labelled. However, this does not mean that

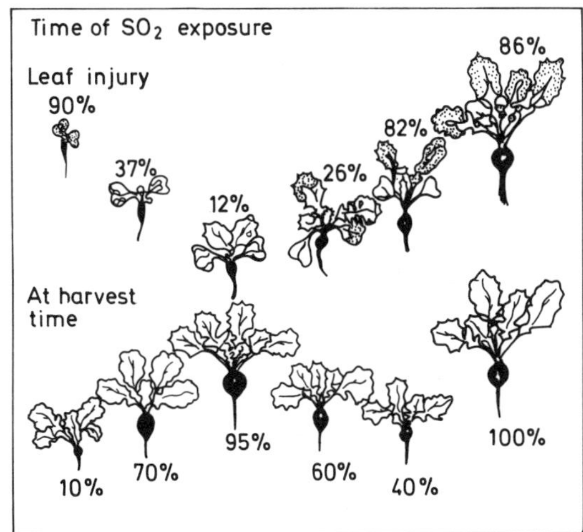

Figure 12.4   Timing of SO_2 exposure, leaf injury and yield loss in radish. The dotted areas are those injured. The values beside each illustration show how percentage leaf injury at various stages in the growth cycle resulted in different yields (as a percentage of the control yield at harvest). From van Haut (1961)

sulphur from SO_2 cannot be utilized by plants at all. Studies in agricultural species with ^{35}S have shown clearly that sulphur from SO_2 can be translocated and metabolized (Thomas *et al.*, 1943; Garsed and Read, 1974). However, questions concerning the extent of translocation and its diversity among plant species remain to be answered.

Very little information is available concerning the concept of a 'critical growth stage' for effects of SO_2. Van Haut (1961) found that yield loss depended on the growth stage of the plant at the time of pollutant exposure (*Figures 12.3, 12.4*). For example, the greatest yield losses in cereals were observed when plants were exposed at the flowering stage, but in radishes when fully mature plants were exposed. Plants showing a 'critical growth stage' apparently do not have sufficient time for recovery and compensation to the stress. Even though Davis (1972) and Haase, Morgan and Salem

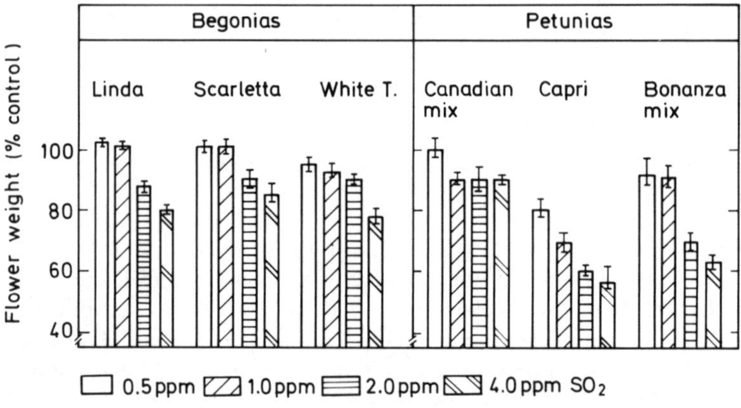

Figure 12.5   Effect of SO_2 on flower weight of cultivars of begonias and petunias. Fumigation for 2 h. From Adedipe, Barrett and Ormrod (1972)

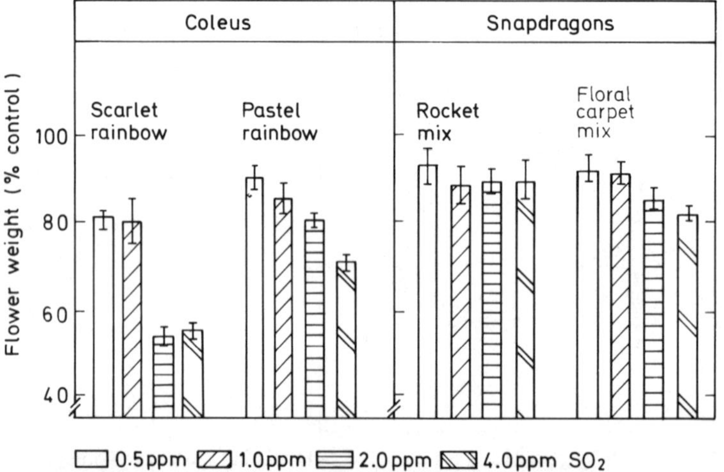

Figure 12.6   Effect of SO_2 on flower weight of cultivars of coleus and snapdragon (antirrhinum). Fumigation for 2 h. From Adedipe, Barrett and Ormrod (1972)

(1980) are of the opinion that the 'critical growth stage' effect cannot be found in the field, Maly (1974) found a yield reduction in broad bean when pollution episodes occurred during flowering.

At this time very little information is available about the growth or biomass effects of SO_2 on herbaceous ornamentals. Adedipe, Barrett and Ormrod (1972) have presented data on the response of coleus, snapdragon, begonia and petunia, showing distinct differences in resistance, between species and cultivars (*Figures 12.5* and *12.6*).

Conclusion

Our knowledge of acute sulphur dioxide injury on vegetation is more extensive than our understanding of the chronic, long-term effects. Growth and yield loss in the field have not been fully examined. Crop-loss models need to be improved and applied in order to evaluate how actual concentrations of sulphur dioxide, varying over periods of hours or minutes, are integrated to produce biomass alteration, with due consideration of the stage of plant growth. Integrated average pollutant concentrations over a growth season do not reflect the relationships between the intensity, time and stress and the effect on crops. Teng and Krupa (1981) have recently developed an air pollutant–crop loss model which is an attempt to evaluate SO_2-induced crop loss on a regional basis by using actual ambient pollutant concentrations and crop response with consideration of the weekly crop-growth stage.

Our knowledge of SO_2 effects on herbaceous ornamentals is limited. Joint effects of SO_2 with other pollutants and parasites also need to be examined in detail, as stressed elsewhere in this book.

Acknowledgement

We are highly grateful to The Northern States Power Company, Minneapolis, Minnesota, for financial support during the preparation of this manuscript.

References

ADEDIPE, N.O., BARRETT, R.E. and ORMROD, D.P. (1972). *Journal of the American Society for Horticultural Science*, **97**, 341–345

DAVIS, C.R. (1972). *Journal of the Air Pollution Control Association*, **22**, 964–966

ENGMAN, F., DAESSLER, H.G. and BOERTITZ, S. (1969). *Archiv für Pflanzenzüchtung*, **5**, 223–232

FALLER, N., HERWIG, K. and KUEHN, H. (1970). *Plant and Soil*, **33**, 177–191

GARBER, K. (1967). *Luftverunreinigungen und ihre Wirkungen*. Gebr. Borntraeger, Berlin

GARSED, G.S. and READ, D.J. (1974). *New Phytologist*, **73**, 299–307

GODZIK, S. (1976). *Uptake of $^{35}SO_2$ and distribution of ^{35}S in some plant species (Pol)*. Polska Akademia Nauk, Instytut Podstaw Inz. Srod., Prace i Studia No. 16, Ossolineum, Wroclaw-Warsaw

GRZESIAK, S. (1979a). *Bulletin de l'Académie polonaise des sciences. Classe II. Série des Sciences biologiques*, **27**, 309–321

GRZESIAK, S. (1979b). *Bulletin de l'Académie polonaise des sciences. Classe II. Série des Sciences biologiques*, **27**, 323–333

GUDERIAN, R. (1966). *Schriftenreihe der Landesanstalt für Immissions, Essen*, **7**, 80–100

GUDERIAN, R. and STRATMANN, H. (1968). *Forschungsberichte des Landes Nordrhein-Westfalen*, No. 1920, p. 114

HAASE, E.F., MORGAN, G.W. and SALEM, J.A. (1980). *Field surveys of sulfur dioxide injury to crops and assessment of economic damage. 73rd Annual Meeting of the Air Pollution Control Association*, Montreal, Abstract 80-26·2. APCA, Pittsburgh

HEAGLE, A.S. and JOHNSTON, J.W. (1978). *Journal of the Air Pollution Control Association*, **29**, 729–732

HECK, W.W. and DUNNING, J.A. (1978). *Journal of the Air Pollution Control Association*, **28**, 241–246

JONES, H.C., NOGGLE, J.C. and WEATHERFORD, F.P. (1978). *Abstracts of papers, Third International Congress of Plant Pathologists, Munich*, p.339. Paul Parey Verlag, Hamburg-Berlin

LAURENCE, J.A. (1979). *Plant Disease Reporter*, **63**, 468–471

McLAUGHLIN, S.B., SHRINER, D.S., McCONATHY, R.K. and MANN, L.K. (1979). *Environmental and Experimental Botany*, **19**, 179–191

MALY, V. (1974). *Scientia Agriculturae Bohemoslovacae*, **6**(XXIII), 147–159

MILLER, J.E., SMITH, H.Y., SPRUGEL, D.G. and XERIKOS, P.B. (1978). *Yield Response of Field-grown Soybeans to an Acute SO_2 Exposure*. ANL-78-65. III. Argonne National Laboratory

NOGGLE, J.C. (1980). In *Atmospheric Sulphur Deposition*, pp. 289–297 (Shriver, D.S., Richmond, C.R. and Lindberg, S.E., Eds), Ann Arbor Science, Michigan

SHIMIZU, H., FURUKAWA, A. and TOLSUKA, T. (1980). *Environmental Control in Biology (Japan)*, **18**, 39–47

SPRUGEL, D.G., MILLER, J.E., XERIKOS, B.P. and SMITH, H.J. (1978). *Effects of Chronic Sulfur Dioxide Fumigation on Development, Yield and Seed Quality of Field-grown Soybeans. Summary of 1977 and 1978 Experiments*. ANL 78-65 P. III. Argonne National Laboratory

SZALONEK, J. and WARTERESIEWICZ, M. (1966a). *Biuletyn ZBN GOP PAN*, **8**, 75–84

SZALONEK, J. and WARTERESIEWICZ, M. (1966b). *Biuletyn ZBN GOP PAN*, **8**, 59–74

TENG, P.S. and KRUPA, S.V. (1981). *Economic Assessment of Air Pollution Damage to Agricultural and Silvicultural Crops in Minnesota*. Minnesota Pollution Control Agency, Roseville, Minnesota

THOMAS, M.D. and HENDRICKS, R.H. (1956). In *Air Pollution Handbook*, pp. 1–44. McGraw-Hill, New York-Toronto

THOMAS, M.D., HENDRICKS, R.H., COLLIER, T.R. and HILL, G.R. (1943). *Plant Physiology*, **18**, 345–371
TINGEY, D.T., HECK, W.W. and REINERT, R.A. (1971). *Journal of the American Society for Horticultural Science*, **96**, 369–371
VAN HAUT, H. (1961). *Staub*, **21**, 52–56
WARTERESIEWICZ, M. (1979). *Archiv Ochrony Środowiska*, **7**, 95–166
WARTERESIEWICZ, M. and SZALONEK, J. (1972). *Mitteilungen aus dem Forstlichen Bundesamt Versuchsaust Wien*, 97/I:209–217
ZAHN, R. (1961). *VDI Berichte*, **53**, 24–28

13

THE EFFECTS OF FLUORIDES ON PLANT GROWTH AND FORAGE QUALITY

ALAN DAVISON
Department of Plant Biology, University of Newcastle upon Tyne

Introduction

In the last decade there have been several reviews of the biological effects of fluorides. Marier and Rose (1971), NAS (1971), Treshow (1971), Chang (1975), Groth (1975), Rose and Marier (1977) and Weinstein (1977) all contain discussions of the effects on plants while NAS (1974) and Suttie (1977) are comprehensive reviews of the effects on animals. Groth (1975) has a useful bibliography on fluorides in the aquatic environment.

Both gaseous and particulate fluorides are deposited on leaf and other plant surfaces (*Figure 13.1*) while gaseous fluoride enters the leaf via the stomata. Some surface-deposited fluoride may penetrate the cuticle, particularly if the leaf is weathered or old, and therefore it may be potentially phytotoxic. Fluoride ions in the leaf are carried in the transpiration stream to the tip and margins where, if the concentration is sufficiently high and the tissues are at a susceptible stage of development, toxic effects may develop. For a number of reasons (Weinstein, 1977), not least

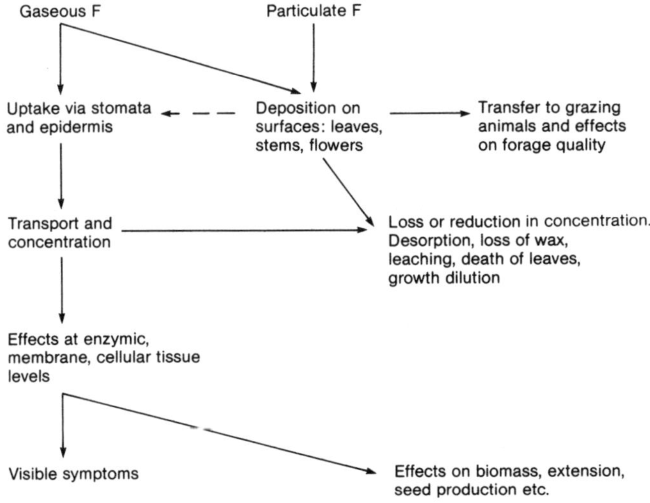

Figure 13.1   Deposition, uptake, loss and effects of airborne fluorides

of which is the problem of defining the biologically active fraction of the leaf fluoride, it is extremely difficult to establish a relationship between the fluoride content of tissues and injury. Weinstein (1977) concluded that the threshold for visible injury of susceptible plants is less than $100\,\mu g\,F\,g^{-1}$ and for some it may be lower than $20\,\mu g\,g^{-1}$. Intermediate or resistant plants can tolerate concentrations in excess of $200\,\mu g\,g^{-1}$. Because there is a better relationship between injury and the concentration of fluoride in the air, environmental standards designed to protect crops are usually based on dose (the product of concentration and time) rather than tissue fluoride levels (McCune, 1969).

Fluoride consumed by animals (*Figure 13.1*) in the form of forage, mineral supplements and water may cause fluorosis. The dietary tolerance of farm animals is well established (NAS, 1974; Suttie, 1977) but much less is known about the tolerance of wild animals. Indeed, there are still few published reports of fluorosis in wild animals (Rose and Marier, 1977). Dietary standards used to protect animals are mostly similar to those proposed by Suttie (1969) in which the main recommendation is that the yearly average fluoride content of the diet should not exceed $40\,\mu g\,g^{-1}$. Some of the difficulties involved in monitoring forage levels are discussed in Davison, Blakemore and Craggs (1979).

The effects of fluoride on the physiology of the plant and on fruiting are covered elsewhere in this volume (Weinstein and Alscher-Herman, Chapter 5; Bonté, Chapter 10). In this review attention is therefore focused on two questions that are of central importance in relation to the economic effects of fluorides: what concentration of fluoride in the air leads to unacceptable levels in forage, and what are the effects on plant growth and production?

Uptake and sorption of fluoride: the effects on forage quality

UPTAKE OF SOIL FLUORIDE

Soil typically contains from 20 to $500\,\mu g\,F\,g^{-1}$ (Weinstein, 1977) although the concentration may be higher where there are deposits of minerals such as fluorspar (Cooke *et al.*, 1976), where certain fertilizers have been added (Robinson and Edgington, 1946; Oelschläger, 1971) or near major emission sources (Thompson, Sidhu and Roberts, 1979). However, the fluoride ion has limited solubility in soil water (Larsen and Widdowson, 1971) so it is not surprising that uptake is relatively low and that there is little or no relationship between uptake and total or water-soluble fluoride (Cooke *et al.*, 1976). A study of uptake in relation to labile fluoride (Larsen and Widdowson, 1971) would probably clarify some of the confusing aspects (Weinstein, 1977) of soil fluoride.

A few species (Weinstein, 1977) can accumulate leaf concentrations in excess of $50\,\mu g\,F\,g^{-1}$ from unenriched soils but, in the majority of species, soil uptake results in concentrations much lower than $20\,\mu g\,F\,g^{-1}$. In most agricultural situations where the airborne fluoride is sufficiently high to make forage unacceptable, the contribution from soil fluoride is relatively small but it may become more significant as the $40\,\mu g\,g^{-1}$ dietary limit is approached, and in areas where soils are acidic and organic, and where there has been long-term deposition.

SORPTION OF AIRBORNE FLUORIDE: THE DOSE-RATE EQUATION

Fumigation of plants with HF under controlled conditions led to the concept that the increase in fluoride content of plant tissue (ΔF, $\mu g\,g^{-1}$) is related to the concentration in air (C, $\mu g\,m^{-3}$) and the duration of exposure (t, days) (McCune and Hitchcock, 1970; Hitchcock *et al.*, 1971). In a summary of extensive experiments on alfalfa (*Medicago sativa*) and orchard grass (*Dactylis glomerata*), Hitchcock *et al.* (1971) showed that the regression of the increase in fluoride on concentration and time accounted for 70% and 54% of the variation in the data, depending upon the species. The regressions were:

$$\Delta F \text{ alfalfa} = 1.89\ Ct + 0.74 \tag{13.1}$$

$$\Delta F \text{ orchard grass} = 1.13\ Ct + 1.17 \tag{13.2}$$

Post-fumigation harvests showed that the concentration in alfalfa fell with a half-life of about eight days and so a modified regression equation was calculated to allow for this. In both publications the authors considered that the regressions did not allow prediction of forage fluoride with sufficient accuracy to be of practical use. Later the regression model of uptake was simplified (NAS, 1971) to produce the dose-rate equation, $\Delta F = KCt$, where K is the accumulation coefficient, rather than a fitted regression coefficient. In NAS (1971) and Weinstein (1977), data from a number of laboratories were used to calculate K for a range of species. Considering the differences in technique used by the different laboratories, the values of K produced were remarkably consistent but it was recognized that the accumulation coefficient is not constant (NAS, 1971; Weinstein, 1977). It varies with species, duration of exposure and the sequence of fumigation (MacLean and Schneider, 1973). There is also evidence that K may be related to the air fluoride concentration and be inversely proportional to the duration of exposure (NAS, 1971). Some examples of K are given in *Table 13.1*.

The attraction of the dose-rate equation is its simplicity but the limitations (NAS, 1971; Weinstein, 1977) are considerable. The most important of these are the variable nature of K, and the fact that although

Table 13.1 ACCUMULATION COEFFICIENTS (K) FOR FORAGE CROPS. Data from Guderian, van Haut and Stratmann (1969) and Weinstein (1977)

Species	*Airborne F concentration and exposure period*				
	0.85	2.6	1.1	2.8 ($\mu g\,m^{-3}$)	*Average*
	16	16	47.7	47.7 (d)	
Lolium multiflorum	4.3	6.3	2.3	4.1	4.2
Lolium perenne	5.1	3.1	4.1	3.0	3.8
Dactylis glomerata	2.2	2.1	2.2	2.0	2.1
Festuca pratensis	3.4	4.7	3.0	3.5	3.6
Phleum pratense	1.3	2.2	2.2	3.7	2.3
Medicago sativa	2.3	3.1	2.0	2.8	2.5
Trifolium pratense	2.0	4.1	3.8	5.4	3.8
Trifolium repens	4.3	4.8	5.9	9.4	6.1

the relation appears to be valid for continuous fumigation with HF under controlled conditions, it has been evaluated only once under field conditions (Israel, 1977).

DEPOSITION VELOCITY

A second approach to the study of fluoride sorption is by means of the deposition velocity, v_g, which may be defined as

$$v_g = \frac{\text{rate of deposition}}{\text{mean concentration at specified reference height above the canopy}}$$

$$= \frac{\mu g\, m^{-2}\, s^{-1}}{\mu g\, m^{-3}} = m\, s^{-1}$$

Velocity of deposition has been extensively used in the study of SO_2, metal aerosols, radionuclides, spores and pollen (Chamberlain, 1967; Chamberlain, 1970; Chadwick and Chamberlain, 1970; Chamberlain and Chadwick, 1972; Garland, Clough and Fowler, 1973; Little, 1977; Little and Wiffen, 1977; Schwela, 1978, 1979). It depends on the properties of the pollutant, on the receptor surface and on wind speed, but it also varies with the reference height above the canopy and the length of time over which measurements are made. It is possible to standardize v_g to any desired reference height (Chamberlain, 1967) but in most of the publications on fluoride deposition insufficient detail is given for this to be done; nonstandardization might therefore be expected to introduce a source of variation. Similarly, differences in the length of time over which authors have made measurements would be expected to affect estimates of v_g because of saturation effects, growth dilution, changes in specific leaf weight and leaf-area index, or loss of pollutant from the canopy. Nevertheless, the few estimates of v_g that are available are remarkably consistent.

The first estimate of the deposition velocity of fluoride was made by Israel (1974a) who recorded gaseous concentrations and the fluoride content of alfalfa at monthly intervals near an aluminium smelter. After assuming a leaf-area index of 5 he calculated v_g on a canopy-area basis to be $6.2\, mm\, s^{-1}$. The work of Schwela (1979) provides estimates for shorter time intervals. He used boxes of *Lolium multiflorum* grown under standard conditions and clipped to 4 cm before being exposed to ambient air for 2 weeks at a number of sites. Gaseous fluoride was sampled continuously during the 2 weeks, then the grass was cropped and the fluoride content measured. The grass fluoride was significantly correlated with gaseous air fluoride, though the values of r were all very low, and v_g showed a weak negative correlation with HF concentration. In 2 out of the 3 years over which v_g was recorded, it varied with season, reaching a peak in May–June and being low during the winter. This was probably due to seasonal variation in the leaf area produced during the 2-week exposure period and in specific leaf weight (Hunt and Cooper, 1967; McColl and Cooper, 1967). The range of v_g was from about $2\, mm\, s^{-1}$ up to about $78\, mm\, s^{-1}$, but a large proportion of the values were in the region of $25\, mm\, s^{-1}$. This is much

Table 13.2 MULTIPLE REGRESSION OF GRASS FLUORIDE, F_{gr} (μg F g^{-1}) ON GRASS FLUORIDE ON THE PREVIOUS DAY, F'_{gr} (μg F g^{-1}), GASEOUS FLUORIDE, F_g (μg m^{-3}) AND PARTICULATE FLUORIDE, F_p (μg F g^{-1}). DEPOSITION VELOCITIES (v_g) ESTIMATED FROM REGRESSION CONSTANTS. Data of Blakemore (1978). * = grass fluoride on the previous week

Experiment	Multiple regression equation	Multiple reg.	Deposition velocity v_g (mm s^{-1})	
			Gaseous F	Particulate F
1. 30 consecutive days, November	$F_{gr} = 0.48\,F'_{gr} + 18\,F_g + 53\,F_p$	8.1	4.2	12.2
2. 30 consecutive days, July	$F_{gr} = 0.71\,F'_{gr} + 21\,F_g + 25\,F_p$	9.3	4.9	5.7
3. 30 consecutive days, December	$F_{gr} = 0.79\,F'_{gr} + 8\,F_g$	8.7	1.8	–
4. 30 consecutive days, March	$F_{gr} = 0.37\,F'_{gr} + 24\,F_g + 59\,F_p$	6.3	5.5	13.6
5. 100 consecutive weeks	$F_{gr} = 0.67\,F'_{gr}* + 281\,F_g + 223\,F_p$	8.8	9.3	7.3

Table 13.3 SPECIFIC LEAF WEIGHT (g m^{-2} BOTH SURFACES) OF TREES, SHRUBS, HERBS AND FORAGE PLANTS. Data of Blakemore (1978)

Trees	g m^{-2}	Shrubs	g m^{-2}	Herbs and forage sp.	g m^{-2}
Acer pseudoplatanus	35	Berberis darwinii	86	Chamaenerion augustifolium	22
Alnus glutinosa	35	Cornus alba	27	Senecio laxifolius	51
Betula pendula	37	Cotoneaster sp.	78	Stachys lanata	39
Fagus sylvatica	25	Crataegus monogyna	28	Lolium perenne leaf	10
Fraxinus excelsior	35	Ligustrum ovalifolium	49	Lolium perenne shoot	56
Ilex aquifolium	118	Mahonia aquifolium	47	Dactylis glomerata leaf	15
Salix fragilis	44	Sambucus nigra	31	Dactylis glomerata shoot	64
Sorbus aucuparia	38	Symphoricarpos rivularis	42	Holcus lanatus leaf	13
Ulmus glabra	37	Viburnum opulus	40	Holcus lanatus shoot	53

higher than the figure quoted for Israel's work because it was based on ground area. With allowance for a leaf-area index of about 3–5, the estimate would be reduced to 5–8 mm s^{-1}. In similar work, Davison and Blakemore (1976, and unpublished work) ran a series of field experiments in which forage and airborne fluorides were sampled daily for four periods of 30 consecutive days and weekly for 2 years. Deposition on alkali papers and in rain gauges was also measured (Davison and Blakemore, 1980). Statistical analysis involving partial correlations and multiple regressions showed that the grass fluoride concentration at any one time was correlated with the grass fluoride concentration at the previous sampling time and with the concentration of gaseous and particulate fluorides in the intervening period. Some data for one 30-day experiment were given by Davison and Blakemore (1976). The multiple regression coefficients (*Table 13.2*) from the various experiments can be used to estimate v_g if the specific leaf weight (SLW = g dry wt m^{-2} of leaf surface, using both sides of the leaf) is known. In the experiments of Blakemore (1978), (*Table 13.3*), the laminae of rye grass had an average SLW of about 10 g m^{-2}, so if an allowance is made for the inclusion of shoot material, then about 20 g m^{-2} should be a reasonable average for a whole tiller. Using this approximation, the regression constants suggest values of v_g ranging from 1.8 to 9.3 mm s^{-1} (*Table 13.2*). Finally, short-term experiments by Hill (1971) with alfalfa give an estimate of 4 mm s^{-1}.

If it is assumed that the velocity of deposition of gaseous fluoride to a forage sward ranges from about 2 to 8 mm s^{-1}, then the rate of deposition from an air concentration of 1 μg m^{-3} would be expected to be 173–691 μg m^{-2} of plant surface per day. The resulting concentration in forage expressed on a dry-weight basis would depend (Little, 1977) on the specific leaf weight of the crop (*Table 13.3*). In the case of a grass with an SLW of 20 g m^{-2}, the concentration on a dry-weight basis would be expected to increase by 9–35 μg g^{-1} d^{-1}.

The velocity of deposition is also suitable for the study of rates of particle deposition (Chamberlain, 1975) and a great deal is known about the factors that affect it. Among the more important are particle size (*Figure 13.2*), wind speed and the properties of the receiving surface. Deposition velocity has a minimum for particles in the size range 0.01–1.0 μm and it increases particularly rapidly with particle diameter between about 5–20 μm. Wind speed increases v_g, but with large particles and some types of surface an increase in wind speed leads to bounce-off and reduced capture efficiency (Little, 1977). McCune *et al.* (1965) fumigated a variety of species with cryolite particles (diameter 2–4 μm) and their data can be used to calculate v_g for corn, tomato and alfalfa (*Table 13.4*). Although the fumigation periods differed for the three species, the estimates are, with one exception, fairly constant at around 0.4 mm s^{-1}. As the wind speed in the greenhouse that they used was probably low, this value is reasonably close to what would be predicted from wind-tunnel experiments (*Figure 13.2*). In contrast, the deposition velocities of 5.7–13.5 mm s^{-1} (*Table 13.2; Figure 13.2*) estimated from Davison and Blakemore's data were higher and more variable. This is probably explained by the fact that the data were recorded at a windy field site where the canopy was wet for long periods and where the particle diameters varied from less than 1 μm to over 5 μm. Only one

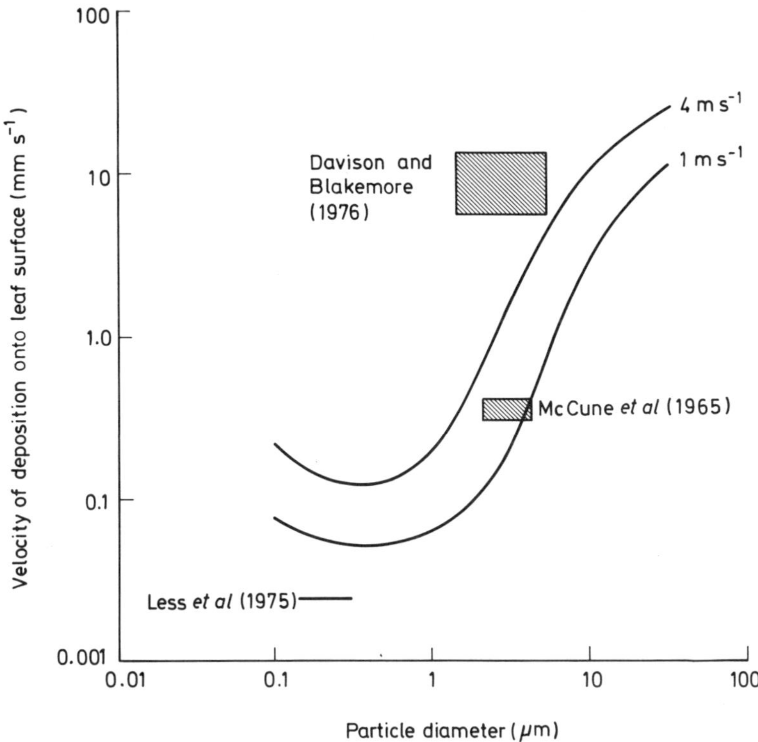

Figure 13.2   Relationship between measured velocity of deposition and particle diameter at wind speeds of $4\,\mathrm{m\,s^{-1}}$ and $1\,\mathrm{m\,s^{-1}}$. Data of Chamberlain (1980) with calculated values of v_g for fluorides superimposed

group has examined the deposition of particulate fluorides less than 1 μm in diameter. Less *et al.* (1975) exposed swards of *Lolium perenne*, for 14-day growth periods, to particles emitted from a heated cell containing cryolite. If an SLW of $20\,\mathrm{g\,m^{-2}}$ is assumed, their data suggest a value for v_g of $0.02\,\mathrm{mm\,s^{-1}}$, but as they did not record the particle size range, it is difficult to compare this estimate with the data in *Figure 13.2*.

Table 13.5 shows some hypothetical rates of fluoride deposition calculated for different particle sizes and SLWs at an air concentration of

Table 13.4   CALCULATED DEPOSITION VELOCITIES FOR CRYOLITE PARTICLES 2–4 μm IN DIAMETER. SPECIFIC LEAF WEIGHT ASSUMED TO BE $20\,\mathrm{g\,m^{-2}}$. From McCune *et al.* (1965)

Species	*Concentration of cryolite* ($\mu\mathrm{g\,m^{-3}}$)	*Exposure* (days)	v_g ($\mathrm{mm\,s^{-1}}$)
Corn, cv. Marcross	3.1	25	0.42
	4.6	25	0.35
Alfalfa, cv. Ranger	2.4	46	0.39
	4.5	46	0.42
Tomato, cv. BonnyBest	13.9	17	0.09
	14.7	17	0.39

Table 13.5 HYPOTHETICAL RATES OF DEPOSITION OF PARTICULATE
FLUORIDE (μg g^{-1} day^{-1}) WITH AIR CONCENTRATION OF 1 μg F m^{-3}. WIND SPEED
4 m s^{-1}

Particle diameter (μm)	Deposition velocity (mm s^{-1})	Specific leaf weight of plant (g m^{-2})		
		20	40	100
0.1	0.2	0.8	0.4	<0.2
1.0	0.2	0.8	0.4	<0.2
5.0	4.0	17	8.6	3.4
10.0	11.5	50	25	10

1 μg m^{-3}. This illustrates the importance of both particle size and SLW but, in particular, it indicates the magnitude of the change in the rate of deposition that occurs in the particle size range from 1 to 10 μm. As particles up to about 20 μm in diameter adhere very strongly (Corn and Stein, 1965) to surfaces (*Figure 13.3*), it is suggested that, where forage quality is concerned, the important range of particle size to consider is approximately 5–20 μm. Below 5 μm the v_g is very low while above 20 μm it is high, but the particles are more easily removed.

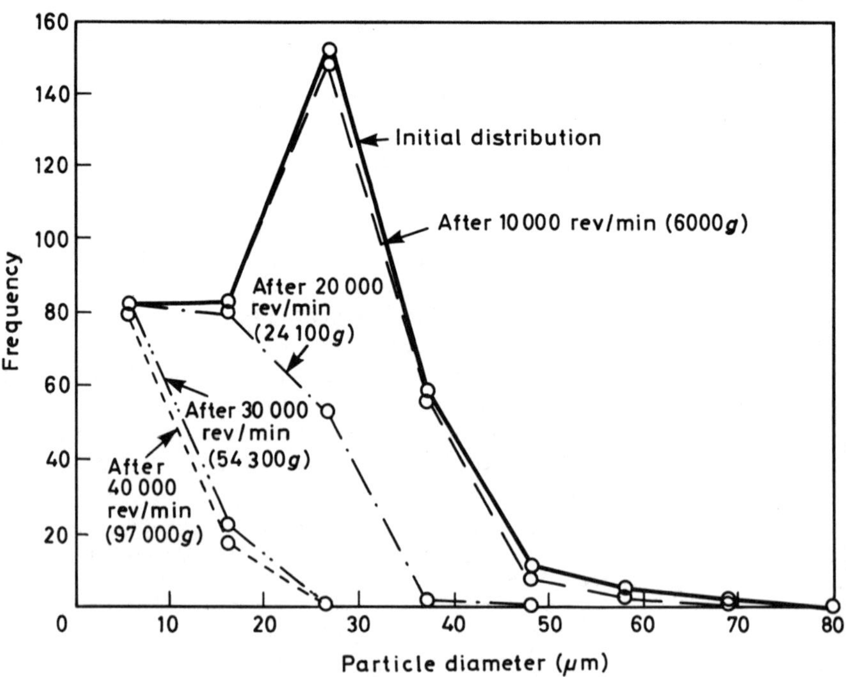

Figure 13.3   Illustration of the adhesion of particles to a surface and the force needed for re-entrainment. Glass beads ranging in diameter from 5 to 80 μm were deposited on a glass slide, subjected to different centrifugal forces, and the frequency distribution subsequently recorded. Even after 24 100 g most particles <20 μm were retained. From Corn and Stein, 1965. Courtesy of the American Industrial Hygiene Association

RESISTANCE ANALOGUES

Electrical resistance analogues have been used extensively in the study of gas exchange and consequently, in recent years, models based on such analogues have been used to study pollutant sorption (Waggoner, 1971; Bennett, Hill and Gates, 1973; Garland, Clough and Fowler, 1973; Fowler and Unsworth, 1974; Bennett and Hill, 1975; Unsworth, Biscoe and Black, 1976; O'Dell, Taheri and Kabel, 1977). None of these studies dealt exclusively with fluoride, although two (Bennett, Hill and Gates, 1973; O'Dell, Taheri and Kabel, 1977) referred to some aspects of the resistances involved in fluoride sorption. The aerodynamic and stomatal resistance to gaseous fluoride movement can be measured in the usual way (Šesták, Čatský and Jarvis, 1971; Unsworth, Biscoe and Black, 1976) using water vapour, replicas and porometry, while the internal resistance is probably negligible (O'Dell, Taheri and Kabel, 1977) because fluoride ions inside the leaf are swept to the tip and margins in the transpiration stream. However, cuticular and surface resistances are both problematical. Both gaseous and particulate fluorides are deposited on surfaces and although short-term measurements (Chamel and Garrec, 1977) using ^{18}F suggest that the intact cuticle may present a high resistance barrier, there is ample evidence that fluoride ions can penetrate the outer layers of the leaf. For example, several workers have used sprays of NaF or other salts and found that the effects were much the same as that of HF (Wander and McBride, 1956; Brewer *et al.*, 1967; Brewer, Sutherland and Guillemet, 1969; Leonard and Graves, 1972). Similarly, experiments on the effects of particulate fluorides (McCune *et al.*, 1965; Keller, 1973; McCune, Silberman and Weinstein, 1977) have shown that sufficient may penetrate the leaf to cause visible symptoms of toxicity. However, no estimates of

Table 13.6   ESTIMATES OF SURFACE RESISTANCE TO HF DEPOSITION FROM SHORT-TERM EXPERIMENTS, $21\,\mu g\,F\,m^{-3}$ FOR 2 h. Davison (unpublished work)

Species	Net surface resistance $(s\,m^{-1})$
Primula ioessa: upper surface	280
lower surface	310
Tiarella cordifolia: upper surface	310
lower surface	200
Polygonum affine: upper surface	850
lower surface	280
Pressure-sensitive tape 1	1380
2	1780
3	1260

cuticular resistance are available. With regard to the resistance to deposition on the leaf surface, Davison (unpublished work) made some preliminary measurements (*Table 13.6*) in a wind tunnel using excised leaves. Leaves were cut, placed in the dark and allowed to wilt in order to close the stomata. Diffusive resistance to water vapour was measured, then the upper or lower surface of each leaf was backed with pressure-sensitive tape cut to the shape of the individual leaf. The leaves were then placed in a wind tunnel and fumigated for 2 h at 21 μg F m^{-3}. After exposure, diffusive resistance was again measured, the backing tape was peeled off and the fluoride content of both leaf and backing tape was measured. The aerodynamic resistance of the leaves was measured using paper replicas and the net surface resistance calculated. Because of the similarity in molecular weight between HF and water vapour, and the magnitudes of other sources of variation, no correction was applied to resistance measurements. The results are very tentative but they suggest (*Table 13.6*) that the resistance to deposition on the surface was of the same order of magnitude as stomatal resistance. However, as the experiments were all short-term there was no indication of whether leaf surfaces have a finite, limited capacity for HF sorption. If there is a limit, then the surface resistance would be expected to increase nonlinearly with time.

COMPARISON OF PREDICTED RATES OF SORPTION

Table 13.7 shows a comparison of hypothetical rates of fluoride sorption by *Lolium* plants using the dose-rate equation, v_g, and a resistance analogue. In order to simplify the calculations in the resistance model, an aerodynamic resistance of 50 s m^{-1} was assumed and it was also assumed that the

Table 13.7   COMPARISON OF RATES OF SORPTION OF GASEOUS FLUORIDE PREDICTED USING THREE METHODS. RYEGRASS PLANTS WITH SPECIFIC LEAF WEIGHT = 20 g m^{-2}. AIR CONCENTRATION = 1 μg F m^{-3}

Method	*Rate of sorption*	
	(μg F m^{-2} d^{-1})	(μg F g^{-1} d^{-1})
1. *Dose-rate equation* with $K = 4$	—	4
2. *Deposition velocity* (v_g) ranging from 2 to 8 mm s^{-1}	173–691	9–35
3a. *Resistance analogue*		
Uptake via stomata 12 h d^{-1} average total resistance 250 s m^{-1}	173	9
Uptake via closed stomata 12 h d^{-1}; average resistance 2000 s m^{-1}	22	1
Surface sorption 24 h d^{-1}; average resistance 350 s m^{-1}	246	12
Total	441	22
3b. As 3a but aerodynamic resistance = 500 s m^{-1}		
12 h open stomata	62	3
12 h closed stomata	17	<1
24 h surface sorption	108	5
Total	187	9

average day-time stomatal resistance (both surfaces) was 200 s m^{-1}. Clearly, some of the assumptions are debatable and fairly small changes in v_g, the resistances or specific leaf weight would alter the picture. Nevertheless, despite this reservation it appears that estimates using v_g and the resistance model agree to a reasonable extent and that both are higher than the estimate predicted from the accumulation coefficient K. The differences in the predicted rates of sorption probably hinge mostly on the fact that the estimate of K was produced from experiments in controlled-environment chambers. In many chambers the wind speed is low enough to create a boundary-layer resistance that is sufficiently high to reduce pollutant sorption significantly (Ashenden and Mansfield, 1977). In support of this idea, when a high aerodynamic resistance was included in the model (*Table 13.7, Method 3b*) the predicted rate of sorption fell to less than half that with a low resistance. Another feature of controlled-environment chambers is that the canopy is usually dry, whereas in the field it is often wet for long periods. This difference would be expected to change the rate of surface deposition by a factor of two or more (Chamberlain and Chadwick, 1972; Fowler and Unsworth, 1974). On the other hand, the relatively high rate of sorption predicted by the resistance model may have been unduly influenced by the low value of the surface resistance that was used. If the leaf surface has a finite capacity, then the rate of sorption would be much lower over periods longer than the 2 h that was used to estimate the surface resistance (*Table 13.6*). However, although it is easy to speculate about the differences in the predicted rates, the comparisons cannot be taken very far because there is still a lack of basic information available about many of the processes involved.

LOSS OF FLUORIDES FROM LEAVES

Reports of rates of postfumigation decrease in fluoride concentration and of loss of fluorides are summarized in *Table 13.8*. In view of the range of species and different experimental conditions represented, it is not surprising that estimates are very variable, but the consensus of the evidence is

Table 13.8 EXAMPLES OF POSTFUMIGATION RATES OF DECREASE OF FLUORIDE CONTENT OF PLANTS

Author	*Observation*
1. Zimmerman and Hitchcock (1946)	First report of loss of fluoride ions. No data quoted.
2. Hitchcock *et al.* (1964)	Corn leaves lost up to 46–70% of fluoride one week or more after treatment
3. Guderian, van Haut and Stratmann (1969)	Old rape plants: [F] decreased from 310 to $105\,\mu g\,g^{-1}$ in 11 days. Young rape plants: [F] decreased from 465 to $62\,\mu g\,g^{-1}$ in 11 days.
4. Knabe (1970)	Spruce needles lost up to about $340\,\mu g\,F\,g^{-1}$ in 5 months. Washed needles lost up to about $200\,\mu g\,F\,g^{-1}$ in 5 months.
5. Hitchock *et al.* (1971)	Alfalfa: [F] decreased by 50% in 8–22 days.
6. Davison and Blakemore (1976)	[F] of washed grass decreased by up to $27\,\mu g\,g^{-1}$ in 1 day
7. Ghiasseddin *et al.* (1978)	Soybean leaves: [F] decreased by up to $180\,\mu g\,g^{-1}$ in 7 days
8. Davison, Blakemore and Craggs (1979)	Many examples of forage [F] decreasing by over $100\,\mu g\,g^{-1}$ in 1 day
9. Bunce (personal communication)	Western hemlock: [F] decreased by $40\,\mu g\,g^{-1}$ in 7 days

that the rate of decrease in fluoride concentration may be substantial and in some cases it appears to be of the same order of magnitude as rates of sorption.

There are several mechanisms that might lead to reduction in the fluoride content or loss of fluoride from forage, including growth dilution, death of leaves, shedding of surface waxes, leaching by rain and volatilization. Forage fluoride levels are frequently observed to be lower in the summer than in other seasons (Allcroft, Burns and Herbert, 1965; Gründer, 1972; Davison, Blakemore and Wright, 1976) and this has led to the belief that dry-matter production can exceed the rate of fluoride sorption and so lead to a reduced concentration. However calculation of rates of dry-matter increase by Hitchcock *et al.* (1971) showed that growth dilution could have accounted for only a part of the postfumigation decrease that they observed in alfalfa and they concluded that weathering was more important. Similarly the data of Guderian, Van Haut and Stratmann (1969) suggest that although growth dilution may have contributed to the observed decreases in concentration, there was an actual loss of fluoride ions from the plants. Knabe (1970) and Davison, Blakemore and Craggs (1979), have reported that the concentration of fluoride fell significantly at times of the year when there was no growth. Recently Craggs and Davison (in press) analysed the data of Blakemore (1978) using Box–Jenkins techniques (Box and Jenkins, 1978), and showed that, although the forage fluoride level did fall in the summer, it was primarily related to trends in airborne fluoride concentrations. It was concluded that, in the summer months, the weather brought about better dispersal of the pollutant and that this led to lower concentrations in the air, and consequently, in forage. However, there are other seasonal changes in swards that might lead to lower concentrations in the spring or summer months. Hunt and Cooper (1967), McColl and Cooper (1967) and Thomas and Norris (1977) have shown that the specific leaf weight of ryegrass is higher during the spring and early summer by a factor of about two compared with other seasons. This would therefore be expected to halve the rate of sorption when expressed on a dry-weight basis. A further seasonal change is the rate of turnover of leaves. Williamson (1976) showed for *Dactylis glomerata* and other grasses that the longevity of leaves depended on the time they were initiated and on subsequent temperature. The rate of turnover was fastest in the summer months with a t_{50} (time of death of 50% of leaves of the same age) of about 30 days compared with up to about 80 days in winter. This faster turnover suggests that fluoride would be more rapidly lost in summer. On the other hand, Davison and Blakemore (1976) found that loss of dry matter during senescence of barley leaves led to an increase in fluoride concentration.

Fluoride ions and particles on the surface of leaves may be lost when wax and other materials are shed during expansion, or when the surface is weathered by mechanical stress, insects and microorganisms. In a discussion of mechanisms of loss of aerosols from grass under field conditions, Chamberlain (1970) considered that weathering was probably the main mechanism involved, but as there appear to be no quantitative estimates available of rates of loss of wax or cuticle, the role of this mechanism remains speculative.

It is frequently observed that forage fluoride concentrations are inversely related to rainfall (NAS, 1971) and this suggests that rain may be an important mechanism for removal. In support of this, Brewer, Sutherland and Perez (1969) found that citrus foliage contained less fluoride when sprinkled or drenched daily with water. However, several considerations cast doubt upon this interpretation and indicate that the effects of rain and 'wetness' may be quite complex. When the relationships was investigated statistically (Davison and Blakemore, 1976; Blakemore, 1978; Craggs, 1980) the zero-order correlation was always low and around -0.3. If this is taken at face value and interpreted as a direct effect, then it suggests that under UK conditions rain accounted for about 9% of the variation in the data. However, even this low correlation may be indirect because the airborne fluoride concentration usually decreases during rainstorms because of 'scrubbing' of the air by falling rain (Thompson and Taylor, 1969), so the rate of sorption may be reduced for that reason. Rain is particularly effective in removing large particles from the air (Gregory, 1973). Further evidence that the correlation is indirect was provided by Craggs and Davison (in press) who did extensive trials with field-grown turf and a rain simulator (described in Clement, Jones and Hopper, 1972) and were unable to find any consistent evidence of removal or leaching. Similarly, Chamberlain (1970) was unable to find any consistent evidence that rain was an important factor in removing aerosols from grass. Some evidence suggests that rain may increase fluoride sorption because wetting the surface of a leaf increases v_g (Fowler and Unsworth, 1974) and because rain in polluted areas contains significant amounts of fluoride (Harriss and Williams, 1969; Garber, 1970; Slanina *et al.*, 1979); there may therefore be an input by wet deposition. There are a number of estimates of fluoride concentration in water collected in standard rain gauges, but there do not appear to be any reliable measurements of rates of wet deposition or of rain fluoride concentration during rainfall, so it is impossible to determine how important wet deposition may be.

Finally, a mechanism of loss that has been proposed or discussed on many occasions (Zimmerman and Hitchcock, 1946; Weinstein, 1961; Hitchcock *et al.*, 1964; Knabe, 1970), but never investigated quantitatively, is volatilization from the leaf surface or substomatal cell walls. This possibility was raised originally by Zimmerman and Hitchcock (1946) who reported that postfumigation losses of fluoride were probably due to the production of volatile compounds. They also noted that fluorides accumulated in the air inside cases in which plants were enclosed that had previously been fumigated. The difficulty with this suggestion has always been in envisaging what type of compound might be involved. The work of Peters and Shorthouse (1971) suggests the possibility of volatile organic compounds such as fluoroketones being formed, whereas the observation of desorption from air filters (Israel, 1974b) suggests that an inorganic form may be volatile. Because field experiments of Davison and Blakemore (1976) and Davison, Blakemore and Craggs (1979) showed that there were such rapid losses of fluoride from grass swards, Nişancioğlu and Davison (unpublished work) investigated volatilization by measuring rates of loss of inorganic fluoride from paper models of leaves. Fluoride as NaF was carefully and evenly spread on paper models that had previously been impregnated with citrate-phosphate buffer. After rapid drying the models were suspended in fluoride-free air for various periods. *Figure 13.4* shows that fluoride was lost at a measurable rate and that it was inversely related to pH. The loss curves were strikingly similar to the well-known curves for the formation of HF in solution so it therefore appeared that, depending on pH, a proportion of the fluoride ions complexed with hydrogen ions to form HF, and a fraction of this was being released. Later experiments showed that the rate of loss was not linear with time, that it was directly

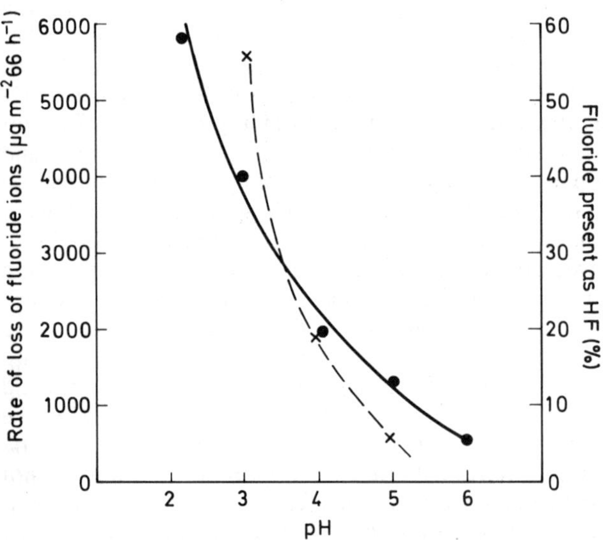

Figure 13.4   Rate of loss of fluoride ions (μg m^{-2} 66 h^{-1}) from paper strips at different pH (solid line), and % of fluoride in solution in the form of HF at different pH (dashed line)

proportional to the concentration on the paper and that it was affected by wind speed, presumably because the boundary layer offered a significant resistance to loss. The effects of concentration and wind speed suggest that loss by this mechanism may be modelled by means of a resistance analogue in the same way as sorption. Although no experiments have been done with real leaves it appears that volatilization of HF is feasible and could account for some of the observed losses.

Returning to the original question of what concentration of fluoride in air leads to unacceptable levels in forage, it is clear that there is no simple answer but the use of deposition velocity and resistance analogues allows the processes of sorption and loss to be analysed in such a way that it may be possible to produce a useful predictive model when more information is available on rates of surface sorption, on the capacity of leaves to sorb fluoride, on the effects of intermittent fumigation and on the mechanisms and rates of loss.

Effects on growth

A number of techniques have been used for assessing the impact of fluorides on vegetation, including field surveys of visible effects and plant performance, tree growth-ring analysis, recording of effects on indicator species and fumigation with controlled concentrations of fluorides or ambient air. As each of these has its own merits and limitations it is useful to consider the effects of growth in terms of the techniques that are available.

FIELD SURVEYS

Field surveys have been done in every country in which there are major fluoride sources. Inspection of the area around a point source usually reveals gradients in the fluoride content of leaves (Kay, 1974; McClenahan and Weidensaul, 1977; Kronberger and Halbwachs, 1978; Mankovska, 1979; Thompson, Sidhu and Roberts, 1979), soil (Thompson, Sidhu and Roberts, 1979) and lichens (Le Blanc, Rao and Comeau, 1972; Perkins, Millar and Neep, 1980) and there may also be variation in the diversity of the lichen flora (Gilbert, 1973), in tree growth (Treshow, Anderson and Harner, 1967; Høgdahl, Karstensen and Virik, 1977; ASC, 1979) and in plant performance. In some cases the composition of vegetation varies (Gilbert, 1975; McClenahan, 1978). Some authors have attributed differences and gradients to the direct effects of fluoride, but interpretation of these correlations is difficult (Weinstein, 1977) because they may represent indirect effects involving other pollutants (most fluoride sources emit other pollutants such as aerosols, SO_2, NO_x, phosphates), variations in soil and climate, and the disturbances that inevitably surround major industrial complexes. Günther (1971) commented on the fact that disturbance of the water table by industrialization often produces visible symptoms very similar to those of pollutant injury. Statistical analysis presents one method of isolating the effects of fluoride in such complex situations, but it has been little used. McClenahan (1978) examined carefully selected stands of vegetation along gradients in pollution levels, and recorded species richness, evenness of stand, diversity index and concentration of three pollutants. Most vegetation characteristics were altered in areas of high pollution. Statistical analysis showed a significant positive correlation between fluorides and shrub density but, overall, chloride appeared to have the greatest impact.

ANALYSIS OF TREE GROWTH

Although analysis of tree growth rings is retrospective and cannot be used to predict future effects, it is a powerful tool for assessing the past effects of pollutants on forest growth. Shaw *et al.* (1951) and Lynch (1951) observed reductions in diameter growth in Ponderosa pine that were attributable to fluoride emission, although variations in precipitation presented a confounding factor that made interpretation more difficult. Treshow, Anderson and Harner (1967) did a similar study near a phosphate works and showed that, at the highest and intermediate fluoride sites, needle length was significantly greater after start-up, but at the highest sites radial growth was reduced. A multidisciplinary approach was used by Høgdahl, Karstensen and Virik (1977) in a study of the effects of smelter emissions on spruce forests. It included a study of lichens, meteorology and plume dispersal, and an inventory of tree deaths that was used to define the area affected. Two statisticians then independently examined growth increments by different methods and mapped zones of growth reduction. The two approaches showed no major differences of opinion and it was concluded that there had been a 20% reduction in growth for about 5 km downwind, and a 10% reduction for about 1 km on either side of that area. In places there was a 5% reduction up to 9 km distant. A similar study was conducted by Reid Collins Associates (Parker, Bunce and Smith, 1974; Bunce, 1978; Bunce, 1979; ASC, 1979) in the Kittimat area of Canada to determine the effects of fluoride emissions from an aluminium smelter. The study is a good illustration of the difficulty of interpreting field observations, because there were several major confounding influences. The smelter was situated in a valley with varying soil and terrain, and was surrounded by mature or overmature climax forest containing a large proportion of dead trees and branches. The low growth potential was a result of the age, the northerly location and the terrain. Between 1960 and 1967, insect infestations involving three species killed a large proportion of the trees growing under the plume path. Various explanations were put forward (*see* Weinstein in ASC, 1979) to explain the association between the outbreak and the plume, but whatever the actual reason it is an inescapable fact that there was some causal interaction between a component or components of the plume and the insects. A further complication arose in 1970 with the building and operation of a pulp mill that released SO_2. The investigators used X-ray densitometry (Parker, Bunce and Smith, 1974) to examine growth increments and they isolated the direct effects of fluoride from the other site factors by comparing periods of growth, for instance before and after smelter operation, before and after insect outbreaks, or before and after pulp-mill operation. It was concluded that the growth loss attributable to fluoride was 28% in the inner zone and 19% in the outer zone around the works, while the total volume loss over 22 years of smelter operation was 2.09×10^6 cubic feet ($5.9 \times 10^4\,\mathrm{m}^3$). This compared with approximately 80×10^6 cubic feet ($2.3 \times 10^6\,\mathrm{m}^3$) lost as a result of the insect infestation. A comment made by the Forest Service (ASC, 1979) on this reduction, was that 19% or 28% loss in an overmature forest of such low potential was of much less significance than the same reduction in a young stand. This is an important point because it illustrates that the impact on growth cannot be judged in simple percentage terms; it has to be judged in context. Other pertinent findings were that there was no relationship between foliar fluoride and growth reduction, that there was satisfactory regeneration within the affected area, that it was impossible to determine whether fluoride was responsible for the deaths of individual trees and that in the outermost zone there may have been some fluoride-induced growth stimulation. In connection with the last finding it is interesting to note that since emissions have been reduced, the British Columbia Forest Service reported that a species trial was showing better growth and survival in the plume area than at other sites (ASC, 1979).

INDICATORS

Indicator species that produce a characteristic visible reaction are extensively used to monitor the occurrence of biologically active doses of pollutants. Posthumus (1976 and this volume, Chapter 2) has discussed the advantages and limitations of the technique and concluded that the main advantage is that the effects are an integration of the varying concentrations and conditions. However, he has also cautioned that it is often difficult to standardize measurements and that there are varying influences of cultural conditions and other biological agents. As the consensus of research suggests (*see below*) that there is little relationship between visible symptoms and effects on growth, it seems that the most effective use of indicators is for preliminary delimiting of areas for further study in the way that Høgdahl, Karstensen and Virik (1977) used lichens, and for monitoring areas where there may be visible effects on plants of aesthetic importance, such as horticultural species or specimen trees in parks and gardens.

A development of the indicator technique is the use of physiological or biochemical measurements on indigenous vegetation or transplants. Keller (1974) used peroxidase activity to map the effects of fluoride on apricot orchards, while Yee-Meiler (1977) recorded differences in phenols in potted plants sited at different distances from a fluoride source. Keller (1977) used the same technique and recorded net assimilation rate. Under standard conditions the assimilation rate decreased in plants sited close to the source, without any associated sign of visible injury. This technique appears to have considerable potential, particularly if coupled with growth measurements and records of atmospheric concentrations.

FUMIGATION STUDIES

Most of the reports on the effects of fluoride involve fumigation in controlled or partly controlled environments. They include a wide variety of environmental conditions and fluoride concentrations, although most have involved constant concentrations of HF and relatively short-term exposures. Weinstein (1977) commented that because of these differences, it is difficult to compare results from different laboratories and that, for a variety of reasons, a proportion of the literature may be of limited relevance to the field situation. Weinstein (1977) listed reports of HF fumigation leading to a decrease in dry weight of roses, citrus, sorghum, alfalfa, orchard grass and lettuce, and to reduced leaf area and leaf size in various citrus crops. He also listed a similar number of reports where there was no effect on growth, and several where there was a positive effect. These latter reports included increases in total linear growth of citrus, rose and tomato, and weight of alfalfa and bean. Because of the range of conditions used by different authors and the limited number of experiments on any one species, it is impossible at present to construct useful dose/growth-response curves of the type that have been produced (McCune, 1969) for visible injury. However, fumigation experiments have consistently demonstrated several important relationships between exposure, symptoms and growth effects. One of these is the relationship between visible symptoms of fluoride injury and growth. Benedict, Ross and Wade (1964) reported that for several crop species there was no significant effect of fluoride on growth in the absence of visible symptoms, while in experiments by Guderian, Van Haut and Stratmann (1969), depression of growth was usually accompanied by chlorosis or necrosis. However, many reports (Brewer *et al.*, 1960, 1960–61; Treshow and Harner, 1968; Pack, 1971; Pack and Sulzbach, 1976; Guderian, 1977; MacLean, Schneider and McCune, 1977) support the view that there may be significant positive or negative effects on the growth of shoots, leaves, or fruit without visible symptoms. Conversely, there may be visible symptoms with no discernible effect on growth (Hitchcock *et al.*, 1971; NAS, 1971). MacLean, Schneider and McCune (1977) showed that, in the case of bean (*Phaseolus vulgaris*) and tomatoes grown at $0.6\,\mu g\,F\,m^{-3}$ throughout their life, there were no visible symptoms on either species but the fresh weight of marketable beans was reduced by 25%. On the basis of foliar injury, tomato is usually classed as being more sensitive than bean and yet the reverse is true when fruiting is considered. It appears that foliar injury is of limited use in predicting effects on growth.

A further finding that has emerged from fumigation experiments, and one that is supported by field observations, is that the reaction of a plant under fluoride stress is not a simple negative or positive one. Different organs may react differently at different phases of growth and the responses may be either independent or integrated (the direct response of one organ causing an effect on another). The result is that the effect of fluoride may be to depress growth of some organs and increase the growth of others on the same plant. For example, Brewer *et al.* (1960–61) showed that fumigation of citrus decreased the size of individual leaves but increased the linear growth of the same trees. In the field, Treshow, Anderson and Harner (1967) recorded decreased radial growth of Douglas fir coupled with longer needles in a high-fluoride area. The work of Pack (1971) and Pack and Sulzbach (1976) has similarly demonstrated the complex positive and negative effects that may occur within a single experiment. The data (Pack and Sulzbach, 1976) in *Table 13.9* shows that exposure of pea (cv. Alaska) to $5-9\,\mu g\,F\,m^{-3}$ resulted in significant increases in the fresh and dry weight of leaves and stems, coupled with significant decreases in fresh weight, dry weight and number of fruit. Pack's work also indicates that effects of this type differ with cultivar and with dose of fluoride. The practical significance is that an effect on one growth characteristic is not a very useful indicator of the potential effect on any other, so a study should have the objectives clearly defined (Tingey, Wilhour and Taylor, 1979) and should include measurements of several

Table 13.9   RESPONSE OF PEA TO HF. Data of Pack and Sulzbach (1976).
* = significantly different from control

HF concentration ($\mu g\,m^{-3}$)	Number of seeds per fruit		Weight per seed (g)		Stem length (cm)		Dry wt, stems and leaves (g)	
	Control	*+ HF*	*Control*	*+ HF*	*Control*	*+ HF*	*Control*	*+ HF*
4.8	3.73	2.52*	0.220	0.177*	102	124*	20.5	26.7*
5.1	3.91	2.24*	0.232	0.180*	96	98	21.4	19.4
9.1	3.40	2.66*	0.215	0.182*	79	97*	16.8	20.4*

different characteristics, preferably those of direct economic, aesthetic or ecological importance.

Studies of the effects of particulate fluorides have been concerned mostly with the toxicity relative to HF, and most have involved cryolite. McCune *et al.* (1965) exposed gladiolus, Milo maize, corn, tomato and alfalfa to cryolite for 9–49 days at concentrations from 1.5 to $15\,\mu g\,F\,m^{-3}$. The only visible injury was some tip necrosis in gladiolus and there was no effect on growth or yield of any species. In later work McCune, Silberman and Weinstein (1977) demonstrated that the presence of free water increased foliar injury by cryolite particles. Keller (1973) compared the effects of NaF, CaF_2, natural cryolite and artificial cryolite on potted plants of birch, Douglas fir and Scots pine. Only NaF produced visible injury, but all of the particles reduced relative photosynthesis of birch. Less *et al.* (1975) did not report any symptoms or growth reduction in ryegrass after exposure to particles less than a micrometre in diameter. In view of the relatively high v_g and strong adhesion of particles in the range 5–20 µm, it would be of interest to compare their toxicity with that of the smaller particles that have been tested to date. Furthermore, the chemical nature of particulate materials is very varied, so studies are needed of materials such as fluoridated alumina, in which the fluoride is chemisorbed rather than chemically combined, and of the complex materials that are released as fugitive emissions from dry scrubbers.

Fumigation of plants with ambient air containing fluorides has not been reported on any scale, probably because of the size and cost of the experiments that are needed and the technical difficulties involved. Thompson and Taylor (1969) found no effects on the growth of citrus exposed to ambient air containing fluoride concentrations from 0 to $1.2\,\mu g\,m^{-3}$. Leonard and Graves (1972) enclosed orange trees in plastic greenhouses and supplied either ambient air, or air that was partly scrubbed by calcium carbonate. Some trees were used as outdoor controls. Leaf size and yields were significantly reduced by the unfiltered air, but complicating factors that seem to be typical of field experiments were a severe frost at the start of the experiment and a mealy-bug infestation that affected only those trees that were in the greenhouses. In Germany, Guderian and Schoenbeck (1970) used test chambers (Van Haut, 1972) ventilated with filtered and unfiltered air to demonstrate that fluorides were a critical source of injury near a chemical factory while in the UK, Brough, Parry and Whittingham (1978) examined the effects of emissions (SO_2 and F) from brickworks on barley growth, using open-top chambers and polythene tunnels. They encountered several difficulties in using chambers in the field and found that enclosing barley plants in open-top chambers had a detrimental effect on yield, largely through a reduction in tiller number (M. Parry, personal communication). In the 6 years that the experiments were repeated, filtration gave variable but usually significant increases in yield. As both SO_2 and fluoride were present in the ambient air, the effects could have been due to either pollutant alone, or to the combination. Further progress with systems using ambient air will probably depend on overcoming the technical problems associated with outdoor chambers and on the development of filtration methods that allow selective removal of individual pollutants. The alternative possibility, of exposing plants to simulated ambient air by continuous monitoring of pollutants and injecting the same concentrations into clean air, is limited by the lack of a fast-response analytical instrument for airborne fluorides.

Interactions between fluorides, pests, pathogens and other pollutants

The effects of fluoride on plant-dwelling insects were reviewed by Weinstein (1977), and its effects on various disease organisms, by Heagle (1973) and Weinstein (1977). Although there is good evidence that the presence of fluoride may alter insect populations and affect the incidence of disease, research has centred on the effects on numbers and performance of insects, and on incidence of disease, so that there is little or no indication whether interactions between fluoride and pest or disease leads to any effect, positive or negative, on plant yield. Interactions between fluoride and other pollutants have received much less attention than some other interactions (for example that of SO_2 and O_3). McCune (1980) listed several publications on the subject (Solberg and Adams, 1956; Hitchcock, Zimmerman and Coe, 1962; Matsushima and Brewer, 1972 and Mandl, Weinstein and Keveny, 1975) and he reviewed a great deal of unpublished work from the Boyce Thompson Institute. Neither the published work nor the previously unpublished work showed any evidence of nonadditive effects on growth when HF was combined with SO_2, NO_2 or O_3. The most consistent effects were on foliar symptoms and fluoride content.

References

ALLCROFT, R., BURNS, K.N. and HERBERT, C.N. (1965). *Fluorosis in Cattle. II. Development and Alleviation: Experimental Studies. Ministry of Agriculture, Fisheries and Food, Animal Disease Surveys Report 2, Part III.* HMSO, London

ASC (1979). *Environmental Effects of Emissions from the Alcan Smelter at Kittimat, BC. Report of the Alcan Surveillance Committee.* Ministry of the Environment, Province of British Columbia, Canada

ASHENDEN, T.W. and MANSFIELD, T.A. (1977). *Journal of Experimental Botany,* **28**, 729–735

BENEDICT, H.M., ROSS, J.M. and WADE, R.H. (1964). *International Journal of Air and Water Pollution,* **8**, 279–289

BENNETT, J.H. and HILL, A.C. (1975). In *Responses of Plants to Air Pollution*, pp. 273–306 (Mudd, J.B. and Kozlowski, T.T., Eds), Academic Press, New York

BENNETT, J.H., HILL, A.C. and GATES, D.M. (1973). *Journal of the Air Pollution Control Association,* **23**, 957–962

BLAKEMORE, J. (1978). *Fluoride Deposition to Grass Swards Under Field Conditions.* PhD thesis, University of Newcastle upon Tyne

BOX, G.E.P. and JENKINS, G.M. (1978). *Time Series Analysis, Forecasting and Control.* Holden-Day, San Francisco

BREWER, R.F., SUTHERLAND, F.H. and GUILLEMET, F.B. (1969). *Environmental Science and Technology,* **3**, 378–381

BREWER, R.F., SUTHERLAND, F.H. and PEREZ, R.O. (1969). *Proceedings of the American Society for Horticultural Science*, **94**, 284–286

BREWER, R.F., CREVELING, R.K., GUILLEMET, F.B. and SUTHERLAND, F.H. (1960–61). *Proceedings of the American Society for Horticultural Science*, **75**, 236–243

BREWER, R.F., GARBER, M.J., GUILLEMET, F.B. and SUTHERLAND, F.H. (1967). *Proceedings of the American Society for Horticultural Science*, **91**, 150–156

BREWER, R.F., SUTHERLAND, F.H., GUILLEMET, F.B. and CREVELING, R.K. (1960). *Proceedings of the American Society for Horticultural Science*, **76**, 208–214

BROUGH, A., PARRY, M. and WHITTINGHAM, C.P. (1978). *Chemistry and Industry*, 51–53

BUNCE, H.W.F. (1978). *Paper 78-24.3, 71st Annual Meeting of the Air Pollution Control Association, Houston, Texas*

BUNCE, H.W.F. (1979). *Journal of the Air Pollution Control Association*, **29**, 642–643

CHADWICK, R.C. and CHAMBERLAIN, A.C. (1970). *Atmospheric Environment*, **4**, 51–56

CHAMBERLAIN, A.C. (1967). *Proceedings of the Royal Society, Series A*, **296**, 45–70

CHAMBERLAIN, A.C. (1970). *Atmospheric Environment*, **4**, 57–78

CHAMBERLAIN, A.C. (1975). In *Heat and Mass Transfer in the Biosphere, Part I, Transfer Processes in the Plant Environment*, pp. 561–582 (de Vries, D.A. and Afgan, N.H., Eds), Scripta Book Co., John Wiley, New York

CHAMBERLAIN, A.C. (1980). In *Air Pollutants and their Effects on the Terrestrial Ecosystem* (Krupa, S.V. and Legge, A.H., Eds), John Wiley, New York (in press)

CHAMBERLAIN, A.C. and CHADWICK, R.C. (1972). *Annals of Applied Biology*, **71**, 141–148

CHAMEL, A. and GARREC, J.P. (1977). *Environmental Pollution*, **12**, 307–310

CHANG, C.W. (1975). In *Responses of Plants to Air Pollution*, pp. 57–95 (Mudd, J.B. and Kozlowski, T.T., Eds), Academic Press, New York

CLEMENT, C.R., JONES, L.H.P. and HOPPER, M.J. (1972). *Journal of Applied Ecology*, **9**, 249–260

COOKE, J.A., JOHNSON, M.S., DAVISON, A.W. and BRADSHAW, A.D. (1976). *Environmental Pollution*, **11**, 9–23

CORN, M. and STEIN, F. (1965). *American Industrial Hygiene Association Journal*, **26**, 325–336

CRAGGS, C. (1980). *A Statistical and Experimental Approach to Problems Arising from Fluoride Contamination of Grassland*. PhD thesis, University of Newcastle upon Tyne

CRAGGS, C. and DAVISON, A.W. (1982). *Environmental Pollution* (in press)

DAVISON, A.W. and BLAKEMORE, J. (1976). In *Effects of Air Pollutants on Plants*, pp. 17–30 (Mansfield, T.A., Ed.), Society for Experimental Biology Seminar Series, Vol I. Cambridge University Press, Cambridge

DAVISON, A.W. and BLAKEMORE, J. (1980). *Environmental Pollution B*, **1**, 305–319

DAVISON, A.W., BLAKEMORE, J. and CRAGGS, C. (1979). *Environmental Pollution*, **20**, 279–296

DAVISON, A.W., BLAKEMORE, J. and WRIGHT, D.A. (1976). *Environmental Pollution*, **10**, 209–215

FOWLER, D. and UNSWORTH, M.H. (1974). *Nature*, **249**, 389–390

GARBER, K. (1970). *Fluoride*, **3**, 22–26

GARLAND, J.A., CLOUGH, W.S. and FOWLER, D. (1973). *Nature*, **242**, 256–257

GHIASSEDIN, M., MASON, J.W., HUGHES, J.M. and DIEM, J.E. (1978). *Paper No. 78-24.4, 71st Annual Meeting of the Air Pollution Control Association, Houston, Texas*

GILBERT, O.L. (1973). In *Air Pollution and Lichens*, pp. 176–191 (Ferry, B.W., Baddeley, M.S. and Hawksworth, D.L., Eds), Athlone Press, University of London, London

GILBERT, O.L. (1975). *Environmental Pollution*, **8**, 113–121

GREGORY, P.H. (1973). *The Microbiology of the Atmosphere*. Leonard Hill, London

GROTH, E. (1975). *Environment*, **17**, 19–38

GRÜNDER, H.D. (1972). *Fluoride*, **5**, 74–81

GUDERIAN, R. (1977). *Air Pollution*. Springer Verlag, Berlin

GUDERIAN, R. and SCHOENBECK, H. (1970). In *Proceedings of the Second International Clean Air Congress*, pp. 266–273. Clean Air Association, Brighton, UK

GUDERIAN, R., VAN HAUT, H. and STRATMANN, H. (1969). *Forschungsberichte des Landes Nordrhein-Westfalen*, No. 2017

GÜNTHER, K.H. (1971). In *Fume Damage to Forests*, p.45. Forestry Commission Research and Development Paper No. 82. Forestry Commission, London

HARRISS, R.C. and WILLIAMS, H.H. (1969). *Journal of Applied Meteorology*, **8**, 299–301

HEAGLE, A.S. (1973). *Annual Reviews of Physiology*, **11**, 365–388

HILL, A.C. (1971). *Journal of the Air Pollution Control Association*, **21**, 341–346

HITCHCOCK, A.E., ZIMMERMAN, P.W. and COE, R.R. (1962). *Contributions. Boyce Thompson Institute for Plant Research*, **21**, 303–344

HITCHCOCK, A.E., McCUNE, D.C., WEINSTEIN, L.H., MACLEAN, D.C., JACOBSON, J.S. and MANDL, R.H. (1971). *Contributions. Boyce Thompson Institute for Plant Research*, **24**, 363–386

HITCHCOCK, A.E., WEINSTEIN, L.H., McCUNE, D.C. and JACOBSON, J.S. (1964). *Journal of the Air Pollution Control Association*, **14**, 503–509

HØGDAHL, B., KARSTENSEN, R. and VIRIK, E. (1977). *Paper No. A77-84, 106th Annual Meeting of the AIME, Atlanta, Georgia*

HUNT, L.A. and COOPER, J.P. (1967). *Journal of Applied Ecology*, **4**, 437–458

ISRAEL, G.W. (1974a). *Atmospheric Environment*, **8**, 1329–1330

ISRAEL, G.W. (1974b). *Atmospheric Environment*, **8**, 159–166

ISRAEL, G.W. (1977). *Atmospheric Environment*, **11**, 183–188

KAY, E. (1974). *Fluoride*, **7**, 7–31

KELLER, T. (1973). *Staub Reinhaltung der Luft* (in English), **33**, 379–381

KELLER, T. (1974). *European Journal of Forest Pathology*, **4**, 11–19

KELLER, T. (1977). *Eidgenössische Anstalt für Das Forstliche Versuchswesen*, **53**, 161–198

KNABE, W. (1970). *Staub Reinhaltung der Luft,* **30**, 29–32

KRONBERGER, W. and HALBWACHS, G. (1978). *Fluoride,* **12**, 129–135

LARSEN, S. and WIDDOWSON, A.E. (1971). *Journal of Soil Science,* **22**, 210–221

LE BLANC, F., RAO, D.N. and COMEAU, G. (1972). *Canadian Journal of Botany,* **50**, 991–998

LEONARD, C.D. and GRAVES, H.B. (1972). *Fluoride,* **5**, 145–163

LESS, L.N., McGREGOR, A., JONES, L.H.P., COWLING, D.W. and LEAFE, E.L. (1975). *International Journal of Environmental Studies,* **7**, 153–160

LITTLE, P. (1977). *Environmental Pollution,* **12**, 293–305

LITTLE, P. and WIFFEN, R.D. (1977). *Atmospheric Environment,* **11**, 437–447

LYNCH, D.W. (1951). *Northwest Science,* **25**, 157–163

McCLENAHAN, J.R. (1978). *Canadian Journal of Forest Research,* **8**, 432–438

McCLENAHAN, J.R. and WEIDENSAUL, T.C. (1977). *Journal of Environmental Quality,* **6**, 169–173

McCOLL, D. and COOPER, J.P. (1967). *Journal of Applied Ecology,* **4**, 113–127

McCUNE, D.C. (1969). *On the Establishment of Air Quality Criteria, With Reference to the Effects of Atmospheric Fluoride on Vegetation.* Air Quality Monograph, 69–3. American Petroleum Institute, New York

McCUNE, D.C. (1980). Terrestrial vegetation-air pollutant interactions: gaseous pollutants: hydrogen fluoride and sulphur dioxide. Paper presented at the *International Conference on Air Pollutants and their Effects on the Terrestrial Ecosystem. May 10–17, Banff, Canada*

McCUNE, D.C. and HITCHCOCK, A.E. (1970). In *Proceedings of the Second International Clean Air Congress*, pp. 289–292. Washington, DC

McCUNE, D.C., SILBERMAN, D.H. and WEINSTEIN, L.H. (1977). In *Proceedings of the 4th International Clean Air Congress*, pp. 116–119. Washington, DC

McCUNE, D.C., HITCHCOCK, A.E., JACOBSON, J.S. and WEINSTEIN, L.H. (1965). *Contributions. Boyce Thompson Institute for Plant Research,* **23**, 1–12

MacLEAN, D.C. and SCHNEIDER, R.E. (1973). *Journal of Environmental Quality,* **2**, 501–503

MacLEAN, D.C., SCHNEIDER, R.E. and McCUNE, D.C. (1977). *Journal of the American Society for Horticultural Science,* **102**, 297–299

MANDL, R.H., WEINSTEIN, L.H. and KEVENY, M. (1975). *Environmental Pollution,* **9**, 133–143

MANKOVSKA, B. (1979). *Biologia,* **34**, 563–570

MARIER, J.R. and ROSE, D. (1971). *Environmental Fluoride.* Publication No. 12226, National Research Council of Canada

MATSUSHIMA, J. and BREWER, R.F. (1972). *Journal of the Air Pollution Control Association,* **22**, 710–713

NAS (1971). *Biologic Effects of Air Pollutants: Fluorides.* National Academy of Sciences, Washington, DC

NAS (1974). *Effects of Fluorides on Animals.* National Academy of Sciences, Washington, DC

O'DELL, R.A., TAHERI, M. and KABEL, R.L. (1977). *Journal of the Air Pollution Control Association,* **27**, 1104–1109

OELSCHLÄGER, W. (1971). *Fluoride,* **4**, 80–84

PACK, M.R. (1971). *Environmental Science and Technology*, **5**, 1128–1132

PACK, M.R. and SULZBACH, C.W. (1976). *Atmospheric Environment*, **10**, 73–81

PARKER, M.L., BUNCE, H.W.F. and SMITH, J.H.G. (1974). The use of X-ray densitometry to measure effects of air pollution on tree growth near Kittimat, British Columbia. Conference on Air Pollution and Forestry, Marianski Lazne, Czechoslovakia

PERKIN, D.F., MILLAR, R.O. and NEEP, P.E. (1980). *Environmental Pollution (Series A)*, **21**, 155–168

PETERS, R.A. and SHORTHOUSE, M. (1971). *Nature*, **231**, 123–124

POSTHUMUS, A.C. (1976). In *Proceedings of the Kuopio Meeting on Plant Damage Caused by Air Pollution*, pp. 115–120 (Kärenlampi, L., Ed.), Kuopio

ROBINSON, W.O. and EDGINGTON, G. (1946). *Soil Science*, **61**, 341–353

ROSE, D. and MARIER, J.R. (1977). *Environmental Fluoride 1977*. Publication No. 16081. National Research Council of Canada

SCHWELA, D. (1978). An estimate of deposition velocities of several air pollutants on grass. Presented at *3rd International Congress of Plant Pathology, Munich*

SCHWELA, D.H. (1979). *Ecotoxicology and Environmental Safety*, **3**, 174–189

ŠESTÁK, Z., ČATSKÝ, J. and JARVIS, P.G. (1971). *Plant Photosynthetic Production: Manual of Methods*. W. Junk, The Hague, Holland

SHAW, C.G., FISCHER, G.W., ADAMS, D.F., ADAMS, M.F. and LYNCH, D.W. (1951). *Northwest Science*, **15**, 156

SLANINA, J., VAN RAAPHORST, J.G., ZIJP, W.L., VERMEULEN, A.J. and ROEF, C.A. (1979). *International Journal of Environmental and Analytical Chemistry*, **6**, 67–81

SOLBERG, R.A. and ADAMS, D.F. (1956). *American Journal of Botany*, **43**, 755–760

SUTTIE, J.W. (1969). *Air Quality Criteria to Protect Livestock from Fluoride Toxicity*. The Aluminium Association, New York

SUTTIE, J.W. (1977). *Journal of Occupational Medicine*, **19**, 40–48

THOMAS, N. and NORRIS, I.B. (1977). *Journal of Applied Ecology*, **14**, 949–964

THOMPSON, C.R. and TAYLOR, O.F. (1969). *Environmental Science and Technology*, **3**, 934–940

THOMPSON, L.K., SIDHU, S.S. and ROBERTS, B.A. (1979). *Environmental Pollution*, **18**, 221–234

TINGEY, D.T., WILHOUR, R.G. and TAYLOR, O.C. (1979). In *Handbook of Methodology for the Assessment of Air Pollution Effects on Vegetation*, pp. 7–1 to 7–35 (Heck, W.W., Krupa, S.V. and Linzon, S.N., Eds), Informative Report No. 3. TE-2 Agricultural Committee. Air Pollution Control Association, Pittsburgh

TRESHOW, M. (1971). *Annual Review of Phytopathology*, **9**, 21–44

TRESHOW, M. and HARNER, F.M. (1968). *Canadian Journal of Botany*, **46**, 1207–1210

TRESHOW, M., ANDERSON, M.K. and HARNER, F. (1967). *Forest Science*, **13**, 114–120

UNSWORTH, H.M., BISCOE, P.V. and BLACK, V. (1976). In *Effects of Air*

Pollutants on Plants, pp. 5–16 (Mansfield, T.A., Ed.), Society for Experimental Biology Seminar Series, Vol. 1. Cambridge University Press, Cambridge

VAN HAUT, H. (1972). *Environmental Pollution*, **3**, 123–132

WAGGONER, P.E. (1971). *Bioscience*, **21**, 455–459

WANDER, F.W. and McBRIDE, J.J. (1956). *Science*, **123**, 933–934

WEINSTEIN, L.H. (1961). *Contributions. Boyce Thompson Institute for Plant Research*, **21**, 215–231

WEINSTEIN, L.H. (1977). *Journal of Occupational Medicine*, **19**, 49–78

WILLIAMSON, P. (1976). *Journal of Ecology*, **64**, 1059–1076

YEE-MEILER, D. (1977). *Eidgenossische Anstalt für das Forstliche Versuchswesen*, **53**, 201–229

ZIMMERMAN, P.W. and HITCHCOCK, A.E. (1946). *American Journal of Botany*, **33**, 233

14

OZONE AND THE GROWTH AND PRODUCTIVITY OF AGRICULTURAL CROPS

JAY S. JACOBSON
Boyce Thompson Institute, Cornell University, Ithaca, NY

Introduction

In recent decades, ozone, generated by photochemical reactions in the lower troposphere, has become an important air pollutant in many countries. It seems likely to continue in importance at least to the end of this century because the reactants in these photochemical reactions, nitrogen oxides and hydrocarbons, are emitted whenever fossil fuels are burned. In countries where dependence on fossil fuels is increasing, photochemical reactions may become more prevalent than they are now. At present, ozone concentrations sufficient to produce foliar symptoms on susceptible vegetation occur in Great Britain and in at least seven European countries (Eastmond and Skarby, 1979). In the USA, more crop plants and forest trees are injured by ozone than by any other pollutant (National Academy of Sciences, 1977; US Environmental Protection Agency, 1978).

The occurrence of ozone in the atmosphere has characteristics that differ from those of pollutants such as sulphur dioxide and fluorides. Ozone is produced by chemical reactions when a polluted air mass is exposed to solar radiation. The formation and occurrence of ozone at any particular place and time are highly dependent on climatic and meteorological conditions. Ozone and its precursors can be transported for hundreds of kilometres and they may extend over thousands of square kilometres. Ozone can persist in the atmosphere overnight in layers above the surface of the earth where there are few scavenging mechanisms. Nonurban air contains smaller amounts of compounds which react with ozone, so ozone can persist for longer periods in rural areas (Coffey, Stasik and Mohnen, 1977; Cleveland *et al.*, 1977; Wolff *et al.*, 1977; Isaksen, Hov and Hesstvedt, 1978).

The sources that burn coal, oil or gasoline, and produce precursors of ozone, are so numerous that no one set can be easily identified as the primary object of control. Emission sources and receptors exposed to ozone are separated in both time and space. Consequently, emission controls in the area of greatest effects do little to reduce ozone concentrations in that area. In fact, these controls will reduce the local concentrations of scavenging compounds that react with and destroy ozone. Strategies to control ozone, therefore, must be imposed over large geographic regions that span political boundaries. The difficulties and expense of controlling emissions of precursors of ozone, and the extensive land areas that are exposed, place greater emphasis on the need for thorough evaluation of the effects of ozone on plant productivity.

A considerable amount of research has been performed on the effects of ozone on vegetation, particularly with agricultural plants. Research in the 1960s and the early 1970s has been reviewed and evaluated previously (Linzon, Heck and Macdowall, 1975; Jacobson, 1977; National Academy of Sciences, 1977; US Environmental Protection Agency, 1978). The purpose of this report is to bring these earlier reviews up to date by summarizing the results of recent investigations. The following questions are considered:

1.   What have recent studies added to our knowledge of the effects of ozone on crop plants?
2.   After all the research of the past several decades, why is there still so much controversy over levels of ozone that should be proscribed to protect agricultural plants from reductions in productivity?
3.   Which new problems have emerged that were not recognized in the past?

This article does not review all research reports of the last decade on effects of ozone on vegetation. Major emphasis has been placed on studies that provide quantitative estimates of effects on growth, yield, development or quantity of plants. This does not imply that other research, carried out in greenhouses and controlled-environment chambers, has not added valuable information. Integration of knowledge from both the field and controlled-environment chambers is needed and certain data can be obtained only in highly controlled experimental situations. It is hoped that the synthesis of information presented here will highlight some of the similarities and contradictions in results that have been obtained by different approaches.

Analysis of recent field experimental data

RELATIONSHIP BETWEEN FOLIAR SYMPTOMS AND EFFECTS ON GROWTH
AND YIELD

As recently as 1977, it was not clear whether susceptibility to foliar symptoms is correlated with yield losses (National Academy of Sciences, 1977). The weight of evidence now indicates that foliar-symptom production is not a reliable index of effects on plant growth or yield (Reinert, 1980). For practical reasons, most evaluations of effects of ozone have been based on the production of foliar symptoms (Larsen and Heck, 1976; National Academy of Sciences, 1977) and, in fact, single exposures to ozone applied at a stage when the harvested product is growing rapidly may produce close correlations between foliar symptoms and yield reductions (Linzon, Heck and Macdowall, 1975). However, the assumption that decreases in growth and yield are a direct result of impairment of leaf function produced by injury to foliage is not valid, because there is uneven competition among the several sinks that receive photosynthate, and compensatory responses to ozone can produce rapid recovery from injury. Recent field studies with soybeans (Tingey *et al.*, 1973), tomatoes (Oshima *et al.*, 1975; Oshima *et al.*, 1977a), alfalfa (Tingey and Reinert, 1975), annual ryegrass and clover (Bennett and Runeckles, 1977), spinach (Heagle, Philbeck and Letchworth, 1979), wheat (Heagle, Spencer and Letchworth, 1979), lettuce (Bennett, 1979) and potatoes (Leone and Green, 1974; Pell, 1980) have demonstrated that foliar-symptom production is not closely correlated with reductions in growth or yield. Recurrent exposures to ozone over extended periods during the development of the crop are unlikely to produce close relationships between foliar symptoms and yield, except when the harvested product is the foliage.

COMPARATIVE SUSCEPTIBILITY OF INDOOR AND FIELD-GROWN PLANTS

Until recently, it has been assumed that effects of ozone on plants grown in the greenhouse or in controlled environmental chambers could be used to predict effects on field-grown plants, even though it has been known for some time that changes in environmental conditions alter plant response to pollutants. In a comparison of susceptibility to ozone of bush beans grown in three different environments, Lewis and Brennan (1977) found that field-grown plants were least susceptible. Beckerson, Hofstra and Wukasch (1979) found that cultivars of bush beans grown in the field differed in susceptibility compared with plants exposed to ozone in laboratory exposure chambers. Results for potatoes (Reinert, 1980) and bush beans (Meiners and Heggestad, 1979) also indicate that field-grown plants are less susceptible to injury and reductions in growth and yield than those grown indoors. General responses to ozone of plants grown in different environments may be similar, but quantitative relationships between dose and response clearly are affected by environmental conditions. A combination of several factors may account for these differences in plant response to ozone.

The direction of air movement in most experimental chambers is vertically through the plant canopy, resulting in substantial airflow around each plant. This situation decreases variability in response from one plant to another. In the field, air movement tends to be horizontal across the canopy and wind speed decreases markedly within the canopy as ground level is approached (Bennett and Hill, 1973). Wind speeds are greater in the field than in experimental chambers during periods of unstable weather which, usually, are not accompanied by increased ozone concentrations. At other times, wind speeds in the field are less than in experimental chambers, where airflow usually is constant throughout the day. Although direct comparisons of ozone flux between field and experimental situations have not been made, it seems reasonable to suspect that ozone flux, on a per unit mass or leaf area basis, frequently is less with field-grown plants than with vegetation exposed in chambers or the greenhouse.

The density of plants in agronomic situations usually is far greater than with most plants grown indoors in chambers or in the greenhouse. The plant canopy in an agricultural field absorbs ozone, lowering the actual dose received by leaves within the canopy (Bennett and Hill, 1973). Consequently, measurements of ozone above a field of plants overestimate the actual dose received by plants within the canopy, whereas measurements in a chamber usually are more accurate representations of the dose received by plants.

Another explanation for differences in results between field and indoor experiments concerns the effects of environmental conditions on plant growth and response to ozone. Conditions within chambers are maintained within ranges conducive to satisfactory plant growth. In the field, plants are exposed to wide fluctuations of temperature, wind, rain and crowding with the effect of producing hardier plants. Water stress in the middle of a sunny summer day is common in agricultural fields, even when soil moisture is adequate. The net effect is that crop plants grown in the field often are less susceptible to yield reduction by ozone than those grown indoors.

There is one important exception to this conclusion. A special combination of environmental conditions can predispose field-grown plants to injury by ozone. A period of poor growing weather with no ozone, followed by excellent growing conditions, can produce rapid gas exchange between leaves and the atmosphere and large amounts of new uninjured foliage. Under these circumstances, field-grown plants can be unusually susceptible (Haas, 1970) and a coinciding period of exposure to ozone could be more injurious to field-grown plants than would otherwise be indicated from results obtained with indoor plants.

OZONE AND THE QUALITY OF CROPS

Emphasis has been placed on effects of ozone on growth and yield but, until recently, little recognition has been given to the potential influence of ozone on crop quality (*Figure 14.1*) (Oshima *et al.*, 1977b). Shrivelled kernels of corn (Oshima, 1973; Thompson, Kats and Cameron, 1976), reduced size of tomatoes (Oshima *et al.*, 1977b), and alterations in chemical composition that affect cooking quality in potatoes and nutritional values of alfalfa (Pell, 1980), and other crops (Pippen *et al.*, 1975) are examples of effects on quality that have economic significance. Howell and Rose (1980) did not find changes in oil or protein content or seed germination, even though ozone reduced soybean yield. Weight-loss measurements alone are clearly insufficient for a thorough evaluation of

GROWTH	DEVELOPMENT	YIELD	QUALITY
Rate	Branching	Number	Appearance: size, shape, colour
Pattern	Flowering	Mass	Storage life
	Fruit set and development		Texture and cooking quality
			Nutrient content
			Viability of seeds

Figure 14.1   Processes and characteristics of crop plants that may be affected by ozone

effects of ozone on crops. Measures of quality such as colour, shape, size, storage life, texture, cooking quality, taste, nutrient content and viability of seeds should be reported when appropriate.

OZONE AND PARTITIONING OF ASSIMILATE

Plants grow in an integrated, sequential pattern controlled by inherited characteristics that are modified by environmental forces (Adams, 1967; Evans, 1975). Ozone and other air pollutants are stresses that can alter the developmental pattern, growth, and yield of plants (*Figure 14.1*). All levels of biological organization can be affected by ozone, beginning with subcellular effects (Oshima and Endress, 1978) and resulting in losses in productivity. The sequence of physiological effects produced by ozone is believed to be: 1) increases in permeability of membranes and leakage of ions; 2) stimulation of stress ethylene production; 3) decreases in photosynthetic carbon dioxide fixation; 4) inactivation of enzymes; 5) alteration of metabolic pools (Tingey, 1977). If repair processes cannot successfully overcome these changes, then foliar symptoms develop, growth may be reduced, leaf senescence accelerated and yield decreased. When photosynthesis is impaired, the supply of metabolites is reduced. Shoots have priority in the utilization of assimilates: consequently, inhibition of root development is one of the initial effects of ozone (Tingey, 1973; Oshima, Bennett and Braegelmann, 1978). Storage capacity of the plant is diminished and transport of nutrients and water to developing fruit decreases. Plants with determinate inflorescences have less opportunity to overcome reductions in yield because no new flowers and fruit are formed (Oshima *et al.*, 1979). Reproductive structures are particularly vulnerable to reduced photosynthetic capacity because they abscise prematurely in response to severe competition with other organs for limited supplies of assimilate (Adams, 1967).

Recent studies have demonstrated clearly that ozone alters the partitioning of assimilates in crop plants (*Table 14.1*). Root development is

Table 14.1 EFFECTS OF OZONE ON PARTITIONING OF DRY MATTER IN CROP PLANTS

Reference	Crop	Vegetative tissues† (Leaves + stems)		Roots	Reproductive tissues† (Fruits + seeds)	
		Number	Mass	Mass	Number	Mass
Tingey (1973)	Radish	–	D	D	–	–
Oshima *et al.* (1976)	Carrot	I	NE	D	–	–
Oshima, Bennett and Braegelmann (1978)	Parsley	–	D-I	D	–	–
Oshima (1973)	Sweet corn	–	D	D	D	D
Heagle, Body and Pounds (1972)	Sweet corn	–	NE	NE	D	D
Oshima and Endress (1978)	Lima bean	–	D	D	–	D
Oshima *et al.* (1979)	Cotton	I	D-I	D	D	NE
Bennett, Oshima and Lippert (1979)	Pepper	I	NE	NE	I	D

†D: decrease; I: increase; D–I: decrease then increase; NE: no effect

reduced, shoot growth is decreased and, in some cases, a decrease in shoot growth is followed by an increase. A new balance is achieved between those organs above and below ground as the plant adjusts to repeated ozone exposures. These effects of ozone on partitioning of assimilate vary with the plant because the degree of competition between organs (pods, seeds, leaves, roots, etc.) varies from one plant species to another. In one study, bush beans and tomatoes were exposed to ambient ozone at a single site. Fresh weight and number of pods and dry weight of leaves and stems were significantly reduced by ozone for beans but for tomatoes only fresh weight of fruit was decreased (MacLean and Schneider, 1976). No doubt, effects of ozone on distribution of dry matter also vary with pattern and dose of ozone and environmental and cultural conditions.

In addition to the studies cited in *Table 14.1*, significant reductions in number or mass of reproductive organs have been reported for bush beans (Maas *et al.*, 1973; Heggestad *et al.*, 1980), soybeans (Heagle, Body and Neely, 1974), tomato (Oshima *et al.*, 1975), and wheat (Shannon and Mulchi, 1974; Heagle, Spencer and Letchworth, 1979).

The change in distribution of dry matter during crop development, caused by ozone, helps to explain the poor association between foliar symptoms and yield reductions and between the response of plants grown in the greenhouse or in controlled environmental chambers and those grown in the field. These observations also may apply to the effects of pollutants such as sulphur dioxide and hydrogen fluoride, because all pollutants apply a stress to the plant even though the initial mechanisms of action are quite different.

DESCRIPTIONS OF OZONE EXPOSURES

The nature and magnitude of effects of ozone on vegetation are not determined by concentration alone. Both the occurrence and severity of effects are dependent on the frequency of ozone exposures, the durations of exposure, the length of time between exposures, the magnitude of fluctuations in concentrations, the time of day of exposures, their sequence and pattern, and the total flux of ozone to the plant as it is affected by canopy characteristics and leaf boundary layers. Far more attention has been paid to the effects of concentration and duration of exposures to ozone than to these other factors.

Table 14.2   DETAILS OF OZONE EXPOSURES
GIVEN IN 23 RECENT STUDIES OF EFFECTS ON CROP
PRODUCTIVITY

Details provided	*Number of publications*
Concentration	23
Duration	18
Frequency	16
Time between exposures	13
Time of day	6
Fluctuation of concentrations	3
Pattern (sequence)	0
Flux	0

A review of recent studies with added ozone indicates that most exposures are performed by adding constant amounts for set periods for a certain number of days per week. Highly artificial exposure patterns are thus produced. Although concentration, duration and frequency of exposure usually are described in published reports, in many cases there is no mention of the exact number of exposures during the course of the experiment, the time of day of the exposures, or the relationship of exposures to the developmental stages of the plant (*Table 14.2*). This effectively vitiates comparisons between studies and limits attempts to explain or understand similarities and differences in results among different studies.

Table 14.3 MEASURES OF OZONE OCCURRENCE USED IN RECENT STUDIES OF EFFECTS ON CROP PRODUCTIVITY

Mean values for 1, 3, 6, 7, or 24 hours
Maximum hourly average concentrations
Diurnal variation of hourly average concentrations
Number of hourly average concentrations exceeding 0.05, 0.10, 0.15 ppm
Frequency distribution of hourly average concentrations
Cumulative dose (ppm-h) greater than 0.10 ppm-h

An additional problem is that ozone exposures in experimental studies performed in the USA are rarely described in terms that can be compared with the US National Ambient Air Quality Standard (NAAQS) for ozone. This problem is perhaps best demonstrated by listing the most popular measures of ozone exposures used by different investigators, presented in their recent publications (*Table 14.3*).

IMPORTANCE OF FREQUENCY OF OZONE OCCURRENCE FOR CROP PROTECTION

Recently, the US Environmental Protection Agency altered both the ozone concentration and frequency of occurrence regulations for the NAAQS for ozone (*Table 14.4*). The current standard is considered to be approximately 50% less stringent than the former standard, because of the increase in limiting concentration from 0.08 to 0.12 ppm. This opinion ignores the change in limitation on frequency of ozone occurrence, which produces an additional relaxation of the standard.

To illustrate this point, several measures of ozone over a 9-year period have been compiled from monitoring data obtained at Yonkers, NY (*Table 14.5*). The previous ozone standard was exceeded between 11 and 50 times as often as the current standard for the places and years compared. Relaxation of the ozone standard is far greater than generally realized.

Table 14.4 MEASURES OF OZONE USED IN THE UNITED STATES NATIONAL AMBIENT AIR QUALITY STANDARDS

Standard	Concentration	Averaging time	Frequency
Former	0.08 ppm	1 h	Not more than 1 h yr^{-1}
Current	0.12 ppm	1 h	Not more than 1 d yr^{-1}

Table 14.5   OZONE CONCENTRATIONS (HOURLY AVERAGE) IN YONKERS, NY, DURING THE GROWING SEASON (MAY TO SEPTEMBER) FOR THE YEARS 1970 TO 1978

Year	Maximum conc.[a] (ppm)	Average daily maximum[a] (ppm)	No. of hours[a] exceeding 0.08[b] ppm	No. of hours[a] exceeding 0.12 ppm	No. of days[a] exceeding 0.08 ppm	No. of days[a] exceeding 0.12[c] ppm
1978	0.254	0.083	329	113	56	29
1977	0.260	0.096	493	144	80	36
1976	0.207	0.081	300	80	59	22
1975	0.150	0.054	128	8	35	5
1974	0.132	0.055	100	3	25	2
1973	0.230	0.077	316	62	59	20
1972	0.151	0.057	123	25	31	11
1971	0.178	0.047	58	9	17	4
1970	0.180	0.046	72	8	27	4

[a]Based on averaging times of 1 hour
[b]Former ambient-air quality standard for oxidants
[c]Current ambient-air quality standard for ozone

One of the aims of the NAAQS—namely to protect crops from ozone—cannot be resolved until ozone exposures in experimental studies can be compared with the concentrations and frequencies of occurrence proscribed by the NAAQS. Modification of the form of the NAAQS may be necessary to achieve this goal. It would also provide an opportunity to relate the standard to the total ozone dose in areas with differing patterns of ozone occurrence. A consensus of scientific opinion concerning whether the ozone standard will protect against yield losses can be obtained only when the NAAQS places more precise limitations on those characteristics of ozone occurrence that determine plant response to this pollutant.

Conclusions

In previous sections of this report, accomplishments of research of the last decade have been reviewed and some remaining gaps in our understanding of the effects of ozone on plant productivity have been identified. Information is, indeed, available at many levels of complexity (*Figure 14.2*). Knowledge at any single level can be used to predict effects at the next higher level of complexity but only with considerable uncertainty. Given specific changes in biochemical and physiological processes, it is difficult to predict whether foliar symptoms will occur, nor can we describe their type or severity. Given a description of the symptoms of leaves after exposure to ozone, we can make only crude estimates of effects on yield. Given effects on yield of one cultivar grown in one field, we can predict only with considerable equivocation the magnitude of effects on yield of other cultivars grown under other soil, cultivation, or climatic conditions. Similar problems exist when attempting to generalize information on yield reductions to effects on market value. When reductions in yield are to be converted into economic effects, shall we consider the direct loss in value to the producer, processor, distributor, consumer, or the national economy? How do we take into account secondary effects and the elasticity of supply and demand?

Our inability to make these extrapolations with confidence has led to decisions to perform studies that are as close as possible to the top of the hierarchy presented in *Figure 14.2*. Much emphasis is now being placed on the exposure of crops to pollutants under conditions that attempt to duplicate actual agronomic situations (Heagle, Riordan and Heck, 1979). There are both advantages to, and difficulties with, this approach.

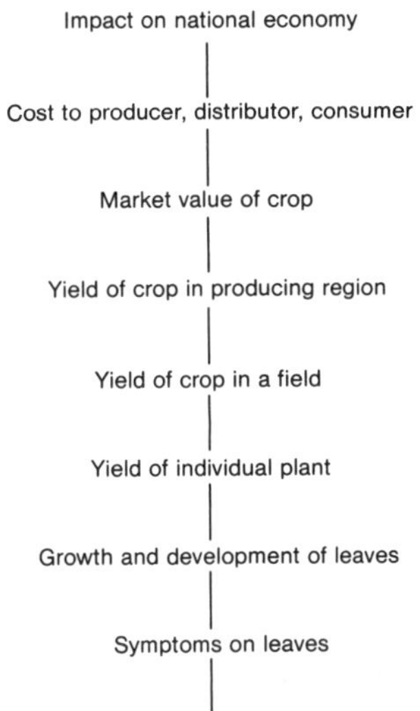

Figure 14.2   Levels of organizational complexity concerning effects of ozone on agricultural crops

Plants are more likely to have realistic levels of susceptibility when they are grown in the field, than when they are grown under controlled conditions. However, wide variations in yield from plant to plant reduce the sensitivity of field tests, and gradients in soil conditions require more complex experimental designs. Yield reductions of 10% or less in the field can be detected only with great difficulty and at great expense. Realistic conditions of pollutant exposure also are obtained in the field, but this is accompanied by loss in the precision of information about the pollutant concentrations. It appears, then, that field studies are necessary but not sufficient to resolve all the issues concerning effects of pollutants on vegetation. They must be accompanied by research in related areas to afford a basis for interpreting the results of these field experiments.

Research in the last decade has expanded our knowledge of the effects of ozone on plants, largely because new systems have been developed for their exposure to ozone under field conditions (Heggestad, 1980) and because new types of experiments have been performed. We have learned that foliar symptoms and effects on growth and yield are not closely associated, that susceptibility of indoor and field-grown plants also are not closely associated, that ozone can alter crop quality as well as productivity, and that ozone changes the distribution of dry matter among vegetative and reproductive organs, frequently producing greater effects on roots than on above-ground organs. Stimulatory effects of ozone on plant growth no longer are viewed as aberrations, nor can they be classified as paradoxical effects. They have a physiological basis and can be explained by the compensatory responses of plants to external stresses.

As we move into the 1980s we view the effects of ozone on crop productivity from a new perspective. Our awareness of the many dimensions of this problem has increased. Consequently, we may once again require improvements in basic approaches and methods in order to resolve the more difficult issues that now are being raised.

Acknowledgement

The author appreciates the research support received from the US Environmental Protection Agency, Corvallis Environmental Research Laboratory and the discussions held with participants in the National Crop Loss Assessment Network during the time this paper was prepared.

References

ADAMS, M.W. (1967). *Crop Science,* **7**, 505–510

BECKERSON, D.W., HOFSTRA, G. and WUKASCH, R. (1979). *Plant Disease Reporter,* **63**, 478–482

BENNETT, J.P. (1979). *Final Report to the California Air Resources Board Research Division.* State of California Air Resources Board, Sacramento, CA

BENNETT, J.H. and HILL, A.C. (1973). *Journal of the Air Pollution Control Association,* **23**, 203–206

BENNETT, J.P. and RUNECKLES, V.C. (1977). *Journal of Applied Ecology,* **14**, 877–880

BENNETT, J.P., OSHIMA, R.J. and LIPPERT, L.F. (1979). *Environmental and Experimental Botany,* **19**, 33–39

CLEVELAND, W.S., KLEINER, B., McRAE, J.E., WARNER, J.L. and PASCERI, R.E. (1977). *Journal of the Air Pollution Control Association,* **27**, 325–328

COFFEY, P., STASIK, W. and MOHNEN, V. (1977). In *Proceedings of the International Conference on Photochemical Oxidant Pollution and Its Control, pp. 89–109 (Dimitriades, B., Ed.) EPA-600(3-77-001a).* US Environmental Protection Agency, Research Triangle Park, North Carolina

EASTMOND, M. and SKARBY, L. (1979). Report from the workshop *Ozone Effects on Vegetation in Europe,* pp. 1–34. Swedish Water and Air Pollution Research Institute, Göteborg, Sweden

EVANS, L.T. (1975). In *Crop Physiology,* pp. 327–356 (Evans, L.T., Ed.), Cambridge University Press, Cambridge

HAAS, J.H. (1970). *Phytopathology,* **60**, 407–410

HEAGLE, A.S., BODY, D.E. and NEELY, G.E. (1974). *Phytopathology,* **64**, 132–136

HEAGLE, A.S., BODY, D.E. and POUNDS, E.K. (1972). *Phytopathology,* **62,** 683–687

HEAGLE, A.S., PHILBECK, R.B. and LETCHWORTH, M.B. (1979). *Journal of Environmental Quality,* **8,** 368–373

HEAGLE, A.S., RIORDAN, A.J. and HECK, W.W. (1979). In *72nd Annual Meeting of the Air Pollution Control Association, Cincinnati, Ohio,* Paper 79-46.6. Air Pollution Control Association, Pittsburgh, Pa

HEAGLE, A.S., SPENCER, S. and LETCHWORTH, M.B. (1979). *Canadian Journal of Botany,* **57,** 1999–2005

HEGGESTAD, H.E. (1980). In *73rd Annual Meeting of the Air Pollution Control Association, Montreal, Canada,* Paper 80-26.1. Air Pollution Control Association, Pittsburgh, Pa

HEGGESTAD, H.E., HEAGLE, A.S., BENNETT, J.H. and KOCH, E.J. (1980). *Atmospheric Environment,* **14,** 317–326

HOWELL, R.K. and ROSE, L.P., Jr (1980). *Plant Disease Reporter,* **64,** 385–387

ISAKSEN, I.S.A., HOV, O. and HESSTVEDT, E. (1978). *Environmental Science and Technology,* **12,** 1279–1284

JACOBSON, J.S. (1977). *VDI Berichte,* **270,** 163–173

LARSEN, R.I. and HECK, W.W. (1976). *Journal of the Air Pollution Control Association,* **26,** 325–333

LEONE, I.A. and GREEN, D. (1974). *Plant Disease Reporter,* **58,** 683–687

LEWIS, E. and BRENNAN, E. (1977). *Journal of the Air Pollution Control Association,* **27,** 889–891

LINZON, S.N., HECK, W.W. and MACDOWALL, F.D.H. (1975). In *Photochemical Air Pollution: Formation, Transport, and Effects,* pp. 89–142. Publication No. 14096, National Research Council of Canada, Ottawa, Canada

MAAS, E.V., HOFFMAN, G.J., RAWLINS, S.L. and OGATA, G. (1973). *Journal of Environmental Quality,* **2,** 400–404

MacLEAN, D.C. and SCHNEIDER, R.E. (1976). *Journal of Environmental Quality,* **5,** 75–78

MEINERS, J.P. and HEGGESTAD, H.E. (1979). *Plant Disease Reporter,* **63,** 273–277

NATIONAL ACADEMY OF SCIENCES (1977). *Ozone and Other Photochemical Oxidants.* Report of the Committee on Medical and Biologic Effects of Environmental Pollutants, Washington, DC

OSHIMA, R.J. (1973). *Plant Disease Reporter,* **57,** 719–723

OSHIMA, R.J. and ENDRESS, A.G. (1978). In *71st Annual Meeting of the Air Pollution Control Association, Houston, Texas,* Paper 78-24.2. Air Pollution Control Association, Pittsburgh, Pa

OSHIMA, R.J., BENNETT, J.P. and BRAEGELMANN, P.K. (1978). *Journal of the American Society for Horticultural Science,* **103,** 348–350

OSHIMA, R.J., TAYLOR, O.C., BRAEGELMANN, P.K. and BALDWIN, D.W. (1975). *Journal of Environmental Quality,* **4,** 463–464

OSHIMA, R.J., POE, M.P., BRAEGELMANN, P.K., BALDWIN, D.W. and VAN WAY, V. (1976). *Journal of the Air Pollution Control Association,* **26,** 861–865

OSHIMA, R.J., BRAEGELMANN, P.K., BALDWIN, D.W., VAN WAY, V. and TAYLOR, O.C. (1977a). *Journal of the American Society for Horticultural Science,* **102,** 286–289

OSHIMA, R.J., BRAEGELMANN, P.K., BALDWIN, D.W., VAN WAY, V. and

TAYLOR, O.C. (1977b). *Journal of the American Society for Horticultural Science*, **102**, 289–293

OSHIMA, R.J., BRAEGELMANN, P.K., FLAGLER, R.B. and TESO, R.R. (1979). *Journal of Environmental Quality*, **8**, 474–479

PELL, E.J. (1980). *Final Progress Report for the US Department of Energy*. Botany Department, Pennsylvania State University

PIPPEN, E.L., POTTER, A.L., RANDALL, V.G., NG, K.C., REUTER, F.W., III, MORGAN, A.I. Jr and OSHIMA, R.J. (1975). *Journal of Food Science*, **40**, 672–676

REINERT, R.A. (1980). In *73rd Annual Meeting of the Air Pollution Control Association, Montreal, Canada*. Paper 80-26.4. APCA, Pittsburgh

SHANNON, J.G. and MULCHI, C.L. (1974). *Crop Science*, **14**, 335–337

THOMPSON, C.R., KATS, G. and CAMERON, J.W. (1976). *Journal of Environmental Quality*, **5**, 410–412

TINGEY, D.T. (1973). In *Proceedings of the Third International Clean Air Congress, Düsseldorf*, pp. A154–A156. VOI-Verlag, Düsseldorf

TINGEY, D.T. (1977). In *Proceedings of the Second International Conference on Photochemical Oxidant Pollution and Its Control*, pp. 601–609 (Dimitriades, B., Ed.). Environmental Protection Agency, Research Triangle Park, North Carolina

TINGEY, D.T. and REINERT, R.A. (1975). *Environmental Pollution*, **9**, 117–125

TINGEY, D.T., REINERT, R.A., WICKLIFF, C. and HECK, W.W. (1973). *Canadian Journal of Plant Science*, **53**, 875–879

US ENVIRONMENTAL PROTECTION AGENCY (1978). *Air Quality Criteria for Ozone and Other Photochemical Oxidants, Volume II*. EPA Office of Research and Development, Washington, DC

WOLFF, G.T., LIOY, P.J., MEYERS, R.E., CEDERWALL, R.T., WIGHT, G.D., PASCERI, R.E. and TAYLOR, R.S. (1977). *Environmental Science and Technology*, **11**, 506–510

IV

Interactions

15

AIR POLLUTANT INTERACTIONS IN MIXTURES

D.P. ORMROD
University of Guelph, Ontario, Canada

Introduction

The urban and industrial regions of the world produce numerous air pollutants. Many of these pollutants, singly and in combination, are known to be toxic to plants and there have been many studies of the effects of selected mixtures of pollutants. More than five years have elapsed since publication of the first extensive literature review on this topic (Reinert, Heagle and Heck, 1975). Much has subsequently been written on the subject of effects of mixtures, helping to increase our understanding of the plant responses to mixtures. We are now aware that the nature of pollutant interactions is extremely dependent on thresholds and upper limits of injury, as well as on duration and number of exposures.

Research with mixtures cannot be separated from that with single gases. Much of the research work on mixtures which has been reported provides useful information on single-gas effects because direct comparisons are usually made with single-gas exposures. As for single gases, a large number of plant and environment variables will affect plant response to mixtures. Plant-to-plant variation due to species, cultivar and individual plant differences, and soil–plant–atmosphere variation due to temperature, water status, irradiation, ventilation, and O_2 and CO_2 status, will affect responses. The current state of knowledge of mixture effects provides only a preliminary understanding of the modes of action of mixtures. Improved experimental protocol (Heck, Krupa and Linzon, 1979) will enhance comparability of results from different laboratories and will hasten the day when mixture interactions are fully understood.

Air pollution mixtures in the atmosphere

SOURCES

Many sources of air pollution emit a combination of two or more major pollutants. In several cases the selection of combinations for study has been based on a particular kind of source. Thus, studies of $SO_2 + NO_2$ have been based on industrial combustion processes which generate SO_2 and NO_x, the high temperatures causing atmospheric O_2 and N_2 to combine to form NO which is then oxidized to NO_2 (Wellburn *et al.*, 1976; Ashenden and Mansfield, 1978; Ashenden, 1979; Mansfield and Law, this volume, Chapter 5). Coal-fired electricity-generating plants may be sources of substantial amounts of $SO_2 + NO_x$ (Hill *et al.*, 1974; White, Hill and Bennett, 1974) as may munitions plants (Skelly, Moore and Stone, 1972). Some research work has also recognized that CO_2, as well as SO_2 and NO_x concentrations, will be increased in the exhaust gases of many industries (Hou, Hill and Soleimani, 1977). In the open air, nitrogen oxides are seldom found alone in sufficiently high concentrations to cause injury, but they may become injurious on interaction with other gases.

Certain types of industrial emissions are likely to contain substantial concentrations of HF and SO_2 (Mandl, Weinstein and Keveny, 1975). Alternate exposures to HF and SO_2 may occur when wind direction fluctuates in an area near several sources of pollutants (Matsushima and Brewer, 1972). Other industrial emissions may contain a predominance of SO_2 and NH_3 (Wellburn *et al.*, 1976), while particulate emissions may add to SO_2 pollution in industrial areas (Williams, Lloyd and Ricks, 1971).

The glasshouse constitutes a specific environment in terms of sources and pollutants. Mixtures of NO, NO_2 and propylene could result from burning propane to form CO_2 (*see* Mansfield and Law, this volume, Chapter 5). The conversion of NO to NO_2 would be slower in the glasshouse than outside because of the lower concentrations of O_3 in a closed glasshouse (Wellburn *et al.*, 1976). An open ventilated glasshouse equipped with an evaporative cooling system may have more vegetation injury, because the forced air movement may increase the absorption of pollutants and the raised humidity may increase plant sensitivity to incoming polluted air (Grosso *et al.*, 1971).

OUTDOOR CONCENTRATIONS

Few investigators have attempted to provide data on outdoor concentrations to support their selection of concentrations for controlled-exposure studies. Data on SO_2 and O_3 concentrations (Tingey *et al.*, 1973a) and on SO_2 and NO_2 concentrations (Tingey *et al.*, 1971) were presented for several places in the USA, to indicate that widespread occurrence of combinations of these gases was possible, and that phytotoxicity would occur if controlled-environment exposure–dose relationships held in the field. However, data were not provided on the frequency with which mixtures of concentrations occurred. Ashenden (1979) provided data on outdoor SO_2 and NO_2 concentrations in the UK. The ratio of SO_2:NO_2:CO_2 concentrations occurring downwind from a power plant burning coal—1 : 0.33 : 326—was used by Hou, Hill and Soleimani (1977) in controlled-exposure studies. Concentrations of SO_2 of an order of magnitude less than that of O_3 in polluted field air were measured by Macdowall and Cole (1971) and formed the basis for controlled-exposure studies. Field concentration data were provided for SO_2 and O_3 in relation to chlorotic dwarf disorder of eastern white pine (Dochinger *et al.*, 1970), and for O_3 but not SO_2 in relation to $SO_2 + O_3$ studies on tobacco cultivars

(Menser and Hodges, 1970). Another approach has been the use of concentrations above and below the US air-quality standards for SO_2 and O_3 to evaluate mixture effects in controlled exposures (Jacobson and Colavito, 1976). In general, investigators have not examined outdoor data before selecting concentrations for mixture studies. In those cases where outdoor data have been examined, the frequency of actual occurrence of particular pollutant combinations has not been recorded

CHEMICAL COMBINATION OF POLLUTANTS

The possibility that gases in mixtures may react chemically before sorption by the plant has been investigated by some researchers but ignored by most. A possible explanation of apparently diminished injury from SO_2 + O_3 mixtures is the chemical combination of these gases in exposure chambers (Costonis, 1973). Preliminary tests of the chamber facilities revealed that, if SO_2 were introduced before O_3, more O_3 production would be required to obtain the desired concentrations than would be necessary if no SO_2 were present. This effect was greater in moist than in dry air. However, Jacobson and Colavito (1976) found that neither SO_2 nor O_3 concentrations changed when they introduced the other pollutant at the concentrations used in the experimental plant exposures, so there was no evidence for reactions that would lower the concentration of either gas. However, direct reaction between O_3 and SO_2 may occur in plant leaves and the direct aqueous oxidation of SO_2 by O_3 has been described (Heagle and Johnston, 1979).

Early research with SO_2 + O_3 mixtures used oxidant analysers based on a reaction with potassium iodide (KI). The interaction of SO_2 and O_3 in such analysers was demonstrated by Menser and Heggestad (1966). Macdowall and Cole (1971) used sufficiently low concentrations of SO_2 to ensure that the analyser interaction did not cause serious errors, while White, Hill and Bennett (1974) used dichromate papers in the air stream to remove SO_2 that would cause negative interference in the KI method for determining NO_2 concentration. The problems of interference in analysers now seem to be resolved.

Experimental design and analysis

SPECIES COMPARISONS

The most useful comparisons of species sensitivities to pollutant mixtures are likely to be based on studies in which more than one species was exposed in the same facilities under the same environmental conditions. However, there have been relatively few of these comparative studies. The effects of SO_2 + O_3 on injury to tobacco, radish, alfalfa, cabbage, broccoli, tomato, onion, brome grass and spinach were compared by Tingey *et al.*, (1973a); on tobacco, alfalfa and radish by Tingey and Reinert (1975); on white bean and tobacco by Jacobson and Colavito (1976); on white bean and soybean by Hofstra and Ormrod (1977); on radish, cucumber and

Table 15.1 ADDITIVE, SYNERGISTIC, ANTAGONISTIC AND MIXED INTERACTIONS IN RESPONSE TO AIR-POLLUTANT MIXTURES

Interaction type	Mixture	Site or characteristic affected	Species	Reference
Additive	$SO_2 + O_3$	Photosynthesis	*Euglena*	De Koning and Jegier (1970)
	$SO_2 + O_3$	Growth	Radish	Tingey, Heck and Reinert (1971)
	$SO_2 + HF$		Orange	Matsushima and Brewer (1972)
	$SO_2 + O_3$		Several	Tingey *et al.* (1973a,b)
	$SO_2 + NO_2$	Photosynthesis	Pea	Bull and Mansfield (1974)
	$SO_2 + NO_2$	Growth	*Poa pratensis*	Ashenden and Mansfield (1978)
	$SO_2 + O_3$	Leaf growth	*Ulmus americana*	Constantinidou and Kozlowski (1979)
Synergistic	$SO_2 + O_3$	Leaf	Tobacco	Menser and Heggestad (1966)
				Heggestad (1969)
				Menser and Hodges (1970)
				Grosso *et al.* (1971)
				Hodges, Menser and Ogben (1971)
				Macdowall and Cole (1971)
	$SO_2 + O_3$	Leaf	Peanut	Applegate and Durrant (1969)
	$SO_2 + O_3$	Needle	Eastern white pine	Dochinger *et al.* (1970)
	$SO_2 + NO_2$	Leaf	Tomato	Matsushima (1971)
	$SO_2 + NO_2$	Leaf	Several	Tingey *et al.* (1971)
	$SO_2 + particulates$	Leaf	*Quercus*	Williams, Lloyd and Ricks (1971)
	$SO_2 + NO_2$	Seedling growth	Eastern white pine	Skelly, Moore and Stone (1972)
	$SO_2 + O_3$	Leaf	Alfalfa, radish, tobacco	Tingey *et al.* (1973a)
	$SO_2 + O_3$	Needle	Eastern white pine	Houston (1974)
	$SO_2 + NO_2$	Photosynthesis	Alfalfa	White, Hill and Bennett (1974)
	$SO_2 + O_3$	Growth	Soybean	Tingey *et al.* (1973b)
	$SO_2 + NO_2$		Radish	Bennett *et al.* (1975)
	$SO_2 + NO_2$	Enzyme activity	Pea	Horsman and Wellburn (1975)
	$SO_2 + O_3$	Leaf	Apple	Kender and Spierings (1975)

	Pollutants	Effect	Plant	Reference
	$SO_2 + F$	Reproduction	Scots pine	Roques, Kerjean and Auclair (1980)
	$SO_2 + O_3$	Root growth	Soybean	Tingey and Reinert (1975)
	$SO_2 + NO_2$	Pollen-tube growth	Lily	Masaru, Syozo and Saburo (1976)
	$NO_2 + O_3$			
	$SO_2 + NO_2$	Growth	Three grasses	Ashenden and Mansfield (1978)
	$SO_2 + HF$		Lichens, mosses	Taylor (1978)
	$SO_2 + O_3$	Leaf	Cucumber, radish	Beckerson and Hofstra (1979b)
	$SO_2 + O_3$	Photosynthesis	Tree species	Carlson (1979)
	$SO_2 + O_3$	Needle	Scots pine	Weidensaul and Darling (1979)
	$SO_2 + NO_2 + O_3$	Photosynthesis	Sunflower	Furukawa and Totsuka (1979)
Antagonistic	$SO_2 + O_3$	Growth	Radish	Tingey, Heck and Reinert (1971)
	$SO_2 + O_3$		Eastern white pine	Costonis (1973)
	$SO_2 + O_3$		Alfalfa	Mandl, O'Neill and Weinstein (1974)
	$SO_2 + O_3$	Fungal infection	Lilac	Hibben and Taylor (1975)
	$SO_2 + O_3$	Leaf and root growth	Alfalfa	Tingey and Reinert (1975)
	$O_3 + PAN$	Leaf	Ponderosa pine	Davis (1977)
	$SO_2 + O_3$	Leaf and leaf growth	White bean, soybean	Hofstra and Ormrod (1977)
	$SO_2 + O_3$	Needle	Scots pine	Neilsen, Terrell and Weidensaul (1977)
	$SO_2 + O_3$		Soybean	Beckerson and Hofstra (1979b)
	$SO_2 + O_3$		Soybean	Heagle and Johnston (1979)
	$SO_2 + O_3$		Apple	Shertz, Kender and Musselman (1980b)
Mixed	$SO_2 + O_3$		Soybean	Heagle, Body and Neely (1974)
	$SO_2 + HF$		Several	Mandl, Weinstein and Keveny (1975)
	$SO_2 + O_3$		White bean, tobacco	Jacobson and Colavito (1976)
	$SO_2 + O_3$		Rieger begonia	Gardner and Ormrod (1977)
	$H_2S + O_3$	Photosynthesis, transpiration	Snap bean	Coyne and Bingham (1978)
	$SO_2 + O_3$		Soybean	Heagle and Johnston (1979)
	$SO_2 + O_3$		Grapevine	Shertz, Kender and Musselman (1980a)
	$SO_2 + O_3$	Nematode infection	Soybean, begonia	Weber, Reinert and Barker (1979)

soybean by Beckerson and Hofstra (1979a); on sugar maple, black oak, and white ash by Carlson (1979); and on white bean, soybean, cucumber and radish by Beckerson and Hofstra (1980). There have been direct comparisons of the SO_2 + NO_2 injury to tomato, radish, oats, tobacco, pinto bean and soybean (Tingey *et al.*, 1971); to a large number of native desert species (Hill *et al.*, 1974); to radish, Swiss chard, oats and peas (Bennett *et al.*, 1975); to *Dactylis glomerata*, *Lolium multiflorum*, *Phleum pratense* and *Poa pratensis* (Ashenden and Mansfield, 1978); and to *Dactylis glomerata* and *Poa pratensis* (Ashenden, 1979). The injurious effects of combined HF and SO_2 have been compared directly on sweet orange and mandarin (Matsushima and Brewer, 1972) and on bean, barley and sweet corn (Mandl, Weinstein and Keveny, 1975).

CULTIVAR AND PLANT COMPARISONS

Most investigators have used one cultivar to represent each species, even though there is ample evidence, from single-gas and mixed-gas studies, that there can be wide variation in cultivar responses within species (Ormrod, 1978). Injury responses of cultivars within species, to mixtures of SO_2 + O_3, have been studied using three tobacco cultivars (Menser and Heggestad, 1966), nine tobacco cultivars (Menser and Hodges, 1970), four cultivars and many breeding lines of tobacco (Hodges, Menser and Ogben, 1971), three tobacco cultivars (Tingey *et al.*, 1973a), two soybean cultivars (Tingey *et al.*, 1973b), two white bean cultivars (Jacobson and Colavito, 1976), six strawberry cultivars (Rajput, Ormrod and Evans, 1977), three petunia cultivars (Elkiey and Ormrod, 1979a, b, c, d; Elkiey, Ormrod and Pelletier, 1979), 33 bean cultivars (Beckerson, Hofstra and Wukasch, 1979), and two soybean cultivars (Heagle and Johnston, 1979). Mandl, Weinstein and Keveny (1975) compared two corn cultivars for sensitivity to injury from HF + SO_2. There were differences among cultivar responses to mixtures in most studies. Most investigators have failed to report the extent of plant-to-plant variation in sensitivity to mixtures. Skelly, Moore and Stone (1972) noted high variability among eastern white pine tree responses to SO_2 + NO_2.

ADDITIVITY, SYNERGISM AND ANTAGONISM

Reports on most studies involving mixtures have included, or implied, an assessment of the interaction of the pollutants in terms of additive effects, greater than additive (synergistic) effects, or less than additive (antagonistic) effects (Tingey and Reinert, 1975). Responses have been expressed in terms of amounts of injury or changes in threshold concentrations causing injury.

Additivity has been demonstrated for a number of diverse plant responses (*Table 15.1*). Many more reports have been of synergistic responses, for example, the effects of SO_2 + NO_2 on cocksfoot growth (*Table 15.2*). Reports of antagonistic responses, for example SO_2 + O_3 on bean leaf injury (*Table 15.3*), have been much more limited. Many of these

Table 15.2 EFFECTS OF SO_2 + NO_2 EXPOSURE[a] ON GROWTH OF *DACTYLIS GLOMERATA*

Treatment	Dry weight of green leaves (g)	Dry weight of roots (g)	Number of tillers	Number of leaves	Leaf area (cm²)
Control	0.25	0.63	11.6	21.7	48.5
NO_2	0.23	0.56	11.5	28.8	58.7
SO_2	0.18	0.40	10.4	22.3	46.2
SO_2 + NO_2	0.04	0.10	7.9	8.3	13.8

[a]68 ppb NO_2 and/or 68 ppb SO_2 for 103.5 h weekly for 20 weeks. After Ashenden (1979)

Table 15.3 EFFECTS OF O_3 + SO_2 EXPOSURES ON % LEAF AREA INJURED ON *PHASEOLUS VULGARIS*

Treatment	Duration (days) 5	10
Control	0	0
O_3	39[a]	89
SO_2	0	0
O_3 + SO_2	0	52

[a]Injury on 2nd trifoliate leaf after 150 ppb O_3 and/or 150 ppb SO_2 for 6 h daily. After Hofstra and Ormrod (1977).

reports of additivity, synergism and antagonism implied that the nature of the interaction was dependent on factors such as pollutant concentration and exposure duration, as well as on the species and gases studied. Thus Mandl, Weinstein and Keveny (1975) noted that a mixture of SO_2 and HF increased foliar injury compared with SO_2 alone, only at low concentrations of SO_2 and HF. The need for systematic testing of a range of concentrations was recognized, having been overlooked by earlier investigators who generally studied only one combination of gas concentrations. General conclusions concerning synergism or antagonism for a particular species require determination of dose–response relationships over a range of concentrations of pollutant mixtures (Jacobson and Colavito, 1976). Heagle, Body and Nealy (1974) found that the concentrations, in relation to the thresholds for injury by separate gases, were critical in determining synergistic effects. In experiments in which no injury was observed for either pollutant singly, visible leaf injury was produced by simultaneous exposure to both pollutants at the same concentrations (Ashenden and Mansfield, 1978; Carlson, 1979). However, Macdowall and Cole (1971) noted, for tobacco-leaf injury, that synergism did not occur below the injury threshold dose of O_3 alone, only below that of SO_2 alone, in SO_2 + O_3 mixtures.

Studies in which several concentrations of gases in mixtures were tested, indicated decreased synergism and finally antagonism as pollutant concentrations were increased. Thus an upper limit for synergism was found for photosynthetic depression in alfalfa by White, Hill and Bennett (1974) and for growth and leaf injury in Rieger begonia by Gardner and Ormrod

Table 15.4 EFFECTS OF O_3 + SO_2 EXPOSURE[a] (AS % INJURED LEAVES) ON BEGONIA PLANTS. After Gardner and Ormrod (1977)

| O_3 concentration (ppb) | 100 | 100 | 100 | 200 | 200 | 200 | 300 | 300 | 300 |
SO_2 concentration (ppb)	600	1200	1800	600	1200	1800	600	1200	1800
Control	0	0	0	0	2	0	1	1	2
O_3	0	1	2	32	23	21	37	40	42
SO_2	0	22	67	4	21	33	3	12	48
O_3 + SO_2	39	44	67	67	50	50	32	41	62

[a]ppb for 4 h daily for 5 days

(1977) (*Table 15.4*). Heagle and Johnston (1979) found additive, synergistic and antagonistic effects of SO_2 + O_3 on soybean, depending on the concentrations of O_3 and SO_2, on the duration of exposure and on the amount of injury caused by each gas singly. Synergism generally occurred when injury by single gases was slight: antagonism was seen when single-gas injury was severe. Statistical analyses of interactions revealed that the type of response to the mixture was related more to single-gas injury, as determined by dose, than to gas concentration, exposure duration or ratio of SO_2 to O_3. Little systematic attention has been given to ratios of gases in mixtures. Middleton, Darley and Brewer (1958) found that an SO_2:O_3 ratio of 4:1 resulted in antagonism in leaf-symptom development in pinto bean, while at 6:1 and higher ratios there was no interference. There was a difference of one order of magnitude between the concentrations of SO_2 and O_3 in a mixture, in their effect on tobacco leaves (Macdowall and Cole, 1971). Clearly, the implications of synergism and antagonism in the field must be viewed in terms of pollutant concentrations and durations likely to occur there. Much of the published research will thus not be directly relevant to the field.

STATISTICAL EVALUATIONS

Studies of pollutant combinations provide opportunities to employ statistical tests in the interpretation of results. Nevertheless, many investigators have not evaluated their data statistically. Most researchers have used analysis of variance of factorial experiments for interpretation of combined effects. Significant interaction of two pollutants is regarded as evidence for a synergistic or an antagonistic effect. Such a technique has been reported by Tingey *et al.* (1973b), Bull and Mansfield (1974), Tingey and Reinert (1975), Wellburn *et al.* (1976), Gardner and Ormrod (1977), Ashenden (1979), and Heagle and Johnston (1979). Separate analyses of variance are usually calculated for each combination with, in some cases, mean separation by Duncan's multiple range test or the least significance test.

More complex methods of statistical analysis have also been used. White, Hill and Bennett (1974) used two statistical methods; the two-sample t-test and the Mann–Whitney–Wilcoxon test to show nonadditivity of pollutant effects at the 99% confidence level. A statistical test of synergism has been made with models derived by polynomial analysis of injury index data (Macdowall and Cole, 1971). Chi-squared analysis with a continuity correction has been used to test the null hypothesis that there was no difference between the incidence of foliar injury in the presence or absence of one of the pollutants (Jacobson and Colavito, 1976).

Transformation of data in probit analysis may help to interpret antagonism and synergism, by the determination of median effective doses (ED_{50}) (*Table 15.5*) (Macdowall and Cole, 1971; Jacobson and Colavito, 1976).

Table 15.5 MEDIAN EFFECTIVE DOSES (ED_{50}) OF O_3 + SO_2 ON TOBACCO. BASED ON PLOTS OF PROBITS ON LOG DOSES

Concn (ppb)	ED_{50} (ppb. h)
200 O_3	3720[a]
1000 SO_2	9780
200 O_3 + 200 SO_2	1660
200 O_3 + 1000 SO_2	1840

[a]Dose that produces some injury on half the leaves. After Macdowall and Cole (1971)

The purpose of probit analysis is to transform the curves of percentage response versus log dose, to straight lines. The dose required to produce some injury on half the leaves can be taken as the ED_{50} value for both individual and combined pollutants. The response to SO_2 + O_3 in terms of percentage of leaves injured was approximately a log function of dose (as concentration of pollutant applied over a 4-hour period) when incidence of injury was converted to probit units (Jacobson and Colavito, 1976).

CONCENTRATIONS, DOSES AND FLUXES

Most research with mixtures has been based on the combination, in controlled-exposure facilities, of certain concentrations of gases for a specified duration. Few investigators have described the exposure in terms of dose—the product of concentration and duration. For most studies, concentration and dose are related because the same duration is used for all treatments. The concept of pollutant flux has barely entered the literature on the effects of air-pollutant mixtures. The flux, which is the rate of movement of pollutant to plants, may be the most important determinant of plant response to pollutant mixtures. The amount of pollutant sorbed by plants is the product of flux and duration of exposure.

The choice of response indicators has ranged from indices of discrete injury to continuous weight measurement, and the concentrations or doses used have ranged widely. Investigators using high concentrations have apparently done so in order to ensure that visible injury symptoms are obtained. Such effects may be referred to as acute and might be found in the direct path of the plume from an industrial plant. Typical of such high-concentration treatments were those of Hou, Hill and Soleimani (1977), Heagle and Johnston (1979). Elkiey and Ormrod (1979a, b, c, d) and Carlson (1979). Others have used concentrations more characteristic of ambient outdoor levels and even below those commonly found during pollution episodes (Jaeger and Banfield, 1970; Menser and Hodges, 1970; Heagle, Body and Neely, 1974; Ashenden, 1979). Such concentrations may result in chronic effects such as growth retardation with little, if any, visible injury.

Dose–response relationships were developed by Macdowall and Cole (1971) who expressed $SO_2 + O_3$ in terms of ppb.h doses for both controlled and field exposures. Linear, quadratic and exponential relationships between plant injury and dose were noted. It may be difficult to establish the threshold dose when the relation between dose and response is sigmoid. The results reported by Jacobson and Colavito (1976) indicated an approximately binomial functional relationship for frequency of foliar injury caused by $SO_2 + O_3$. They found that interpretation was difficult because of (1) a variation in foliar response with pollutant concentration; and (2) a narrow range over which injury increased from 0 to 100%.

DURATION OF EXPOSURE

Mixture studies have been conducted over a wide range of exposure durations. Many experiments have been based on one exposure of a few hours or less (Grosso *et al.*, 1971; Hodges, Menser and Ogben, 1971; Tingey *et al.*, 1971, 1973a; Hill *et al.*, 1974; White, Hill and Bennett, 1974; Bennett *et al.*, 1975; Jacobson and Colavito, 1976). Such exposure durations were usually associated with acute injury symptoms. They do not provide an opportunity to observe acclimatization effects, nor do they permit estimation of growth and development effects or of visual symptom development during exposure. Some investigators have exposed plants for a few days, either a few hours per day or continuously, to elicit responses that might be characteristic of outdoor pollution episodes. Examples of such experiments are found in the reports of Horsman and Wellburn (1975), Hofstra and Ormrod (1977), Beckerson and Hofstra (1979a, b, c) and Elkiey and Ormrod (1979a, b, c, d). Such studies enable changing visual symptoms to be noted, together with some effects on growth and development. Long-term exposures have been used to characterize seasonal or growth-cycle responses to pollutant mixtures, usually supplied at low concentrations continuously or discontinuously for up to several months (Jaeger and Banfield, 1970; Tingey, Heck and Reinert, 1971; Tingey *et al.*, 1973b; Heagle, Body and Neely, 1974; Mandl, Weinstein and Keveny, 1975; Ashenden and Mansfield, 1978; Ashenden, 1979; Carlson, 1979). A few investigators have directly contrasted exposures of short or medium and of long duration in studies of mixtures.

EXPOSURE PATTERNS

It is unlikely that steady concentrations of mixed pollutants will be found outdoors: nevertheless, almost all investigations to date have used simultaneous exposures to nonvarying concentrations. In the natural environment, peak concentrations of pollutants may occur at different times for different pollutants. Such patterns may have considerable impact, with the preconditioning of plants by one pollutant affecting their response to a later peak concentration of another pollutant. Some investigators have called for studies of sequence effects (Tingey and Reinert, 1975; Wellburn *et al.*, 1976) but there have been few reports of such studies. Matsushima and Brewer (1972) investigated sequential reciprocal exposure of orange to SO_2 and HF to see whether one gas influenced the subsequent response of plants to the other gas, but found little effect. Costonis (1973) studied certain sequences of $SO_2 + O_3$ on eastern white pine. A sequence of O_3 followed by SO_2 was more toxic, in terms of severity and rate of lesion development, than exposure to both gases simultaneously. White, Hill and Bennett (1974) used sequential studies in which plants were exposed during the first hour to either SO_2 or NO_2, followed by $SO_2 + NO_2$ during the second hour. This did not produce a significantly greater effect than did simultaneous application of both pollutants. Sulphur dioxide exposure before exposure to O_3 or to $SO_2 + O_3$ can markedly alter plant sensitivity to further pollutant episodes (Hofstra and Beckerson, 1981). Many investigators have subjected plants to intermittent exposures within an overall long-term exposure, for example, 5 days per week, 8 hours per day. Tingey, Heck and Reinert (1971) and Tingey *et al.* (1973b) noted that this might allow the plants time to repair injury or regain normal metabolic functions during the period when no exposure took place.

COMPARISONS BETWEEN CONTROLLED ENVIRONMENTS AND THE FIELD

Most studies of effects of mixtures have been conducted in controlled-environment cabinets or in glasshouses. A few studies have been conducted entirely in the field, and a few have included both field and controlled-exposure components to allow direct comparisons. Menser and Hodges (1970) compared tobacco cultivar sensitivity to $SO_2 + O_3$ in the field directly with controlled exposures in the glasshouse. A major shift in sensitivity of one cultivar, between greenhouse and field, illustrated the differences sometimes observed when comparing natural and simulated environments. Hodges, Menser and Ogben (1971) also compared chamber responses with field responses of tobacco to $SO_2 + O_3$. Beckerson, Hofstra and Wukasch (1979) compared bean cultivar sensitivity to $SO_2 + O_3$ in controlled environments, with injury development on outdoor exposure. Outdoor responses correlated more strongly with sensitivity to O_3 than with sensitivity to mixtures.

Outdoor-exposure chambers, with environmental conditions more characteristic of ambient conditions, have been used by some researchers. Heagle, Body and Neely (1974) used outdoor-exposure chambers for long-term SO_2 and O_3 treatments of soybean. Hill *et al.* (1974) used a portable field chamber to expose desert plants to $SO_2 + NO_2$ at concentrations measured downwind of a large coal-fired power plant.

The lack of good correlations of responses in the laboratory and in the field has been a cause of concern. The amount of injury in the field may be underestimated because of (1) inadequate air movement in some controlled exposures, resulting in high diffusion resistance in the boundary layer;

(2) the fluctuation of air-pollution concentrations in the field, with up to 10 times the daily mean occurring occasionally in industrial areas, compared with controlled exposures near the daily mean; and (3) the impact on plants in the field, of several pollutants not present in laboratory experiments (Ashenden and Mansfield, 1978).

BIOINDICATORS

The use of test plants known to be particularly sensitive to single pollutants, as indicators of the presence of these pollutants, is a well-developed practice. However, only two recommendations of plant indicators of mixtures have been made to date. Menser and Hodges (1970) suggested that the distinct shift from O_3 insensitivity to mixed-gas susceptibility shown by 'Burley 49' tobacco would make this cultivar a useful indicator of the simultaneous presence of $SO_2 + O_3$. Grosso *et al.* (1971) noted that *Nicotiana glutinosa* reacted synergistically to a low-concentration mixture of $SO_2 + O_3$, and suggested that it could be a useful bioindicator to detect $SO_2 + O_3$ injury after low O_3 concentrations have been recorded.

Responses to pollutant mixtures

VISIBLE INJURY SYMPTOMS

Most investigators have applied a rating system for extent of visible leaf injury and used the data obtained for interpretation of mixture effects. Few have assessed differences in appearance by providing descriptions of the injury: in some cases, experiments did not last long enough to permit the development of stable characteristic injury symptoms. In general, O_3-injury symptoms have dominated in studies of mixtures of $SO_2 + O_3$ (Menser and Heggestad, 1966; Menser and Hodges, 1970; Hodges, Menser and Ogben, 1971; Tingey, Heck and Reinert, 1971; Tingey *et al.*, 1973b; Heagle, Body and Neely, 1974; Elkiey and Ormrod, 1979d). There were exceptions and, in some cases, different symptoms developed in response to the combination. Grosso *et al.* (1971) considered that injury caused by $SO_2 + O_3$ could be distinguished from that attributable to O_3 alone, as the mixture produced a general flecking or diffuse bleaching of upper leaf surfaces of all tobacco cultivars, whereas O_3 alone produced punctate flecking. Leaf injury from combined $SO_2 + O_3$ on apple trees cv. Golden Delicious is characterized by large greyish-green water-soaked areas in the midshoot leaves, a symptom that differs from SO_2 or O_3 injury (Kender and Spierings, 1975; Shertz, Kender and Musselman, 1980b). Undersurface glazing, a symptom usually attributed to peroxyacetyl nitrate, has been found on petunia leaves exposed to $SO_2 + O_3$ (Lewis and Brennan, 1978). Injury caused by $SO_2 + O_3$ in radish has symptoms similar to those caused by O_3 alone, but cucumber, in the same experiment, had an additional symptom of interveinal necrosis (Beckerson and Hofstra, 1979b). Visible injury to three woody species caused by simultaneous exposure to SO_2 + O_3 included symptoms characteristic of each pollutant (Carlson, 1979). Acute responses of eastern white pine to SO_2 + O_3 indicated different symptoms from either SO_2 or O_3 alone, according to Costonis (1973), but Dochinger *et al.* (1970) reported that separate or combined SO_2 + O_3 produced the same initial symptoms—mottling of current needles and premature dwarfing.

The importance of the duration of exposure in the development of symptoms of injury can be illustrated by the experiments, with white bean, of Hofstra and Ormrod (1977). On white bean leaves, O_3 by itself caused early development of bronzing: on the other hand, a mixture of SO_2 + O_3 caused yellow interveinal chlorosis on corresponding leaves, after several days' additional exposure. Thus, for short-term exposure of white bean, while SO_2 + O_3 appears to have antagonistic effects compared with O_3 alone (Jacobson and Colavito, 1976), the effects may later be considered to be synergistic with regard to leaf chlorosis. Soybean, in the same experiment, developed the characteristic symptoms of O_3 injury in the SO_2 + O_3 mixture, but required a longer exposure than that needed for development of corresponding injuries with O_3 alone.

The symptoms of injury from mixtures of SO_2 + NO_2 differed greatly, in several species, from those caused by single gases (Tingey *et al.*, 1971). The injuries caused by mixtures were chlorotic and necrotic flecking on the upper surface of the interveinal areas of tomato, radish, oats and tobacco. In contrast, the symptom on pinto bean and soybean was reddish-brown stipple, which easily could be confused with O_3 injury. Hill *et al.* (1974) found that, on many species, the symptoms caused by SO_2 + NO_2 and by SO_2 alone were identical. Matsushima and Brewer (1972) found that leaves of orange exposed to SO_2 + HF developed chlorosis patterns which were indistinguishable from those produced by individual gases, indicating that chlorosis patterns would not be useful indicators in the field.

GROWTH AND YIELD RESPONSES

Many studies of growth responses to mixtures have been reported. In general, such studies have involved long-term exposures, and growth usually has been evaluated by weight measurements and, occasionally, by leaf-area determination. Effects on economic yield have been obtained less frequently. Growth and yield effects have been demonstrated in various ways. Mixtures of SO_2 + O_3 caused the following symptoms: defoliation and reduced growth and yield of soybean, compared with the effects of single gases (Heagle, Body and Neely, 1974); reductions in fresh weights of the tops, and in fresh and dry weights of the roots of young soybean plants (Tingey *et al.*, 1973b); reductions in growth and yield of radishes (Tingey, Heck and Reinert, 1971); growth reduction in three species at high but not at low humidity (Carlson, 1979); growth reduction in tobacco and alfalfa but not in radish (Tingey and Reinert, 1975); and reductions in plant and flower weights of petunia (Elkiey and Ormrod, 1979d). Combinations of SO_2 + O_3 had antagonistic weight effects in soybean (Heagle and Johnston, 1979) and in white bean and soybean (Hofstra and Ormrod, 1977).

Long-term exposure of four grass species to SO_2 + NO_2 caused major reductions in dry weight, associated with reductions in leaf area (Ashenden and Mansfield, 1978). The reductions in leaf areas and dry weights of *Dactylis glomerata* and *Poa pratensis* were associated with decreases in the initiation of tillers and leaves, not with their enhanced senescence (Ashenden, 1979). Mixtures of SO_2 + HF decreased linear growth and leaf area of oranges (Matsushima and Brewer, 1972) but Mandl, Weinstein and Keveny (1975) found no significant effects on growth of several species, even though there were visible symptoms of injury. Fruiting effects were not studied.

Several investigators have expressed concern over the apparent lack of correlation of visible symptoms and growth effects in some experiments (Tingey, Heck and Reinert, 1971; Tingey *et al.*, 1973b; Mandl, Weinstein and Keveny, 1975). Tingey, Heck and Reinert (1971) noted that growth reduction was far greater than leaf injury in radish, while in soybean Tingey *et al.* (1973b) observed that growth reductions were not well related to foliar injury, and that cultivars differed in foliar-injury response but not in growth response. They suggested that an air pollutant would not be expected to reduce plant growth unless it directly or indirectly affected some growth-limiting system. Tingey and Reinert (1975) noted that the effects of mixed gases were additive with regard to growth reduction but more than additive with regard to foliar injury.

Roques, Kerjean and Auclair (1980) have reported deleterious effects of combined SO_2 + F on reproduction in Scots pine. There were fewer seeds per cone, more abortions, delays in lignification and smaller cones, with variation from tree to tree. Mixed air pollutants also seem to interfere with reproduction in lily: combinations of SO_2 + NO_2, O_3 + NO_2 and NO_2 + formaldehyde markedly inhibited pollen-tube elongation on a culture medium (Masaru, Syozo and Saburo, 1976).

ROOT RESPONSES

Root growth has seldom been measured in studies of gas mixtures. Although it is unlikely that the pollutants penetrate the rooting media and exert a direct effect on roots (Tingey, Heck and Reinert, 1971), root growth may be a sensitive indirect indicator of the physiological status of plant shoots. Fresh weight of radish leaves was reduced by a smaller percentage than was root fresh weight after the same SO_2 + O_3 treatment, indicating either that the plants had sufficient photosynthate for normal leaf growth but insufficient reserves for normal root growth, or that there had been a change in partitioning (Tingey, Heck and Reinert, 1971). Root growth of soybeans was reduced more than top growth by SO_2 + O_3 mixtures, also probably as a result of reduced translocation of photosynthate to the roots (Tingey *et al.*, 1973b). Greater reductions of root growth than top growth of soybean indicated that root growth was more sensitive than top growth (Tingey and Reinert, 1975).

INTERACTIONS WITH THE ENVIRONMENT

Irradiance, temperature, water supply and other environmental factors are known to affect plant responses to air pollutants and have been studied extensively for single gases (Ormrod, 1978). However, studies of the interactions of gas mixtures with the environment have been more limited. Sugar maple and white ash showed greater visible injury and more marked photosynthetic reduction after exposure to $SO_2 + O_3$ in conjunction with a combination of high irradiance and high humidity (Carlson, 1979). Under conditions of low irradiance and low humidity there were no visible symptoms, but photosynthesis was reduced. Thus the presence of visible symptoms was not necessarily a good indicator of pollutant-induced reductions in photosynthesis. Environmental conditions that led to full stomatal opening predisposed leaf tissue to more severe injury than did suboptimal conditions.

Table 15.6 EFFECTS OF CO_2 CONCENTRATION ON PERCENTAGE INHIBITION OF PHOTOSYNTHESIS BY $SO_2 + NO_2$ IN ALFALFA

CO_2 concn[a]	*Time* (h)		
(ppm)	1.2	2	3
315	38	50	54
645	10	18	25

[a]Plants were also exposed to about 1.00 ppm $SO_2 + 0.33$ ppm NO_2. After Hou, Hill and Soleimani (1977)

Much more injury was noted after exposure of eastern white pine to $SO_2 + O_3$ during three successive days of high humidity than after periods of low humidity (Jaeger and Banfield, 1970). Dochinger *et al.* (1970) also noted that $SO_2 + O_3$ sensitivity of eastern white pine is influenced by environment, but they did not study environmental factors directly. Humidity was found to have a marked effect on the diffusive-resistance response of petunia leaves to $SO_2 + O_3$ (Elkiey and Ormrod, 1979b). The impact of CO_2 concentration on $SO_2 + NO_2$ effects was studied by Hou, Hill and Soleimani (1977), who found that doubling the CO_2 concentration increased net rates of photosynthesis in the presence of $SO_2 + NO_2$, even though $SO_2 + NO_2$ decreased net rates of photosynthesis at ambient CO_2 levels (*Table 15.6*). There have been no reports of any possible effects of temperature or of nutritional status on the response of plants to gas mixtures.

INFLUENCE OF LEAF AGE AND GROWTH STAGE

Few studies of mixtures have considered the impact of leaf age or of growth stage in determining sensitivity to pollutants. Menser and Heggestad (1966) noted that leaves of tobacco were most sensitive to $O_3 + SO_2$ exposure when mature. Elkiey and Ormrod (1979b) found that middle-aged leaves of three petunia cultivars were generally more sensitive to $O_3 + SO_2$ than younger leaves. Growth stage did not affect the form of response to the mixture. The midshoot leaves of grape and apple are most sensitive to $SO_2 + O_3$ (Shertz, Kender and Musselman, 1980a, b).

DOMINANT COMPONENT IN MIXTURES

In many studies of air-pollutant mixtures the results indicate that one component plays a dominant role. Many investigators have attempted to identify that component, on the basis of the expression of injury or of concentration effects. In studies of $SO_2 + O_3$, the dominant component was usually O_3 (Menser and Hodges, 1970; Hodges, Menser and Ogben, 1971; Macdowall and Cole, 1971; Tingey *et al.*, 1973b; Heagle, Body and Neely, 1974; Beckerson and Hofstra, 1979a, b, c; Elkiey and Ormrod, 1979a, b, c, d). In contrast, Dochinger *et al.* (1970) noted that SO_2 seemed to be more reactive than O_3 in producing chlorotic dwarf of eastern white pine. In studies of mixtures of $SO_2 + NO_2$, SO_2 was considered to be the dominant component by Bennett and Hill (1974) and Bennett *et al.* (1975), while Skelly, Moore and Stone (1972) considered that NO_x was dominant over SO_2. White, Hill and Bennett (1974) attempted to separate the effects of NO and NO_2 in mixtures of $NO_x + SO_2$ and concluded that NO had little, if any, effect in a mixture, compared with NO_2. Few investigators have considered whether the mere presence or the amount of a second component in a mixture affects plant response to the other component. In the studies of Hofstra and Ormrod (1977) with white bean, the antagonistic effect of SO_2 on O_3 response was similar over the SO_2 concentration range of 75–600 ppb, suggesting that the presence of SO_2 was the determining factor in the antagonistic response.

Modes of action

The frequent occurrence of synergism, and occasionally antagonism, in plant responses to gas mixtures within certain dose ranges makes it imperative that the mechanism of pollutant interactions in the plant should be more fully understood. Strong synergistic responses are of concern in agricultural and horticultural plant production, because very substantial crop losses occur when gases are present in concentrations that, individually, would not be very damaging.

If the effects of the gas mixture are equal to the additive effects of the single gases, the gases may be acting independently to produce their responses (Tingey and Reinert, 1975). If the effects are not equal to the additive effects of the single gases, this indicates that the plant response to one gas is dependent on the presence, and possibly the amount, of the second gas. The mechanisms of synergism and antagonism may be related to (1) direct reactivity between pollutant gases in plant leaves; (2) an effect of either gas on the stomata; (3) competition for reaction sites; (4) a change in the sensitivity of reaction sites; or (5) a combination of these (Heagle and Johnston, 1979). Antagonism suggests either that there is competition between the two gases for an active site, or that one gas may give protection from the injurious effects of the other (Tingey and Reinert, 1975). Synergism suggests that the two gases augment each other's action.

Interactions of pollutants, leading to synergism or antagonism, may have a basis in biochemistry or physiology or both. Biochemical changes may be mitigated by additional protective mechanisms within the cell (Wellburn *et al.*, 1976). Foliar injury occurs only when the expenditure of energy and material on such alleviation exceeds the input of energy and materials by these additional mechanisms. Retardation of critical anabolic processes, or the acceleration of particular catabolic processes, or both, may occur without destroying the biochemical mechanisms responsible for them (Tingey, Heck and Reinert, 1971).

Close relationships between plant growth and foliar injury are not necessarily to be expected, because plant growth is a composite of many reactions, any one of which could be growth limiting (Tingey *et al.*, 1973b). Substantial growth reductions, with or without visible leaf injury, may be the result of more than a simple reduction in photosynthetic leaf area. Reduction in numbers of leaves, branches, tillers and other plant parts, attributable to a decrease in their formation, will also have a secondary effect on plant growth (Ashenden, 1979). A reduction in the number of plant parts is more likely to be caused by a decreased supply of assimilates than by a direct interference in the formation of these parts, but this in turn will result in substantial decreases in plant growth.

Reductions in root growth in response to gas mixtures may also have a secondary effect on the whole plant. Root growth is reduced, probably as a result of reduced translocation of photosynthate to the roots (Tingey *et al.*, 1973b) caused either by diminished photosynthesis or by interference with translocation (Tingey, Heck and Reinert, 1971). A reduction in root size would decrease the absorption of water and nutrients and affect photosynthetic rate.

Stomata may also play a major part in response to gas mixtures, and this is closely related to environmental conditions. Adequate soil water, optimal mineral nutrition, high relative humidity, and sufficient irradiance will lead to full stomatal opening, subjecting the leaf tissue to maximal pollutant entry (Carlson, 1979). Synergism and antagonism may be explained readily if the pollutants exert direct effects on stomatal opening. However, there is considerable evidence that other factors in the plant, in addition to stomata, have an important role in response to pollutants.

ACTION ON ENZYMES AND METABOLISM

There have been few studies of biochemical responses to gas mixtures (*but see* Wellburn, this volume, Chapter 8). One pollutant may cause a change in metabolic reactions which alters the susceptibility of the plant to the other pollutant (Jacobson and Colavito, 1976). Studies of enzyme activity

Table 15.7   EFFECTS OF SO_2 + NO_2 ON PEROXIDASE ACTIVITY IN 14-DAY-OLD PEA SEEDLINGS. [a]Activity as percentage of control after 6 days pollutant exposure. After Horsman and Wellburn (1975)

NO_2 (ppm)	SO_2 (ppm)				
	0	0.2	1.0	1.5	2.0
0	100[a]	100	112	127	114
0.1	100	124	149	182	205
1.0	99	151			

have indicated that, in some cases, the threshold concentration of a pollutant required to produce a change in enzyme activity was lowered when combinations of pollutants were used (Horsman and Wellburn, 1975). Peroxidase had the greatest response to NO_2 + SO_2 (*Table 15.7*). Examination of numerous enzyme responses to SO_2 + NO_2 or to SO_2 + NH_3 showed that many were affected in a different manner by the combination than by the single gases (Wellburn *et al.*, 1976). Studies of enzyme activities may be a useful method to determine the sensitivity of higher plants to mixtures of pollutants. Also, the study of critical enzyme activities of insensitive and sensitive species or cultivars may provide a better understanding of injury development in plants and of the way in which plants counteract such injury (Horsman and Wellburn, 1975).

ACTION ON CHEMICAL COMPOSITION OF PLANTS

Studies of the influence of mixtures on the chemical composition of plants have been largely concerned with chlorophyll concentration. Horsman and Wellburn (1975) found that SO_2 + NO_2 mixtures decreased the chlorophyll content of plant tissue even though NO_2 alone increased chlorophyll. More chlorophyll *a* than chlorophyll *b* was lost. Beckerson and Hofstra (1979c) studied the antagonistic response of SO_2 + O_3 in white bean in which exposure to the mixture caused a decrease in chlorophyll *a* and *b* which occurred 2 days later than did the response to O_3 alone. After a 5-day exposure period, chlorophyll levels were lower than in leaves treated with O_3 alone. Interveinal chlorosis observed was not due to phaeophytinization of the chlorophyll molecules, nor were the changes typical of normal leaf senescence. Protein and RNA levels were not affected. The fact that chlorosis, but no lesions, developed in leaves exposed to the mixture, suggested that the chloroplasts were adversely affected but that the membranes remained intact as a result of some protective action of SO_2 (Hofstra and Ormrod, 1977).

ACTION ON PHOTOSYNTHESIS

Impairment of photosynthesis by mixtures has been studied by several investigators. Carlson (1979) found that photosynthetic rates of sugar maple and white ash leaves exposed to SO_2 + O_3 decreased more than additively during the first 2 days of exposure. He considered that the mechanism might be a result of a direct effect of the pollutants on stomata. The reduction in photosynthesis was least when irradiance was optimal for photosynthesis. Carlson (1979) also noted that large reductions in photosynthesis could occur before, or in the absence of, visible injury.

Interactions of SO_2 + NO_2 affecting photosynthesis were studied by Bull and Mansfield (1974) who observed combined effects in the pea over a concentration range up to 250 ppb for each gas. Initially, exposure to these pollutants increased photosynthesis, but substantial reductions soon followed. No visible injury accompanied the depression in photosynthesis.

White, Hill and Bennett (1974) and Bennett and Hill (1974) found synergism of SO_2 + NO_2 at low concentrations, but synergism decreased as concentrations increased, until at 500 ppb of each gas there was no synergism. The inhibition of photosynthesis was complete in about 20 min and recovery to control levels was complete within 30 min to 2 h, provided that tissue injury did not occur. Net photosynthesis rates in sunflower leaves were depressed by all combinations of SO_2, NO_2 and O_3 (Furukawa and Totsuka, 1979). The inhibition of apparent photosynthesis by SO_2 + NO_2 was less than half at higher than ambient CO_2 concentrations than at ambient concentrations (Hou, Hill and Soleimani, 1977). Depression of CO_2 uptake was slower at raised levels of CO_2 than at ambient levels, and the overall effect of increased CO_2 was increased photosynthesis compared with that found at ambient CO_2 concentrations. The high CO_2 level may have caused the stomata to close slightly, may have reduced SO_2 uptake and, by increasing photosynthesis, may have inactivated more leaf sulphite. Addition of O_3 to H_2S resulted in greater reductions in apparent photosynthetic rates of snap beans than those caused by H_2S alone (Coyne and Bingham, 1978).

ACTION ON TISSUE PERMEABILITY

Tests of tissue permeability are intended to determine the effect of mixtures on membrane integrity. Total electrolyte leakage was more sensitive than potassium ion leakage from petunia leaf tissue exposed to SO_2 + O_3 (Elkiey and Ormrod, 1979c). Beckerson and Hofstra (1980) found a temporary increase in solute leakage, indicating increased membrane permeability, in cucumber and radish, two species with a synergistic response to SO_2 + O_3, but only after visible symptoms appeared. The SO_2 + O_3 mixture did not increase solute leakage above the control level in white bean or soybean, two species with antagonistic responses, whereas there was increased leakage with O_3 alone.

ACTION ON PLANT WATER STATUS AND STOMATA

Plant water status, described by water potential, not only affects the responses to mixtures but also is affected by plant exposure to mixtures. Plant water potential decreased quickly on exposure of petunia to SO_2 + O_3, but there was a delay compared with the effect of exposure to O_3 alone, suggesting that some protective action against water potential decrease was exerted by SO_2 (Elkiey and Ormrod, 1979a).

The limitation on water movement from leaf surfaces may be evaluated in terms of leaf diffusive resistance. Much of the leaf diffusive resistance is assumed to be stomatal resistance. In mixture studies, one pollutant may inhibit or promote foliar uptake of the other pollutant by altering stomatal resistance (Jacobson and Colavito, 1976). Such interaction may involve effects on stomatal function or altered physiological or metabolic processes that indirectly affect stomata. A mixture of SO_2 + O_3 caused an increase in resistance which was much greater than that caused by O_3 alone, for three species, two of which have synergistic responses in terms of leaf injury, and one of which has antagonistic responses (Beckerson and Hofstra, 1979b). The differences in species responses suggest that differences in injury response must be due to some further physiological or biochemical mechanisms associated with the leaf, which alter the effects of the pollutant mixture and result in either synergistic or antagonistic effects. Similarly, the effect of SO_2 pretreatments on plant sensitivity to O_3 and to $SO_2 + O_3$ could not be explained entirely on the basis of stomatal function (Hofstra and Beckerson, 1981).

Exposure of petunia plants to $SO_2 + O_3$ at about 50% relative humidity caused marked and rapid increases in leaf diffusive resistance—an indication of stomatal closure—regardless of cultivar sensitivity, while at about 90% relative humidity there was a slight increase in resistance, but this was seen in a sensitive cultivar only (Elkiey and Ormrod, 1979b). Exposure to $SO_2 + O_3$ apparently caused rapid stomatal closure, but normal resistance values were restored within 2 h of the end of exposure. Diffusive resistance of both leaf surfaces of white bean leaves was increased much more by $SO_2 + O_3$ than by SO_2 or O_3 alone (Beckerson and Hofstra, 1979a). The increased resistance began on the first day of exposure and was not related to visible injury. Initially, the resistance of the upper surface was about double that of the lower surface, corresponding to differences in stomatal numbers. The increased resistance could help to explain the antagonism in injury response to the $SO_2 + O_3$ mixture in white bean, but other physiological or metabolic factors must also be involved. Menser and Heggestad (1966) took impressions of tobacco stomata during exposure to $SO_2 + O_3$ and found that some had partly closed, particularly those on the lower surface.

Stomata became closed in response to $SO_2 + NO_2$, resulting in reduced transpiration and less uptake of pollutant (Ashenden, 1978). The closure may be a response to physiological injury within the leaf: for example interference with respiration may increase CO_2 concentrations within the leaf and thus reduce the amount of stomatal opening. It was assumed that stomata had closed, in the $SO_2 + HF$ studies of Mandl, Weinstein and Keveny (1975), because less fluoride accumulated in corn leaves exposed to the mixture. Pollutant exposure could have induced stomatal closure or reduced the duration of stomatal opening. There was greater stomatal closure in snap beans when O_3 was added to H_2S, than was seen in response to H_2S alone (Coyne and Bingham, 1978). Williams, Lloyd and Ricks (1971), in the evaluation of interactions between SO_2 and particles, noted that particles accumulated in stomatal pores, probably keeping them open, increasing permeability and admitting more SO_2.

INTERACTION WITH HISTOLOGY, ANATOMY AND MORPHOLOGY

While many studies of tissue and cell injury have been conducted with single pollutant gases, very little has been done with mixtures. Solberg and Adams (1956) exposed apple leaves to $SO_2 + HF$ and found no difference in microscopic injury resulting from exposure to the single gases or the mixture. Evans and Miller (1975) compared the effects of separate and mixed $SO_2 + O_3$ on pine needles and noted that the site of injury differed according to the gas responsible, whether exposed to only one pollutant or in combination treatments. Some anatomical features of a petunia cultivar which was found to be insensitive to $SO_2 + O_3$ included more raised cuticle around the stomata, larger epidermal cells, rougher cuticle surface, and many more trichomes than on the leaves of a sensitive cultivar (*Table 15.8*) (Elkiey, Ormrod and Pelletier, 1979). Cultivars did not differ in guard-cell shape, stomatal-pore area or trichome shape. The relative amounts of stomatal absorption and surface adsorption of SO_2 and O_3 differed among cultivars (*Table 15.9*).

Table 15.8 SUMMARY OF STOMATAL AND LEAF-SURFACE FEATURES IN RELATION TO $SO_2 + O_3$ SENSITIVITY OF PETUNIA CULTIVARS. After Elkiey, Ormrod and Pelletier (1979)

Variable		*Cultivar sensitivity*	
	Sensitive	*Intermediate*	*Insensitive*
Guard-cell cuticle	Not raised	Raised	Raised
Stomatal density	Lowest	Intermediate	Highest
Stomatal pore length	Longest	Intermediate	Shortest
Epidermal cells	Largest	Intermediate	Smallest
Cuticle surface	Smooth	Intermediate	Rough
Trichome number	Least	Intermediate	Most

Table 15.9 STOMATAL ABSORPTION AND SURFACE ADSORPTION OF $SO_2 + O_3$ BY THREE PETUNIA CULTIVARS. After Elkiey and Ormrod (1980)

		O_3		SO_2
Cultivar	*Alone*	*In mixture*	*Alone*	*In mixture*
Absorption				
Capri	6.2[a]	4.0	10.5	6.0
White Magic	8.0	6.0	13.5	10.5
White Cascade	7.0	4.5	13.5	13.2
Adsorption				
Capri	4.2	4.6	14.2	13.8
White Magic	1.4	1.4	11.2	12.4
White Cascade	1.0	0.8	1.6	1.6

[a]$\mu l\,dm^{-2}\,min^{-1} \times 10^{-2}$

Examination of hybrid poplar leaves exposed to $O_3 + SO_2$ revealed crystalline-like inclusions within bundle-sheath extension cells, distorted and injured chloroplasts, and globules in mesophyll cells (Krause and Jensen, 1978). These symptoms of injury were not present in leaves exposed to SO_2 or O_3 separately. Hybrid poplar leaf surfaces were injured by the mixture. Epidermal cells with smooth cuticles deteriorated to an emaciated condition, but single gases had no such effect (Krause and Jensen, 1979).

Research needs

In contrast to studies with single gases, there has been little research with air-pollutant mixtures and less is known of most plant responses to mixtures. The effects of mixtures of pollutants seem to be complex, differing wih species, cultivar and experimental conditions (Hofstra and Ormrod, 1977). The results of the research that has been conducted suggest a number of ways in which our knowledge of the effects of mixtures can be improved. The general expectation is for outdoor plants to be exposed to several pollutants simultaneously and sequentially, with intermittent and fluctuating concentrations. Experimental exposures should reflect such outdoor patterns (Tingey and Reinert, 1975). Fluctuating concentrations of pollutants may cause more stress than steady amounts, to which a plant may easily adapt at the cellular and physiological level (Wellburn *et al.*, 1976). Systematic experimental evaluations of dose –response relationships are needed to determine the effects of various patterns of exposure on appearance, growth and yield. Special attention should be directed to the study of mixtures in which concentrations of individual gases do not violate the recommended air-quality standards (Bennett and Hill, 1974; Jacobson and Colavito, 1976). Effects of other important phytotoxic air pollutants, such as ethylene, should also be examined in mixtures (Bennett and Hill, 1974). Long-term studies of growth, yield and quality of plants exposed to mixtures are needed to develop a more complete understanding of effects on crop production (Tingey *et al.*, 1971, 1973a). Many mixture effects seem to be species-dependent (Tingey and Reinert, 1975) and studies are therefore needed for each major crop species. The effects of ratios of pollutants are not well known, so studies are needed in which ratios are studied systematically (Heagle, Body and Neely, 1974). The extent of genetic control of plant responses to mixtures needs definition for a range of species. Strong genetic control has been demonstrated in eastern white pine (Houston and Stairs, 1973).

Economic evaluation of the effects of pollutant mixtures on crop production will be required if we are to know whether such mixtures are causing substantial losses (Ashenden, 1979). In addition, if laboratory experiments are to provide information applicable to field conditions, much more care must be taken in the choice of exposure conditions (Ashenden and Mansfield, 1978). Interactions among pollutants, and between pollutants and the environment, may represent sources of error in going from controlled exposures to the field.

The lack of significant effects and relationships in some past research may have resulted from the high plant-to-plant variability sometimes encountered, and from insufficient replication because of the large number of treatments in studies of mixtures. Lack of significance does not prove the lack of an effect or relationship. Future experiments will have to combine adequate replication with control of extraneous random variation.

Finally, the diversity and variability of responses of plants to pollutant mixtures suggest that much more attention should be devoted to the factors that affect these responses (Heagle and Johnston, 1979). To date, few patterns of pollutant interaction in mixtures have been found. Changing conditions have not resulted in consistent responses in uptake rates or in sensitivity of plant tissue. Much research needs to be done at the physiological and biochemical levels (Dugger and Ting, 1970) if alleviation of the effects of mixtures is to take place efficiently and effectively, based on knowledge of specific sites of action. The ambient outdoor environment is generally characterized by mixtures of pollutants rather than by single gases. Future air-pollution research should be a reflection of this 'real world' situation in crop production.

References

APPLEGATE, H.G. and DURRANT, L.C. (1969). *Environmental Science and Technology,* **3**, 759–760

ASHENDEN, T.W. (1978). *Environmental Pollution,* **18**, 45–80

ASHENDEN, T.W. (1979). *Environmental Pollution,* **18**, 249–258

ASHENDEN, T.W. and MANSFIELD, T.A. (1978). *Nature,* **273**, 142–143

BECKERSON, D.W. and HOFSTRA, G. (1979a). *Atmospheric Environment,* **13**, 533–535

BECKERSON, D.W. and HOFSTRA, G. (1979b). *Atmospheric Environment,* **13**, 1263–1268

BECKERSON, D.W. and HOFSTRA, G. (1979c). *Canadian Journal of Botany,* **57**, 1940–1945

BECKERSON, D.W. and HOFSTRA, G. (1980). *Canadian Journal of Botany,* **58**, 451–457

BECKERSON, D.W., HOFSTRA, G. and WUKASCH, R. (1979). *Plant Disease Reporter,* **63**, 478–482

BENNETT, J.H. and HILL, A.C. (1974). In *Air Pollution Effects on Plant Growth,* American Chemical Society Symposium Series 3, pp. 115–127 (Dugger, M., Ed.), American Chemical Society, Washington, DC

BENNETT, J.H., HILL, A.C., SOLEIMANI, A. and EDWARDS, W.H. (1975). *Environmental Pollution,* **9**, 127–132

BULL, J.N. and MANSFIELD, T.A. (1974). *Nature,* **250**, 443–444

CARLSON, R.W. (1979). *Environmental Pollution,* **18**, 159–170

CONSTANTINIDOU, H.A. and KOZLOWSKi, T.T. (1979). *Canadian Journal of Botany,* **57**, 170–175

COSTONIS, A.C. (1973). *European Journal of Forest Pathology,* **3**, 50–55

COYNE, P.I. and BINGHAM, G.E. (1978). *Journal of the Air Pollution Control Association,* **28**, 1119–1123

DAVIS, D.D. (1977). *Plant Disease Reporter,* **61**, 640–644

DE KONING, H.W. and JEGIER, Z. (1970). *Atmospheric Environment,* **4**, 357–361

DOCHINGER, L.S., BENDER, F.W., FOX, F.O. and HECK, W.W. (1970). *Nature,* **225**, 476

DUGGER, W.M. and TING, I.P. (1970). *Annual Review of Plant Physiology,* **21**, 215–233

ELKIEY, T. and ORMROD, D.P. (1979a). *Zeitschrift für Pflanzenphysiologie,* **91**, 177–181

ELKIEY, T. and ORMROD, D.P. (1979b). *Journal of the Air Pollution Control Association,* **29**, 622–625

ELKIEY, T. and ORMROD, D.P. (1979c). *Atmospheric Environment,* **13**, 1165–1168

ELKIEY, T. and ORMROD, D.P. (1979d). *Scientia Horticulturae,* **11**, 269–280

ELKIEY, T. and ORMROD, D.P. (1980). *Journal of Environmental Quality,* **9,** 93–95

ELKIEY, T., ORMROD, D.P. and PELLETIER, R.L. (1979). *Journal of the American Society for Horticultural Science,* **104,** 510–514

EVANS, L.S. and MILLER, P.R. (1975). *American Journal of Botany,* **62,** 416–421

FURUKAWA, A. and TOTSUKA, T. (1979). *Environmental Control in Biology,* **17,** 161–166

GARDNER, J.O. and ORMROD, D.P. (1977). *Scientia Horticulturae,* **5,** 171–181

GROSSO, J.J., MENSER, H.A., HODGES, G.H. and MacKINNEY, H.H. (1971). *Phytopathology,* **61,** 945–950

HEAGLE, A.S. and JOHNSTON, J.W. (1979). *Journal of the Air Pollution Control Association,* **29,** 729–732

HEAGLE, A.S., BODY, D.E. and NEELY, G.E. (1974). *Phytopathology,* **64,** 132–136

HECK, W.W., KRUPA, S.V. and LINZON, S.N. (1979). *Methodology for the Assessment of Air Pollution Effects on Vegetation.* Air Pollution Control Association, Pittsburgh

HEGGESTAD, H.E. (1969). *Journal of the Air Pollution Control Association,* **19,** 424–426

HIBBEN, C.R. and TAYLOR, M.P. (1975). *Environmental Pollution,* **9,** 107 –114

HILL, A.C., HILL, S., LAMB, C. and BARRETT, T.W. (1974). *Journal of the Air Pollution Control Association,* **24,** 153–157

HODGES, G.H., MENSER, H.A. and OGBEN, W.B. (1971). *Agronomy Journal,* **63,** 107–111

HOFSTRA, G. and BECKERSON, D.W. (1981). *Atmospheric Environment,* **15,** 383–389

HOFSTRA, G. and ORMROD, D.P. (1977). *Canadian Journal of Plant Science,* **57,** 1193–1198

HORSMAN, D.C. and WELLBURN, A.R. (1975). *Environmental Pollution,* **3,** 37–49

HOU, L.Y., HILL, A.C. and SOLEIMANI, A. (1977). *Environmental Pollution,* **12,** 7–16

HOUSTON, D.B. (1974). *Canadian Journal of Forest Research,* **4,** 65–68

HOUSTON, D.B. and STAIRS, G.R. (1973). *Forest Science,* **19,** 267–271

JACOBSON, J.S. and COLAVITO, L.J. (1976). *Environmental and Experimental Botany,* **16,** 277–285

JAEGER, J. and BANFIELD, W. (1970). *Phytopathology,* **60,** 577 (Abstract)

KENDER, W.J. and SPIERINGS, F.H.D.F. (1975). *Netherlands Journal of Plant Pathology,* **81,** 149–151

KRAUSE, C.R. and JENSEN, K.F. (1978). *Scanning Electron Microscopy,* **2,** 755–758

KRAUSE, C.R. and JENSEN, K.F. (1979). *Scanning Electron Microscopy,* **3,** 77–80

LEWIS, E. and BRENNAN, E. (1978). *Phytopathology,* **68,** 1011–1014

MACDOWALL, F.D.H. and COLE, A.F.W. (1971). *Atmospheric Environment,* **5,** 553–559

MANDL, R.H., O'NEILL, M.C. and WEINSTEIN, L.H. (1974). *Plant Physiology (Supplement),* **30,** (Abstract)

MANDL, R.H., WEINSTEIN, L.H. and KEVENY, M. (1975). *Environmental Pollution,* **9**, 133–143

MASARU, N., SYOZO, F. and SABURO, K. (1976). *Environmental Pollution,* **11**, 181–187

MATSUSHIMA, J. (1971). *Industrial and Environmental Pollution,* **7**, 218 –224

MATSUSHIMA, J. and BREWER, R.F. (1972). *Journal of the Air Pollution Control Association,* **22**, 710–713

MENSER, H.A. and HEGGESTAD, H.E. (1966). *Science,* **153**, 424–425

MENSER, H.A. and HODGES, G.H. (1970). *Agronomy Journal,* **63**, 265–269

MIDDLETON, J.T., DARLEY, E.F. and BREWER, R.F. (1958). *Journal of the Air Pollution Control Association,* **8**, 9–15

NIELSEN, D.G., TERRELL, L.E. and WEIDENSAUL, T.C. (1977). *Plant Disease Reporter,* **61**, 699–703

ORMROD, D.P. (1978). *Pollution in Horticulture.* Elsevier, Amsterdam

RAJPUT, C.B.S., ORMROD, D.P. and EVANS, W.D. (1977). *Plant Disease Reporter,* **61**, 222–225

REINERT, R.A., HEAGLE, A.S. and HECK, W.W. (1975). In *Response of Plants to Air Pollution,* pp. 159–177 (Mudd, J.B. and Kozlowski, J.J., Eds), Academic Press, New York

ROQUES, A., KERJEAN, M. and AUCLAIR, D. (1980). *Environmental Pollution (Series A),* **21**, 191–201

SHERTZ, R.D., KENDER, W.J. and MUSSELMAN, R.C. (1980a). *Scientia Horticulturae,* **13**, 37–45

SHERTZ, R.D., KENDER, W.J. and MUSSELMAN, R.C. (1980b). *Journal of the American Society for Horticultural Science,* **105**, 594–598

SKELLY, J.M., MOORE, L.D. and STONE, L.L. (1972). *Plant Disease Reporter,* **56**, 3–6

SOLBERG, R.A. and ADAMS, R.F. (1956). *American Journal of Botany,* **43**, 755–760

TAYLOR, R.J. (1978). *Water, Air and Soil Pollution,* **10**, 199–213

TINGEY, D.T. and REINERT, R.A. (1975). *Environmental Pollution,* **9**, 117 –125

TINGEY, D.T., HECK, W.W. and REINERT, R.A. (1971). *Journal of the American Society for Horticultural Science,* **96**, 369–371

TINGEY, D.T., REINERT, R.A., DUNNING, J.A. and HECK, W.W. (1971). *Phytopathology,* **61**, 1506–1511

TINGEY, D.T., REINERT, R.A., DUNNING, J.A. and HECK, W.W. (1973a). *Atmospheric Environment,* **7**, 201–208

TINGEY, D.T., REINERT, R.A., WICKLIFF, C. and HECK, W.W. (1973b). *Canadian Journal of Plant Science,* **53**, 875–879

WEBER, D.E., REINERT, R.A. and BARKER, K.R. (1979). *Phytopathology,* **69**, 624–628

WEIDENSAUL, T.C. and DARLING, S.L. (1979). *Phytopathology,* **69**, 939–941

WELLBURN, A.R., CARRON, T.M., CHAN, H.S. and HORSMAN, D.C. (1976). In *Effect of Air Pollution on Plants,* pp. 105–114. (Mansfield, T.A., Ed.), Cambridge University Press, Cambridge

WHITE, K.L., HILL, A.C. and BENNETT, J.H. (1974). *Environmental Science and Technology,* **8**, 574–576

WILLIAMS, R.J.H., LLOYD, M.M. and RICKS, G.R. (1971). *Environmental Pollution,* **2**, 57–68

16

INTERACTIONS BETWEEN AIR POLLUTANTS AND PARASITIC PLANT DISEASES

ALLEN S. HEAGLE
Agricultural Research Service, US Department of Agriculture and Plant Pathology Department, North Carolina State University, Raleigh

Introduction

Parasitic plant diseases are influenced by many factors. Among the most important are the genetics of host and parasite, climate, and the presence or absence of other disease-inducing agents which may be either biotic or abiotic. Plant disease, as influenced by genetics and climate, has long been a favourite subject of pathologists. More recently, pathologists have begun to study cases where one organism affects disease caused by another (Powell, 1971).

Air pollution is a relatively new environmental factor that can injure plants and decrease growth and yield (Mudd and Kozlowski, 1975). It can also affect parasitic plant disease (Heagle, 1973; Laurence, 1980). Reports of decreased incidence of foliar disease caused by fungi near sources of severe sulphur dioxide pollution are relatively common. Other pollutants, such as ozone (O_3) and fluorides also affect parasitism. Leaves injured by ambient levels of O_3 are more susceptible to attack by *Botrytis*, but diseases caused by rust fungi or other organisms may be inhibited by O_3.

The subject of interactions between pollutants and plant parasites was thoroughly reviewed in 1972 (Heagle, 1973). For this reason the present review will primarily consider results published since then. Recent work will be discussed in relation to previous evidence, in an attempt to delineate our current knowledge of the subject.

Effects of SO_2 on parasitism

FUNGI

Most reports published before 1973 described field observations near point sources of SO_2. Several genera of rust and needle-cast fungi were less prevalent near the sources. The occurrence of *Armillaria mellea* in trees, and of needle-cast fungi on spruce, was increased near point sources of SO_2. Other fungi were not affected (Heagle, 1973). Thus, the variable sensitivity of fungi to SO_2 was known but there was no knowledge of the mechanisms involved. Reports of direct results from controlled experiments were sparse before 1973, although controlled doses of SO_2 were known to decrease parasitism of roses by *Diplocarpon rosae*. There were no reported cases of increased parasitism from known doses of SO_2.

Controlled experiments have been of increasing importance since 1973: the effects on parasitism by obligate and nonobligate fungi have been studied. The results generally substantiate evidence from field observations that parasitism by obligate fungi is more sensitive to SO_2 than is that by other fungi. For example, parasitism of bean leaves by the rust fungus *Uromyces phaseoli* was inhibited by SO_2 (Weinstein *et al.*, 1975). Continuous exposures to 0.13 ppm of SO_2 on the 8 days before inoculation or on the 7 days after inoculation caused foliar injury and decreased the number of pustules and the size and percentage germination of spores produced on exposed leaves.

Sulphur dioxide inhibited parasitism of wheat by *Puccinia graminis*, but the type of wheat resistance to the rust was critical in the response to SO_2 (Laurence *et al.*, 1979). With Thatcher wheat, which contains horizontal resistance to race 15-B, the fungus produced fewer pustules on leaves exposed after inoculation (0.10 ppm for 100 h starting 2 days after inoculation). The rust was not inhibited on Prelude wheat, which has vertical resistance to race 15-B. When Thatcher leaves were exposed before inoculation, the tendency was for decreased numbers of pustules, but the effect was not statistically significant.

Parasitism of lilac leaves by *Microsphaera alni* was decreased by SO_2 (Hibben and Taylor, 1975). Exposure of inoculated leaves during spore germination and penetration (0.40 ppm: continuous for 24–72 h) caused decreased germination, penetration and hyphae production. Penetration was decreased by an acute SO_2 dose (1.0 ppm for 6 h) when exposures occurred during penetration. However, exposure of developing or mature conidia did not affect their ability to cause disease.

Recent reports of the effects of SO_2 on parasitism by nonobligate fungal parasites show cases of inhibition, stimulation, or no response. Exposures that decreased parasitism of bean by *U. phaseoli* did not affect parasitism of tomato leaves by *Alternaria solani* (Weinstein *et al.*, 1975). However, the number of *Helminthosporium maydis* lesions was decreased by 38% when maize plants were exposed to SO_2 on the 8 days before inoculation (0.15 ppm for 14 h daily); if the exposures occurred on the 8 days before and on the 2 days after inoculations, the number of lesions decreased by only 13–16% (Laurence *et al.*, 1979). Maize leaves were not visibly injured by SO_2, although sulphur levels in the leaves were increased.

One fungus is known to benefit from controlled exposure to SO_2: more lesions of *Scirrhia acicola* occurred on needles of Scots pine seedlings exposed at 5 days after inoculation, than on plants not exposed (0.20 ppm for 6 h) (Weidensaul and Darling, 1979). The effect was similar, but not significant, when the 6 h exposure ended 30 min before inoculation. A mixture of SO_2 and O_3 (0.20 ppm each for 6 h) caused a greater than additive amount of needle injury, but the number of *S. acicola* lesions was similar to that on plants treated with SO_2 only.

The incidence of sycamore tar spot, caused by *Rhytisma acerinum*, has been suggested as a biological indicator of SO_2 pollution in Britain (Bevan and Greenhalgh, 1976; Greenhalgh and Bevan, 1978). Detailed field studies have shown good negative correlations between the number of tar spot lesions per unit of leaf area and mean annual (Bevan and Greenhalgh, 1976), mean daily (May–June), or mean seasonal (May–June) concentrations of SO_2 (Greenhalgh and Bevan, 1978). *Rhytisma acerinum* appears to be sensitive to SO_2 influence during spore germination or penetration, when free water is required on leaf surfaces. Once infection occurs, SO_2 does not affect lesion size (Greenhalgh and Bevan, 1978).

BACTERIA, VIRUSES AND NEMATODES

A few recent articles have begun to fill the void in our knowledge of SO_2 effects on parasitism by bacteria (Laurence and Aluisio, 1981), viruses (Laurence *et al.*, 1980) and nematodes (Weber, Reinert and Barker, 1979). Results with bacterial diseases indicate that exposure to SO_2 can reduce the rate of lesion development and maximum lesion size. The effects occurred when the SO_2 dose was sufficient to increase the sulphur content of leaves. In some cases the timing of exposure in relation to inoculation did appear to be important. *Corynebacterium nebraskense* was inhibited when maize plants were exposed continuously to SO_2 (0.20 ppm for 24 h daily) on 1) the 5 days before inoculation; 2) the 2 days after inoculation; or 3) both before and after inoculation: however, the greatest inhibition was caused by the 2-day postinoculation exposures (Laurence and Aluisio, 1981). Inhibition of *Xanthomonas phaseoli* occurred when soybean plants were exposed to SO_2 (0.10 ppm for 24 h daily) on the 5 days before, 5 days after, or 5 days both before and after inoculation (Laurence and Aluisio, 1981). With *X. phaseoli*, exposures both before and after inoculation caused inhibition that was greater than the additive effects of either exposure period alone.

Sulphur dioxide can increase viruses in bean and maize (Laurence *et al.*, 1981). Again, the timing of exposure was not important. The titre of southern bean mosaic virus and sulphur content in bean leaves was increased by continuous SO_2 exposure either before, after, or both before and after inoculation (0.10 ppm for 24 h daily for 7 d). A trend toward increased virus titre, more infections and more severe symptoms, without an increase in sulphur content, was shown for maize dwarf mosaic virus (MDMV) in plants exposed to SO_2 continuously before or after inoculation.

Noninjurious doses of SO_2 failed to affect adversely four nematode parasites of soybean (*Pratylenchus penetrans*, *Belonolaimus longicaudatus*, *Heterodera glycines*, and *Paratrichodorus minor*) or one nematode parasite in leaves of begonia (*Aphelenchoides fragariae*) (Weber, Reinert and Barker, 1979). However, reproduction of *P. penetrans* was increased when soybeans were exposed to SO_2 on the 4 days before inoculation and on 3 days per week for 13 weeks after inoculation (0.25 ppm for 4 h daily). This dose of SO_2 was not enough to injure leaves or significantly to affect plant growth.

Effects of ozone on parasitism

FUNGI

Reports before 1973 generally supported the theory that O_3 would inhibit obligate fungus parasitism, whereas facultative parasitism might be increased, inhibited, or not affected (Heagle, 1973). The responses were known to depend on the dose of O_3, the parasite, the host, and even the host tissue. Evidence for an indirect effect of O_3 on parasitism by rusts was strong, while powdery mildew seemed to be affected directly during conidia formation, conidia germination and penetration. Mature spores of all fungi were known to be resistant to doses of O_3 likely to occur in ambient air. Research had begun to identify relationships between the O_3 dose and the rate of infection, invasion and sporulation. However, there was no direct evidence of how these relationships would affect the course of epidemics under field conditions. *Botrytis cinerea* was the only fungus known to benefit substantially from the presence of O_3: it caused more disease on potato when leaves were first injured by ambient O_3. There were very few reports on the effects of O_3 on soil-borne disease (Heagle, 1973).

Apparently, there are no new studies to measure the effects of O_3 on obligate parasitism, although work in Canada with *Uromyces phaseoli* on bean has been reported (Resh and Runeckles, 1973). The results suggest both a negative and a positive effect. Ozone inhibited the size of primary uredia but there were more of them, often with secondary uredia at their margins when plants were exposed on the 12 days after inoculation (0.10 ppm for $14\,h\,d^{-1}$). The decreased size of primary uredia was greatest in leaf areas between veins, where O_3 injury was also greatest. Mature conidia of *Microsphaera alni* were very resistant to O_3: germination of conidia or infection of lilac was not affected when conidia were exposed to O_3 before incubation (1.0 ppm for 6 h, or 0.25 ppm for 72 h) (Hibben and Taylor, 1975). Powdery mildew of barley is known to be sensitive to O_3 during development of conidia, and during germination or penetration (Heagle, 1973). However, sensitivity of *M. alni* at these stages was not tested.

The effects of O_3 on facultative parasitism has received significant attention in the past 8 years. Previous evidence that disease caused by *Botrytis* is increased on plants injured by O_3, has been supported by recent work with onions (Wukasch and Hofstra, 1977a). Onions grown in open-top field chambers receiving nonfiltered air were injured by O_3, while those in chambers with filtered air were not injured. When both groups were inoculated with conidia of *Botrytis squamosa*, the injured plants developed twice as many *Botrytis* lesions as did noninjured plants. Field-grown onions, partly protected from O_3 injury by an antioxidant chemical (DPX-4891, Dupont Chemical Co.) had less natural infection by *Botrytis* sp. than noninjured onions (Wukasch and Hofstra, 1977b). This is further evidence that O_3 injury predisposes plants to *Botrytis* infection but, in this study, the chemical may have directly changed the sensitivity of onions to infection or invasion by *Botrytis* sp.

Although O_3 can predispose plants to infection by *Botrytis*, it appears that O_3 can decrease the inoculum potential of *B. cinerea* if exposure occurs during sporulation. Conidia of *B. cinerea*, produced during one or two exposures to O_3 (0.15 ppm for 6 h), germinated poorly and produced shorter germ tubes, causing fewer and smaller infections on detached geranium leaves than did nonexposed spores (Krause and Weidensaul, 1978a). At an O_3 concentration of 0.30 ppm, fewer conidia were produced: those that managed to mature, germinated poorly and were unable to infect detached geranium leaves. The result was similar, whether conidia were produced during exposure on V8 juice agar or on infected geranium leaves, indicating a direct effect of O_3. Scanning electron microscopy revealed that conidia produced during exposure to 0.15 ppm of O_3 were unable to reinfect geranium leaves through stomata (Krause and Weidensaul, 1978b). A flocculent substance was present on exposed geranium leaves when *B. cinerea* colonies were exposed to O_3; this substance was not present in the absence of either O_3 or *B. cinerea*. At first, these results might appear contradictory to previously summarized reports that O_3 causes increased disease by *Botrytis* (Heagle, 1973). However, neither of these recent studies examined whether O_3 injury to leaves would affect severity of subsequent *Botrytis* infection.

Present evidence strongly indicates that several plant species are more susceptible to invasion by *Botrytis* when leaves are injured by O_3. However, *Botrytis* spores can be inactivated by direct effects of relatively low levels of O_3 if exposure occurs during sporulation. The net effect of O_3 on *Botrytis* epidemiology under field conditions remains unknown. The result will probably depend on the frequency and timing of O_3 exposure in relation to various stages in the *Botrytis* disease cycle.

The reaction to O_3 of *Helminthosporium maydis* on maize leaves was negative or positive, depending on the time of exposure in relation to the stage of fungus development, as well as on the concentration of O_3 (Heagle, 1977). When maize plants were exposed to O_3 for 6 h d^{-1} on the 6 days before inoculation, sporulation was increased at 0.06 and 0.12 ppm but not at 0.18 ppm. When exposures occurred on the 6 days after inoculation, sporulation was decreased by increasing O_3 concentrations. Exposures for 6 days, both before and after inoculation, tended to decrease sporulation. Apparently, the effects on maize plants of pre-inoculation exposures that indirectly caused increased sporulation were offset by postinoculation exposures that caused decreased sporulation.

Studies of interactions between air pollutants and fungus parasites of roots are rare, but two reports have recently been published. Ponderosa pine trees in the San Bernadino mountains 70 miles east of Los Angeles, California have been suffering from oxidant air pollution for the past 30 years. Tree stands have been severely reduced and bark beetles have been implicated in hastening death of trees injured by O_3 (Heagle, 1973). Now it appears that *Heterobasidion annosum* (syn, *Fomes annosus*) can also cause the death of trees injured by O_3 (James *et al.*, 1980). Trees with different degrees of sensitivity to O_3 were inoculated with *H. annosum* to determine whether O_3 injury was correlated with the rate of infection and invasion by the fungus. One year later, trees with good needle retention had less infection and colonization by *H. annosum* than those suffering from premature needle abscission caused by O_3. These results were corroborated with controlled exposures in greenhouse experiments.

A relatively large dose of O_3 can inhibit the association between *Glomus fasciculatus*, a vesicular-arbuscular (VA) endomycorrhizal fungus, and 'Troyer' citrange roots (McCool, Menge and Taylor, 1979). Infection was decreased by O_3 at 0.90 ppm when exposures occurred for 6 h each week for 19 weeks. At this dose, root growth and total sporulation per plant were decreased, but sporulation per gram of root tissue was not affected. Exposures to O_3 (0.45 ppm for 3 h on 2 days per week for 19 weeks) did not affect root growth or sporulation of *G. fasciculatus*, but root infections were significantly decreased. Preliminary evidence indicates that sporulation of the VA endomycorrhizal fungus, *Glomus macrocarpus*, in soybean roots is inhibited by chronic exposures of plants to O_3 in open-top field chambers (Feicht, 1981).

BACTERIA AND NEMATODES

Decreased *Rhizobium* nodulation of legumes from exposure to O_3 has been discussed in the previous review, but the effects of O_3 on foliar bacterial diseases had not then been reported (Heagle, 1973). Recent studies indicate that O_3 can reduce the development of foliar lesions caused by some bacteria but not by others. These variable results may relate to the O_3 dose, the timing of exposure relative to inoculation, the bacterial species, or the host. For example, the number of lesions of *Pseudomonas glycinea* on soybean leaves was decreased when leaves were exposed to O_3 (0.08 or 0.25 ppm for 4 h) from 16 days to 1 h before inoculation, or from 1 h to 1 day after inoculation, but not when exposures occurred 2 days after inoculation (Laurence and Wood, 1978a). This report was particularly interesting because the inhibition of *P. glycinea* occurred with or without the development of foliar injury due to O_3. The mechanism for inhibition was not shown, but the authors suggested the involvement of some bacteriostatic compounds (isoflavenoids or peroxidases) produced by plants in response to O_3. Ozone (0.20 ppm for 4 h) decreased the number of lesions caused by *Xanthomonas fragariae* on wild strawberry (*Fragaria virginiana*) (Laurence and Wood, 1978b). The results with *X. fragariae* were similar to those for *P. glycinea*: exposures before inoculation caused greater disease inhibition than did exposures after inoculation.

The responses of four soybean nematodes (*H. glycines, P. minor, B. longicaudatus* and *P. penetrans*) to O_3 and SO_2, singly or in a mixture, has been reported (Weber, Reinert and Barker, 1979). Exposures occurred for 4 h d^{-1} on 4 days during the week preceding inoculation and for 4 h d^{-1} on 3 days per week for 6–13 weeks after inoculation (0.25 ppm for each pollutant). In all tests, O_3 alone or in the mixture caused severe foliar injury and decreased growth of soybean plants; SO_2 singly caused relatively mild effects. Reproduction of *H. glycines* and *P. minor* was decreased by O_3, alone and in the mixture, but that of *B. longicaudatus* and *P. penetrans* was not usually altered. In a similar series of exposures, doses of O_3 alone or in a mixture with SO_2 severely injured leaves of begonia plants and curtailed reproduction of *A. fragariae* (Weber, Reinert and Barker, 1979). Reproduction of *P. penetrans* on tomato was slightly suppressed by O_3 at 0.20 ppm but not by SO_2 at 0.80 ppm (3 h d^{-1} on 2 days per week, for 8 weeks) (Brewer, 1979). For some reason, however, the mixture of 0.20 ppm of O_3 and 0.80 ppm of SO_2 caused increased reproduction of *P. penetrans*.

Effects of hydrogen fluoride on parasitism

Work on the effects of HF on fungus parasitism indicates that most fungi tested were inhibited if F levels in leaves were increased significantly (McCune *et al.*, 1973). Exposures of bean plants to HF before, after, or both before and after inoculation with *U. phaseoli*, generally inhibited the number and growth of uredia in leaves where F levels were increased to approximately 600–1300 ppm. However, the importance of exposure timing in relation to inoculation was not clear: in one experiment, pustule numbers were reduced by exposure before but not after inoculation. The reverse was true in a second experiment, while in a third experiment both pre-inoculation and postinoculation exposure caused inhibition of rusts.

The effects of HF on parasitism of tomato by *Alternaria solani* and *Phytophthora infestans* were mixed (McCune *et al.*, 1973). Pre-inoculation exposures of tomato plants to HF caused decreased development of *A. solani*, mainly on young leaves where F levels were in the 200–500 ppm range, but postinoculation exposures had little effect (McCune *et al.*, 1973; D.C. McCune, personal communication). With *P. infestans* on tomato, pre-inoculation exposures did not change disease development, whereas the effects of postinoculation exposure were inconsistent (McCune *et al.*, 1973). Disease caused by *Pseudomonas phaseolicola* in bean leaves was not affected by exposure, either before or after inoculation, to HF that caused an accumulation of 700 ppm of F in the unifoliate leaves.

Effects of 'acid rain' on parasitism

Decreases in the pH of precipitation have been charted in several areas of the world and attempts to quantify potential effects on biological systems have increased. Preliminary work on the effects of simulated acid rain has been completed for several diverse types of parasitic organisms.

Simulated acid rain caused inhibition of fusiforme rust (*Cronartium fusiforme*) on leaves of willow oak (*Quercus phellos*) inoculated with aeciospores. The numbers of infections and telia were decreased when 'rain' acidified to pH 3.2 with sulphuric acid was applied on each of 14 days before and after inoculation (0.63 cm 'rainfall' applied over 10 minutes per day). The number of *Helminthosporium maydis* (race T) lesions on maize with normal cytoplasm was increased when conidia were incubated in water at pH 3.5 and leaves were subsequently treated with 'rain' at pH 3.5 (0.63 cm over 10 minutes per day for 21 days). This effect did not occur for maize with Texas male-sterile cytoplasm, or when conidia were incubated in water at pH 4.5. Disease symptoms caused by *Pseudomonas phaseolicola* on kidney beans were increased when host plants were stressed by simulated acid rain at pH 3.2 before inoculation (0.63 cm over 10 minutes per day for 10 days). However, disease symptoms were inhibited when the pH 3.2 'rain' treatments occurred on the 11 days after inoculation. The progress of natural epidemics of *Uromyces phaseoli* and *Meloidogyne hapla* was measured on field-grown red kidney beans treated three times weekly with simulated rain at pH 6.0 or 3.2. The results suggest that 'rain' at pH 3.2 caused a temporary inhibition in rust severity and decreased root infection and reproduction of the nematode (Shriner, 1978).

'Rain' at pH 3.2 decreased the incidence of *Rhizobium* nodulation of greenhouse-grown red kidney beans and soybeans and of field-grown red kidney beans (Shriner, 1977). Preliminary evidence at our field site in Raleigh, North Carolina indicated no effect of biweekly 'rain' at pH 2.8 on *Rhizobium* nodulation of soybeans, although sporulation of *Glomus macrocarpus* in soybean roots was decreased (Feicht, 1981).

Results to date suggest that acid rain at levels that can occur in the field might significantly affect parasitism. However, there are no reports of dose–response studies using rainfall acidities, durations, and intensities that span those likely to occur in the field. Such work is urgently needed.

Effects of parasitism on injury caused by pollutants

Plants infected with fungi, bacteria or viruses often suffer less from O_3 than noninfected plants (Heagle, 1973). With fungi and bacteria, the protection is usually confined to the margins of disease lesions. With viruses, the effect can be more generalized. Several new articles have appeared which add to our knowledge of this phenomenon with respect to bacteria and viruses.

BACTERIA

Some research indicates that bacteria protect leaves from O_3 injury: other work indicates the lack of a protective effect. The variation in results suggests that bacterial invasion must be fairly advanced for protection to occur.

Neither *Pseudomonas glycinea* (Laurence and Wood, 1978a) nor *Xanthomonas fragariae* (Laurence and Wood, 1978b) afforded protection of infected plants from O_3 injury, as has so often been reported for other parasites. Perhaps bacterial invasion was not sufficiently developed when exposures occurred (less than 2 days after inoculation). *Xanthomonas phaseoli* provided some protection of white bean leaves from ambient oxidants in Canada, although the effect was not commercially significant; the degree of defoliation was similar for infected and noninfected plants (Temple and Bisessar, 1979). In a controlled experiment, *X. phaseoli* did not protect leaves of navy bean from O_3 injury (0.24 ppm over 8 h for 1, 2 or 4 days after inoculation) (Olson and Saettler, 1979).

A preliminary report indicates a generalized protection of soybeans by *P. glycinea* when O_3 exposures occurred 1–3 days after inoculation (Pratt and Krupa, 1979). When exposures occurred after lesion development, the protection was limited to the lesion margins and was visible only when chlorophyll was removed.

An experiment was recently performed to determine whether acute foliar injury caused by a *Pseudomonas* sp. (hypersensitive response) would protect soybean leaves from acute injury caused by O_3 and vice versa (Pell *et al.*, 1977). Bacterial infection afforded some protection from O_3 injury when plants were inoculated 1 day before, but not 4 h before, exposure to 0.35 ppm for 2 h. Exposure to O_3 (0.25 ppm for 3 h) at any time 1–6 days before inoculation, caused a decreased plant response to the bacterium, but the effect was absent if exposure occurred 7 days before inoculation. It has been suggested that O_3 decreased the injury of alfalfa caused by *X. alfalfae* and that infection by the bacterium caused decreased O_3 injury (Howell and Graham, 1977). However, in this experiment the injury caused by either insult singly was severe, probably being too extensive to allow room for additional injury when plants were treated with both.

VIRUSES

Reports since 1973 have expanded our understanding of the effects of virus infection on plant response to O_3. Pinto bean leaves were partly protected from O_3 (0.25 ppm for 4 h) when inoculated with bean common mosaic virus 4, 5, or 6 days before exposure (Davis and Smith, 1974). The effect did not occur if plants were inoculated 3 days before exposure, apparently because virus titre in leaves was too low. Several other viruses (tobacco ringspot, tomato ringspot, alfalfa mosaic and tobacco mosaic) also provided some protection from O_3 injury of primary leaves of pinto beans when plants were inoculated 5 days before exposure (0.25 ppm for 4 h) (Davis and Smith, 1976). The protection was confined to the area around the lesions for all viruses except tobacco ringspot virus which provided a more general protection.

Tobacco mosaic virus in one primary leaf of pinto bean provided some degree of protection of the opposite primary leaf from injury caused by O_3 at 0.20 ppm but not at 0.25 ppm (4 h) (Davis and Smith, 1976). A similar effect occurred with tobacco ringspot virus in soybean (Vargo, Pell and Smith, 1978). Noninoculated primary leaves were partly protected from O_3 when the 'companion' primary leaves were inoculated 8 or 10 days before exposure (0.35 ppm for 4 h) and apical necrosis caused by the virus was present. The effect did not occur if leaves were inoculated 4 or 6 days before exposure.

There are several new reports of virus-induced protection from O_3 in tobacco. Field-grown tobacco infected with tobacco mosaic virus had 60% less injury from ambient oxidants than noninfected tobacco (Bisessar and Temple, 1977). Tobacco etch virus protected tobacco leaves from O_3 (0.25 ppm for 4 h) when plants were exposed 9 days after inoculation (Moyer and Smith, 1975). However, one virus has been shown to increase sensitivity to O_3 (Reinert and Gooding, 1978). Systemic infection by tobacco streak virus caused tobacco plants to suffer significantly more injury than the noninoculated controls when exposed to O_3 (0.30 ppm for 3 h) on 1 or 2 days at 3 weeks after inoculation.

The mechanisms for virus-induced changes in plant response to O_3 are not known. Decreased stomatal conductance has been suggested as a mechanism for protection, but this was not found to be true for tobacco ringspot virus in soybean (Vargo, Pell and Smith, 1978) or for tobacco mosaic virus in tobacco or pinto bean (Brennan, 1975). Although the mechanisms remain unknown, the importance of virus titre has been recognized: protection generally increases with increased time between inoculation and exposure. Plant age and season of the year have also been identified as important factors.

Concluding remarks

A small list of noteworthy generalities can be gleaned from research on interactions between air pollutants and parasitic plant disease. Generalities that first come to mind might have been predicted from other research in plant pathology. Perhaps the most obvious is that the majority of parasitic associations are inhibited by doses of air pollution that injure host tissues (*Tables 16.1–16.4*). As might have been expected, obligate parasitism is generally inhibited by air pollutants. However, the picture with regard to facultative parasitism is less clear. Facultative parasites which thrive on senescing plant tissue, e.g. *Botrytis cinerea* and *Heterobasidion annosum*, are known to benefit when they find tissues injured by air pollutants. For other facultative parasites, e.g. *Helminthosporium maydis* or *Alternaria solani*, the impact of pollutants can depend upon the period in the disease cycle during which pollutant exposure occurs, the pollutant dose, the genetics of host or parasite, and many other factors.

Most pollutant-induced alterations of parasitism seem to be caused indirectly, through effects on the host. This is certainly the case for most bacteria, viruses and nematodes, which rarely contact pollutants directly. Mature fungus spores and hyphae are resistant, even to very large doses of pollutant. However, fungi are known to be sensitive to the direct effects of relatively small pollutant doses during spore formation (on surfaces) or during germination and penetration phases.

Almost any foliar parasite can confer some degree of protection of leaf tissue from O_3 injury. While this observation may not be of great economic importance at present, relationships of this kind may one day be used to sort out the mechanisms involved in O_3 injury. Curiously, cases of a similar nature have never been reported for SO_2.

There is strong evidence that air pollution can affect parasitism in the field. This is certainly true for areas near point sources of pollution where plants have suffered severe injury. But can the same be said for parasitism in areas where chronic doses of pollution cause subtle plant injury? Oxidant air pollution (primarily O_3) causes subtle foliar injury to crop plants in many areas of the United States. Scientists are beginning to recognize similar problems in other parts of the world. Has the long absence of a severe wheat stem-rust epidemic in the United States been caused by advances in plant pathology or by a decline in air quality? Is the occurrence and severity of disease caused by *Botrytis* on potatoes or onions more or less severe than that found 30 years ago? Does air pollution alter parasitism by *Rhizobium* or mycorrhizal fungi to a degree that will require short-term or long-term changes in fertilizer requirements for crops? Of course, the answers to these questions are not known; obtaining them will not be easy but attempts should be made.

Table 16.1 EFFECTS OF SULPHUR DIOXIDE ON PARASITIC PLANT DISEASE REPORTED SINCE 1972[a]

Organism and host	Effects	Sulphur dioxide dose and timing[b]	References
Fungi			
Uromyces phaseoli on bean leaves	Decreased pustule numbers, spore size, spore germination	0.13 ppm for 24 h d^{-1} (on the 8 days before or 7 days after inoculation)	Weidensaul and Darling (1979)
Microsphaera alni on lilac leaves	Decreased conidia germination and penetration of leaves	0.30–0.40 ppm for 24–72 h (before incubation of conidia)	Hibben and Taylor (1975)
Scirrhia acicola on Scots pine needles	Increased lesion numbers	0.20 ppm for 6 h (5 days after inoculation)	Weber, Reinert and Barker (1979)
Rhytisma acerinum on sycamore leaves	Decreased incidence	Ambient levels in England	Bevan and Greenhalgh (1976); Greenhalgh and Bevan (1978)
Helminthosporium maydis on maize leaves	Decreased lesion numbers	0.15 ppm for 14 h d^{-1} (on the 8 days before inoculation)	Laurence *et al.* (1979)
Puccinia graminis on wheat leaves	Decreased lesion numbers	0.10 ppm for 100 h (starting 2 days after inoculation)	Laurence *et al.* (1979)
Bacteria			
Corynebacterium nebraskense on corn leaves	Decreased lesion enlargement	0.20 ppm for 24 h over a period of 5 days (before or after inoculation)	Laurence and Aluisio (1981)
Xanthomonas phaseoli on soybean leaves	Decreased lesion enlargement	0.10 ppm for 24 h d^{-1} over 5 days (before or after inoculation)	Laurence and Aluisio (1981)
Viruses			
Southern bean mosaic virus in bean	Increased virus titre in leaves	0.10 ppm for 24 h d^{-1} over 7 days (before or after inoculation)	Laurence *et al.* (1981)
Maize dwarf mosaic virus in maize	Increased symptom development on leaves	0.20 ppm for 24 h d^{-1} over 7 days (before or after inoculation)	Laurence *et al.* (1981)
Nematodes			
Pratylenchus penetrans	Increased reproduction per plant and per gram of root tissue	0.25 ppm for 4 h d^{-1} (on the 4 days before inoculation and 3 days per week for 13 weeks after inoculation)	Vargo, Pell and Smith (1978)

[a]Effects of sulphur dioxide on parasitic plant disease, reported before 1973, were reviewed by Heagle (1973).
[b]Timing in relation to stage in the disease cycle is shown in parentheses.

Table 16.2 EFFECTS OF OZONE ON FUNGUS PLANT PARASITES, REPORTED SINCE 1972[a]

Fungus and host	Effects	Ozone dose and timing[b]	References
Botrytis cinerea on geranium leaves	Decreased conidia germination and stomatal penetration	0.15 ppm for 6 h d^{-1} over a period of 2 days (during sporulation)	Krause and Weidensaul (1978b)
Botrytis cinerea on geranium leaves	Decreased conidia germination, germ-tube length, infection	0.15 or 0.30 ppm for 6 h d^{-1} over 2 days (during sporulation)	Krause and Weidensaul (1978a)
Helminthosporium maydis on maize leaves	Increased sporulation	0.06 or 0.12 ppm for 6 h d^{-1} over 6 days (before inoculation)	Heagle (1977)
Helminthosporium maydis on maize leaves	Decreased sporulation	0.18 ppm for 6 h d^{-1} over 6 days (after inoculation)	Heagle (1977)
Botrytis squammosa on onion leaves	Increased lesion numbers	Ambient air in open-top field chambers	Weinstein *et al.* (1975)
Uromyces phaseoli on bean leaves	Decreased pustule size, increased pustule numbers and secondary pustules	0.10 ppm for 14 h d^{-1} over 12 days (after inoculation)	Reinert and Gooding (1978)
Heterobasidion annosum (syn. *Fomes annosus*) on pine roots	Increased infection and colonization of trees injured by oxidants	Ambient air	James *et al.* (1980)
Heterobasidion annosum (syn. *Fomes annosus*) on pine roots	Increased infection and colonization of 6-year-old seedlings	0.23 ppm for 12 h d^{-1} (58 days after inoculation, or 37 days before and 55 days after inoculation)	James *et al.* (1980)
Glomus fasiculatus[c] on 'Troyer' citrange roots	Decreased sporulation per root system	0.90 ppm for 6 h d^{-1} on 1 day per week for 19 weeks (starting 7 days after inoculation)	McCool, Menge and Taylor (1979)

[a]Effects of ozone on fungus plant parasites reported before 1973 were reviewed by Heagle (1973).
[b]Timing shown as in *Table 16.1*.
[c]Mycorrhizal fungus.

Table 16.3 EFFECTS OF OZONE ON BACTERIAL AND NEMATODE PARASITES OF PLANTS

Organism and host	Effects	Ozone dose and timing[a]	References
Bacteria			
Pseudomonas glycinea on soybean leaves	Decreased lesion numbers	0.08 or 0.25 ppm for 4 h (before or after inoculation)	Laurence and Wood (1978a)
Xanthomonas fragariae on wild strawberry	Decreased lesion numbers	0.08 or 0.20 ppm for 4 h (before or after inoculation)	Laurence and Wood (1978b)
Pseudomonas sp. in soybean leaves	Decreased hypersensitive response caused by bacterial infection	0.25 ppm for 3 h (1–6 days before inoculation)	Olson and Saettler (1979)
Nematodes			
Heterodera glycines in soybean roots	Fewer cysts per plant; more cysts per gram of root tissue	0.25 ppm for 4 h d^{-1} (4 days before inoculation and 3 days per week for 10 weeks after inoculation)	Vargo, Pell and Smith (1978)
Paratricodorus minor in soybean roots	Decreased reproduction per plant and per gram of root tissue	0.25 ppm for 4 h d^{-1} (4 days before inoculation and 3 days per week for 7 weeks after inoculation)	Vargo, Pell and Smith (1978)

[a]Timing shown as in *Table 16.1*.

Table 16.4 EFFECTS OF FLUORIDE AND SIMULATED ACID RAIN ON PLANT PARASITES

Pollutant and disease affected	Effects	Pollutant dose and timing[a]	References
Fluoride			
Uromyces phaseoli on bean leaves	Fewer pustules	HF exposures producing accumulation of 800–1200 ppm of F in leaf tissue	McCune *et al.* (1973)
Alternaria solani on tomato leaves	Decreased disease severity on young leaves	HF exposures producing accumulation of 200–500 ppm of F in leaf tissue	McCune *et al.* (1973)
Acid 'rain'			
Cronartium fusiforme on willow oak leaves	Decreased infection and telia production	pH 3.2 in 0.63 cm 'rainfall' during 10 min d^{-1} for 14 days (before and after inoculation)	Shriner (1977)
Pseudomonas phaseolicola on bean leaves	Increased disease	pH 3.2 in 0.63 cm 'rainfall' during 10 min d^{-1} for 10 days before inoculation	Shriner (1977)
Pseudomonas phaseolicola on bean leaves	Decreased disease	pH 3.2 in 0.63 cm 'rainfall' during 10 min d^{-1} for 10 days after inoculation	Shriner (1977)
Rhizobium on kidney bean and soybean roots	Decreased nodulation	pH 3.2	Resh and Runeckles (1973)
Glomus macrocarpa in soybean roots	Decreased sporulation	pH 2.8	Feicht (1981)

[a]Timing shown as in *Table 16.1*.

Our present knowledge of pollutant–parasite interactions stems largely from short-term experiments performed mostly in the greenhouse or laboratory. Although this type of experiment has provided results that have served to arouse our interest, many have been lacking in several important aspects. Few experiments have embodied the important concepts of dose–response: future work should employ at least four pollutant doses (three plus a control) that span those likely to occur in ambient air. With this approach, threshold doses for effects can be determined and the prediction of the magnitude of effects at other doses can be made. Many studies have dealt with only one stage in the parasitic cycle. More work needs to be done to differentiate the effects of exposures that occur before, during and after inoculation. There are no reports of the influence of different levels of pollution stress on disease epidemics. New developments in field exposure systems, recently reviewed (Heagle and Philbeck, 1979), make studies of this kind more feasible.

References

BEVAN, R.J. and GREENHALGH, G.H. (1976). *Environmental Pollution*, **10**, 271–285

BISESSAR, S. and TEMPLE, P.V. (1977). *Plant Disease Reporter*, **61**, 961–963

BRENNAN, E. (1975). *Phytopathology*, **65**, 1054–1055

BREWER, B.J. (1979). *Interactions of Ozone, Sulfur Dioxide, and* Pratylenchus *species on Tomato and Soybean*. MS thesis, North Carolina State University, Raleigh, North Carolina

DAVIS, D.D. and SMITH, S.H. (1974). *Phytopathology*, **64**, 383–385

DAVIS, D.D. and SMITH, S.H. (1976). *Plant Disease Reporter*, **60**, 31–34

FEICHT, P.G. (1981). *Effects of Ozone and Simulated Acidic Rain on Interactions between* Glomus macrocarpus *and Soybeans*. MS thesis, North Carolina State University, Raleigh, North Carolina

GREENHALGH, G.N. and BEVAN, R.J. (1978). *Transactions of the British Mycological Society*, **71**, 491–523

HEAGLE, A.S. (1972). *Annual Review of Phytopathology*, **11**, 365–388

HEAGLE, A.S. (1977). *Phytopathology*, **67**, 616–618

HEAGLE, A.S. and PHILBECK, R.B. (1979). In *Methodology for the Assessment of Air Pollution Effects on Vegetation*, Chapter 6, pp. 6.1–6.19 (Heck, W.W., Krupa, S.V. and Linzon, S.N., Eds). Agriculture Committee, Air Pollution Control Association, Pittsburgh, Pa, USA

HIBBEN, C.R. and TAYLOR, M.P. (1975). *Environmental Pollution*, **9**, 107–114

HOWELL, R.H. and GRAHAM, J.H. (1977). *Plant Disease Reporter*, **61**, 565–567

JAMES, R.L., COBB, F.W. Jr, MILLER, P.R. and PARMETER, J.R. Jr (1980). *Phytopathology*, **80**, 560–563

KRAUSE, C.R. and WEIDENSAUL, T.C. (1978a). *Phytopathology*, **68**, 195–198

KRAUSE, C.R. and WEIDENSAUL, T.C. (1978b). *Phytopathology*, **68**, 301–307

LAURENCE, J.A. (1981). *Zeitschrift für Pflanzenkrankheiten, Pflanzenpathologie und Pflanzenschutze*, **87**, 156–172

LAURENCE, J.A. and ALUISIO, A.L. (1981). *Phytopathology*, **71**, 445–448

LAURENCE, J.A. and WOOD, F.A. (1978a). *Phytopathology*, **68**, 441–445

LAURENCE, J.A. and WOOD, F.A. (1978b). *Phytopathology*, **68**, 689–692

LAURENCE, J.A., WEINSTEIN, L.H., McCUNE, D.C. and ALUISIO, A.L. (1979). *Plant Disease Reporter*, **63**, 975–978

LAURENCE, J.A., ALUISIO, A.L., WEINSTEIN, L.H. and McCUNE, D.C. (1981). *Environmental Pollution*, **24**, 185–191

McCOOL, P.M., MENGE, J.A. and TAYLOR, O.C. (1979). *Journal of the American Society for Horticultural Science*, **104**, 151–154

McCUNE, D.C., WEINSTEIN, L.H., MANCINI, J.F. and VAN LEUKEN, PAUL (1973). In *Proceedings of the 3rd International Clean Air Congress*, pp. A146–149. Intern. Union of Air Pollution Prevention Association, VDI-Verlag. Dusseldorf, Germany

MOYER, J.W. and SMITH, S.H. (1975). *Environmental Pollution*, **9**, 103–106

MUDD, J.B. and KOZLOWSKI, T.E. (1975). *Responses of Plants to Air Pollution*. Academic Press, New York

OLSON, B. and SAETTLER, A.W. (1979). In *Proceedings of the IX International Congress of Plant Protection, Washington, DC, Aug. 5–11, 1979*. Abstract 192

PELL, E.J., LUKEZIC, F.L., LEVINE, R.G. and WEISSBERGER, W.C. (1977). *Phytopathology*, **67**, 1342–1345

POWELL, N.T. (1971). *Annual Review of Phytopathology*, **9**, 253–274

PRATT, G.C. and KRUPA, S.V. (1979). In *Proceedings of the IX International Congress of Plant Protection, Washington, DC, Aug. 5–11, 1979*. Abstract 196

REINERT, R.A. and GOODING, G.V. Jr (1978). *Phytopathology*, **68**, 15–17

RESH, H.M. and RUNECKLES, V.C. (1973). *Canadian Journal of Botany*, **51**, 725–727

SHRINER, D.S. (1977). *Water, Air and Soil Pollution*, **8**, 9–14

SHRINER, D.S. (1978). *Phytopathology*, **68**, 213–218

TEMPLE, P.H. and BISESSAR, S. (1979). *Phytopathology*, **69**, 101–103

VARGO, R.H., PELL, E.J. and SMITH, S.H. (1978). *Phytopathology*, **68**, 715–719

WEBER, D.E., REINERT, R.A. and BARKER, K.R. (1979). *Phytopathology*, **69**, 624–628

WEIDENSAUL, T.C. and DARLING, S.L. (1979). *Phytopathology*, **69**, 939–941

WEINSTEIN, L.H., McCUNE, D.C., ALUISIO, A.L. and VAN LEUKEN, P. (1975). *Environmental Pollution*, **9**, 145–155

WUKASCH, R.T. and HOFSTRA, G. (1977a). *Phytopathology*, **67**, 1080–1084

WUKASCH, R.T. and HOFSTRA, G. (1977b). *Journal of the American Society for Horticultural Science*, **102**, 543–546

17

MINERAL NUTRITION AND PLANT RESPONSE TO AIR POLLUTANTS

D.W. COWLING*
Grassland Research Institute, Hurley, Maidenhead

M.J. KOZIOL
Botany School, University of Oxford

Introduction

In this review two main aspects of the relationship between the nutrition of plants and their response to air pollutants are discussed. First, attention is given to the role of nutrition in determining the response of cultivated plants to air pollutants, especially their sensitivity to chronic and acute injury. Second, the possibility that pollutant gases containing elements essential for plant growth may sometimes fulfil a role in nutrition is examined.

The trend of increasing yield of cultivated crops measured over the last hundred years owes much to improvements in their breeding, nutrition and management. It would be unfortunate if, in seeking increased yield of crops through improved nutrition, the potential gain was partly offset by crops becoming more sensitive to injury from air pollutants. Conversely, it is important to understand whether adjustment of crop nutrition may be used as a strategy to avoid injury by pollutants.

When Thomas (1951) reviewed the early evidence about the relationship between nutrition and plant response to pollutants he found contradictory results, but made the general statement that 'environmental factors conducive to high physiological activity . . . were found to cause maximum toxicity of sulphur dioxide'. In a later review of injury to plants by air pollutants, Brandt and Heck (1968) observed that 'Experimental research has largely neglected the area of soil factors and their influence on the sensitivity of plants to phytotoxic air pollutants . . .'. They also stated that plants are more sensitive when grown under conditions of low fertility, but that differing results have been reported with various levels of nitrogen. They concluded that there was uncertainty as to the role of plant nutrition in influencing plant sensitivity. Guderian (1977) emphasizes such uncertainty and comments that 'the number of experimental investigations is quite small and presents an incomplete and controversial picture'.

Before examining the present state of knowledge about the two main aspects already mentioned, it is worth while to consider some of the principles of plant nutrition that are relevant to the effects of air pollutants.

*Now deceased

Mineral nutrition and fertilizer practice

The mineral elements that are essential for the nutrition of plants are usually divided into two groups: the major elements—nitrogen, phosphorus, potassium, calcium, magnesium and sulphur; and the trace elements—iron, manganese, copper, zinc, molybdenum, boron, chlorine and, for nitrogen-fixing legumes, cobalt. The major elements are required in greater quantities than the trace elements and each usually represents from 0.1 to 3.0% of tissue dry weight; each trace element may be present at concentrations less than 0.01%.

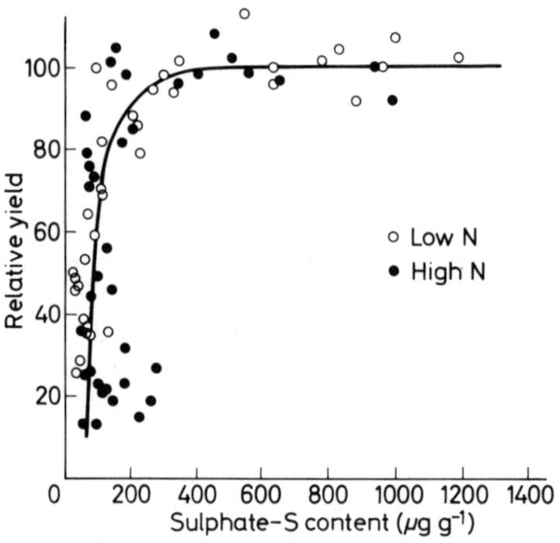

Figure 17.1   The relationship between the concentration of sulphate-S and relative yield of the shoots of perennial ryegrass (*Lolium perenne* L.) cv. S23, based on data for four harvests of plants grown on nine soils at two rates of added nitrogen. Relative yield = (yield without added S/yield with added S) × 100

A simple concept has been developed to define the nutritional status of plants. This depends on the relationship between the dry weight of the plant and the concentration of a specific element in the tissue of leaves or of a leaf of a specific age (Macy, 1936; Ulrich, 1952). As an example, *Figure 17.1* shows the relationship between the concentration of sulphate-S in the shoots of perennial ryegrass (*Lolium perenne* L.) and their relative yield when grown in soil without added sulphur (Cowling and Jones, 1978). The relative yield is calculated on the basis that maximum yield (100) was obtained from plants grown with added sulphate but otherwise similarly treated. The data are for plants grown on nine soils at two rates of added nitrogen, and harvested on four occasions. The fitted curve may be divided into a nearly horizontal part representing a zone of adequacy, a steeply descending part representing a zone of deficiency and, between these, a transition zone. The 'critical concentration' of sulphate-S which lies in the transition zone may be defined as the concentration which is just deficient for maximum growth, in this case about 0.025%. In the zone of adequacy, an increasing concentration without an increase in yield represents 'luxury consumption'.

The availability to plants of nutrients present in the soil, or added through fertilizers, is influenced by many soil factors including pH, water status and content of organic matter, as well as by plant factors and by the activity of microorganisms. Nutrients in soil may become unavailable to plants through the processes of precipitation as insoluble compounds, strong adsorption by clay minerals and organic matter (Russell, 1973), or through microbiological activity. Both nitrogen and sulphur in the soil may be incorporated into bacterial tissue and become unavailable to plant roots (Freney, Barrow and Spencer, 1962; Broadbent and Clark, 1965). Nutrients may also be lost from soils through leaching by water, weakly adsorbed ions like nitrate being particularly prone to this process (Gardner, 1965). Sulphate, an ion of interest in the present paper, is adsorbed strongly by clay minerals at pHs below 6.0 (Harward and Reisenauer, 1966) but is subject to leaching in soil with a higher pH. Under some conditions, nitrogen may be lost from soils in gaseous form as N_2, nitrous oxide (N_2O), or ammonia (NH_3) (Focht and Verstraete, 1977; Terman, 1979).

From the beginning of the present century, the use of farmyard manure to sustain crop yield has been supplemented by ever-increasing amounts of inorganic fertilizer. Over the period 1950–75 alone, and on a world-wide basis, fertilizer consumption of nitrogen, phosphorus and potassium (NPK) increased elevenfold, fourfold and fivefold, respectively (FAO, 1953, 1977). The fertilizers often provided these elements in the form of ammonium sulphate, potassium sulphate and superphosphate (made by the treatment of rock phosphate with sulphuric acid). However, a post-1950 trend that is relevant to the present review is towards the use of high-analysis fertilizers, low in sulphur. This trend was partly a response to the increased price of sulphuric acid which made chemicals other than sulphates of ammonia and potassium attractive as fertilizers. Urea, ammonium nitrate and ammonia, besides di-ammonium phosphate in compound fertilizers, have gained favour as sources of nitrogen. Rock phosphate is treated with phosphoric acid to produce triple superphosphate (19.6% phosphorus compared with 7.4% in superphosphate) or with nitric acid to produce nitric phosphate, to give sources of phosphorus, and potassium chloride has become the main source of potassium (Sauchelli, 1950). The nutritional value of sulphur, which seemed to have been taken for granted (Alway, 1940), was highlighted with this trend towards high-analysis fertilizers (Sauchelli, 1950) as many crops began to show signs of sulphur deficiency (Bertramson, Fried and Tisdale, 1950; Coleman, 1966).

It is worth noting the particular relationships between sulphur and nitrogen nutrition because these elements may be found in atmospheric pollutants. Both elements are present in the soil mainly in organic matter and they are released through its degradation by microorganisms as the available forms sulphate and nitrate, the process being known as mineralization (Freney, Barrow and Spencer, 1962; Broadbent and Clark, 1965). The ratio of the amounts of nitrogen to sulphur in soils is about 10:1; for available forms of the elements it is somewhat higher and close to 14:1–17:1 which is the ratio required for healthy growth by a range of plants (Dijkshoorn and Van Wijk, 1967). Thus, when soil supplies of nitrogen are deficient, those of sulphur will almost always be deficient as well (Stewart, 1967) and the deficiency will be exacerbated by large additions of nitrogen in fertilizers or through symbiotic fixation. However, in industrial regions there are often substantial additions of sulphur to crops through rainfall, the absorption of SO_2 by leaves, and the deposition of SO_2 on to soils (Terman, 1978). These additions must largely account for the absence of sulphur deficiency in crops in many of the industrialized areas of the world (Alway, Marsh and Methley, 1937; Jones, Cowling and Lockyer, 1972; Korkman, 1973). The adequacy of supplies of sulphur to plants may be assessed from the ratio of nitrogen to sulphur in their shoots (values above 17 indicating deficiency) or, as previously seen, from the concentration of sulphate in their shoots (Metson, 1973).

A deficiency in the supply of any one of the essential elements may be reflected in a reduction of growth rate and, in turn, of the final dry weight of the plant. In addition, there may be visible signs of deficiency such as chlorosis, discolouration and necrosis of the leaves, together with a reduction in the concentration of the element in the tissue. In some circumstances there is evidence that stomatal conductance is affected by nutrient supplies (Drew, 1967; Ryle and Hesketh, 1969; Brag, 1972; Leone and Brennan, 1972b). It therefore becomes of interest to assess plant response to atmospheric pollutants under various regimes of nutrition.

Mineral nutrition and sensitivity of plants to air pollutants

In this section, the review and discussion centre on three phytotoxic air pollutants, namely sulphur dioxide (SO_2), hydrogen fluoride (HF), and ozone (O_3), there being no substantial information about the relevant effects as far as other pollutants are concerned.

SULPHUR DIOXIDE

A summary of the studies which have examined the influence of mineral nutrition on the sensitivity of plants to SO_2 is given in *Table 17.1*. Generally, plants that are given an adequate supply of nutrients are less sensitive to injury than plants with a deficient supply, although there are some exceptions. Notably, Leone and Brennan (1972a) working with tobacco (*Nicotiana tabacum* L.) and tomato (*Lycopersicon esculentum* Mill.) found that sensitivity was greatest in plants with an adequate supply of nitrogen, and that it decreased in plants with either a deficient or a 'luxury' supply of this element. Decreased sensitivity appeared to be linked with increased stomatal resistance.

The concentrations of SO_2 to which plants were exposed in most of the studies given in *Table 17.1* were high in relation to those of the air in rural areas of most countries, and resulted in considerable leaf necrosis. However, Bleasdale (1952) conducted a number of experiments in which perennial ryegrass was exposed to filtered air or ambient air in which SO_2 was the principal air pollutant. He investigated the effect of variation in nutrient status of soil-grown plants on their response to pollution (*Table 17.2*). The experiments (1–3) were conducted in unheated glasshouses

Table 17.1 INFLUENCE OF MINERAL NUTRITION ON PLANT SENSITIVITY TO SULPHUR DIOXIDE

Plant species	Exposure Concn ($\mu g\,m^{-3}$)	Time (h)	Nutritional regime and culture medium	Effects on sensitivity	Source
Lucerne	2490	5	Complete: low and high (sand and 'soil')	Increased with low supply	Setterstrom and Zimmerman (1939)
Buckwheat	3490	6			
Lucerne Buckwheat	2490–3030	4–6	S: 70–1650 mg ℓ^{-1} (sand and soil)	Unchanged	Setterstrom and Zimmerman (1939)
Lucerne	286	6–7 (daily: over 2 years)	Complete: low and high (sand)	More chlorosis and leaf drop with low supply	Thomas *et al.* (1943)
Lucerne	286	6–7 (daily: over 2 years)	S: 0, 0.8, 1.5, 10 mg ℓ^{-1} (sand)	Unchanged	Thomas *et al.* (1943)
Rape	2290–4380	9–81	N: 0, 120 kg ha^{-1} (field, soil)	Increased with low N	Zahn (1963)
Rape, Spinach, Oat Barley	5720	1.5	NPK: 3 rates ('soil')	Increased with low N Decreased with low N Unchanged	Zahn (1963)
Red clover Barley	2500	60	CaCO$_3$: 0–4000 mg kg^{-1} (soil) P: 0–350 mg kg^{-1} (soil) K: 0–660 mg kg^{-1} (soil)	Increased with low Ca Decreased with low P Increased with low K	Guderian (1971)
Wheat	2500	38	N: 0, 500 mg kg^{-1} (soil)	Increased with low N	Guderian (1971)
Rape	2000	20	N: 0, 700 mg kg^{-1} (soil)	Increased with low N	
Sunflower	2000	30	N: 0, 1400 mg kg^{-1} (soil)	Increased with low N	
Tobacco	5720	5	N: 14, 280, 1120 mg ℓ^{-1} (sand)	Decreased with deficiency or with excess	Leone and Brennan (1972a)
Tomato	5720	5	N: 14, 112, 336 mg ℓ^{-1} (sand)		
Tobacco	5720	5	S: 0, 1.5, 96, 384 mg ℓ^{-1} (sand)	Decreased with deficiency	Leone and Brennan (1972b)
Tomato	5720	5	S: 0, 1.5, 64, 224 mg ℓ^{-1} (sand)	Decreased with deficiency	

through winter, spring and summer periods, respectively, and the concentration of SO_2 was monitored on a daily basis either outside the glasshouses (Experiments 1 and 2) or inside the glasshouse with polluted air (Experiment 3). In each experiment the addition of nutrients to soil increased the rate of plant growth and, apart from the winter period, growth was reduced in polluted air to a lesser extent with plants having the high rather than the low supply of nutrients. No leaf lesions were observed in any of these experiments but an increased rate of leaf senescence was usually recorded for plants grown in polluted air.

Table 17.2 DRY WEIGHT OF SHOOTS OF PERENNIAL RYEGRASS GROWN IN FILTERED AND IN POLLUTED AIR AT DIFFERENT LEVELS OF SOIL FERTILITY. YIELD OF PLANT IN POLLUTED AIR RELATIVE TO THAT IN FILTERED AIR (100) IS SHOWN IN BRACKETS. After Bleasdale (1952)

Experiment	Air treatment	Nutrient level	Harvest 1[a]	Harvest 2[a]
			(mg per plant)	
1	Filtered	Low[b]	73 (100)	35 (100)
	Polluted[c]		59 (80)	22 (73)
	Filtered	High	99 (100)	57 (100)
	Polluted[c]		72 (73)	33 (58)
2	Filtered	Low	542 (100)	–
	Polluted[d]		345 (64)	–
	Filtered	High	817 (100)	–
	Polluted[d]		825 (101)	–
3	Filtered	Low	673 (100)	–
	Polluted[e]		287 (43)	–
	Filtered	High	1084 (100)	–
	Polluted[e]		728 (67)	–

[a]1) Harvest 1 and 2 at 115 and 197 d after sowing in autumn, respectively
  2) One harvest only at 78 d after sowing in spring
  3) One harvest only at 65 d after sowing in summer
[b]Addition to soil. Low: Nil; High: (kg ha^{-1} equivalent): 600 calcium (as $Ca(OH)_2$), 190 nitrogen, 200 potassium, 100 phosphorus
[c]Approximate mean SO_2 concentration (final 52 d): 108 µg m^{-3} (range 28–252)
[d]Approximate mean SO_2 concentration: 104 µg m^{-3} (range 28–504)
[e]Mean SO_2 concentration: 32 µg m^{-3} (range 0–252)

Although a high concentration of SO_2 (380 µg m^{-3}) was used in a recent study with perennial ryegrass, no visible signs of injury were observed (Ayazloo, Bell and Garsed, 1980). Nevertheless, there were indications that the reduction in yield through exposure to SO_2 was less in plants grown with high nitrogen than with low nitrogen. This supports the finding that, with SO_2 close to ambient (55 µg m^{-3}), the only evidence of injury in perennial ryegrass—an increase in the proportion of dead leaves—was observed with low but not with adequate addition of nitrogen to the plant culture (Cowling and Lockyer, 1978).

There is also a difference among findings about the effect of the prolonged exposure of perennial ryegrass to SO_2 at 40–300 µg m^{-3} (*see* Bell, this volume, Chapter 11). Thus, Cowling and colleagues (Cowling, Jones and Lockyer, 1973; Cowling and Lockyer, 1976; Lockyer, Cowling and Jones, 1976; Cowling and Koziol, 1978), consistently found little evidence of reduced plant growth, whereas others (Bell and Clough, 1973; Bell, Rutter and Relton, 1979; Ashenden and Mansfield, 1977; Crittenden and Read, 1978) have found substantial reductions in growth. This difference may arise because of a generally higher level of nutrition and a higher growth rate in the first than in the second group of studies. For example, the daily rate of accumulation of shoot weight was an order of magnitude greater ($18\,\mathrm{g\,m^{-2}\,d^{-1}}$ versus $1.6\,\mathrm{g\,m^{-2}\,d^{-1}}$) in an experiment reported by Lockyer, Cowling and Jones (1976) than in an experiment reported by Bell, Rutter and Relton (1979). Other direct comparisons are complicated by differences in the methods of plant culture used, i.e. close planting in contrast to spaced single plants, but the indications are that plants may be more sensitive under conditions that give a low rather than a high rate of growth (Davies, 1980).

The special case of sulphur nutrition is interesting; it is referred to only briefly here as a more detailed examination is given later. In a report by Thomas *et al.* (1943) the level of sulphur nutrition had no effect on the sensitivity of plants to SO_2. On the other hand, Leone and Brennan (1972b) found increasing sensitivity to SO_2 in tobacco and tomato with an increase in the concentration of sulphate in the culture solution. Although it was not possible to discriminate critically between sulphur taken up from the two sources, solution and air, estimates based on differences in uptake by plants grown in solution with low and high concentrations of sulphur suggested a reduced uptake from the air in sulphur-deficient plants. The conclusion that this arose through increased stomatal resistance was supported by preliminary measurements of stomatal aperture (Leone and Brennan, 1972b). The exposure of ryegrass to low concentrations of SO_2 has produced no evidence that sulphur nutrition influences sensitivity to injury as measured by the weight and proportion of dead and senescent leaves (Cowling and Lockyer, 1978). In the same study, the lack of a consistent and significant effect of sulphur nutrition on the transpiration coefficient of ryegrass does not suggest major changes in stomatal resistance. However, in a series of studies with ryegrass, estimates of uptake of SO_2 based on the 'difference' method described above indicate a higher rate in plants with a low rather than an adequate supply of sulphate to the roots, and hence point to a lower stomatal resistance in sulphur-deficient plants (Cowling *et al.*, 1981). Two studies in which ^{35}S has been used to discriminate between uptake from culture solution containing sulphate and from air containing SO_2 have given conflicting results on this point. In studies with poppy (*Papaver somniferum* L.) and tobacco, Furrer (1967) found that the uptake of SO_2 following prolonged exposure to $26\,\mu\mathrm{g\,m^{-3}}$ was greater with low than with high concentrations of sulphate in the culture solution. In contrast, Faller (1972a) found that the uptake of SO_2 by tobacco was unaffected by the supply of sulphate in solution following exposure to SO_2 at $1500\,\mu\mathrm{g\,m^{-3}}$ for $14\,\mathrm{d}$ ($12\,\mathrm{h}$ daily). One factor possibly contributing to this difference is that in the second study, but not in the first, there was evidence that the plants were deficient in sulphur when grown with the low supply of sulphate alone.

It is difficult to reconcile all these findings. In the studies of Leone and Brennan (1972b) the high concentration of SO_2 may have decreased stomatal resistance to a greater extent than occurs at the lower concentrations generally used elsewhere. It would be helpful if complete data were available for: 1) the plant material, e.g. specific leaf area, leaf weight; 2) the mean concentration and duration of plant exposure; 3) the resistance to uptake; and 4) the environmental conditions during exposure. Additionally, the content of sulphur in the plants is of basic importance. Even without the use of labelled sulphur, this would permit an estimate of uptake, and thus confirm measurement of uptake based on changes in concentration of SO_2 in exposure chambers.

HYDROGEN FLUORIDE

A number of workers have pointed out the complex nature of nutrient deficiencies and the interactions that may occur among nutrients as well as with air pollutants (Applegate and Adams, 1960; MacLean *et al.*, 1969). In particular, 'when HF injury occurs in plants with a deficiency of a nutrient element the response may be due to an imbalance peculiar to that element rather than to the deficiency of that element *per se*' (MacLean *et al.*, 1969). The response then, may be an indirect rather than a direct effect of the nutrient deficiency.

A summary of the experiments in which nutrition has modified the response of plants to HF is given in *Table 17.3*. Among the earliest reports is that by Brennan, Leone and Daines (1950), who found that tomatoes grown with a low or with a 'luxury' supply of nitrogen or of calcium were less sensitive to injury by HF than plants with a supply of either element close to optimum. With phosphorus, however, plants became increasingly sensitive to HF as the supply increased but even the highest level of supply of phosphorus may not have represented 'luxury'. The mechanisms involved were not discussed but the degree of injury was not obviously related to the content of fluoride in the leaves. McCune, Hitchcock and Weinstein (1966) also found that sensitivity to injury decreased under conditions of nitrogen deficiency as well as of calcium deficiency, while other investigators have, in contrast, found that with deficiency of nitrogen, phosphorus, potassium, magnesium or calcium there was increased sensitivity to injury (*Table 17.3*). Adams and Sulzbach (1961) reported that, following an accidental 15 h exposure to 43 µg HF m^{-3}, only beans that were nitrogen-deficient showed leaf necroses. Chlorosis and necrosis of leaves due to HF were particularly marked with calcium-deficient or magnesium-deficient tomato, whilst some necrosis of apical leaves was observed in potassium-deficient plants (MacLean *et al.*, 1969). It was concluded that leaf injury by HF was dependent on the relative solubilities of the fluorides of the cations that were deficient, and on the metabolic roles of these cations. Thus, relatively insoluble calcium or magnesium fluorides decreased the effective supply of these elements within plant tissue to a greater extent than did potassium, which has a more soluble fluoride salt. Deficiency of potassium, required as an activator of enzyme reactions, is less critical than deficiency of the other two cations, which are required as co-factors (*see* Weinstein and Alscher Herman, this volume, Chapter 7).

Further studies with tomato (MacLean, Schneider and McCune, 1976) showed that magnesium deficiency and excess respectively increased and decreased the phytotoxic effects of HF, compared with plants with an

Table 17.3 INFLUENCE OF MINERAL NUTRITION ON PLANT SENSITIVITY TO HYDROGEN FLUORIDE

Plant species	Exposure Concn (μg m^{-3})	Time (h)	Nutritional regime[a] and culture medium	Effect on sensitivity	Source
Tomato	420	3.5	N: 14, 56, 112, 448 mg ℓ^{-1} (sand) Ca: 10, 40, 80, 240 mg ℓ^{-1} P: 0.8, 16, 62 mg ℓ^{-1}	Greatest at medium N Greatest at 40 mg ℓ^{-1} Decreased with P deficiency	Brennan, Leone and Dainer (1950)
Bean	2	>240	N, P, K, Ca, or N deficient (soil)	No injury: uptake greater with P or K deficiency	Applegate and Adams (1960)
Bean	43	15	N deficient (sand)	Increased with N deficiency	Adams and Sulzbach (1961)
Tomato	3, 6	3696	Ca: 40, 200 mg ℓ^{-1}	Increased with Ca deficiency	Pack (1966)
Gladiolus	2	24	P, K, or Mg deficient (sand) N or Ca deficient	Increased with deficiency Decreased with deficiency	McCune, Hitchcock and Weinstein (1966)
Tomato	5–12 1–15 3	120 120 1536	Mg deficient (sand) Ca deficient P deficient	Increased with deficiency Increased with deficiency Increased with deficiency	MacLean et al. (1969)
Tomato	5–10	168	Mg: 2, 12, 48, 192, 384 mg ℓ^{-1} (sand)	Increased with deficiency, decreased with excess	MacLean, Schneider and McCune (1976)
Barley	12	192	Various rates of P, K, or Ca (soil)	Increased with deficiency of P, K, and more so of Ca	Guderian (1977)
Spinach	Not given		N deficient (soil)	Increased with deficiency	Guderian (1977)

[a]The regimes almost always include a treatment providing adequate nutrition of plants as a basis for comparison

adequate supply of this nutrient. It was postulated that the effects were associated with detoxification of fluoride through the formation of MgF_2, so that more fluoride was precipitated in plants with excess magnesium. In plants with low magnesium, less fluoride would be thus detoxified and, in addition, there would be an acute deficiency of magnesium for essential biochemical reactions. It was also noted that accumulation of fluoride was reduced in magnesium-deficient plants through leaf necrosis; a reduction of fluoride in plants with excess magnesium could not be explained.

Accumulation of fluoride by barley (*Hordeum vulgare* L.) was shown to increase almost linearly with increasing supply of either phosphorus or potassium when plants were exposed to HF (Guderian, 1977). However, the sensitivity to injury either decreased, or was unchanged, as the supply of each nutrient increased. On the other hand, both accumulation of F and injury were higher in spinach grown with a poor rather than a good supply of nitrogen (Guderian, 1977). In a study with bean (*Phaseolus vulgaris* L.), there were indications that the uptake of fluoride from air containing $2\,\mu g\,HF\,m^{-3}$ was increased in plants that were deficient in phosphorus or potassium but was not significantly altered with deficiency of nitrogen or calcium; none of the plants displayed signs of injury (Applegate and Adams, 1960).

It is clear that the interpretation of the effects of nutrition on the response of plants to HF, or indeed to other air pollutants, is complicated. This may largely account for the fact that experiments on the topic are rare (MacLean *et al.*, 1969). Nevertheless, it is important to obtain a clearer insight into the mechanisms involved. For example, it is interesting to note that one of the main approaches to the control of fluorosis in cattle in Great Britain is to encourage increased yields of grass through good nutrition (Allcroft, Burns and Herbert, 1965). This has been found to dilute the concentration of fluoride in forage presumably because, leaving aside the question of the influence of nutrients on F accumulation, resistances to uptake in a dense, closed canopy are sharply increased. The protective action of calcium carbonate sprayed on to the leaves of plants, appears to operate by inactivation of HF as calcium fluoride.

OZONE

A summary of the experiments in which nutrition has modified the sensitivity of plants to O_3 is given in *Table 17.4*. The earliest indications of such effects came, however, from studies with ozonated hexene (O_3 equivalent, $600\,\mu g\,m^{-3}$) which showed that oat (*Avena sativa* L.), barley, spinach (*Spinacea oleracea* L.) and lettuce (*Lactuca sativa* L.) were more sensitive when grown with the equivalent of $50\,kg\,ha^{-1}$ of added nitrogen than with a lower supply (Middleton, 1956). The response of mangel (*Beta vulgaris* L.) to ozonated hexene was also examined in plants grown in a factorial experiment with treatments that provided three rates of nitrogen, and two each of phosphorus and potassium (Brewer, Guillemet and Creveling, 1961). The exposure involved two periods of 6 h with O_3 at $400\,\mu g\,m^{-3}$, together with some organic oxidants. Spinach, grown with similar nutrient treatments, was exposed on each of three successive days

Table 17.4 INFLUENCE OF MINERAL NUTRITION ON PLANT SENSITIVITY TO OZONE

Plant species	Exposure Concn ($\mu g\,m^{-3}$)	Time (h)	Nutritional regime[a] and culture medium	Effect on sensitivity	Source
Mangel	430[b]	2×6	N: 0, 30, 90 mg kg^{-1} (soil) P: 20, 150 mg kg^{-1}	Increased with N and with P + K	Brewer, Guillemet and Creveling (1961)
Spinach	535	3×6	K: 5, 40 mg kg^{-1}	Increased with N, decreased with P and high NPK	
Tobacco	Ambient (field)		N: 30, 45, 60 mg kg^{-1} (soil)	Decreased with N supply	Menser and Street (1962)
Tobacco	750	48	N: 15, 84, 168, 840 mg ℓ^{-1} (sand or soil)	Increased with deficient and excess N	MacDowall (1965)
Tobacco	200–350 130	4 6	N: 14, 280, 560, 1120 mg ℓ^{-1} (sand)	Decreased with deficient and excess N	Leone, Brennan and Daines (1966)
Tobacco	190–770	2–5	N: 15, 80, 160, 800 mg ℓ^{-1} (sand)	Decreased with deficient and excess N	Menser and Hodges (1967)
Bean	640	1	Low, medium, high nutrients (sand)	Decreased with increased nutrient supply	Heck, Dunning and Hindawi (1965)
Tomato	320–960	3	P: 1.5, 15, 62 mg ℓ^{-1} (sand)	Decreased with P deficiency	Leone and Brennan (1970)
Bean	1070	2×2	S: 1.3, 32 mg ℓ^{-1} (sand)	Increased with S deficiency	Adedipe, Hofstra and Ormrod (1972)
Radish	535	4	N: 60, 300 mg ℓ^{-1} (sand) P: 30, 150 mg ℓ^{-1} (sand)	Increased with N supply Unchanged with P supply	Ormrod, Adedipe and Hofstra (1973)
Bean, soybean	353–1820	0.5–4	K: 4.1, 28 g ℓ^{-1} ('soil')	Increased with K supply	Dunning, Heck and Tingey (1974)
Soybean	1285	1.5	4 rates of NPK (sand or 'soil')	Decreased with deficient and excess nutrient supply	Heagle (1979)

[a]The regimes almost always include a treatment providing adequate nutrition of the plants as a basis for comparison
[b]Total oxidant from ozonated hexene

for 9 h to O_3 alone at 500 µg m^{-3}. With both plants, injury increased with addition of nitrogen and, with mangel, to a greater extent when the higher rather than lower rates of phosphorus and potassium were given together. When either one of these two nutrients was given alone at the high rate, injury was reduced. With spinach grown with a low rate of nitrogen, an increase in the supply of potassium gave increased injury when phosphorus was low but not when it was high. At high nitrogen rates, increasing the potassium supply had the opposite effect, injury being reduced. The interaction was believed to have practical implications in that injury to spinach might be avoided by ensuring that both potassium and phosphorus are given in substantial amounts. With mangel there was little evidence that fertilizer treatment might be manipulated to avoid injury; decreased sensitivity was linked with treatments that gave reduced yields. Some problems arise in the interpretation of these results because, at least with potassium and phosphorus, the low levels of supply did not always result in deficiency.

The influence of nutrition on the sensitivity of tobacco has been the subject of a number of reports. Menser and Street (1962) conducted a field experiment in Maryland, where ambient O_3 frequently causes injury to this species. Among the factors investigated, three rates of addition of nitrogen fertilizer influenced the response of the tobacco crop to air pollution. The incidence and severity of leaf fleck were greatest with low nitrogen and declined as the supply increased. The additions of nitrogen were sufficient to produce an increase in yield, the increment declining with the highest addition. It was suggested that these changes in sensitivity arose through delaying and shortening the period when leaves were at the most sensitive stage i.e. recently matured. In the exposure-chamber study reported by MacDowall (1965), injury was greatest with either the lowest or the highest rates of nitrogen, which represented, respectively, deficiency and excess in the plants, but there was no clear conclusion as to the mechanism involved in this pattern of response. A somewhat different effect of nitrogen on response of tobacco to O_3 was found in another series of experiments, again conducted in exposure chambers (Leone, Brennan and Daines, 1966). The plants were grown in sand with additions of nitrogen in solution to give a supply ranging from low to 'luxury' (14, 280, 560 and 1120 mg ℓ^{-1}) and exposed to O_3 in 15 independent experiments. Injury to leaves (necrosis and leaf fleck) was always more severe in plants with an optimum supply of nitrogen (as defined by the concentration of nitrogen in leaf 3) than with either a low or a 'luxury' supply. Carbohydrates were measured in the leaves before and after exposure treatments, and the rate of respiration after exposure was noted. These measurements supported the conclusion that there may be a particular level of carbohydrate at which leaves are most sensitive to ozone injury, and that the effect of nitrogen nutrition on sensitivity operates through this. In contrast to many studies of this sort, careful attention was given to defining the nutritional status of the plants rather than the level of nutrient supply *per se*. This is an important point that should be followed in all studies of this kind, if the confusion that appears to exist at present is to be avoided in future.

Menser and Hodges (1967) also examined the effect of O_3 on tobacco that was grown in sand culture with additions of nitrogen in solution (15,

80, 160 and 800 mg ℓ^{-1}) giving successive increases in yield. Leaf injury was greatest in plants given 160 mg N ℓ^{-1} and least in nitrogen-deficient plants (15 mg ℓ^{-1}). The protective effect against injury of nitrogen at the highest rate was again suggested to arise through the delaying of leaf maturity. These findings support those of Leone, Brennan and Daines (1966); the difference between them and the findings of MacDowall (1965) was believed to be due to a cultivar difference.

The influence of the concentration of nitrogen in the leaf of tobacco on the sensitivity to O_3 was examined more closely by Lee (1966) who studied the effect of various chemical treatments on detached leaves. He showed that sensitivity to injury induced by treatments with chemicals, including urea and potassium nitrate, could be reversed by treatment with sucrose. The positive correlation between sensitivity and the concentration both of total soluble nitrogen and of ammonia does not provide a clear explanation for the reduced sensitivity found in plants grown with excessive supplies of nitrogen (Leone, Brennan and Daines, 1966; Menser and Hodges, 1967). It does explain the low sensitivity of nitrogen-deficient plants in those studies, especially as their content of sucrose was likely to have been high.

The influence of phosphorus nutrition on the sensitivity of tomato to O_3 was examined by Leone and Brennan (1970). The plants, grown in sand culture, were given phosphorus in solution at 1.5, 15 and 62 mg ℓ^{-1} which produced deficient and healthy plants, and plants with an excess content of phosphorus, respectively. Exposure to ozone at 320–960 $\mu g\,m^{-3}$ for 3 h resulted in leaf injury that decreased with deficiency of phosphorus. In phosphorus-deficient plants, which showed little or no injury, it was observed that total carbohydrates in the leaves decreased sharply after exposure; before exposure, this component had not been greatly affected by phosphorus nutrition. However, it was suggested that carbohydrate content, which represented a relatively greater reserve in the phosphorus-deficient plants, was linked with their low sensitivity to injury.

When bean plants, grown in sand culture with a varied supply of sulphur, were exposed to O_3 for 2 h, leaf injury was greater in the plants given a low rate of sulphur (Adedipe, Hofstra and Ormrod, 1972). Injury was associated with increased concentration of chlorophyll and carbohydrate in the leaf, and it was accompanied by reduced growth of the plant, especially of the root. It was concluded that sensitivity was reduced through a higher content of sulphydryl groups in plants grown with a high rate of sulphur, and not through carbohydrate content. It is known that plants may be protected from oxidant injury by sprays containing sulphydryl compounds.

A study of the influence of temperature on the sensitivity of radish (*Raphanus sativus* L.) included varying nitrogen and phosphorus nutrition in sand culture (Ormrod, Adedipe and Hofstra, 1973). Exposure to O_3 gave a greater reduction in yield of plants grown at high rather than at low nitrogen rates, irrespective of the temperature in the range 20–30 °C; the plants showed a marked response to nitrogen. Variation in phosphorus nutrition had no effect on sensitivity or on plant yield, but was involved in an unexplained interaction with the effect of temperature. Thus, with plants grown with high nitrogen, exposure to O_3 had no effect on yield at either rate of phosphorus at 20 °C, but yield was reduced at 30 °C. No explanation was given for these findings: they serve to indicate the complexity of interactions between nutritional and physical factors in determining plant sensitivity.

The sensitivity of bean and soybean (*Glycine max* (L.) Merr.) to a range of doses of O_3 was examined in a factorial experiment which included two rates of supply of potassium (Dunning, Heck and Tingey, 1974). Leaf damage was higher at low than at normal rates of potassium, and damage was inversely related to the concentration of potassium in the leaf. The concentrations of reducing sugars and sucrose were unaffected by potassium nutrition, and it was suggested that the low rate of supply had its effect through interference with stomatal closure or through a poorer enzymatic repair system.

Soybean was also exposed to O_3 in another factorial experiment involving various growth media and different cultivars as well as four rates of addition (zero, low, medium, high) of a fertilizer containing nitrogen, phosphorus, potassium and micronutrients (Heagle, 1979). Those plants given no added fertilizer showed signs of nutrient deficiency and (except on high addition of fertilizer with two of the media used) plant weight increased with fertilizer rate. Plants grown without added fertilizer were always less sensitive to O_3 than plants grown with the low rate, and their shoot weight was reduced proportionately less; effects with the medium and high rates were generally intermediate.

There are clearly discrepancies between the various findings about the effect of nutrition on the sensitivity of plants to O_3. However, there is sufficient evidence to support the hypothesis that changes in sensitivity are linked to changes in soluble carbohydrate status. It is therefore important that this component be measured in future studies. With this information, it should be possible to establish a concept unifying the effects of nutrients on sensitivity, that would amplify the findings of Dugger *et al.* (1962). They showed that, if the content of carbohydrate was above or below a certain optimum, then the leaves were predisposed to injury by O_3, but they also noted that other factors were involved.

Role of air pollutants in plant nutrition

Two of the problems that have arisen with changes in agricultural practices are the deficiency in many soils of sulphur for maximum plant growth and the loss from fertilized soils of nitrogen, either by leaching of nitrates, denitrification, or by the volatilization of NH_3. It therefore becomes of economic interest, both in terms of crop yield and of fertilizing practice, to investigate whether certain gaseous compounds containing elements necessary for plant growth may have some beneficial effect by compensating for deficiency in soil supplies.

SULPHUR DIOXIDE AND HYDROGEN SULPHIDE IN SULPHUR NUTRITION

Sulphur is important in nutrition: for instance, the formation of disulphide bonds between specific cysteine residues in proteins helps to maintain the particular steric configurations required for enzymatic activity. Further, highly reactive sulphydryl groups play an integral part in the functioning of coenzyme A and dihydrolipoyl transacetylase, while thioether bridges are essential to maintaining the structure of cytochrome c. Sulphur is also a constituent of the amino acids cystine and methionine, and of other metabolites such as glutathione, thiamin and biotin; it also has a role in chlorophyll synthesis and in nodule formation in leguminous plants. As the biochemistry of sulphur is closely linked with that of nitrogen, it is not surprising that the symptoms of deficiency of either element are similar (Treshow, 1970) and that the plants require specific S:N ratios for healthy growth (Dijkshoorn and Van Wijk, 1967).

Although plants normally obtain their sulphur as sulphate absorbed from the soil, the possibility that they may use atmospheric sources—particularly when soil supplies are insufficient—has attracted the attention of various investigators since the 1930s. The experiments that have been conducted to investigate the nutritive potential of atmospheric SO_2 and H_2S are summarized in *Table 17.5*.

In the midst of the controversy concerning plant response to prolonged exposure to SO_2 at concentrations not causing visible injury, Setterstrom, Zimmerman and Crocker (1938) observed that 'None of these (earlier) studies, however, attempted to determine the "invisible" effects of sulphur dioxide under conditions unfavourable to plant growth such as frequently encountered in actual practice'. Further, it was recognized that 'Investigators who found stimulation with SO_2 treatment may have grown their plants in sulphur-deficient soil . . . those who found no effect on yield may have used plants with an adequate sulphur supply . . . plants growing in soil containing an excess of sulphur may actually have been retarded on treatment with sulphur dioxide as would be the case with an over-supply of fertilizer'. They therefore investigated plant response to SO_2 under various regimes of water, sulphur, and other nutrient supply that simulated conditions likely to be encountered in the field. Their first experiment (summarized in *Table 17.5*) showed that the exposure of lucerne to about $270 \, \mu g \, SO_2 \, m^{-3}$ from 30 d after sowing increased the yield of plants given 70–$260 \, mg \, S \, \ell^{-1}$, but decreased yield in plants given 0 and 330–$790 \, mg \, S \, \ell^{-1}$. The experiment was repeated at a higher SO_2 concentration, studying in addition the effect of plant age on the response; the 'young' and 'old' plants were exposed at 10 and 59 d, respectively, after sowing. Except for an anomaly in plants given $130 \, mg \, S \, \ell^{-1}$, the yield of young plants exposed to SO_2 was greater than that of control plants. With old plants, yield was also greater in exposed than in control plants, except where sulphur was given at 0 and $790 \, mg \, \ell^{-1}$; furthermore, the increase in yield in exposed compared with control plants was greater than in young plants. It is curious, however, that without added sulphur the old plants as well as the plants in the first experiment showed *decreases* in yield when exposed to SO_2. In another study with lucerne, the plants were given 130, 330 and $830 \, mg \, S \, \ell^{-1}$ and a complete nutrient solution either at a sufficient or at a low rate. Exposure to SO_2 resulted in a slight increase in dry weight which was generally greater in plants grown with the low nutrient supply. It was thus shown that not only sulphur, but also the general nutrient supply, could affect the yield response to SO_2 exposure. In the last of these experiments, four cruciferous species (*Table 17.5*), each having a high requirement for

Table 17.5 INFLUENCE OF SULPHUR DIOXIDE AND HYDROGEN SULPHIDE ON THE SULPHUR NUTRITION OF PLANTS

Plant species	Exposure Concn ($\mu g\,m^{-3}$)	Time (h)	Nutritional regime and culture medium	Effect on plant weight	Source
a) Sulphur dioxide					
Lucerne	276	650	Complete; S: 0, 70, 130, 200, 260, 330, 560, 790 mg ℓ^{-1} (sand)	Nonsignificant increase with up to 260 mg S ℓ^{-1}	Setterstrom, Zimmerman and Crocker (1938)
Lucerne	409	402	As above	Increased: greater effect with older plants	
Lucerne	268	600	Complete, at low and high nutrient concentrations; S: 130, 330, 830 mg ℓ^{-1} (sand)	Nonsignificant increases at all S rates: greater effects shown with low nutrient supply	
Cabbage Rutabaga Mustard Turnip	539	117	Complete; S: 130, 330, 830 mg ℓ^{-1} (sand)	Nonsignificant increases at low and middle S rates, decrease at high S rate: increase in turnip at all S rates	Setterstrom, Zimmerman and Crocker (1938)
Wheat	373	300 (3–6 h daily to maturity)	Complete; S: 200 mg ℓ^{-1} (solution)	Increase in grain, slight decrease in whole plant	Swain and Johnson (1936)
Lucerne	286	6–7 (daily over 2 years)	Complete, at low and high nutrient concentrations; S: 0, 0.8, 1.5, 10 mg ℓ^{-1} (sand)	Increased at two low S rates, slightly decreased at two higher S rates	Thomas *et al.* (1943)

Tobacco	200, 500, 1000, 1500	12 (daily for 9 d)	Complete; S: 0.12 mg ℓ^{-1} (solution)	Increased: maximum effect at 1500 µg m^{-3}	Faller, Herwig and Kuhn (1970)
Sunflower	200, 500, 1000, 1500	12 (daily for 15 d)	Complete; S: 0.5 mg ℓ^{-1} (solution)	Increased: maximum effect at 1000 µg m^{-3}	
Maize	200, 500, 1000, 1500	12 (daily for 13 d)	Complete; S: 0.25 mg ℓ^{-1} (solution)	Increased: maximum effect at 500 µg m^{-3}	
Radish	200, 500, 1000, 1500	12 (daily for 20 d)	Complete; S: 2.98 mg ℓ^{-1} (solution)	Increased: maximum effect at 1500 µg m^{-3}	
Tobacco	1500	12 (daily for 14 d)	Complete; S: 0, 80 mg ℓ^{-1} (solution)	Increased in plants with deficient supply of S	Faller (1972a)
Ryegrass	131	408, 504, 504 (3 harvests)	Complete with added NPK; S: 0, 10 mg kg^{-1} (soil)	Increased in plants with deficient supply of S	Cowling, Jones and Lockyer (1973)
Ryegrass	55	1584, 696	Complete; S: 0,10 mg kg^{-1}; N: 10, 100 mg kg^{-1} (soil)	Increased in high N–low S treatment	Cowling and Lockyer (1978)
b) Hydrogen sulphide					
Sunflower	1518–303 600 (variable)	504	Complete (solution)	Increased	Faller (1972b)
Lucerne	42, 139, 417	672–840 (8 harvests)	Complete (soil)	Increased at 42 µg m^{-3}, no effect at 139, and decreased at 417 µg m^{-3}	Thompson and Kats (1978)
Lettuce	42, 139, 417	1416–2304	Complete (soil)	Increase at 42 and sometimes at 139 µg m^{-3}, decreased at 417 µg m^{-3}	
Sugar beet	42, 139, 417	2952	Complete (soil)	Increased in leaves and roots at 42 and 139 µg m^{-3}	
Grapes	42, 139, 417	3480	Complete (soil)	Increased in leaves at 42 and 139 µg m^{-3}, decreased in canes at all exposure concentrations	

sulphur, were given 130, 330 and 830 mg S ℓ^{-1}. Increases in yield with exposure to SO_2 were variable: cabbage showed an increase at the intermediate rate of sulphur, rutabaga at the intermediate and high rates, and turnip at all rates of sulphur. Once again, it is interesting that with low S the dry weights of rutabaga, cabbage and mustard were reduced through exposure to SO_2.

In five of their solution-culture experiments with wheat, Swain and Johnson (1936) found dry weight increased in plants that had been exposed to 480–1226 μg SO_2 m^{-3} for 4 or 6 h daily for 28 d, but increases were not statistically significant. In a further experiment in which plants were grown to maturity, it was found that exposure to 373 μg SO_2 m^{-3} for 3–6 h daily (total exposure time = 300 h) reduced the dry weight of the plant but increased that of grain (Swain and Johnson, 1936).

Interest in the possible role of SO_2 in plant nutrition continued, and detailed experiments on the utilization of sulphate and SO_2 by lucerne were conducted by Thomas *et al.* (1943) in sand culture. Included in the experimental design were refinements to assess over a 2-year period the interactions between exposure to SO_2 and low and high nutrient concentrations, each with different levels of sulphur: 0, 0.8, 1.5 and 10 mg S ℓ^{-1}. The data for crop yield showed that, with both low and high nutrient supply, exposure to SO_2 raised the yield of plants given the two lower rates of sulphur; exposure to SO_2 slightly reduced the yield of plants given the two higher rates of sulphur. These results support the hypothesis of Setterstrom, Zimmerman and Crocker (1938), that exposure to SO_2 of plants with a poor supply of sulphur to their roots compensates for the deficiency but has an adverse effect on yield when root supply is sufficient.

Faller, Herwig and Kühn (1970) exposed tobacco, sunflower (*Helianthus annuus* L.), maize (*Zea mays* L.) and radish growing in solution culture to 200, 500, 1000 and 1500 μg SO_2 m^{-3} for 12 h daily for 9–20 d; sulphate concentrations in the nutrient solution were adjusted for each species. Despite the availability of sulphur to the roots and the rather high exposure concentrations, each species showed increases in yield with SO_2, although the maximum increase was found at different SO_2 concentrations for each species. Thus, the maximum increase in yield was at 500 μg m^{-3} with maize, at 1000 μg m^{-3} with sunflower, and at 1500 μg m^{-3} with tobacco and radish. Faller (1972a) grew tobacco in solution culture using a complete nutrient solution minus sulphur, and demonstrated that atmospheric SO_2 could supply the whole of the plant's requirement; exposure to SO_2 doubled the yield after the supply of sulphur to the roots was stopped. Furthermore, no difference in dry weight was found between plants grown without sulphur in the nutrient solution and exposed to SO_2, and plants grown in a nutrient solution with sulphate added at a rate of 80 mg S ℓ^{-1}.

Jones, Cowling and Lockyer (1972) reported that ryegrass, grown in a chamber with the air filtered to remove SO_2 and including 16 soils from England and Wales, soon developed signs of sulphur deficiency, whereas deficiency was not observed in plants in the field. In a subsequent experiment, ryegrass was grown in soil either with or without sulphate added at a rate of 10 mg S kg^{-1} soil, and otherwise with adequate nutrients. Exposure to SO_2 at 131 μg m^{-3} had a beneficial effect on growth and increased the yield of plants grown without added sulphate; it had no effect on the yield of plants given sulphate (Cowling, Jones and Lockyer, 1973).

The interaction between sulphur and nitrogen nutrition and the effects of SO_2 on ryegrass were investigated by Cowling and Lockyer (1978). The plants were grown in soil with adequate nutrients, apart from sulphur and nitrogen which were given at 0 or 10, and 10 or 100 mg kg^{-1} respectively, in factorial combination. The high rate of added N increased yield fourfold where sulphate was also added, but resulted in symptoms of S deficiency in plants grown without added sulphate. Exposure of otherwise similarly treated plants to 55 µg SO_2 m^{-3} for 1584 and 696 h (two successive harvests) was found to alleviate the signs of sulphur deficiency in plants grown without added sulphate, and was without effect on the yield of plants given sulphate. These results are illustrated in *Figure 17.2*. Exposure to SO_2 and the addition of sulphate to the soil had no effect on the yield of plants grown with low N.

Figure 17.2   The effect of SO_2 on the growth of perennial ryegrass (*Lolium perenne* L.) cv. S23 grown on a soil low in available sulphur. N2O and N2S: plants grown without or with added sulphate, respectively; the two left-hand pots were housed in filtered air and the other two in air containing SO_2 at 55 µg m^{-3} during a 29 d growth period. (Photograph: Grassland Research Institute)

Comparatively little attention has been paid to the possible role of H_2S in plant nutrition. Faller (1972b) found that exposure to a variable high concentration of H_2S (1.5–303 mg m^{-3}) increased the yield of sunflower grown in solution culture low in sulphur; surprisingly, he reported no foliar necrosis but the experimental method was not rigorous. Thompson and Kats (1978) examined the effect of prolonged exposure to H_2S at 42, 139 and 417 µg m^{-3} on lucerne, lettuce, sugar beet (*Beta vulgaris* L.) and grape (*Vitis vinifera* L.) (*Table 17.5*). Except with grape, exposure at the lowest concentration resulted in an increase in yield; the yield of grape leaves increased, but that of vines decreased. The yield of sugar beet leaves and roots, and of grape leaves, increased with exposure to the intermediate concentration of H_2S, but the yield of all plants decreased at the highest exposure concentration. At low concentrations, H_2S was believed to act as a source of sulphur.

The data presented in *Table 17.5* indicate that in laboratory experiments SO_2 and H_2S can compensate for deficiencies in the supply of sulphur to plant roots. Furthermore, it has been shown that exposure to low concentration of these gases of plants with an adequate supply of sulphur does not always result in decreases in yield or in foliar damage. It is interesting to consider the effects of the relationship between the absorption of SO_2 through leaves and of sulphate through roots. It would seem reasonable to assume that if plants obtain some of the sulphur required for metabolism from the atmosphere less will be required from the root medium. Such a 'sparing effect' on root uptake was indeed observed in the data collected by Faller (1972a), that are presented in *Table 17.6*. Tobacco

Table 17.6 UPTAKE OF SULPHUR (μg S PER PLANT) BY TOBACCO EXPOSED TO SO_2 FOR 12 h DAILY OVER 14 DAYS. (a) FROM SULPHATE IN CULTURE SOLUTION AND (b) FROM LABELLED SULPHUR DIOXIDE. After Faller (1972a)

| Solution SO_4-S | SO_2 in air | | | |
| ($mg\ \ell^{-1}$) | Nil | | $1.5\,mg\,m^{-3}$ | |
	a	b	a	b
0	1892	–	1932	22738
80	19115	–	16301	22697

was grown with and without sulphate in the culture solution and exposed to air containing 0 and $1.5\,mg\,SO_2\,m^{-3}$ in a factorial experiment, the SO_2 being isotopically labelled to discriminate between the two sources of sulphur. Uptake of sulphur from solution containing adequate sulphate was found to be reduced, through exposure to SO_2, from 19 115 to 16 301 μg S per plant, representing a 'sparing effect' of SO_2 on root uptake of 15%. On the other hand, the supply of sulphur to roots had no effect on the absorption of SO_2 through the leaves (22 738 and 22 697 μg S per plant). In a similar study (Furrer, 1967) the 'sparing effect' of SO_2 was 29 and 36% with Egyptian clover (*Trifolium alexandrinum* L.) and rape, respectively, but with tomato the opposite effect was noted, i.e. root uptake of sulphur was enhanced through exposure to SO_2. It remains to be established to what extent a 'sparing effect' occurs among different crop species growing on soils in the field.

Assuming a mean value of $8\,mm\,s^{-1}$ for the velocity of deposition of SO_2 to wheat and grass crops on a ground area basis (Garland, 1977; Fowler, 1978), and SO_2 in ambient air at $30\,\mu g\,m^{-3}$, the annual input of sulphur to these crops would be around $38\,kg\,ha^{-1}$. Thus, leaving aside inputs of sulphur through rainfall and from the soil *per se*, the deposition of SO_2 may meet the sulphur requirements of these crops, namely 12–$23\,kg\,ha^{-1}$, and probably many other crops, apart from crucifers which have a higher requirement (Whitehead, 1964).

AMMONIA AND NITROGEN DIOXIDE IN NITROGEN NUTRITION

Perhaps because of the relatively low cost of nitrogenous fertilizers and their widespread use, there are few experiments that have given attention to the possible beneficial effect of air pollutants containing nitrogen on the nutrition of plants. There is however, recent unequivocal evidence that NH_3 and NO_2 may be taken up through leaves and metabolized by the plant (Porter, Viets and Hutchinson, 1972; Rogers, Campbell and Volk, 1979). A summary of other available data is presented in *Table 17.7*.

Faller (1972b) exposed sunflower, grown in solution culture with a full nutrient medium, to a range of concentrations of NH_3 continuously for 504 h. Compared with plants grown in NH_3-free air, total plant weight was increased by exposure to NH_3 at 400, 800 and 1200 $\mu g\,m^{-3}$, but not at 1600 $\mu g\,m^{-3}$ which resulted in plant injury. Total nitrogen content in the plants increased linearly with concentration of NH_3 up to 1200 $\mu g\,m^{-3}$. It was concluded that ammonia absorbed through the leaves provided a source of nitrogen for growth.

Sunflower, grown in a similar way, was also exposed to a range of concentrations of NO_2 (*Table 17.7*) continuously for 504 h (Faller, 1972b): total plant weight as well as nitrogen content increased with NO_2 concentration. The effect of exposure to NO_2 on tomato, grown at two rates of nitrogen in sand culture, was investigated by Troiano and Leone (1977). With 470 $\mu g\,NO_2\,m^{-3}$, the dry weight of leaves and stems increased in plants given the high but not the low rate of nitrogen: with 750 $\mu g\,NO_2\,m^{-3}$ there were no significant effects on plant weight. The results are interesting in that they represent the reverse of what might have been expected—that growth might have been enhanced through exposure to NO_2 in plants given *low* nitrogen. In contrast to these findings, Matsumaru *et al.* (1979) found that the dry weight of tomato plants of a similar age, and grown at three rates of nitrogen, decreased with exposure to NO_2 at 565 $\mu g\,m^{-3}$ (*Table 17.7*). Under similar conditions, the dry weight of sunflower and of maize plants increased slightly at the low and intermediate rates of nitrogen respectively, with exposure to NO_2. With all three species, a substantial proportion (13–46%) of the total nitrogen in the plant was derived from NO_2, but it was not possible to differentiate between direct absorption by leaves and uptake through roots after absorption in the culture solution. In subsequent studies, six species were exposed to NO_2 at 615 $\mu g\,m^{-3}$ for 10 h daily, either in the light or dark period, during 14 d (Yoneyama *et al.*, 1980). The effects of NO_2 varied between species and with timing of the exposure period. Thus, plant weight was increased by exposure in the light in cucumber (*Cucumis sativus* L.) and in the dark in cucumber, bean and sunflower. It was decreased by exposure during the dark period in maize, and unaffected by exposure in tomato and Swiss chard (*Beta vulgaris* L.). The effects on plant weight could not be attributed positively to benefits to nitrogen nutrition, although absorbed NO_2 represented from 2–12% of total plant nitrogen.

Table 17.7 INFLUENCE OF AMMONIA AND NITROGEN DIOXIDE ON THE NITROGEN NUTRITION OF PLANTS

Plant species	Exposure Concn (μg m^{-3})	Time (h)	Nutritional regime and culture medium	Effect on plant weight	Source
Ammonia					
Sunflower	400, 800, 1200, 1600	504	Complete (solution)	Increased: maximum effect at 800 μg m^{-3}	Faller (1972b)
Nitrogen dioxide					
Sunflower	1500, 3000, 4500, 6000	504	Complete (solution)	Increased: maximum effect at 6000 μg m^{-3}	Faller (1972b)
Tomato	470	80	Complete; N: 6.3, 31.6 mg ℓ^{-1} (sand)	Increased in stems and leaves at high N, no effect at low N	Troiano and Leone (1977)
Tomato	750	164	Complete; N: 6.3, 31.6 mg ℓ^{-1} (sand)	Nonsignificant increases in stems and leaves at high N, no effect at low N	
Tomato	565	336	Complete; N: 26, 105, 260 mg ℓ^{-1} (carbonized rice hull)	Decreased	Matsumaru *et al.* (1979)
Sunflower	565	336	Complete; N: 26, 105, 260 mg ℓ^{-1} (carbonized rice hull)	Slightly increased at low N	
Maize	565	336	Complete; N: 26, 105, 260 mg ℓ^{-1} (carbonized rice hull)	Slightly increased at middle N	

Uptake of pollutants and mineral nutrition

A number of estimates have been made of the rate of uptake of gaseous air pollutants by plants including:

1. Direct measurement of the concentration in the plant of the elements involved, sometimes using isotopic labelling (Faller, 1972a; Matsumaru *et al.*, 1979).
2. Measurements of changes in concentration of the gas across a well-stirred chamber (Rogers *et al.*, 1977; Rogers, Campbell and Volk, 1979).
3. Indirect measurements of the flux to plants in the field by the gradient method (Garland, 1977; Platt, 1978; Fowler and Unsworth, 1979).

The total resistance to uptake is made up of components that include aerodynamic resistance in the turbulent layer, bluff-body resistance in the viscous layer, and surface resistance. When the surface is represented by plants, the resistance may be further subdivided into stomatal and cuticular resistance; with growing crops, soil resistance may act in parallel with these resistances to influence canopy resistance. Plant nutrition may have an important role in total resistance, either by affecting stomatal resistance in individual plants, or by modifying both this resistance and the external resistances that apply to plants forming a canopy.

The influence of nutrition on stomatal resistance has been the subject of many investigations but it is not yet possible to make a broad generalization about the topic. Desai (1937) observed that whereas the size and frequency of stomata were unchanged in plants that were deficient in nitrogen, phosphorus or potassium, stomatal movements were slower and their degree of opening was less than in healthy plants. Some support for these findings comes from studies by Shimshi (1970) working with transpiration in bean. He found a lower degree of opening and closing of stomata in nitrogen-deficient plants than in healthy plants. Ryle and Hesketh (1969) noted increased resistance to CO_2 transfer with nitrogen deficiency in three crop species, but there was evidence, as in the previous study, that increased mesophyll resistance was the main factor involved. Farquhar (1979) found that stomatal conductance increased (i.e. resistance decreased) with the content of nitrogen in leaves, which reflects the nitrogen status of the plant. However, the supply of nitrogen, phosphorus or potassium had little effect on the stomatal resistance to the diffusion of water vapour in cotton (*Gossypium hirsutum* L.) (Longstreth and Nobel, 1980). An increased supply of potassium was found to reduce markedly the transpiration rate in wheat and bean (Brag, 1972) by reducing stomatal aperture, but the effect was specific for potassium deficiency and was not found for sodium. On the other hand, potassium deficiency was found to reduce photosynthetic CO_2 exchange, partly through increased stomatal resistance but mainly through increased metabolic resistance (Peaslee and Moss, 1968; Terry and Ulrich, 1973). In an examination of the dosage kinetics of SO_2, the lack of correlation between the effects of the gas on rates of photosynthesis and transpiration suggested that changes in stomatal resistance were not of major importance in determining sensitivity to this pollutant (McLaughlin *et al.*, 1979).

As well as having effects on stomatal resistance, mineral nutrition may affect the uptake of gaseous pollutants through the density of leaves in the canopy. Adequate nutrition invariably results in much denser canopies than may be found with an insufficient supply. This will increase the resistance of individual leaves to uptake, because of increased aerodynamic resistance within the canopy; uptake by the whole canopy on a ground-area basis does not increase linearly with leaf area (Milne, Roberts and Williams, 1979). Increased canopy resistance may certainly result in marked gradients in the concentration of gaseous pollutants found in canopies, a decrease being found with depth into the canopy (Bennett and Hill, 1973; Horsman *et al.*, 1979).

It is likely that the sensitivity to injury of isolated individual plants will be greater than that of plants within canopies. The greater potential for injury in isolated plants was referred to by Guderian (1977) and, leaving aside the question of differences in resistance to uptake of pollutants by isolated plants compared with plant communities, there are also almost certainly compensatory effects by the growth of non-sensitive individuals in communities of crop plants given genotypic variation within the species (Cowling and Koziol, 1978; Horsman, Roberts and Bradshaw, 1978). The success of these resistant genotypes in compensating for any reduction in growth sustained by the sensitive genotypes within a community is likely to be favoured by a high supply of nutrients and hindered by a deficiency of nutrients.

References

ADAMS, D.F. and SULZBACH, C.W. (1961). *Science,* **133**, 1425–1426

ADEDIPE, N.O., HOFSTRA, G. and ORMROD, D.P. (1972). *Canadian Journal of Botany,* **50**, 1789–1793

ALLCROFT, R., BURNS, K.N. and HERBERT, C.N. (1965). *Fluorosis in Cattle: Animal Disease Surveys Report 2, Part II.* HMSO, London

ALWAY, F.J. (1940). *Journal of American Society of Agronomy,* **32**, 913–921

ALWAY, F.J., MARSH, A.W. and METHLEY, W.J. (1937). *Proceedings of the Soil Science Society of America,* **2**, 229–238

APPLEGATE, H.G. and ADAMS, D.F. (1960). *Phyton,* **14**, 111–120

ASHENDEN, T.W. and MANSFIELD, T.A. (1977). *Journal of Experimental Botany,* **28**, 729–735

AYAZLOO, M., BELL, J.N.B. and GARSED, S.G. (1980). *Environmental Pollution,* **22**, 295–307

BELL, J.N.B. and CLOUGH, W.S. (1973). *Nature,* **241**, 47–49

BELL, J.N.B., RUTTER, A.J. and RELTON, J. (1979). *New Phytologist,* **83**, 627–643

BENNETT, J.H. and HILL, A.C. (1973). *Journal of the Air Pollution Control Association,* **23**, 203–206

BERTRAMSON, B.R., FRIED, M. and TISDALE, S.L. (1950). *Soil Science,* **70**, 27–41

BLEASDALE, J.K. (1952). *Atmospheric Pollution and Plant Growth,* PhD thesis, University of Manchester

BRAG, H. (1972). *Physiologia Plantarum,* **26**, 250–257

BRANDT, C.S. and HECK, W.W. (1968). In *Air Pollution,* Vol. 1, pp. 401–444. 1st Edn. (Stern, A.C., Ed.). Academic Press, New York

BRENNAN, E.G., LEONE, I.A. and DAINES, R.H. (1950). *Plant Physiology,* **25**, 736–747

BREWER, R.F., GUILLEMET, F.B. and CREVELING, R.K. (1961). *Soil Science,* **92**, 298–301

BROADBENT, W.V. and CLARK, F.E. (1965). In *Soil Nitrogen,* pp. 287–306 (Broadbent, W.V. and Clark, F.E., Eds), American Society of Agronomy, Madison, NJ

COLEMAN, R. (1966). *Soil Science,* **101**, 230–239

COWLING, D.W. and JONES, L.H.P. (1978). In *Sulphur in Forages: Sulphur Institute Symposium.* An Foras Taluntais, Dublin

COWLING, D.W. and KOZIOL, M.J. (1978). *Journal of Experimental Botany,* **29**, 1029–1036

COWLING, D.W. and LOCKYER, D.R. (1976). *Journal of Experimental Botany,* **27**, 411–417

COWLING, D.W. and LOCKYER, D.R. (1978). *Journal of Experimental Botany,* **29**, 257–265

COWLING, D.W., JONES, L.H.P. and LOCKYER, D.R. (1973). *Nature,* **243**, 479–480

COWLING, D.W., LOCKYER, D.R., CHAPMAN, P.F. and KOZIOL, M.J. (1981). *Environmental Pollution,* **26**, 1–13

CRITTENDEN, P.D. and READ, D.J. (1978). *New Phytologist,* **80**, 49–62

DAVIES, T. (1980). *Nature,* **284**, 483–485

DESAI, M.C. (1937). *Plant Physiology,* **12**, 253–283

DIJKSHOORN, W. and VAN WIJK, A.L. (1967). *Plant and Soil,* **26**, 129–157

DREW, D.H. (1967). *Plant and Soil,* **27**, 92–102

DUGGER, W.M., TAYLOR, O.C., CARDIFF, E.A. and THOMPSON, C.R. (1962). *Proceedings of the American Society of Horticultural Science,* **81**, 304–315

DUNNING, J.A., HECK, W.W. and TINGEY, D.T. (1974). *Water, Air and Soil Pollution,* **3**, 305–313

FALLER, N. (1972a). In *Isotopes and Radiation in Soil–Plant Relationships Including Forestry,* pp. 403–407. International Atomic Energy Agency, Vienna

FALLER, N. (1972b). *Zeitschrift für Pflanzenernahrung und Bodenkunde,* **131**, 120–130

FALLER, N., HERWIG, K. and KUHN, H. (1970). *Plant and Soil,* **33**, 177–191

FARQUHAR, G.D. (1979). In *Photosynthesis and Plant Development,* pp. 321–328 (Marcelle, R., Clijsters, H. and Van Poucke, M., Eds). W. Junk, The Hague

FOCHT, D.D. and VERSTRAETE, W. (1977). *Advances in Microbial Ecology,* **1**, 135–214

FOOD AND AGRICULTURE ORGANIZATION (1953). *Annual Fertilizer Review, 1952.* FAO, Rome

FOOD AND AGRICULTURE ORGANIZATION (1977). *Annual Fertilizer Review, 1976.* FAO, Rome

FOWLER, D. (1978). *Atmospheric Environment,* **12**, 369–373

FOWLER, D. and UNSWORTH, M.H. (1979). *Quarterly Journal of Royal Meteorological Society,* **105**, 767–783

FRENEY, J.R., BARROW, N.J. and SPENCER, K. (1962). *Plant and Soil,* **17**, 295–308

FURRER, O.J. (1967). In *Isotopes in Plant Nutrition and Plant Physiology*, pp. 51–57. International Atomic Energy Agency, Vienna

GARDNER, W.R. (1965). In *Soil Nitrogen*, pp. 555–572 (Broadbent, W.V. and Clark, F.E., Eds), American Society of Agronomy, Madison, NJ

GARLAND, J.A. (1977). *Proceedings of the Royal Society, A.,* **354**, 245–268

GUDERIAN, R. (1971). *Schriftenreihe der Landesanstalt für Immissions – und Bodennutzungsschutz des Landes Nordrhein – Westfalen,* **23**, 51–57

GUDERIAN, R. (1977). *Air Pollution.* Springer Verlag, Berlin

HARWARD, M.E. and REISENAUER, H.M. (1966). *Soil Science,* **101**, 326–335

HEAGLE, A.S. (1979). *Environmental Pollution,* **18**, 313–322

HECK, W.W., DUNNING, J.A. and HINDAWI, I.J. (1965). *Journal of the Air Pollution Control Association,* **15**, 511–515

HORSMAN, D.C., ROBERTS, T.M. and BRADSHAW, A.D. (1978). *Nature,* **276**, 493–494

HORSMAN, D.C., ROBERTS, T.M., LAMBERT, M. and BRADSHAW, A.D. (1979). *Journal of Experimental Botany,* **30**, 485–493

JONES, L.H.P., COWLING, D.W. and LOCKYER, D.R. (1972). *Soil Science,* **114**, 104–114

KORKMAN, J. (1973). *Journal of the Scientific Agricultural Society of Finland,* **45**, 121–215

LEE, T.T. (1966). *Canadian Journal of Botany,* **44**, 487–496

LEONE, I.A. and BRENNAN, E. (1970). *Phytopathology,* **60**, 1521–1524

LEONE, I.A. and BRENNAN, E. (1972a). *Journal of the Air Pollution Control Association,* **22**, 544–547

LEONE, I.A. and BRENNAN, E. (1972b). *Atmospheric Environment,* **6**, 259 –266

LEONE, I.A., BRENNAN, E. and DAINES, R.H. (1966). *Journal of the Air Pollution Control Association,* **16**, 191–196

LOCKYER, D.R., COWLING, D.W. and JONES, L.H.P. (1976). *Journal of Experimental Botany,* **27**, 397–409

LONGSTRETH, D.J. and NOBEL, P.S. (1980). *Plant Physiology,* **65**, 541–543

MacDOWALL, F.D.H. (1965). *Canadian Journal of Plant Science,* **45**, 1–12

MacLEAN, D.C., SCHNEIDER, R.E. and McCUNE, D.C. (1976). *Journal of the American Society for Horticultural Science,* **101**, 347–352

MacLEAN, D.C., ROARK, O.F., FOLKERTS, G. and SCHNEIDER, R.E. (1969). *Environmental Science and Technology,* **3**, 1201–1204

McCUNE, D.C., HITCHCOCK, A.E. and WEINSTEIN, L.H. (1966). *Contributions. Boyce Thompson Institute for Plant Research,* **23**, 295–299

McLAUGHLIN, S.B., SHRINER, D.S., McCONATHY, R.K. and MANN, L.K. (1979). *Environmental and Experimental Botany,* **19**, 179–191

MACY, P. (1936). *Plant Physiology,* **11**, 749–764

MATSUMARU, T., YONEYAMA, T., TOTSUKA, T. and SHIRATORI, K. (1979). *Soil Science and Plant Nutrition,* **25**, 255–265

MENSER, H.A. and HODGES, G.H. (1967). *Tobacco Science,* **11**, 151–154

MENSER, H.A. and STREET, O.E. (1962). *Tobacco Science,* **6**, 167–171

METSON, A.J. (1973). *Sulphur in Forage Crops*, Technical Bulletin 20. The Sulphur Institute, Washington

MIDDLETON, J.T. (1956). *Journal of the Air Pollution Control Association,* **6**, 7–9

MILNE, J.W., ROBERTS, D.B. and WILLIAMS, D.J. (1979). *Atmospheric Environment,* **13**, 373–379

ORMROD, D.P., ADEDIPE, N.O. and HOFSTRA, G. (1973). *Plant and Soil,* **39**, 437–439

PACK, M.R. (1966). *Journal of the Air Pollution Control Association,* **16**, 541–544

PEASLEE, D.E. and MOSS, D.N. (1968). *Crop Science,* **8**, 427–430

PLATT, U. (1978). *Atmospheric Environment,* **12**, 363–367

PORTER, L.K., VIETS, F.G. and HUTCHINSON, G.L. (1972). *Science,* **175**, 759–761

ROGERS, H.H., CAMPBELL, J.C. and VOLK, R.J. (1979). *Science,* **206**, 333–335

ROGERS, H.H., JEFFRIES, H.E., STAHEL, E.H., HECK, W.W., RIPPERTON, L.A. and WITHERSPOON, A.M. (1977). *Journal of the Air Pollution Control Association,* **27**, 1192–1197

RYLE, G.J.A. and HESKETH, J.D. (1969). *Crop Science,* **9**, 451–455

RUSSELL, E.W. (1973). *Soil Conditions and Plant Growth,* 10th Edn. Longmans, London

SAUCHELLI, V. (1950). *Soil Science,* **70**, 1–8

SETTERSTROM, C. and ZIMMERMAN, P.W. (1939). *Contributions. Boyce Thompson Institute for Plant Research,* **10**, 155–181

SETTERSTROM, C., ZIMMERMAN, P.W. and CROCKER, W. (1938). *Contributions. Boyce Thompson Institute for Plant Research,* **9**, 179–198

SHIMSHI, D. (1970). *New Phytologist,* **69**, 405–412

STEWART, B.A. (1967). In *Soil Chemistry and Fertility,* pp. 131–138 (Jacks, G.V., Ed.), Transactions of Meeting of International Society of Soil Science (1966), Aberdeen

SWAIN, R.E. and JOHNSON, A.B. (1936). *Industrial and Engineering Chemistry,* **28**, 42–47

TERMAN, G.L. (1978). *Atmospheric sulphur—the agronomic aspects, Technical Bulletin 23.* The Sulphur Institute, Washington

TERMAN, G.L. (1979). *Advances in Agronomy,* **31**, 189–223

TERRY, N. and ULRICH, A. (1973). *Plant Physiology,* **51**, 783–786

THOMAS, M.D. (1951). *Annual Review of Plant Physiology,* **2**, 293–322

THOMAS, M.D., HENDRICKS, R.H., COLLIER, T.P. and HILL, G.R. (1943). *Plant Physiology,* **18**, 345–371

THOMPSON, C.R. and KATS, G. (1978). *Environmental Science and Technology,* **12**, 550–553

TRESHOW, M.G. (1970). *Environment and Plant Response,* 1st Edn. McGraw-Hill, New York

TROIANO, J.J. and LEONE, I.A. (1977). *Phytopathology,* **67**, 1130–1133

ULRICH, A. (1952). *Annual Review of Plant Physiology,* **3**, 207–228

WHITEHEAD, D.C. (1964). *Soils and Fertilizers,* **27**, 1–8

YONEYAMA, T., TOTSUKA, T., HAYAKAWA, N. and YAZAKI, J. (1980). *Research Report National Institute of Environmental Studies (Japan),* **11**, 31–50

ZAHN, R. (1963). *Zeitschrift für Pflanzenkrankheiten, Pflanzenpathologie und Pflanzenschutz,* **70**, 81–95

V

Pollution in Perspective

18

EVOLUTION OF RESISTANCE TO GASEOUS AIR POLLUTANTS

M.L. ROOSE
A.D. BRADSHAW
Department of Botany, University of Liverpool

T.M. ROBERTS
Central Electricity Research Laboratories, Leatherhead

Introduction

Populations often have the capacity to evolve resistance to environmental stresses: indeed, this capacity forms the basis of much plant breeding. It should not, therefore, be surprising that evolution of resistance also occurs in natural populations, but the speed with which evolutionary responses can take place is often surprising.

There is now abundant evidence that evolution can occur within a few years or generations when plant populations are subject to new stress factors. This has been most apparent for resistance to high concentrations of heavy metals in soils (Antonovics, Bradshaw and Turner, 1971; Bradshaw, 1980) but also occurs in populations treated with herbicides (Ryan, 1970; Bandeen and McLaren, 1976) and in response to variation in soil nutrient status (Snaydon, 1970; Snaydon and Davies, 1976). In this chapter we try to assess the likelihood and importance of similar evolutionary responses to air pollutants.

Because the term has been used in a variety of ways, it is necessary to define resistance as used in this chapter. Resistance refers to the relative ability of a genotype to maintain normal growth and remain free from injury in a given polluted environment. It is quantitative, rather than qualitative, as resistance need not be complete. Our use of the term includes both stress avoidance (exclusion of pollutants) and stress tolerance (avoidance, repair or compensation of injury resulting from pollutant uptake) in the sense of Levitt (1972) (*cf* Taylor, 1978a).

Theory: should evolution of resistance occur?

Evolution (change in the genetic constitution of a population) will occur when a population contains individuals with heritable differences in characters which affect fitness, where fitness is defined as the net reproduction of offspring, including, for perennial plants, vegetative reproduction. Thus to assess the likelihood of evolution of a particular character, such as resistance to air pollutants, we must establish that genetic variation for the character is present and that this variation influences fitness.

GENETIC VARIATION IN RESISTANCE TO STRESS

Genetic variation in resistance to stresses occurs at three different levels of organization (Bradshaw, 1976). In evolution, the most significant of these is that which exists *between individuals* of a single population because it is this variation which, when subject to directional selection, results in better adaptation. Measurement of this variation is thus crucial in assessing the potential for evolution of resistance in a population. Differences *between populations* may be due to either adaptations to environmental differences between the populations, or genetic drift, or natural selection for resistance if the populations have been subject to different levels of the stress. *Interspecific differences* in resistance are an incidental byproduct of the many (genetically determined) morphological, physiological and biochemical differences between the species and are not a result of selection for resistance unless adaptation preceded speciation (most unlikely for new stresses such as air pollutants). The differences between species which cause differential resistance may or may not provide insights into mechanisms likely to arise within species, because they have not been subject to natural selection and thus may or may not add to fitness.

Why does genetic variation in resistance to stresses occur in populations? All variation originates via mutation, and is thereafter influenced by a number of factors (*Table 18.1*). Alleles causing differences in resistance

Table 18.1 EVOLUTIONARY FORCES AND THEIR EXPECTED ROLES IN EVOLUTION OF AIR-POLLUTION RESISTANCE

Force	*General effect*	*Probable role in air-pollution resistance*
Mutation	Origin of genetic variation	Direct role minimal because extensive variation in resistance already present in most populations
Gene flow	1. Decreases differentiation between populations	Minimal–pollution widespread in comparison with gene-flow distances.
	2. Allows spread of new genes and gene combinations	Could be important for spread of resistance within polluted area
Genetic drift	Random changes in gene frequency due to sampling effects	Unlikely to be important in large populations or over short times. Could account for variation in resistance in unselected populations
Mating system	Inbreeding decreases variability	Outbreeding species will be more likely to evolve resistance
Selection	Decreases frequency of genes which reduce fitness	Leads to increase in resistance in population

can be maintained in an unstressed population by mutation and random genetic drift, particularly if the genes have significant effects on morphological and physiological characters only when the stress is present (e.g. a hypothetical gene which closes stomata in the presence of air pollutants, but otherwise has no expression). Variation affecting resistance to one stress may be maintained as adaptations to other stresses. For example, Winner and Mooney (1980a,b) suggest that SO_2 resistance may be related, in part, to intrinsic photosynthetic rates and drought tolerance. Intrapopulation variation in these characters might be maintained by selection if their adaptive value differed on north-facing and south-facing slopes. Where selection is important in maintaining variation, evolution of resistance may be limited by conflicting selection pressures, but resistance will increase in frequency if it increases fitness. In such cases, population adaptation to some other environmental factor would decrease as resistance increases. Understanding the reasons for the occurrence of variation in resistance in previously unstressed populations is therefore important in evaluating net fitness of various genotypes and in predicting the consequences of evolution for other ecological characteristics of the population (not an easy task).

EVOLUTIONARY RESPONSES TO HEAVY METALS AND HERBICIDES

We know little about evolution of resistance to air pollutants, but much more about evolutionary responses to other new stresses imposed by human activity. Therefore, before discussing evolution of resistance to air pollutants, we will consider some of the factors influencing evolution of tolerance to heavy metals and herbicides.

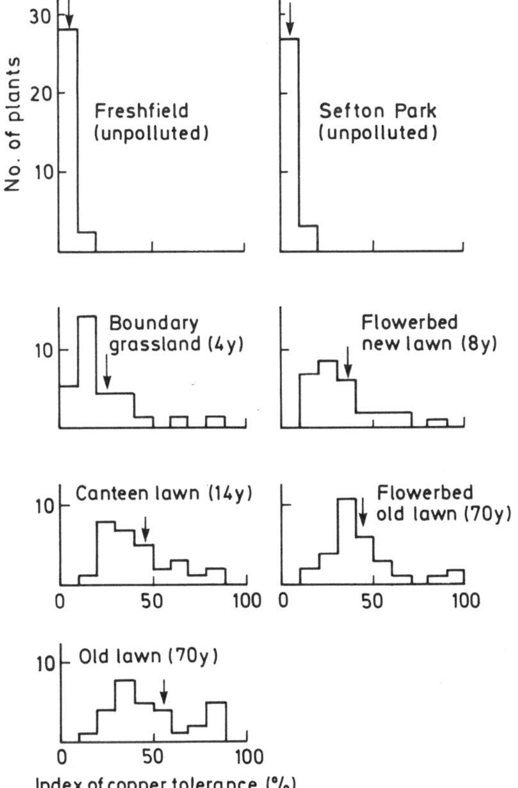

Figure 18.1   The distribution of copper tolerance in *Agrostis stolonifera* populations of different ages around a copper refinery near Liverpool. Arrows indicate mean value. After Wu, Bradshaw and Thurman (1975)

At least 70 cases of plants evolving resistance to various soils contaminated by heavy metals are now known (Antonovics, Bradshaw and Turner, 1971; Bradshaw, 1980). Identification of resistant individuals was greatly aided by the development of a simple rooting test for tolerance: root growth in solutions containing the metal ion is compared with that in control solution (Wilkins, 1957; Jowett, 1964).

Studies of the evolution of heavy-metal tolerance have been critical in changing views on rates of evolution in natural populations because it has been possible to estimate the time within which evolution has occurred. This is clearly demonstrated in lawns near a copper refinery in Liverpool (Wu, Bradshaw and Thurman, 1975). The lawns were sown at different times and now consist of *Agrostis stolonifera* and *A. tenuis*. The distribution of copper tolerance was studied in plants sampled from these lawns and nearby unpolluted areas (*Figure 18.1*). The mean copper tolerance was greater in the older lawns, which generally had better cover, but even the 4-year-old boundary grassland had significantly greater copper tolerance than the samples from unpolluted populations.

On contaminated soils, selection for metal tolerance is often intense, with fitness of nontolerant plants being 0.05 or less relative to tolerant plants (Jain and Bradshaw, 1966). When seeds from normal populations are sown on mine waste mixed with soil, as the proportion of mine waste increases, the percentage of plants surviving after 4 months declines from about 28% for a 3:1 waste:soil mix to <1% for a 48:1 mix (Walley, Khan and Bradshaw, 1974). Selection on metal-contaminated soils is continuous: plants experience the selective agent throughout growth and development, although effective metal concentrations do vary in time and space.

Interspecific competition is probably relaxed on contaminated soils because few species are present. This is possibly important in allowing plants with high-cost (or low-benefit) resistance mechanisms to colonize the soils, thus increasing the probability and rate of evolution. The importance of competition in determining fitness is clearly shown by the experiment of Hickey and McNeilly (1975) on competitive abilities of metal-tolerant and nontolerant populations on normal soil. Clones of tolerant and nontolerant populations of four species were planted into swards of S23 or S24 ryegrass. At subsequent periods, over the course of a year, the number of plants surviving was counted and dry weight of green matter estimated by clipping at 4 cm. Relative fitness was estimated as the ratio (dry weight tolerant population/dry weight normal population). In every case (*Figure 18.2*) relative fitness of the tolerant plants declined over time and the number of surviving tolerant plants, relative to the normal plants, also declined. There was clearly strong selection against tolerant plants. On a logarithmic scale, relative fitness declined linearly with time, showing that the effects of selection are cumulative (*Figure 18.3*). Precisely what increased costs or reduced benefits are responsible for the inability of tolerant plants to compete effectively on normal soils awaits elucidation of the mechanism of tolerance.

Evolution of heavy-metal tolerance appears to be limited by the amount of genetic variation for tolerance available in a population. Large samples of a number of species from uncontaminated soils have been tested for heavy-metal tolerance (Gartside and McNeilly, 1974a; Ingram, personal

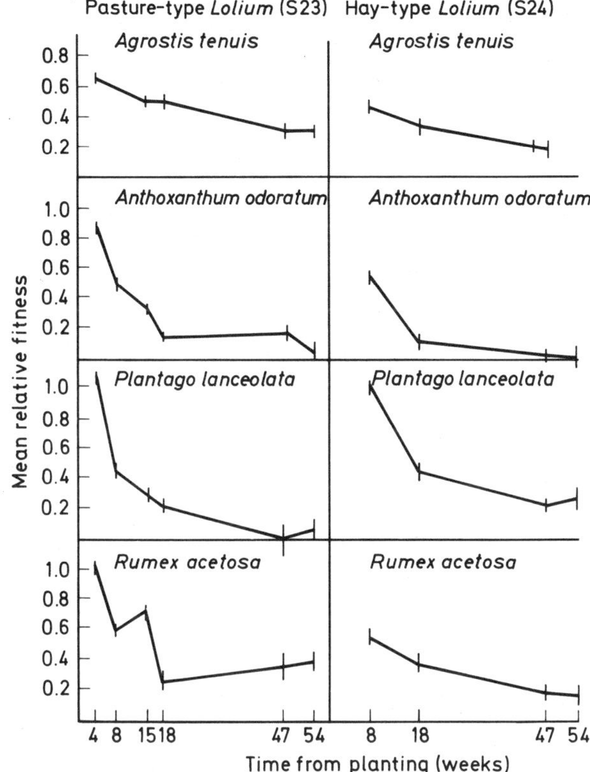

Figure 18.2   Plots of mean relative fitness (dry weight tolerant population/dry weight normal population) against time for various species grown in competition with two different varieties of ryegrass. Standard deviations are also shown. After Hickey and McNeilly (1975)

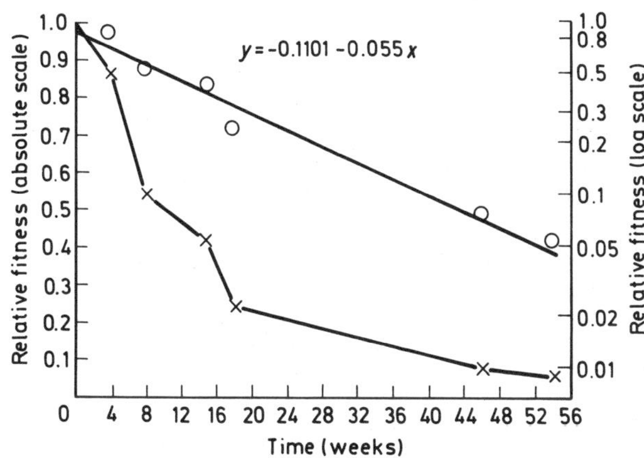

Figure 18.3   Change in relative fitness of metal-tolerant populations with time. × = values plotted on absolute scale. o = values plotted on log scale. Inset is the equation of the regression line of $\log_e$ (relative fitness) on time. Courtesy of Hickey and McNeilly (1975)

communication). In general, tolerant plants are found only in those species which have been observed to evolve tolerant populations. Why some species apparently do not possess genetic variation for heavy-metal tolerance is not known. Tolerance is usually inherited as a quantitative character (Urquhart, 1971; Gartside and McNeilly, 1974b,c) but in some cases only a few genes seem to be involved (Macnair, 1979).

Evolution of resistance to herbicides provides additional examples of population changes in response to Man's activities and in some ways is more comparable to air pollution as a selective agent. At present, herbicide resistance appears to be less common than heavy-metal tolerance, but herbicides have been in common use for only about 30 years. Mortality from herbicides can range from rapid and episodic for a nonpersistent herbicide applied at levels sufficient to kill most of the weed population, to slow and continuous for a persistent herbicide applied at levels which only reduce weed growth. Such episodic selection will tend to reduce the rate of evolution because resistance may be selected against, when the herbicide is not present (Conard and Radosevich, 1979). Most reported cases of evolution of resistance to herbicides (e.g. Ryan, 1970; Bandeen and McLaren, 1976) concern herbicides which persist in the soil and hence exert selection pressures over long periods. The model, by Gressel and Segel (1978), of the evolution of herbicide resistance, indicates that critical factors controlling the rate of evolution are selection against resistant genotypes when the herbicide has decayed, escape of susceptible genotypes as buried seed, and the strength of selection by the herbicide. In addition, fields are often treated with more than one herbicide; thus survivors of one treatment may be eliminated by a second (Holliday and Putwain, 1980). Herbicides are commonly used at concentrations giving 90–95% mortality in the weed species: potential for selection is therefore strong, but less intense than on metal wastes. The intensity of competition varies depending on the crop–weed combination and the effectiveness of the herbicide but, in general, competition is more intense than on metal-contaminated soils.

FACTORS AFFECTING EVOLUTIONARY RESPONSES

What do these cases suggest about evolution of resistance to air pollutants? Response of a population to each of the three situations is influenced by a unique combination of characteristics (*Table 18.2*).

Genetic variability may limit evolution, and this seems to be the case for resistance to heavy metals and herbicides, because resistance is absent in most species. In contrast, variation in resistance to air pollutants appears to be common in most species (*see below*). In part, this difference probably reflects the severity of stress imposed: resistance to heavy metals and herbicides is identified by survival at concentrations which kill a large proportion of the population, whereas resistance to air pollutants is studied at dosages causing differential injury or growth reduction. The mode of inheritance of variation influences the rate of evolution (*discussed below*) but the effect is small compared with the effects of selection intensity if resistance occurs at frequencies greater than about 0.10. The mode of

Table 18.2 FACTORS AFFECTING EVOLUTION OF RESISTANCE TO METAL-CONTAMINATED SOILS, HERBICIDES AND GASEOUS AIR POLLUTANTS

Factor		Stress	
	Metal wastes	*Herbicides*	*Air pollutants*
Variability	Rare in most species	Rare	Common
Inheritance	Usually polygenic	Often monogenic	Usually polygenic
Life history	Usually perennial	Annual	Annual or perennial
Gene flow	Small–medium	Small–medium[1]	Small
Selection			
In 'polluted' environment			
Fitness[2]	0.01–0.10	0.01–0.10	?
Mortality	Rapid	Rapid	Usually slow
Competition	Low	Low	High
Duration	Continuous	Episodic	Episodic
In normal environment			
Fitness[3]	<0.5	<0.5?	?
Mortality	Slow	Slow	Slow
Competition	High	Low–high	High
Duration	Continuous	Episodic	Episodic

[1]Includes plants which 'escape' treatment
[2]Fitness of sensitive genotype relative to resistant genotype in polluted environment
[3]Fitness of resistant genotype relative to sensitive genotype in normal environment

inheritance is somewhat more important in an annual plant, where recombination alters the genetic composition of the entire population every year, than in a perennial plant, where considerable evolution may occur through differential growth and survival of genotypes without any sexual reproduction.

Gene flow from unselected 'populations' can be a significant factor in retarding evolution. Gene flow, in the normal sense (movement of pollen or seeds), is confined to short distances in most species (Levin and Kerster, 1974); thus, it does not prevent differentiation of populations over even short distances. However, plants which escape herbicides, by late germination, substantially retard evolution of resistance (Gressel and Segal, 1978), and this process is analogous to gene flow. In comparison with herbicides and heavy metals, where affected areas may be small and boundaries between affected and unaffected populations are sharp, air pollutants affect large areas with diffuse boundaries. In addition, pollution episodes are generally more frequent than herbicide treatments. Thus it seems unlikely that gene flow from unselected populations will greatly retard evolution of resistance to air pollutants.

Selection is of critical importance in determining the rate of evolution, and it is perhaps here that comparison of air pollution with heavy metals and herbicides is most valuable. Because plants are exposed continuously to heavy metals, fitness differences accumulate over time and eventually may become quite large. In contrast, selection by herbicides and air pollutants is episodic. Consequently, growth and survival of genotypes depends upon their responses to both polluted and normal environments, thus slowing evolution of resistance. Acute injury from air pollution resembles that from herbicides, in that selection for resistance occurs only during a short period, and therefore net fitness will depend, to a great extent, on the fitness of resistant genotypes in normal air. Chronic air pollution is more similar to soil contaminated with heavy metals in that genotypes experience the polluted environment for a considerable portion of their lives and hence fitness in unpolluted air is expected to be less important than in evolution of resistance to acute injury. As judged from obvious mortality, air pollution seems unlikely to lead to fitness differences as large as those observed on metal wastes or in fields treated with herbicides. It should be noted, however, that it makes little difference to the rate of evolution if selection occurs as mortality at germination, as mortality on a single day or over several months or years prior to reproduction, or as reduced fecundity.

Interspecific and intraspecific competition are critical components of natural selection. In the absence of competition, many species can be grown far outside their normal range, but adaptations which are adequate for survival in the physical environment do not necessarily allow successful acquisition of resources when other species are present. In only a few areas is air pollution so severe that only a few species persist; competitive ability must, therefore, generally be maintained during evolution of resistance. On the other hand, growth reductions which are small at low density may rapidly lead to large differences in fitness in the presence of other species. Thus competition increases selection *for* resistance in polluted conditions and selection *against* resistance in less polluted conditions.

GENETIC VARIATION IN RESISTANCE TO GASEOUS AIR POLLUTANTS

Cultivar differences

The most abundant evidence for variation in resistance is from comparisons of resistance among cultivars of crop plants. Nearly all of these studies have been concerned with estimating foliar injury resulting from brief exposures to high concentrations of pollutants (greater than $400\,\mu g\,O_3\,m^{-3}$ or $1000\,\mu g\,SO_2\,m^{-3}$) with the aim of providing guidelines for choosing cultivars suited to polluted areas. When large numbers of cultivars are compared, foliar injury often ranges from slight to severe (e.g. O_3 injury in *Nicotiana*, *Petunia*, *Phaseolus*, *Poa* and *Spinacia* (*Table 18.3*) and SO_2 injury in *Poa* and *Glycine* (*Table 18.4*)). While there are fewer studies of long-term cultivar responses to low or ambient pollution levels, significant differences again appear common (e.g. injury or yield reductions from O_3 in *Allium*, *Lycopersicon*, *Nicotiana*, *Phaseolus* and *Zea* (*Table 18.3*), from NO in *Lycopersicon* (Anderson and Mansfield, 1979), and from SO_2/O_3 mixtures in *Phaseolus* (*see Table 18.6*)).

Cultivar differences in resistance to air pollutants may arise in several different ways: conscious selection of resistant types by the breeder (Howell, Devine and Hanson, 1971; Aycock *et al.*, 1977); correlations (due to genetic linkage or pleiotropy) between resistance and some other character selected by the breeder; new mutations (Wilton *et al.*, 1972); differences between the base populations from which the cultivars were selected; or, as Heggestad and Heck (1971) suggested, unconscious selection for resistance as a component of yield when breeding occurred in polluted areas. In general, the origin of resistance differences between cultivars reflects past events which are difficult or impossible to reconstruct.

Clonal and family differences

Differences between clones provide further evidence for heritable variation in pollution resistance, particularly in forest trees. Differences in percentage leaf injury of more than 30% have been found between clones exposed to O_3 in *Pinus sylvestris*, *Pinus strobus* and *Populus tremuloides* (*Table 18.3*); SO_2 in *Larix decidua*, *Pinus strobus*, *Pinus sylvestris* and *Populus tremuloides* (*Table 18.4*); and HF or mixtures of SO_2 and O_3 in *Larix decidua* (*Tables 18.5* and *18.6*). Families collected from different areas show significant differences in response to O_3 in *Acer*, *Fraxinus* and *Pinus sylvestris* (*Table 18.3*) and in response to SO_2 in *Pinus contorta* (*Table 18.4*).

Variation within and between natural populations

Variation in pollution resistance within and between natural populations grown in a common environment is known in *Lolium perenne* (Bell and Clough, 1973; Bell and Mudd, 1976; Horsman, Roberts and Bradshaw, 1978, 1979; Ayazloo, 1979), other British grasses (Ayazloo, 1979), *Geranium carolinianum* (Taylor and Murdy, 1975), *Lepidium virginianum* (Murdy, 1979), *Populus tremuloides* (Karnosky, 1977), *Pinus sylvestris* and *Picea abies* (Huttunen, 1978), and *Pinus contorta* (Lang, Neumann and Schutt, 1971). Variation within populations is critical in the evolution of resistance, but most studies have emphasized comparisons between populations. Histograms of the frequency in populations of genotypes with a given degree of injury or yield loss suggest that considerable variation exists (*Figure 18.4*; Taylor and Murdy, 1975; Horsman, Roberts and Bradshaw, 1979; Horsman *et al.*, 1979; Ayazloo, 1979) but how much of this variation has a genetical basis is not clear.

Inheritance of resistance

The simplest, but least rigorous method of demonstrating that variation in resistance is genetically determined is clonal repeatability analysis which partitions the total phenotypic variation into within-clone (environmental) and between-clone (primarily genetic) components (Falconer, 1960). Repeatability (R) is the ratio of between-clone variance to total variance and is usually similar to broad-sense heritability (H_B = genetic variance/total variance). Repeatability and broad-sense heritability of response to air pollutants ranges from 0.46 to 0.83 (*Table 18.7*), implying that a large proportion of the observed variation is genetically determined.

More detailed studies in a number of species include estimates of narrow-sense heritability (h^2)—the proportion of phenotypic variation

Table 18.3 CULTIVAR, CLONAL AND FAMILY VARIATION IN RESISTANCE TO OZONE

Species	Concentration[1] ($\mu g\,m^{-3}$)	Time[2]	Comparison[3]	Results[4]	Reference[5]
Acer rubrum L. (red maple)	1470	3 d (7 h d^{-1})	4 F	24–48% F*	Townsend and Dochinger (1974)
	490	6 w (56 h w^{-1})	3 P	7–28% F*	Dochinger and Townsend (1979)
Agrostis palustris Huds. (creeping bentgrass)	450	6 h	3 CV	5–9 r F	Brennan and Halisky (1970)
Agrostis tenuis Sibth. (colonial bentgrass)	450	6 h	3 CV	1–7 r F	Brennan and Halisky (1970)
Allium cepa L. (onion)	A	s	4 CV	0–8 r F	Engle and Gabelman (1966)
	1960	4 h	4 CV	6–16% F, ns; 16–47% Y*	Ormrod, Adedipe and Hofstra (1971)
Avena sativa L. (oat)	608	4 h	9 CV	20–90% F	Brennan, Leone and Daines (1964)
Carthamus tinctorus L. (safflower)	490	2 h	12 CV	3–7 r F*	Howell and Thomas (1972)
Coronilla varia L. (crown vetch)	588	4 h	2 CV	0 r F	Brennan, Leone and Halisky (1969)
Cucumis sativus L. (cucumber)	1960	4 h	4 CV	11–30% F*; 11–37% Y*	Ormrod, Adedipe and Hofstra (1971)
Festuca rubra L. (red fescue)	588	6 h	2 CV	2–6 r F	Brennan and Halisky (1970)
Fraxinus americana L. (white ash)	490	6 h	2 F	25–38% F	Steiner and Davis (1979)
Fraxinus pennsylvanicus Marsh. (green ash)	490	6 h	10 F	2–33% F*	Steiner and Davis (1979)
Glycine max (L.) Merr. (soybean)	1372	1.5 h	14 CV	17–57% F*	Tingey, Reinert and Carter (1972)
	980	2 h	21 CV	40–74% F*	Miller, Howell and Caldwell (1974)
	1176	1.5 h	4 CV	47–56% F*	Heagle (1979)
Lolium perenne L. (perennial ryegrass)	588	6 h	2 CV	6 r F	Brennan and Halisky (1970)
Lycopersicon esculentum Mill. (tomato)	98	14 d	22 CV	Sensitive to tolerant }	Gentile *et al.* (1971)
	490	3 h			
	176	45 d (8 h d^{-1})	2 CV	20–50% Y	Manning and Feder (1976)
	784	1.5 h	12 CV	9–62% F*	Reinert, Tingey and Carter (1972)
Medicago sativa L. (alfalfa)	392	4 h	14 CV	1–5 r F*	Howell, Devine and Hanson (1971)
	294	4 h	2 CV	1–2 r F	Brennan, Leone and Halisky (1969)
Nicotiana tabacum L. (tobacco)	196	6 h	3 CV	58–80% F*	Menser and Hodges (1968)
	588	2 h	9 CV	47–72% F*	Menser and Hodges (1970)
	196–490	2 h	14 CV	11–45% F*	Menser and Hodges (1972)
	A	s	4 CV	10–71% F***	Turner, Rich and Tomlinson (1972)
Petunia hybrida Vilm. (petunia)	294–1176	3 h	65 CV	1–6 r F	Cathey and Heggestad (1972)
	1960	1 h	14 CV	1–38% F*	Feder *et al.* (1969)
	2940	4–7 h	7 CV	2–7 r F	Hanson, Addis and Thorne (1976)

Species	[1]	[2]	[3]	[4]	Reference
Phaseolus vulgaris L. (bean)	98	3 w (40 h w⁻¹)	2 CV	4–13% F*	Tingey *et al.* (1973)
	A	s	387 CV	0–9 r F	Meiners and Heggestad (1979)
	687	3 h	20 CV	4–8 r F*	
	A	s	30 CV	Ranked only	Beckerson, Hofstra and Wukasch (1979)
	294	5 d (6 h d⁻¹)	32 CV	0–68% F	
	490	4 h	10 CV	1–33% F	Davis and Kress (1974)
	2744	1 h	18 CV	29–82% F*	Butler and Tibbits (1979a)
Pinus strobus L. (e.white pine)	194	8 w (22 h w⁻¹)	2 G	0–3% F	Dochinger *et al.* (1970)
	882	6 h	8 CV + G	0–7 r F	Genys and Heggestad (1978)
Pinus sylvestris L. (Scots pine)	1764	6 h	5 G	7–9 r F	Genys and Heggestad (1978)
	1960	66 h	17 G	1–38% F	Bialobok, Karolewski and Oleksyn (1977)
	1960	66 h	17 F	8–45% F*	Karolewski and Bialobok (1978)
Pinus taeda L. (loblolly pine)	490	8 h	18 F	5–60% F*	Kress and Skelly (1977)
Poa pratensis L. (Kentucky bluegrass)	588	2,4 h	3 CV	F**	Wilton *et al.* (1972)
	588	6 h	2 CV	4–6 r F	Brennan and Halisky (1970)
	784	2 h	17 CV	1–9 r F*	Murray, Howell and Wilton (1975)
Populus tremuloides Michx. (trembling aspen)	392	3 h	5 G	6–37% F*	Karnosky (1976)
	392	3 h	10 G	7–56% F*	Karnosky (1977)
Solanum tuberosum L. (potato)	1960	3 h	4 CV	F*, 0–40% Y	Ormrod, Adedipe and Hofstra (1971)
Spinacia oleracea L. (spinach)	294	4 h	6 CV	0–9 r F*	Manning, Feder and Perkins (1972)
Trifolium pratense L. (red clover)	294	4 h	3 CV	2–4 r F	Brennan, Leone and Halisky (1969)
Triticum aestivum L. (bread wheat)	118	27 d	11 CV	8–42% Y*	Heagle, Spencer and Letchworth (1979)
Zea mays L. (sweet corn)	A	s	9 CV	1–8 r F*	Cameron (1975)
Zoysia japonica Steud. (japanese lawngrass)	588	6 h	2 CV	0 r F	Brennan and Halisky (1970;

[1]Gas concentration in μg m⁻³. Where several concentrations were used in a study, that giving the largest differences is reported. Conversion factors used:
ppm SO₂ × 2610 = μg m⁻³ (25 °C); ppm O₃ × 1960 = μg m⁻³; ppm HF × 817 = μg m⁻³.
Abbreviations: A = ambient exposure—mean listed if reported
[2]Duration of exposure; abbreviations: h = hours, d = days, w = weeks, s = season; unless otherwise noted, days means 24 h d⁻¹ exposure
[3]Number of classes and types compared. Abbreviations: CV = cultivars, G = genotypes or clones, F = families, P = populations
[4]Range of percentage foliar injury (% F), foliar injury ratings on scale from 0–9 (r F), or percentage decrease in dry weight yield (% Y) where given. Significance shown where given: * $P <0.05$, ** $P <0.01$, *** $P <0.001$
[5]These tables include most, but not all comparisons of intraspecific variation in response to air pollutants.

Table 18.4 CULTIVAR, CLONAL AND FAMILY VARIATION IN RESISTANCE TO SULPHUR DIOXIDE

Species	Exposure Concentration[1]	Time[2]	Comparison[3]	Results[4]	Reference[5]
Cucumis sativas L. (cucumber)	7830	16.7 h	2 CV	5–40% F*	Bressan, Wilson and Filner (1978)
Cucurbita pepo L. (squash)	20880	16.7 h	2 CV	35–45% F*	Bressan, Wilson and Filner (1978)
Festuca rubra L. (red fescue)	1960	6 h	2 CV	2.5–4 r F	Brennan and Halisky (1970)
Glycine max (L.) Merr. (soybean)	5220	1.75 h	19 CV	10–50% F**	Miller, Howell and Caldwell (1974)
Larix decidua Miller (European larch)	5220	4 d (5 h d⁻¹)	21 G	3–65% F*	Bialobok, Karolewski and Oleksyn (1977)
	10440	5 h	21 F	3–67% F*	Karolewski and Bialobok (1978)
Lolium perenne L. (perennial ryegrass)	2220	6 h	2 CV	0–2 r F	Brennan and Halisky (1970)
	600	84 d	3 CV	0–36% Y*	Bell (1980)
	600	35 d	2 CV	0% Y	Bell (1980)
Pelargonium × *hederaefolium*	26100	1 h	3 CV	27–29% F	Bonte *et al.* (1977)
Pelargonium × *hortorium*	26100	1 h	8 CV	44–60% F	Bonte *et al.* (1977)
Petunia hybrida Vilm.	6525	1 h	14 CV	1–19% F*	Feder *et al.* (1969)
Phaseolus vulgaris L. (bean)	2610	5 d (6 h d⁻¹)	32 CV	0–18% F	Beckerson, Hofstra and Wukasch (1979)
Pinus contorta Dougl. (lodgepole pine)	3900	109 h	4 P	r F**	Lang, Neumann and Schutt (1971)
Pinus strobus L. (e.white pine)	A-86	150 d	2 G	0–63% Y	Roberts (1976)
	2358	6 h	8 G	1–9 r F	Genys and Heggestad (1978)
Pinus sylvestris L. (Scots pine)	5240	6 h	5 G	7–9 r F	Genys and Heggestad (1978)
	5240	18 h	17 G	1–53% F*	Bialobok, Karolewski and Oleksyn (1977)
	2610	24 h	17 F	0–40% F*	Karolewski and Bialobok (1978)
Poa pratensis L. (Kentucky bluegrass)	520	2 h	17 CV	3–8 r F*	Murray, Howell and Wilton (1975)
	2220	6 h	2 CV	3–4 r F	Brennan and Halisky (1970)
Populus deltoides Bartr. × *P. trichocarpa* Torr. and Gray (poplar hybrid)	650	6 w (40 h w⁻¹)	2 G	F*	Dochinger and Jensen (1975)
Populus tremuloides Michx. (trembling aspen)	1305	3 h	5 G	0–46% F	Karnosky (1976)
	1305	3 h	10 G	1–34% F*	Karnosky (1977)
Triticum aestivum L. (bread wheat)	1566	100 h	7 CV	F*; 20–40% Y*	Laurence (1979)
Zea mays L. (sweet corn)	1300	100 h	4 CV	F, NS; Y, NS	Laurence (1979)
Zoysia japonica Steud. (japanese lawngrass)	4700	6 h	2 CV	0 r F	Brennan and Halisky (1970)

[1–5]See notes to *Table 18.3*

Table 18.5 CULTIVAR AND CLONAL VARIATION IN RESISTANCE TO HYDROGEN FLUORIDE

Species	Exposure Concentration[1]	Time[2]	Comparison[3]	Results[4]	Reference[5]
Citrus limon Burm. (lemon)	25–98	78 w	2 CV	13–15% Y; 16% F	Brewer *et al.* (1960)
Citrus paradisi Macf. (grapefruit)	25–98	78 w	2 CV	16–18% Y; 25–33% F	Brewer *et al.* (1960)
Citrus sinensis Osbeck (orange)	25–98	78 w	3 CV	+15–33% Y; 13–20% F	Brewer *et al.* (1960)
Gladiolus hortulanus Bailey (gladiolus)	A	s	110 CV	4–83% F	Hendrix and Hall (1958)
Larix decidua Miller (European larch)	817	5 h	20 G	5–70% F*	Karolewski and Bialobok (1978)
Phaseolus vulgaris L. (bean)	14	s	2 CV	11–52% Y**	Pack (1971)
Pinus sylvestris L. (Scots pine)	408	12 h	17 G	30–50% F*	Karolewski and Bialobok (1978)
Zea mays L. (sweet corn)	8	7 d	6 CV	0–8 r F	Hitchcock *et al.* (1964)
	3	7 d	8 CV	0–40% F	

[1-5]See notes to *Table 18.3*

Table 18.6 CULTIVAR AND CLONAL VARIATION IN RESISTANCE TO GAS MIXTURES

Species	Gas and conc.[1] (μg m^{-3})	Time[2]	Comparison[3]	Results[4]	Reference[5]
Larix decidua Miller (European larch)	O_3–980 + SO_2–5220	15 h	20 G	7–65% F*	Karolewski and Bialobok (1978)
Nicotiana tabacum L. (tobacco)	O_3–59 + SO_2–1044	4 h	9 CV	9–27% F*	Menser and Hodges (1970)
Phaseolus vulgaris L. (bean)	O_3–294 + SO_2–392	5 d (6 h d^{-1})	32 CV	0–30% F	Beckerson, Hofstra and Wukasch (1979)
	O_3–98 + SO_2–131	3 w (40 h w^{-1})	2 CV	3–16% F*	Tingey *et al.* (1973)
Pinus strobus L. (e.white pine)	O_3–194 + SO_2–256	4–8 w (20–40 h w^{-1})		0–16% F	Dochinger *et al.* (1970)
Pinus sylvestris L. (Scots pine)	O_3–980 + SO_2–1300	36 h	17 G	5–25% F*	Karolewski and Bialobok (1978)
Populus tremuloides Michx. (trembling aspen)	O_3–98 + SO_2–914	3 h	10 G	0–20% F	Karnosky (1977)

[1-5]See notes to *Table 18.3*

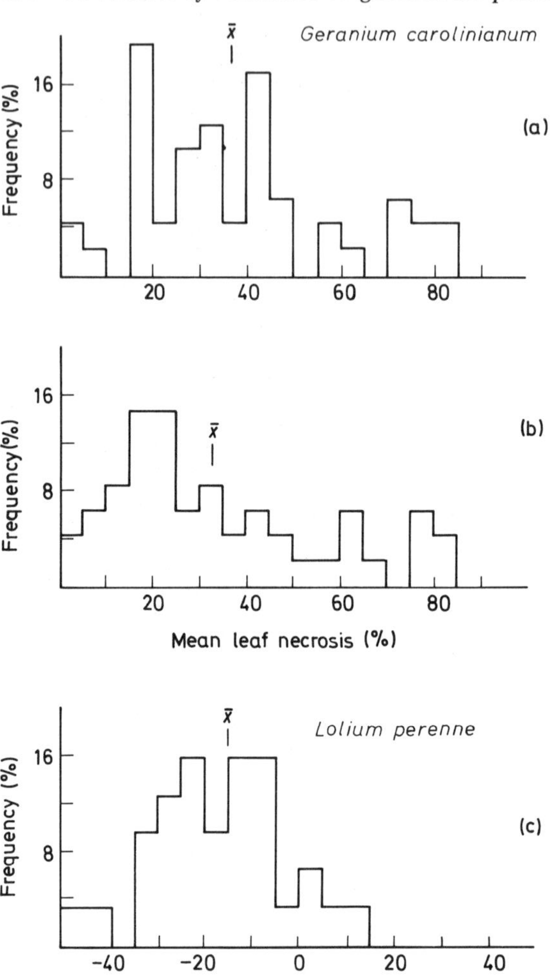

Figure 18.4 (a and b)   Frequency distribution of percentage leaf necrosis in two populations of *Geranium carolinianum* collected from unpolluted sites following 12-h exposure to 2088 µg SO$_2$ m^{-3}. Reprinted from *Botanical Gazette* by Taylor and Murdy (1975) by permission of University of Chicago Press. (c) Frequency distribution of percentage reduction in living shoot dry weight of *Lolium perenne* cv. S23 following 35-d exposure to 650 µg SO$_2$ m^{-3}. Mean percentage change from control is the change in yield of each individual clone due to exposure to the SO$_2$-polluted air expressed as a percentage of its growth in control air. $\bar{x}$ is arithmetic mean of frequency histogram. After Horsman, Roberts and Bradshaw (1979)

contributed by the additive genetic component. Estimates of h^2 are quite high in *Petunia* (0.80) and *Nicotiana* (0.95), but lower in *Geranium* (0.50), *Pinus strobus* (0.64) and *Pinus taeda* (0.33–0.41) (*Table 18.7*). Like R and H_B, h^2 is a ratio and thus its value is greatly influenced by the success of the investigator in reducing the environmental variance. Therefore, the values obtained apply only to the population and environment in which the measurement was made. Estimates of heritability in field-grown plants are

Table 18.7 INHERITANCE OF POLLUTION RESISTANCE IN VARIOUS SPECIES

Species	Gas	Level[1] ($\mu g\,m^{-3}$)	Time[2]	Results[3]	Reference
Allium cepa L. (onion)	O_3	A	s	Single dominant gene for resistance	Engle and Gabelman (1966)
Geranium carolinianum L.	SO_2	2090	12h	$h^2 = 0.50$, no dominance	Taylor (1978b)
Nicotiana tabacum L. (tobacco)	O_3	A	s	$h^2 = 0.95$	Aycock (1972)
	O_3	A	s	GCA large, heterosis not significant	Huang, Aycock and Mulchi (1975)
	O_3	98	4 h	GCA large, heterosis signif. for susceptibility	Huang, Aycock and Mulchi (1975)
Petunia hybrida Vilm.	A	A	4 h	Signif. additive & dominance components, some maternal effect $h^2 = 0.80$, $H_B = 0.93$	Hanson, Addis and Thorne (1976)
Phaseolus vulgaris L. (bean)	O_3	725	11 h	Significant additive variation	Butler, Tibbits and Bliss (1979)
	PAN	426–593	1–2.5 h	$H_B = 0.83$; resistance recessive	
Pinus strobus L. (e. white pine)	SO_2	2607	1 h	$h^2 = 0.64$ for needle colour; $h^2 = 0.32$ for needle length	Thor and Gall (1978)
	A		s		
	SO_2 &	65	6 h	$R = 0.81$ for needle damage; $R = 0.53$ for needle elongation	Houston and Stairs (1973)
	O_3	98			
Pinus taeda L. (loblolly pine)	O_3			$h^2 = 0.33$–0.41	Anon. (1973), cited in Karnosky (1977)
Populus tremuloides Michx. (trembling aspen)	O_3	392	3 h	$R = 0.62$	Karnosky (1977)
	SO_2	1305	3 h	$R = 0.64$	
	SO_2 &	913	3 h	$R = 0.46$	
	O_3	98			
Zea mays L. (sweet corn)	O_3	A	s	Highly heritable, resistance partly recessive	Cameron (1975)

[1] Gas concentration in $\mu g\,m^{-3}$. A = ambient
[2] Duration of exposure. h = hours, s = season
[3] h^2 = narrow sense heritability; H_B = broad sense heritability; R = clonal repeatability; GCA = general combining ability

thus most directly relevant to evolutionary and breeding studies because this is the environment in which selection will take place.

The mode of inheritance of pollution resistance has been studied most intensively in *Nicotiana tabacum* where several genes with additive effects are involved (Sand, 1960; Provilaitis, 1967; Aycock, 1972; Huang, Aycock and Mulchi, 1975). SO_2 resistance in *Geranium carolinianum* was also controlled by genes with additive effects (Taylor, 1978b), as was O_3 resistance in *Petunia hybrida* (Hanson, Addis and Thorne, 1976) where there was some evidence for partial dominance of resistance. In beans, O_3 resistance was recessive and determined by a few major genes (Butler, Tibbits and Bliss, 1979). In contrast to these cases in which resistance was a complex character influenced by variation at several gene loci, O_3 resistance in onion was apparently determined by two alleles at one locus with resistance dominant to susceptibility (Engle and Gabelman, 1966).

Because genes for susceptibility which act in an additive manner will decrease in frequency if there is selection against susceptibility, the presence of large amounts of additive variation for resistance to pollutants probably reflects either the recent advent of pollution stress or the use of between-cultivar crosses in the inheritance studies.

Evaluation of genetic variation

While, at face value, the above evidence indicates that there is an abundance of genetic variation in resistance to air pollutants, and that the differences between genotypes are often large, there are several qualifications to this inference. First, most of the variation is in foliar injury from high concentrations of pollutants, and there is increasing evidence that resistance to acute injury is, at best, weakly correlated with resistance to long-term, ambient exposures (Ayazloo, 1979; Horsman *et al.*, 1979; Meiners and Heggestad, 1979). A related problem is that many studies have used only one or two dosages and it is not clear over what range the observed differences extend. Where response to several dosages has been evaluated, differences are often smaller at both lower and higher dosages (Brennan and Halisky, 1970; Menser and Hodges, 1970; Manning, Feder and Perkins, 1972; Karnosky, 1976, 1977; Bressan, Wilson and Filner, 1978). Such a response is expected as the dosage is increased from one with small effects to that lethal to all plants, but studies at only a single dosage give no indication of where, on the dose–response curves, the genotypes are being tested. Selection for resistance is effective only over the range of dosages at which the genotypes differ in response; thus, high pollutant concentrations are not necessarily the most effective in selection.

A more disturbing finding is that cultivar rankings are sometimes not the same at different dosages (Tingey, Reinert and Carter, 1972) although this appears to be the exception (Menser and Hodges, 1970; Ormrod, Adedipe and Hofstra, 1971; Bressan, Wilson and Filner, 1978). Cultivar rankings may also be influenced by environmental conditions before and during exposure (Heagle, 1979). This should not be surprising since resistance may be influenced by a variety of physiological and morphological characters that are themselves influenced by genotype and environment.

At present there is not enough information to generalize about the importance of environmental effects on cultivar or genotype rankings. While differences in resistance between genotypes which experience fluctuating pollution levels in a variable environment may be somewhat less than those measured in controlled environments, variation is so pervasive and of such magnitude that abundant variation in resistance still seems likely to be expressed in natural populations of most species.

RESISTANCE AND FITNESS

Differences in resistance to air pollutants among the genotypes in a population can lead to evolution only if these differences affect fitness (technically, the net reproduction of offspring). This has yet to be demonstrated for any plant in response to air pollution: thus, we must relate information on injury, growth and yield differences to fitness in natural populations. Where visible injury occurs, it is usually inversely correlated with dry-weight yields (Hill and Thomas, 1933; Heagle, Body and Pounds, 1972; Heagle, Body and Neely, 1974; Markowski, Grzesiak and Schramel, 1975; Heagle, 1979) and reproduction (Brisley and Jones, 1950; Shannon and Mulchi, 1974). Within a population, differential growth reductions are likely to lead to differences in fitness for two reasons: first, growth is related to fitness because larger plants produce more seeds and thus have a greater chance of contributing to the next generation; second, in many populations, mortality is enormous because reproduction is far in excess of the number of plants which reach reproduction in the next generation. At least in monocultures, those plants which die tend to be smaller than those which survive (Black, 1958; Ford, 1975; Naylor, 1976) and in many cases relative growth rate is positively correlated with plant size. It is inferred that competition for light is responsible for these effects (Koyama and Kira, 1956; Ford, 1975). Therefore, genotypic differences in growth rate, even for a short time, will generate size differences which can lead to decreased competitive ability and reduced fecundity or death of the smaller plant. These effects are most intense during early growth, but many populations continue to experience mortality during later stages (Harper, 1977). It is not clear to what extent this mortality is determined by genetic variation, but the potential for selection is clearly present.

The primary concern in most studies of air-pollution effects on plants has been with reductions in growth. However, sexual reproduction is critical to the persistence of many natural populations and to the economic value of many crops. Air pollutants can influence reproduction indirectly, through effects on growth, or directly by influencing flowering, pollen germination, fruit and seed development (Feder, 1968; Feder and Sullivan, 1969; Sulzbach and Pack, 1971; Masuru, Syozo and Saburo, 1976; Houston and Dochinger, 1977; Roques, Kerjean and Auclair, 1980; Bonte, this volume, Chapter 10). There is clearly genetic variation in resistance of reproductive characters to air pollutants (Feder and Sullivan, 1969; Manning and Feder, 1976; Murdy, 1979) so the potential for evolutionary response is present. Considering their potential importance, effects of air pollutants on these components of fitness have not been well characterized.

COSTS OF RESISTANCE

Known mechanisms of resistance to air pollution involve characteristics which are also adaptively important in unpolluted environments and therefore resistance is likely to be based on genetic variation subject to selection by factors other than pollution. Evolution of resistance is thus governed by *net* fitness in an environment characterized by varying pollution intensity and often by conflicting selection pressures. Some idea of the net fitness of genotypes with differing levels and mechanisms of resistance can be gained by cost-benefit analysis. Costs and benefits of a particular character are ultimately measured as fitness, but energy is commonly used as a more easily obtained estimate (Penning de Vries, 1975). To illustrate this approach, we consider some of the costs and benefits of various resistance mechanisms known to occur within species, in the hope that this will stimulate more detailed and quantitative analyses.

Avoidance of pollution stress often involves lower stomatal conductance of resistant genotypes, even in the absence of pollutants (Turner, Rich and Tomlinson, 1972; Thorne and Hanson, 1976; Braun, 1977a; Butler and Tibbits, 1979b). While some protection from pollution injury is provided by such a mechanism, photosynthesis and growth may be reduced at pollution levels which do not cause injury. A second mechanism reducing pollution uptake is closure of stomata in the presence of pollution (Engle and Gabelman, 1966; Bonte *et al.*, 1977; Butler and Tibbits, 1979b). Such genotypes pay only the energy cost needed to reopen stomata and the cost of decreased photosynthesis while stomata are closed. However, the pollution level at which closure occurs is crucial to the adaptive significance of this mechanism because closure at too low a pollution level would decrease photosynthesis more than exposure to pollution. Tolerance of pollution stress may involve increased buffering capacity (Braun, 1977b) with costs including synthesis of additional buffer ions and the perhaps detrimental effects of increased ionic strength on enzyme function. Increased ability to convert pollutants to less toxic forms (Miller and Xerikos, 1979) may require increased synthesis or altered properties of relevant enzymes, whereas decreased sensitivity of an enzyme to pollutants may alter other functional properties of the enzyme. Such changes may have effects on cellular metabolism and growth rates.

A direct approach to assessing costs of resistance is to compare growth in filtered and polluted air of several sensitive and resistant genotypes derived from the same population. Significant costs of resistance will be reflected in reduced growth of resistant genotypes (relative to sensitive) in filtered air. Such comparisons are not informative if the genotypes derive from different populations or cultivars because many other genetic differences in addition to those controlling air-pollution resistance are likely to affect growth responses of such plants. No adequate data are yet available which permit this comparison to be made.

That resistance often has costs means that some of the genetic variation in resistance discussed above may not add to fitness under the variable pollution levels which affect real populations, or that differences in net fitness may be smaller than suggested by pollution resistance alone. Because the frequency of costly resistance mechanisms should be kept low by natural selection in unpolluted populations (as apparently happens for resistance to heavy metals and herbicides), resistance mechanisms which occur at reasonably high frequencies in such populations are unlikely to reduce fitness greatly in *that* environment. However, a character with low costs in one environment may have much higher costs in another; thus, we cannot conclude that costs of resistance are always small. The analogous argument for cultivars of crop plants is that resistance has minimal costs in the environments for which the cultivar was developed (otherwise the cultivar would not be used), but may have much higher costs in other environments.

THE RATE OF EVOLUTION

Rates of evolution for resistance controlled by single dominant and recessive genes are shown in *Figure 18.5*. An initial frequency of 0.04 for resistance is assumed as this appears reasonable from the comparative studies considered above. Evolution of resistance is more rapid if resistance is dominant than if it is recessive, because a higher proportion of

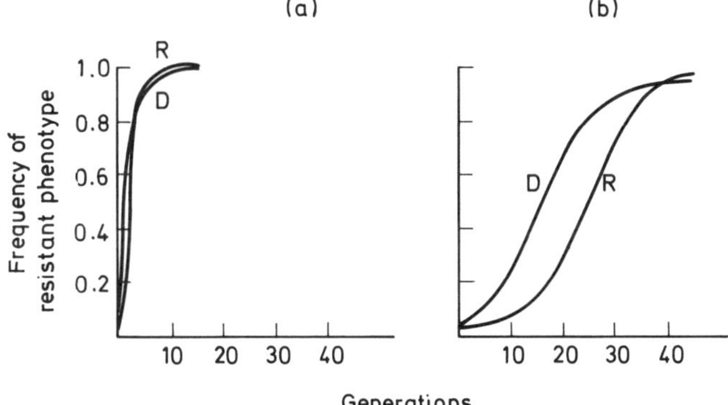

Figure 18.5 (a)   Changes in frequency of resistance under selection where fitness of sensitive genotype is 0.20 relative to resistant genotype. R and D indicate changes when resistance is controlled by recessive and dominant genes respectively. (b) As in (a) but fitness of sensitive genotype is 0.80 relative to tolerant (or 0.20 for 25% of the time and 1.0 for 75% of the time)

individuals carrying the gene actually express the character. Evolution is slow when resistance is rare, but increases rapidly when the frequency of resistance is above 0.10. Clearly, substantial changes in resistance can occur in 10–20 generations when the fitness of sensitive genotypes is 0.80 relative to that of resistant genotypes. More intense selection results in more rapid change.

Heritability does not provide enough information about the genetic control of resistance to compute rates of evolution, but if heritability of resistance is high, rates broadly comparable to those shown in *Figure 18.5* are reasonable.

Evolutionary response to environmental stresses is determined by the magnitudes of differences in fitness which the stress induces among the genotypes in the population, and by the genetic architecture of fitness in the population. Setting aside genetic details, it is clear that the amount of genetic variation and the severity of the stress interact in determining the rate of evolution. However, their relative importance seems likely to depend on severity of stress. When stress (measured as percentage change .from mean fitness in the absence of the stress) is severe, genetic variation limits evolution, whereas, when stress is slight, the rate of evolution *must* be slow because fitness differences are likely to be small.

Reproductive biology may also play a part in limiting the rates of evolution. Species in which each individual produces a large number of offspring relative to population size can evolve more rapidly, because selection can act on a wider range of progeny genotypes, and a well-adapted genotype can contribute a larger proportion of the next generation. In many species, seed can be buried in the soil for several years before germinating. This reduces evolutionary responses to annual environmental variation, because the seeds germinating in any season reflect genotypes successful during several years. Longer generation times also tend to reduce the rate of evolution.

What then is likely to govern the rate of evolution to air pollutants? In situations where pollution has been so severe that many species are eliminated, for example, around metal smelters such as Copper Basin, Tennessee (Hursh, 1948), Wawa, Ontario (Gordon and Gorham, 1963), and Sudbury, Ontario (Linzon, 1978), and around gas refineries (Winner and Bewley, 1978), stress is clearly severe and, by analogy with similar situations on wastes from metal mines, genetic variation seems likely to determine the ability of species to survive in these environments. In less extreme situations, which currently prevail in many urban fringe areas, the situation is less clear. We do not know enough about long-term effects of ambient air pollution to predict fitness reductions, but unless growth or fecundity reductions are greater than 5–10% per year for sensitive genotypes, evolution of resistance will occur on a time-scale of decades rather than years. When several pollutants are present, evolution is likely to be slower because resistance to one may not confer resistance to others, or to interactions. Pollutant mixtures are potentially a serious problem; how much genetic variation is present to cope with mixtures is not yet clear, but available evidence (*Table 18.6*) suggests that there is a considerable amount.

Evidence: cases of evolution of resistance

Evolution of resistance to air pollutants has been demonstrated in several populations by correlating degree of resistance with the pollution level to which the population had been exposed (Bell and Clough, 1973; Taylor and Murdy, 1975; Bell and Mudd, 1976; Horsman, Roberts and Bradshaw, 1978, 1979; Murdy, 1979). Because response to air pollution depends on climatic (Heagle, Heck and Body, 1971; Ashenden and Mansfield, 1977; Davies, 1980) and edaphic (Cowling and Lockyer, 1978; Heagle, 1979) conditions, it is probable that adaptation to natural variation in these factors will influence pollution resistance. Therefore, it is critical that these other environmental factors should be as similar as possible in the populations compared for resistance to air pollutants. The evidence is also more convincing if many populations are sampled from each pollution level, because this gives an estimate of the variation in resistance among populations from each pollution level. These caveats are less important in severely polluted environments where pollution-induced mortality is clearly apparent, because in these cases air pollution is likely to be the primary selective agent.

The most thoroughly studied case of evolution of air-pollution resistance is in populations of *Geranium carolinianum* L. from the SO_2-polluted area around a power plant in Georgia (Taylor and Murdy, 1975; Taylor, 1978b). Populations collected from within 0.8 km of the plant sustained an average of about 20% less injury from 12 h exposures to 2130 μg m^{-3} than populations from adjacent, unpolluted sites. Visible injury was observed in the field, although SO_2 concentrations were not reported. Taylor (1978b) further showed that consistent differences in SO_2 resistance could be obtained by artificial selection in five generations, and that these differences were inherited as a quantitative character. The power plant had been operating for 31 years when the populations were sampled; thus evolution has occurred within 31 generations (*G. carolinianum* is a winter annual).

Evolution of resistance to SO_2 was reported in *Lolium perenne* L. by Bell and Clough (1973) when yield of two clones indigenous to Helmshore, a polluted area in the Pennines, was more resistant to SO_2 than yield of a commercial cultivar, S23. Subsequently a larger sample from the Helmshore population indicated that mean resistance was greater than that of S23 (Horsman, Roberts and Bradshaw, 1979). As the Helmshore population had been exposed to pollution for a number of years, it is reasonable to infer that its resistance results from selection by SO_2. However, it also differs from S23 in having been subject to selection for a variety of additional adaptations (winter dormancy, grazing regime) which might influence SO_2 resistance. Comparison with a population native to a more similar but unpolluted environment would increase confidence that selection for SO_2 resistance is responsible for the differences between the populations. Two such comparisons have since been made and in both cases evidence for evolution of resistance has been found, as described below.

Ryegrass populations were sampled from polluted and unpolluted sites subject to similar management regimes and situated at similar altitudes within a few kilometres of one another in Merseyside (Horsman, Roberts and Bradshaw, 1979). Thirty-six clones from each of four populations were grown in wind tunnels for 8 weeks at either 35 μg SO_2 m^{-3} or 650 μg SO_2 m^{-3}, and shoot dry-matter yields were determined. The populations from polluted sites showed smaller reductions in yield from SO_2 treatment than the populations from the less polluted sites, and had a lower frequency of highly sensitive genotypes (*Figure 18.6*). These results are precisely those expected from selection for increased resistance to SO_2. There was no suggestion that resistance decreased yield of the Newsham and Wavertree populations when they were grown in low SO_2 concentrations, because their yield was greater than or equal to that of the West

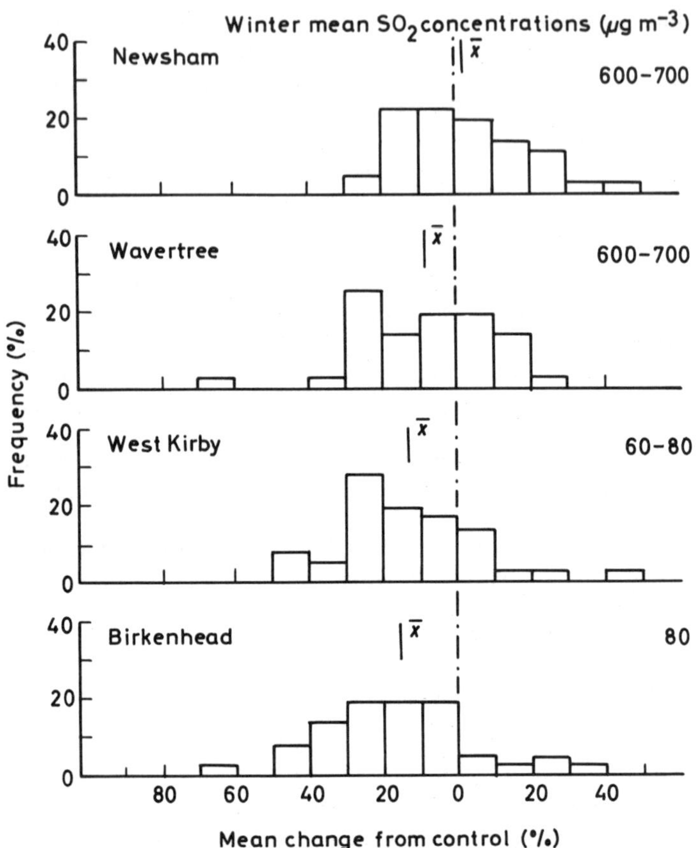

Figure 18.6   Differential response in living dry-matter yield of four Merseyside ryegrass populations to an 8-week exposure to $650\,\mu g\,SO_2\,m^{-3}$. $\bar{x}$ is arithmetic mean of frequency histogram. Typical late 1950s–early 1960s winter mean SO_2 level of each population is also shown. After Horsman, Roberts and Bradshaw (1979)

Kirby and Birkenhead populations. Furthermore, in the SO_2 treatment, the resistant populations had significantly greater yields than the others, suggesting that evolution of resistance can maintain yields in polluted environments. The extent to which yields are depressed *during* evolution of resistance cannot be determined from these experiments.

Further evidence that evolution of SO_2 resistance is common in the UK is provided by the studies of Ayazloo (1979). She collected *Dactylis glomerata, Festuca rubra, Holcus lanatus, Lolium perenne* and *Phleum pratense* from three polluted sites in northern England and compared their responses to acute and chronic fumigations with those of populations from unpolluted sites or with those of commercial cultivars. When compared with commercial cultivars, Helmshore populations of *L. perenne, F. rubra, D. glomerata* and *P. pratense* all showed significantly smaller reductions in leaf dry weight or total dry weight when exposed to 320–$480\,\mu g\,SO_2\,m^{-3}$ for 72–180 days. *D. glomerata, F. rubra, H. lanatus* and *L. perenne* collected from low-SO_2 sites at Askern showed greater reductions in leaf weight, root weight and total weight than populations from nearby high-SO_2 sites, although only in *L. perenne* was the gas × population interaction significant. *L. perenne* sown in 1959 in a polluted park in Manchester showed significantly less reduction in leaf dry weight from exposure to 346 µg SO_2 m^{-3} for 61 days than the commercial cultivar S23. In most cases, plants from the polluted sites showed greater resistance to acute injury (5320 µg m^{-3} for 6 h) than plants from unpolluted sites or commercial cultivars, but there was no correlation between acute injury and yield reduction from chronic fumigations for individual clones. These studies significantly extend the evidence for evolution of SO_2 resistance by demonstrating that it occurs in a number of species, and may be detected within 17–25 years.

Plants of the annual weed *Lepidium virginicum* L. from the SO_2-polluted Copper Basin showed significantly less flower sterility after exposure of inflorescences to 2130 µg SO_2 m^{-3} for 9 h than populations from outside the basin (13.8% and 24.5% sterility respectively), whereas in the absence of SO_2 the populations did not differ in sterility (6.3 and 8.1% respectively) (Murdy, 1979). This could be an ideal case to examine further, using reciprocal transplant experiments, because (unlike yield) sterility is relatively easy to score and is quite directly related to fitness.

Genotypes of eastern white pine (*Pinus strobus* L.) differ in sensitivity to a number of air pollutants including SO_2, O_3 and $SO_2 + O_3$ mixtures (Costonis, 1970; Dochinger and Seliskar, 1970; Dochinger *et al.*, 1970). As far as we know, changes in the genotypic composition of populations subject to pollution have not been directly demonstrated, but several results suggest that such changes are occurring. Progenies from a large number of white pine stands showed symptoms of air-pollution damage when grown near a point source, but progenies from stands in the vicinity of power plants and smeltering operations had less damage than those from unpolluted sites (Thor and Gall, 1978). Dochinger and Seliskar (1970) reported that 'In 10- to 15-year-old plantations, typical chlorotic dwarf trees are usually not more than 2 to 3 feet tall'. It appears that fairly strong selection is occurring, but it is not clear that sensitive genotypes form a large proportion of the population. As it also appears that air pollution reduces the growth of pines in the absence of visible injury (Farrar, Relton and Rutter, 1977; Phillips, Skelly and Burkhart, 1977a,b), evolutionary responses to long-term exposures may be quite significant.

An early report of possible evolution of resistance to O_3 injury is that of Dunn (1959). In experimental gardens, significantly less vegetative and floral injury from photochemical smog was observed in the *Lupinus bicolor* subspecies from Los Angeles than in others. Given the severity of the air-pollution problem in southern California, it is rather surprising that other species have not been studied.

There are considerable opportunities for further studies of evolution of air-pollution resistance. Miller and McBride (1975) present excellent descriptions of forests in Europe and North America in which extensive injury, mortality and growth reductions due to air pollution are occurring. In view of the ubiquitous genetic variation in pollution resistance which occurs in most species, it is almost certain that evolution is taking place. The most definite evidence for evolution of resistance to air pollutants would involve long-term study of natural populations from the beginning of their exposure to pollution, and would include estimates of growth responses, mortality, fertility and fecundity changes.

Significance of evolutionary responses to air pollutants

From the foregoing evidence there is no doubt that breeding for resistance to air pollutants is possible. In some species, specific breeding programmes may be unnecessary if cultivars with sufficient resistance are available. Resistance to air pollutants has been consciously included in plant-breeding programmes in tobacco (Menser and Hodges, 1972; Huang, Aycock and Mulchi, 1975), onions (Gabelman, cited in Ryder, 1973), eastern white pine (Thor and Gall, 1978), alfalfa (Howell, Devine and Hanson, 1971) and other species (Ryder, 1973; Houston, 1976). It must be an unconscious part of all breeding programmes carried out in polluted areas. Success of breeding programmes is dependent on the existence of genetic variation for resistance which improves quality and yield. Evolution of resistance in natural populations provides further evidence for the abundance of genetic variation. Genetic studies allow breeding programmes to progress more rapidly and, in view of the variety of resistance mechanisms already known, generalizations about the mode of inheritance of resistance are unlikely. The construction and maintenance costs and effect on growth in clean air of different resistance mechanisms may vary considerably: more physiological information on mechanisms of resistance would therefore allow selection for physiological characteristics which confer resistance at minimum cost. Strictly empirical breeding methods—selecting plants with adequate resistance, yield and quality—may succeed, but may not be efficient.

The abundance of genetic variation in resistance within and between populations indicates that adequate sampling of a species is essential when attempting to compare resistance of different species. The relative magnitudes of intrapopulation, intraspecific and interspecific variation in resistance to air pollutants are difficult to estimate because responses have not been measured under comparable conditions, but clearly all are large. Intraspecific variation must be estimated if interspecific comparisons are to be meaningful.

In some crops it may not be necessary to carry out breeding programmes. Natural, unassisted evolution of resistance may maintain yields in genetically variable crop species which normally experience mortality before final harvest, although this mechanism is not available in uniform, inbred cultivars. Maintenance of yields is possible in variable crops because, over time, growth of many plants obeys the equation $w = cp^{-1.5}$ where w is dry weight per plant, c is a constant, and p is the density of surviving plants. This means that, as mortality decreases density, dry weight lost is more than compensated by growth of survivors (growth of survivors is believed to be responsible for mortality which is mainly confined to small plants) (Yoda *et al.*, 1963; White and Harper, 1970). Eventually the population reaches 'ceiling' yield, which is often independent of initial density (*reviewed in* Harper, 1977) and thereafter mortality is just compensated by growth of survivors (i.e. $w = cp^{-1.0}$). Both of these

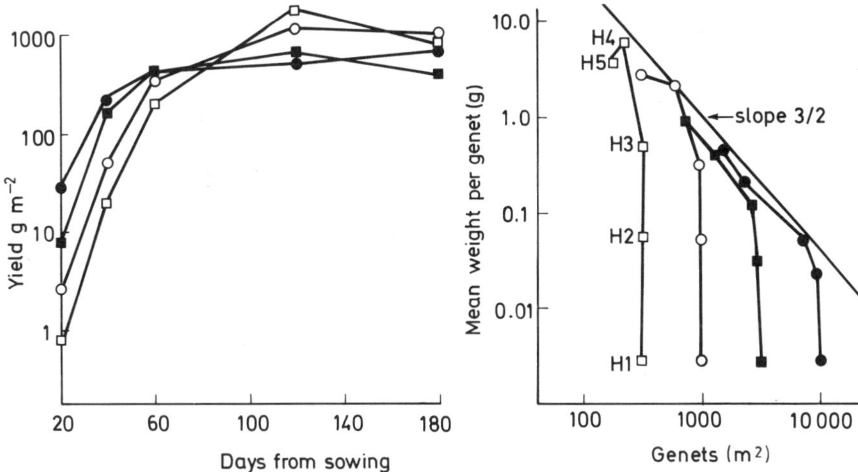

Figure 18.7 (a)   Changes in dry-matter yield of ryegrass populations sown at four initial densities: $\square$ = 320 seeds m^{-2}; $\bigcirc$ = 1000 seeds m^{-2}; $\blacksquare$ = 3200 seeds m^{-2}; $\bullet$ = 10000 seeds m^{-2} (note log scale). (b) The relationship between changing genet (genotype) weight and genet density in the ryegrass populations. H1–H5 represent five successive harvests (note log scales). After Kays and Harper (1974)

processes are illustrated in an experiment of Kays and Harper (1974), in which ryegrass was sown at four different densities and the growth and survival of individual plants was followed over 180 days (*Figure 18.7*).

We can infer that when a variable population is stressed by air pollution, and the growth of sensitive genotypes is reduced more than that of resistant genotypes, a hierarchy of size results which leads to further reductions in size of sensitive genotypes and their eventual death. Decreased growth and mortality of sensitive genotypes frees resources (primarily light) for resistant genotypes which respond by increased growth, thereby maintaining overall yield. The extent to which this can occur depends on sowing density, frequency of resistant genotypes and costs of resistance. Experiments evaluating air-pollution effects on species in which evolution of resistance is possible, should employ realistic densities and allow time for compensation of yield losses to occur.

However, it would be an error to assume that evolution or breeding can always compensate for air-pollution effects. Most resistance mechanisms themselves have costs which will reduce potential yields or alter the niche of a species in ecosystems.

Problems

At present we know very little about the amount of genetic variation in resistant to air pollutants which exists *within* natural populations before exposure to pollution. Studies of clonal repeatability and of between-family versus within-family variation in resistance are thus much needed. If yield or reproductive success of the plants in clean air is reported, these studies can also be useful in assessing fitness of resistance genotypes in the absence of pollutants. Eventually, we need to determine whether genes causing susceptibility to air pollution are adaptively significant in unpolluted environments: if they are, then evolution of resistance to air pollution may lead to changes in species distribution.

A primary difficulty in assessing the importance of evolution of resistance to air pollutants is that only a few cases of the phenomenon have been demonstrated. Therefore, while theoretical evidence suggests that evolution should be occurring in most species suffering injury or yield reductions from pollution, there is not yet sufficient empirical evidence to support this hypothesis. Of particular importance are studies in which changes in populations can be observed over several years, and reproduced in controlled environments.

The costs and benefits of various resistance mechanisms need to be explored more fully and testable models developed to estimate the range of environments over which a particular mechanism will increase fitness or yield. This information is important in predicting resistance mechanisms likely to evolve in natural populations, and in allowing breeders to select for biochemical and physiological characteristics which confer resistance with minimal reductions in yield.

The hypothesis that natural evolution of resistance can maintain yields of some crop species should be tested experimentally and supported more fully by means of modelling. To a great extent, the practical significance of this hypothesis depends on the rate at which resistance evolves, but at present we know only that it can be detected after about 25 years; thus it is important to obtain more information on rates of evolution.

Conclusions and summary

Comparisons of resistance to gaseous air pollutants among cultivars of crop plants and among clones of forest trees indicate that most species possess a large amount of genetic variability for resistance. There also appears to be considerable variation in resistance within populations, but only a few cases have been studied. It therefore seems likely that where air pollution affects plant populations, evolution of resistance will occur. Indeed, evolution of resistance to chronic or acute injury by air pollutants has now been demonstrated in at least a dozen populations. Furthermore, significant differences in resistance evolve within 25–30 years and perhaps faster.

However, resistance to air pollutants is likely to have costs which may be large or small depending on the resistance mechanism and environment. These costs may reduce growth of resistant plants and alter the adaptation of populations in which resistance evolves.

Resistance to air pollutants can be artificially selected, and breeding for resistance may be important in reducing yield losses in some crops. Abundant variability in resistance makes artificial selection relatively easy. In some crops it may allow natural selection to bring about evolution of resistance and this may be important in reducing yield losses.

References

ANDERSON, L.S. and MANSFIELD, T.A. (1979). *Environmental Pollution*, **20**, 113–121

ANTONOVICS, J., BRADSHAW, A.D. and TURNER, R.G. (1971). *Advances in Ecological Research*, **7**, 1–85

ASHENDEN, T.W. and MANSFIELD, T.A. (1977). *Journal of Experimental Botany*, **28**, 729–735

AYAZLOO, M. (1979). PhD Dissertation, University of London

AYCOCK, M.K., Jr. (1972). *Crop Science*, **12**, 672–674

AYCOCK, M.K. Jr., SKOOG, H.A., MORGAN, O.D., McKEE, C.G., HOYER, T.J.H. and MUCHI, C.L. (1977). *Performance of Maryland Tobacco Varieties and Breeding Lines, 1975 and 1976*. Publication 915, Maryland Agric. Expt. Station

BANDEEN, J.D. and McLAREN, R.P. (1976). *Canadian Journal of Plant Science*, **56**, 411–412

BECKERSON, D.W., HOFSTRA, G. and WUKASCH, R. (1979). *Plant Disease Reporter*, **63**, 478–482

BELL, J.N.B. and CLOUGH, W.S. (1973). *Nature*, **241**, 47–49

BELL, J.N.B. and MUDD, C.H. (1976). In *Effects of Air Pollution on Plants*, pp. 87–104. (Mansfield, T.A., Ed.), Cambridge University Press, Cambridge

BELL, R.M. (1980). PhD Dissertation, University of Liverpool

BIALOBOK, S., KAROLEWSKI, P. and OLEKSYN, J. (1977). In *Third Annual Report of the Institute of Dendrology*, pp. 13–30, Kornik, Poland

BLACK, J.N. (1958). *Australian Journal of Agricultural Research*, **9**, 299–318

BONTE, J., BONTE, C., CORMIS, L. and LOUGUET, P. (1977). *Physiologie Végétale*, **15**, 15–27

BRADSHAW, A.D. (1976). In *Effects of Air Pollution on Plants*, pp. 135–159. (Mansfield, T.A., Ed.), Cambridge University Press, Cambridge

BRADSHAW, A.D. (1980). In *Herbicide Resistance in Plants*. (Le Baron, H.M. and Gressel, J., Eds), Wiley, New York

BRAUN, G. (1977a). *European Journal of Forest Pathology*, **7**, 129–152

BRAUN, G. (1977b). *European Journal of Forest Pathology*, **7**, 236–249

BRENNAN, E. and HALISKY, P.M. (1970). *Phytopathology*, **60**, 1544–1546

BRENNAN, E., LEONE, I. and DAINES, R.H. (1964). *Plant Disease Reporter*, **48**, 923–924

BRENNAN, E., LEONE, I. and HALISKY, P.M. (1969). *Phytopathology*, **59**, 1458–1459

BRESSAN, R.A., WILSON, L.G. and FILNER, P. (1978). *Plant Physiology*, **61**, 761–767

BREWER, R.F., CREVELING, R.K., GUILLEMET, F.B. and SUTHERLAND, F.H. (1960). *Proceedings of the American Society for Horticultural Science*, **75**, 236–243

BRISLEY, H.R. and JONES, W.W. (1950). *Plant Physiology*, **25**, 666–681

BUTLER, L.K. and TIBBITS, T.W. (1979a). *Journal of the American Society for Horticultural Science*, **104**, 208–210

BUTLER, L.K. and TIBBITS, T.W. (1979b). *Journal of the American Society for Horticultural Science*, **104**, 213–216

BUTLER, L.K., TIBBITS, T.W. and BLISS, R.A. (1979). *Journal of the American Society for Horticultural Science*, **104**, 211–213

CAMERON, J.W. (1975). *Journal of the American Society for Horticultural Science*, **100**, 577–579

CATHEY, H.M. and HEGGESTAD, H.E. (1972). *Journal of the American Society for Horticultural Science*, **97**, 695–700

CONARD, S.G. and RADOSEVICH, S.R. (1979). *Journal of Applied Ecology*, **16**, 171–177

COSTONIS, A.C. (1970). *Phytopathology*, **60**, 994–999

COWLING, D.W. and LOCKYER, D.R. (1978). *Journal of Experimental Botany*, **29**, 257–265

DAVIES, T. (1980). *Nature*, **284**, 483–485

DAVIS, D.D. and KRESS, L. (1974). *Plant Disease Reporter*, **58**, 14–16

DOCHINGER, L.S. and JENSEN, K.F. (1975). *Environmental Pollution*, **9**, 219–229

DOCHINGER, L.S. and SELISKAR, C.E. (1970). *Forest Science*, **16**, 46–55

DOCHINGER, L.S. and TOWNSEND, A.M. (1979). *Environmental Pollution*, **19**, 229–237

DOCHINGER, L.S., BENDER, F.W., FOX, F.L. and HECK, W.W. (1970). *Nature*, **225**, 476

DUNN, D.B. (1959). *Ecology*, **40**, 621–625

ENGLE, R.L. and GABELMAN, W.H. (1966). *Proceedings of the American Society for Horticultural Science*, **89**, 423–430

FALCONER, D.S. (1960). *Introduction to Quantitative Genetics*. Ronald Press, New York

FARRAR, J.F., RELTON, J. and RUTTER, A.J. (1977). *Environmental Pollution*, **14**, 63–68

FEDER, W.A. (1968). *Science*, **160**, 1122

FEDER, W.A. and SULLIVAN, F. (1969). *Phytopathology*, **59**, 399

FEDER, W.A., FOX, F.L., HECK, W.W. and CAMPBELL, F.J. (1969). *Plant Disease Reporter*, **53**, 506–510

FORD, E.D. (1975). *Journal of Ecology*, **63**, 311–333

GARTSIDE, D.W. and McNEILLY, T. (1974a). *Heredity*, **32**, 335–348

GARTSIDE, D.W. and McNEILLY, T. (1974b). *Heredity*, **32**, 287–297

GARTSIDE, D.W. and McNEILLY, T. (1974c). *Heredity*, **32**, 303–308

GENTILE, A.G., FEDER, W.A., YOUNG, R.E. and SANTNER, Z. (1971). *Journal of the American Society for Horticultural Science*, **96**, 94–96

GENYS, J.B. and HEGGESTAD, H.E. (1978). *Plant Disease Reporter*, **62**, 687–691

GORDON, A.G. and GORHAM, E. (1963). *Canadian Journal of Botany*, **41**, 1063–1078

GRESSEL, J. and SEGEL, L.A. (1978). *Journal of Theoretical Biology*, **75**, 349–371

HANSON, G.P., ADDIS, D.H. and THORNE, L. (1976). *Canadian Journal of Genetics and Cytology*, **18**, 579–592

HARPER, J.L. (1977). *Population Biology of Plants*. Academic Press, London

HEAGLE, A.S. (1979). *Environmental Pollution*, **18**, 313–322

HEAGLE, A.S., BODY, D.E. and NEELY, G.E. (1974). *Phytopathology*, **64**, 132–136

HEAGLE, A.S., BODY, D.E. and POUNDS, E.K. (1972). *Phytopathology*, **62**, 683–687

HEAGLE, A.S., HECK, W.W. and BODY, D. (1971). *Phytopathology*, **61**, 1209–1212

HEAGLE, A.S., SPENCER, S. and LETCHWORTH, M.B. (1979). *Canadian Journal of Botany*, **57**, 1999–2005

HEGGESTAD, H.E. and HECK, W.W. (1971). *Advances in Agronomy*, **23**, 111–145

HENDRIX, J.W. and HALL, H.R. (1958). *Proceedings of the American Society for Horticultural Science*, **72**, 503–510

HICKEY, D.A. and McNEILLY, T. (1975). *Evolution*, **29**, 458–464

HILL, G.R. Jr and THOMAS, M.D. (1933). *Plant Physiology*, **8**, 223–245

HITCHCOCK, A.E., WEINSTEIN, L.H., McCUNE, D.C. and JACOBSON, J.S. (1964). *Journal of the Air Pollution Control Association*, **14**, 503–508

HOLLIDAY, R.J. and PUTWAIN, P.D. (1980). *Journal of Applied Ecology*, **17**, 779–791

HORSMAN, D.C., ROBERTS, T.M. and BRADSHAW, A.D. (1978). *Nature*, **276**, 493–494

HORSMAN, D.C., ROBERTS, T.M. and BRADSHAW, A.D. (1979). *Journal of Experimental Botany*, **30**, 495–501

HORSMAN, D.C., ROBERTS, T.M., LAMBERT, M. and BRADSHAW, A.D. (1979). *Journal of Experimental Botany*, **30**, 485–493

HOUSTON, D.B. (1976). In *Proceedings 4th North American Forest Biology Workshop*, pp. 56–72. State University of New York College of Environmental Science and Forestry, Syracuse, NY

HOUSTON, D.B. and DOCHINGER, L.S. (1977). *Environmental Pollution*, **12**, 1–5

HOUSTON, D.B. and STAIRS, C.R. (1973). *Forest Science*, **19**, 267–271

HOWELL, R.K. and THOMAS, C.A. (1972). *Plant Disease Reporter*, **56**, 195–197

HOWELL, R.K., DEVINE, T.E. and HANSON, C.H. (1971). *Crop Science*, **11**, 114–115

HUANG, T.R., AYCOCK, M.K., Jr. and MULCHI, C.L. (1975). *Crop Science*, **15**, 785–789

HURSH, C.R. (1948). *Local Climate in the Copper Basin of Tennessee as Modified by Removal of Vegetation*, pp. 1–35. Circular 774, US Dept. of Agriculture

HUTTUNEN, S. (1978). *Silva Fennica*, **12**, 1–16

JAIN, S.K. and BRADSHAW, A.D. (1966). *Heredity*, **21**, 407–441

JOWETT, D. (1964). *Evolution*, **18**, 70–80

KARNOSKY, D.F. (1976). *Canadian Journal of Forest Research*, **6**, 166–169

KARNOSKY, D.F. (1977). *Canadian Journal of Forest Research*, **7**, 437–440

KAROLEWSKI, P. and BIALOBOK, S. (1978). In *Fourth Annual Report of the Institute of Dendrology*, pp. 7–17. Kornik, Poland

KAYS, S. and HARPER, J.L. (1974). *Journal of Ecology*, **62**, 97–105

KOYAMA, H. and KIRA, T. (1956). *Journal of the Institute and Polytechnic of Osaka City University*, **7**, 73–94

KRESS, L.W. and SKELLY, C.M. (1977). In *Proceedings of the International Conference on Photochemical Oxidant Pollution and its Control*, pp. 655–662, EPA 600/3-77-001B. Environmental Protection Agency, Research Triangle Park, North Carolina

LANG, K.J., NEUMANN, P. and SCHUTT, P. (1971). *Flora*, **160**, 1–9

LAURENCE, J.A. (1979). *Plant Disease Reporter*, **63**, 468–471

LEVIN, D.A. and KERSTER, H.W. (1974). *Evolutionary Biology*, **7**, 139–220

LEVITT, J. (1972). *Response of Plants to Environmental Stress*. Academic Press, New York

LINZON, S. (1978). In *Sulfur in the Environment. Part II. Ecological Impacts*, pp. 109–162. (Nriagu, J.K., Ed.), Wiley, New York

MACNAIR, M.R. (1979). *Genetics*, **91**, 553–563

MANNING, W.J. and FEDER, W.A. (1976). In *Effects of Air Pollutants on Plants*, pp. 47–60, (Mansfield, T.A., Ed.), Cambridge University Press, Cambridge

MANNING, W.J., FEDER, W.A. and PERKINS, I. (1972). *Plant Disease Reporter*, **56**, 832–833

MARKOWSKI, A., GRZESIAK, S. and SCHRAMEL, M. (1975). *Bulletin de L'Académie Polonaise des Sciences, Cl. II*, **23**, 637–653

MASURU, N., SYOZO, F. and SUBURO, K. (1976). *Environmental Pollution*, **11**, 181–187

MEINERS, J.P. and HEGGESTAD, H.E. (1979). *Plant Disease Reporter*, **63**, 273–277

MENSER, H.A. and HODGES, G.H. (1968). *Agronomy Journal*, **60**, 349–352

MENSER, H.A. and HODGES, G.H. (1970). *Agronomy Journal*, **62**, 265–269

MENSER, H.A. and HODGES, G.H. (1972). *Agronomy Journal*, **64**, 189–192

MILLER, J.E. and XERIKOS, P.B. (1979). *Environmental Pollution*, **18**, 259–264

MILLER, P.R. and McBRIDE, J.R. (1975). In *Responses of Plants to Air Pollution*, pp. 195–235. (Mudd, J.B. and Kozlowski, T.T., Eds), Academic Press, New York

MILLER, V.L., HOWELL, R.K. and CALDWELL, B.E. (1974). *Journal of Environmental Quality*, **3**, 35–37

MURDY, W.H. (1979). *Botanical Gazette*, **140**, 299–303

MURRAY, J.J., HOWELL, R.K. and WILTON, A.C. (1975). *Plant Disease Reporter*, **59**, 852–854

NAYLOR, R.E.L. (1976). *Journal of Applied Ecology*, **13**, 513–521

ORMROD, D.P., ADEDIPE, N.O. and HOFSTRA, G. (1971). *Canadian Journal of Plant Science*, **51**, 283–288

PACK, M.R. (1971). *Environmental Science and Technology*, **5**, 1128–1132

PENNING DE VRIES, F.W.T. (1975). In *Photosynthesis and Productivity in Different Environments*, pp. 459–480. (Cooper, J.P., Ed.), Cambridge University Press, Cambridge

PHILLIPS, S.O., SKELLY, J.M. and BURKHART, H.E. (1977a). *Phytopathology*, **67**, 716–720

PHILLIPS, S.O., SKELLY, J.M. and BURKHART, H.E. (1977b). *Phytopathology*, **67**, 721–725

PROVILAITIS, B. (1967). *Canadian Journal of Genetics and Cytology*, **9**, 327–334

REINERT, R.A., TINGEY, D.T. and CARTER, H.B. (1972). *Journal of the American Society for Horticultural Science*, **97**, 149–151

ROBERTS, B.R. (1976). *Environmental Pollution*, **11**, 175–180

ROQUES, A., KERJEAN, M. and AUCLAIR, D. (1980). *Environmental Pollution*, **21**, 191–201

RYAN, G.F. (1970). *Weed Science*, **18**, 614–616

RYDER, E.J. (1973). In *Air Pollution Damage to Vegetation, Advances in Chemistry Series* **122**, 75–84. (Naegele, J.A., Ed.). American Chemical Society, Washington

SAND, S.A. (1960). *Tobacco Science,* **4**, 137–146

SHANNON, J.G. and MULCHI, C.L. (1974). *Crop Science,* **14**, 335–337

SNAYDON, R.W. (1970). *Evolution,* **24**, 257–269

SNAYDON, R.W. and DAVIES, M.S. (1976). *Heredity,* **37**, 9–25

STEINER, K.C. and DAVIS, D.D. (1979). *Canadian Journal of Forest Research,* **9**, 106–109

SULZBACH, C.W. and PACK, M.R. (1971). *Phytopathology,* **61**, 413

TAYLOR, G.E., Jr. (1978a). *New Phytologist,* **80**, 523–534

TAYLOR, G.E., Jr. (1978b). *Botanical Gazette,* **139**, 362–368

TAYLOR, G.E., Jr. and MURDY, W.H. (1975). *Botanical Gazette,* **136**, 212–215

THOR, E. and GALL, W.R. (1978). *Metropolitan Tree Improvement Alliance (METRIA) Proceedings,* **1**, 80–86

THORNE, L. and HANSON, G.P. (1976). *Journal of the American Society for Horticultural Science,* **101**, 60–63

TINGEY, D.T., REINERT, R.A. and CARTER, H.B. (1972). *Crop Science,* **12**, 268–270

TINGEY, D.T., REINERT, R.A., WICKLIFF, C. and HECK, W.W. (1973). *Canadian Journal of Plant Science,* **53**, 875–879

TOWNSEND, A.M. and DOCHINGER, L.S. (1974). *Atmospheric Environment,* **8**, 957–964

TURNER, N.C., RICH, S. and TOMLINSON, H. (1972). *Phytopathology,* **62**, 63–67

URQUHART, C. (1971). *Heredity,* **26**, 19–33

WALLEY, K., KHAN, M.S.I. and BRADSHAW, A.D. (1974). *Heredity,* **32**, 309–319

WHITE, J. and HARPER, J.L. (1970). *Journal of Ecology,* **58**, 467–485

WILKINS, D.A. (1957). *Nature,* **180**, 37–38

WILTON, A.C., MURRAY, J.J., HEGGESTAD, H.E. and JUSKA, F.V. (1972). *Journal of Environmental Quality,* **1**, 112–114

WINNER, W.E. and BEWLEY, J.D. (1978). *Oecologia,* **33**, 311–325

WINNER, W.E. and MOONEY, H.A. (1980a). *Oecologia,* **44**, 290–295

WINNER, W.E. and MOONEY, H.A. (1980b). *Oecologia,* **44**, 296–302

WU, L., BRADSHAW, A.D. and THURMAN, D.A. (1975). *Heredity,* **34**, 165–187

YODA, K., KIRA, T., OGAWA, H. and HOZUMI, K. (1963). *Journal of Biology of Osaka City University,* **14**, 107–129

19

FUTURE DIRECTIONS IN AIR POLLUTION RESEARCH

WALTER W. HECK
Agricultural Research Service, US Department of Agriculture, and Botany Department, North Carolina State University

Introduction

It is time that we in agricultural research started to think in terms of the effects of *air quality* on agricultural production rather than the more narrow term of *air pollution*. Air quality considers all aspects of the interactions between atmospheric chemicals and the biosphere, and gives a more positive direction to our responsibility to understand this complex system. In addition, we are interested in an objective understanding of the effects of substances in the atmosphere on crop production, whether they are beneficial or adverse. Additionally, it is difficult to separate the relative importance of anthropogenic (human) sources (air pollutants) and natural sources for some components of air quality.

Air pollutants are the components of air quality that are released by human activities. The 32nd Nottingham School in Agricultural Science was developed to focus on the air-pollutant component of air quality and thus the remainder of this chapter will be concerned with that air-pollution component. Although both managed and unmanaged biological systems are affected by air pollutants, this paper will cover only the effects of air pollutants on agricultural systems (agronomic and horticultural). However, the major part of this chapter will be pertinent to research involving effects of air pollutants on vegetation of all types.

Agricultural production is significantly affected by air pollution in all industrial nations. The problem may be ubiquitous, affecting large areas of a country, or localized around industrial sources. Thus, regardless of the relative importance of individual pollutants, all pollutants (alone or as mixtures) must be of concern to the agricultural scientist.

The papers in this book cover a breadth of topics that will complement other review articles such as those by Mudd and Kozlowski (1975), Heck and Brandt (1977), Heck, Mudd and Miller (1977), Guderian (1977), National Research Council (1977a,b; 1979), Galloway *et al.*, (1978), Heck, Krupa and Linzon (1979) and Krupa and Legge (1982). All of these review articles, whatever their emphasis, help to keep us aware of the basic

Cooperative investigations of the US Department of Agriculture and North Carolina State University, Raleigh, NC.

information and, by implication, suggest future directions for research programmes.

Two papers (Heck, 1973; Heck, Taylor and Heggestad, 1973) remain relevant to the subject of research priorities and future research direction. The first paper gives a conceptual overview of the biological system at several organizational levels and discusses possible future effects of air pollutants on agricultural systems. The second paper focuses on research needs by first developing the historical background and reviewing the state of knowledge at that time. The needs outlined in that paper are still relevant. The concepts developed in both papers fit into a systems approach to research on the effects of air pollution.

Research into the effects of air pollution has two primary goals. First, to develop scientific criteria for use in setting ambient air-quality standards or guidelines. Such criteria or data would meet the needs of regulatory agencies. In the United States, such information would support the mandate of the Department of Energy to protect the environment while developing new energy technologies. Second, to understand the effects of air pollution on agricultural production. This should include the mechanisms of pollutant action in plants and the effects on growth processes. The two goals are complementary and should lead to the development of predictive techniques that can be used both in setting ambient air-quality standards and in quantifying agricultural losses. A supplementary objective would be to improve our knowledge of fundamental plant processes and the relationships between plant systems and their environments.

This chapter will: 1) develop a systems approach to air-pollution research; 2) briefly discuss the main subsystems; 3) discuss research approaches and facilities; 4) highlight the importance of external factors on plant response; 5) discuss biological organization and the types of effects to study; 6) identify future directions of research; and 7) put forward modified concepts of research management.

Air pollution research—a systems approach

The major challenge to the agricultural specialist over the next decade is to treat air pollution and its effects as an integrated system. It is no longer sufficient to understand the effects of given pollutants or combinations of pollutants on specific biological systems; the entire air-pollution system must be understood. The agricultural scientist must accept this integration, because many physical scientists and administrators must be educated to understand that the effects of air pollutants on biological systems serve as the primary rationale for continuing to study the air-pollution system.

The air-pollution system, in relation to agricultural production, is diagrammatically represented in *Figure 19.1* and is discussed below. The challenge to understanding the entire system lies in understanding the intricacies of the subsystems. An understanding of effects is the central core around which all the other subsystems are built. The system is cyclic, with feedback at critical steps. It involves legal and economic aspects, source technology, emission and control technology, meteorology and

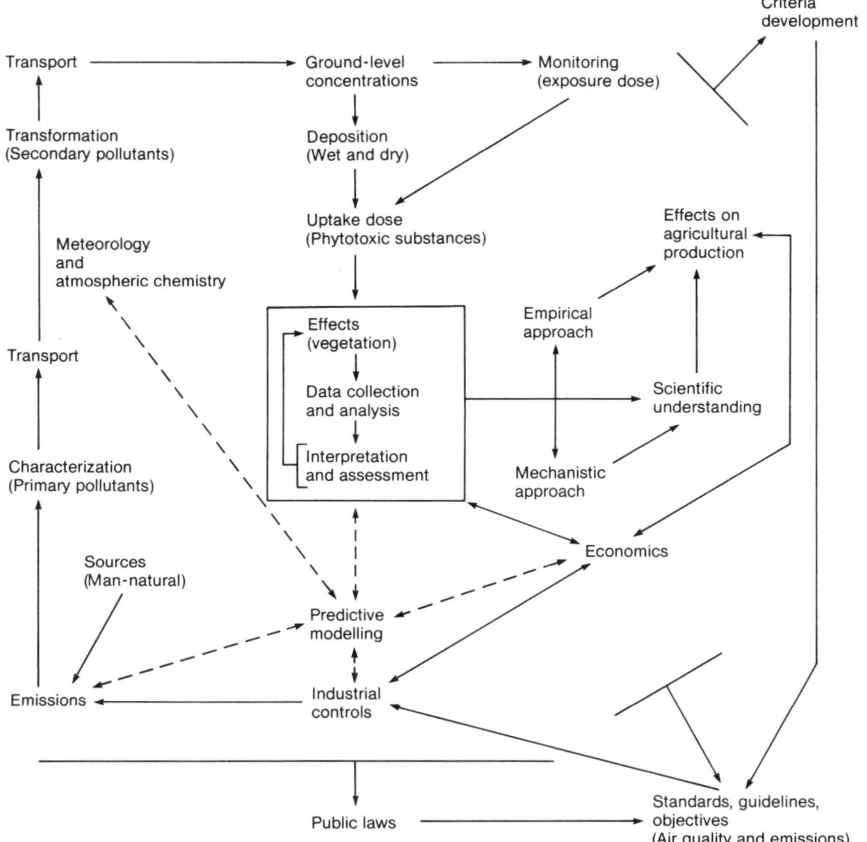

Figure 19.1   A model of the air-pollution system and its major components. The solid lines suggest direct linkages; the dotted lines are areas of uncertainty associated with predictive concepts. The lines with an arrow for Criteria, Standards and Public Laws show that all aspects of the system are included in developing these three pivotal parts. The central effects box refers to all effects, but this paper deals specifically with vegetation

transport processes, atmospheric chemistry and transformation processes, and deposition and uptake processes, all of which help shape a research programme on pollution effects.

The historical development of the air-pollution system was as follows. Effects were initially observed and possible sources identified; cause-and-effect relationships were developed; causative agents (primarily pollutants) were identified; public concern was voiced; additional research studies were started and contributions of meteorology and atmospheric chemistry were recognized; monitoring was begun; differences between primary and secondary pollutants were characterized and more quantitative assessments were made; laws of various strength were enacted as knowledge grew; criteria and standard protocols were developed; economic aspects were given greater consideration and predictive efforts were intensified; and the 'cycle' continues. The agricultural scientist should understand this historical development in order to appreciate and understand the air-pollution system and to deal with air-pollution problems that affect crops.

The legal, criteria-development, standard-setting, and economic aspects of the air-pollution system are not discussed further because their influence on research is indirect. However, these subsystems *should* be understood, for they are critical parts of the entire system. Additional discussion of sources, air pollutants, meteorology and atmospheric chemistry is included, because a basic understanding of the role of these components of the air-pollution system is essential in planning for an effective programme of research into biological effects. Such research is discussed in a separate section and, in this chapter, will be limited to effects on vegetation.

AIR POLLUTANTS AND THEIR SOURCES

Anthropogenic sources of air pollutants that are of national and international importance are usually associated with the use of fossil fuel for energy production. Smelting of ores and various industrial processes also release many different chemicals into the atmosphere. Air pollutants emitted directly from industrial technology are primary pollutants, while those formed from primary pollutants through atmospheric transformations are secondary pollutants. *Table 19.1* lists some of the more important

Table 19.1 LIST OF PHYTOTOXIC AIR POLLUTANTS IN ORDER OF IMPORTANCE TO PLANT SYSTEMS[a]

Pollutant	Primary or secondary	Form	Major source(s)
O_3	Secondary	Gas	Atmospheric transformation (associated with automotive emissions, NO_2, hydrocarbons)
SO_2	Primary	Gas	Power generation, smelter operations
NO_2	Primary and secondary	Gas	From direct release and atmospheric transformation (high-temperature combustion, from NO); fertilizer production
HF	Primary	Gas–particulate	Superphosphate, aluminium smelters
$C=C$[b]	Primary	Gas	Combustion, natural
PAN-Oxid.[c]	Secondary	Gas	Atmospheric transformation (automotive emissions, NO_2, hydrocarbons)
NO	Primary	Gas	Combustion, natural
Cl_2	Primary	Gas	Spills, manufacture
HCl	Primary	Gas	Burning of plastics
Toxic elements	Primary	Particulate	Smelters, combustion processes
NH_3	Primary	Gas	Feedlots, natural
SO_4^{2-}	Secondary	Aerosol	Atmospheric transformation (SO_2)
NO_3^-	Secondary	Aerosol	Atmospheric transformation (NO_2)
H_2S	Primary	Gas	Paper production, natural, geothermal
CO_2	Primary	Gas	Combustion, natural

[a]This list is not meant to be complete but represents the most important air pollutants with respect to terrestrial plant systems. Several of those low on the list have been poorly studied and may be more important than thought.
[b]Ethylene
[c]Peroxyacetyl nitrate-oxidant

air pollutants, designates whether they are primary or secondary, and identifies their major sources.

The photochemical oxidants (primarily ozone—O_3) are secondary pollutants. They are the most important pollutants in the United States and Canada, and possibly in the international community. Sulphur dioxide (SO_2) and nitrogen dioxide (NO_2), in themselves and as the precursors to acid aerosols, are the next most important pollutants. For certain industrialized nations (primarily those European nations at higher latitudes), they are probably *the* most important pollutants.

There is some interest in understanding effects in relation to a specific industrial technology as a means of assessing the environmental impact of that technology. This technology-orientated assessment has been encouraged to aid in comparing the environmental effects of different energy technologies. This approach to environmental assessment requires detailed knowledge of each technology and ignores the fact that many energy technologies release similar chemicals but in different relative amounts. Thus, it seems reasonable for the agricultural scientist to adopt a generic approach and just to study those chemicals associated with emissions from all or most energy technologies. The results of such studies can be used to assess the environmental impact of specific technologies when emission data are available.

Vegetation is more sensitive to gaseous pollutants than to aerosols. Some scientists discount the direct effects of aerosols because they do not enter the plant system, unless small enough to act as gases. Research with acidic precipitation and other chemicals that may be deposited on the leaf in a soluble form suggests considerable cuticular or stomatal absorption of the soluble components. Additional research is needed to understand the significance of wet and dry deposition of soluble chemicals.

TRANSPORT, TRANSFORMATION AND DEPOSITION

Transport, transformation and deposition are the most important aspects of the air-pollution system (after emission data) in relation to agricultural effects. An understanding of the transport and deposition processes is basically the responsibility of meteorologists, whereas the transformation processes are the responsibility of atmospheric chemists. *Figure 19.2* is an overall view of these subsystems, starting with the sources and showing deposition of both primary and secondary pollutants on vegetation. These subsystems must be well understood by the atmospheric chemist in the development of materials-balance equations for air pollutants (emissions versus deposition). The subsystems also comprise the basic aspects of the atmospheric portion of biogeochemical cycling.

Transport

This phenomenon is associated with meteorological processes, although mechanistically it cannot be separated from transformation and deposition processes. Meteorologists have used dispersion modelling and predictive

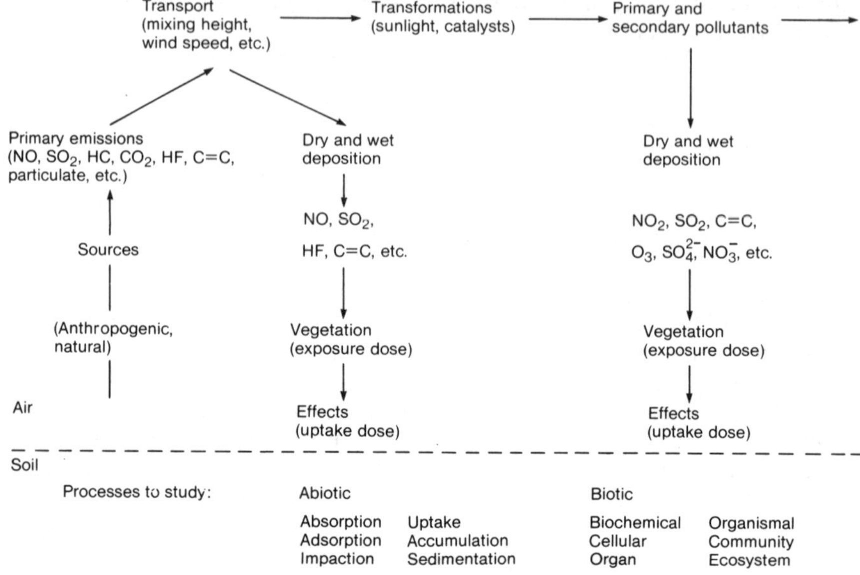

Figure 19.2 The emission, transport, transformation and deposition subsystems. The atmospheric portion of biogeochemical cycling. (C=C is ethylene)

concepts as a basic approach in describing atmospheric transport (National Research Council, 1977a; Nriagu, 1978). The modelling approach requires a knowledge of selected meteorological parameters, source emission data and terrain. The models predict expected ground-level concentrations over time at various distances and directions from a source. Most such modelling efforts have been developed for single sources, while a few have considered transport from multiple sources or small regional sources (cities) within relatively short distances (<50 km). The significance of long-range transport did not receive serious consideration until the mid-1970s with the identification of large rural ozone concentrations and the increased concern about acidic deposition (National Research Council, 1977a; Nriagu, 1978). We now recognize that pollutants may be carried hundreds and even thousands of kilometres before they impact on biological systems. Some researchers have studied air-mass trajectories to identify the origin of air masses associated with high concentrations of pollutants at a specific site.

Transformations

Transformation processes (chemical reactions in the atmosphere) have been a major area of atmospheric sciences since the photochemical oxidants were first recognized in the Los Angeles basin. Numerous reviews discuss the complex chemistry of atmospheric transformations (National Research Council, 1977a).

Hundreds of chemical reactions have been reported, primarily in relation to photochemical oxidants. The significance of most of these reactions, in terms of phytotoxic effects, is not known. The most interesting group of substances include the strong oxidants and the free radicals. In general, the stronger the oxidant the more rapidly it will react in the atmosphere and the less chance it will have to cause a biological effect. This is also true for free radicals where half-lives are usually less than $1\,\mu s$. Ozone is the most stable phytotoxic oxidant yet identified. Free radicals probably have too short a half-life to cause direct injury to the vegetation.

During the 1970s, interest developed in the long-range transport of SO_2 and NO_2, and their transformation to SO_4^{2-} and NO_3^- by oxidation and hydration processes. The rate of oxidation is dependent on environmental factors and the presence of catalytic amounts of certain trace elements. The transformation of SO_2 to SO_4^{2-} takes place at an average estimated rate of 1–3% per hour under a variety of conditions (Nriagu, 1978).

The agricultural scientist needs to keep in contact with the atmospheric chemist in order to understand better the transformation processes. This contact would also serve as a means of identifying other potentially harmful chemicals.

Deposition

Deposition is a process that links the interests of the meteorologist, the chemist and the biologist. Deposition can be dry or wet and concerns both gases and aerosols. Aerosols can be deposited dry, by impaction or sedimentation, or wet, by incorporation into some form of precipitation.

Gaseous deposition to vegetation is by sorption processes that include surface adsorption, cuticular absorption, and absorption through plant stomata. Stomatal sorption is by diffusive processes and is dependent upon stomatal resistance, the atmospheric concentration of the gas, the internal concentration of the gas and the duration of the exposure. The actual amount of uptake may be called the uptake dose. The exposure dose (Heck, Larsen and Heagle, 1981) is an important consideration for developing predictive models and should include some expression for the frequency of occurrence of concentrations over some threshold level (in some instances this will approach zero). To understand gaseous deposition we must understand the relationships between stomatal opening, total uptake and uptake rates, and the biotic and abiotic factors which affect these rates.

Research approach

Air-quality research on plant systems has both philosophical and pragmatic overtones. The former concerns the basic goals of such research and whether the research should be primarily for regulatory purposes or to understand the effects. Both of these influence how the scientist conducts the research. The research methodology involves both the facilities to be used and the experimental approaches.

Some standardization of methods must occur across laboratories without sacrificing the ability to develop innovative techniques and approaches. A handbook on methodology (Heck, Krupa and Linzon, 1979) was an initial attempt to encourage standardization of certain procedures and to suggest guidelines that could be followed. We believe that scientists should follow these guidelines whenever possible.

Air-pollution research has been conducted using four different approaches: 1) chambers or other devices in the field; 2) when possible, field studies using ambient conditions without experimental devices; 3) greenhouse chambers; and 4) chambers in controlled-environment facilities.

EXPERIMENTAL FACILITIES

The facilities for air-pollution research can be both complex and expensive because of the nature of the atmosphere being studied. In most cases, special exposure chambers are required to handle and contain the test atmospheres. Individually, these may be relatively inexpensive, but the complete system can be expensive, especially when the dispensing and monitoring portions of the system are automated. Various chamber designs have been recommended for greenhouse and laboratory (Heck, Philbeck and Dunning, 1978), and for controlled field exposures (Heagle *et al.*, 1979). Chamber concepts and design were also discussed in the handbook edited by by Heck, Krupa and Linzon (1979).

Field

The most versatile systems are open-top chambers (Heagle *et al.*, 1979) that permit control of the atmospheric composition around the test plants. These chambers have been built to simulate natural conditions to the maximum extent possible, while still permitting control of the pollutant dose. They are essential for the study of ubiquitous pollutants such as ozone (they permit the use of a control treatment and several concentrations of ozone) and are of value in the study of any pollutant. In certain cases and for specific objectives the zonal air pollution system (ZAPS) as adapted by Miller for crop studies (Miller, 1979) and the linear gradient system (LGS) (Shinn, Clegg and Stuart, 1977) are relevant and should be considered.

Ambient air concentration gradients

These designs are of value only when studying point or area sources of pollution when a concentration gradient (dose) exists with distance from the source. Experimental designs using distance to obtain dosage levels have been broadly used in most studies of point-source pollution problems. This approach has been tried with regional problems when the region was not too large (Oshima *et al.*, 1976) but as the distance increases it becomes increasingly difficult to separate the confounding effects of environmental variables.

Greenhouse

Greenhouses permit year-round experimental work and can be used for initial verification of controlled-environment results. They also permit initial screening of numerous responses associated with various experimental designs, so field experiments can be more selective in the plant species or varieties, and in the exposure regimes and response measures used. Results may be indicative of results from field exposures, but field verification is necessary. Greenhouses may be designed to give from good to medium control of environmental variables. Environmental fluctuations in greenhouse exposure chambers are usually similar to those in the greenhouses themselves.

Controlled environment

These facilities are, to varying degrees, capable of controlling the environmental conditions under which plants are grown and exposed. Research in these facilities is essential if models are to be developed that include an understanding of environmental factors. These facilities also permit the study of plant response under various steady-state environmental conditions. Thus, physiological processes can be studied and changes associated with alterations in single or multiple environmental factors can be measured. Uptake rates of pollutants can also be studied, and the physiological resistance related to stomatal or biochemical control of uptake can be measured under different conditions. The effects of pollutants on biochemistry and cellular organelles should first be studied under controlled conditions, so steady-state conditions can be used. Once responses have been noted and measured under various environmental conditions, the response should be verified under field conditions.

EXPERIMENTAL APPROACHES

Research is often started without an understanding of the overall air-pollution system. Thus, experimental approaches may be used which produce data that is incompatible with the system (i.e. use of an exceptionally high single concentration of gas). This subsection focuses on the basic experimental approaches that will produce useful data on the effects of air pollution. These approaches can be used to study plant communities, the whole plant or any part of the plant. The approaches discussed are: 1) modelling; 2) dose–response; 3) sensitive versus resistant varieties; and 4) acute versus chronic responses.

Modelling

This term elicits a variety of responses from plant scientists. Some refuse to discuss the concept, pointing to extravagant claims of earlier modelling enthusiasts and their lack of success. Others think that modelling is the answer to all societal problems. The truth lies somewhere between these extremes but no serious scientific student of air pollution can afford to ignore the value or utility of modelling approaches.

Models have been, and will continue to be, valuable tools to aid in the understanding of air-pollutant effects on vegetation. Models can be simple conceptual schemes (*Figure 19.1*) that define a problem or present a hypothesis, or they can be complex mathematical descriptions that describe or predict the functioning of biological systems. The plant scientist in air-pollution research must understand the basic concepts of models, what they attempt to portray and their inherent weaknesses. Such an appreciation is essential before the scientist can correctly use them to understand the response of plant systems.

Modelling is used regularly by scientists to develop experimental designs and enunciate hypotheses. Modelling concepts are also used to predict cause–effect relationships. Such models may be subjective predictions based on the intuitive judgment of the scientist, a prediction based on some quantitative assessment of effects (i.e. a mathematical model) or a combination of the two. At present the subjective prediction is still our best model for use in predicting effects of air pollutants on vegetation.

The ultimate in predictive modelling is a mathematical expression that will predict a happening with a high degree of confidence. Such a mathematical expression becomes harder to develop as the system becomes more complex. Thus, the degree of fit for a mathematical model is a function of the complexity of the system being modelled and how well the functioning of the system is understood. The accuracy of a subjective estimate is also a function of the complexity and functional knowledge of the system.

A reasonable approach in order to understand the response of plants to air pollutants would be to study a single plant type in some detail. Hypotheses could then be put forward concerning which were the most critical responses to study in plant systems. These critical responses could then be studied for a number of other species, and generalities across species and varieties could be developed.

Modelling of plant response and the development of predictive capabilities should involve a number of steps. The following steps could lead to the development of usable predictive models.

1. Envisage the response system and determine what must be studied.
2. Short-term results should consider empirical models. These can be divided into acute and chronic air-pollution studies:
    (*a*) The quickest approach would be to use short-term acute exposures within a single day. Responses could be related to the length of exposure and the gas concentration. With sufficient data, empirical dose–response models can be developed (Heck, Larsen and Heagle, 1980). These do not focus on mechanism but some mechanistic concepts can be derived from the results. Foliar injury may be the primary response studied. They can give an indication of expected injury from short-term acute ambient exposures.
    (*b*) Growth and yield studies may be undertaken in the field or greenhouse and related to ambient concentrations of the pollutant(s) of interest (Heagle and Heck, 1980). They can give an

adequate description of what happened and, once verified, may be used to predict yield effects at other locations where concentrations of pollutants, over some time, are known.

3. The long-term results require extensive research. They are based on mechanistic modelling concepts that call for an understanding of the effects of air pollutants on plant processes. Such studies require the development of a growth model based on healthy plants, and then the addition of the pollutant stress and, eventually, other stresses.

Modelling, in its broadest concept, should be the basic approach to air-pollution research; the primary approach is conceptual.

Dose–response designs

Whereas modelling should be the broad basic approach to air pollution research, the dose–response concept should be the primary experimental approach for all research on the studies of the effects of atmospheric chemicals on vegetation. Dose is the level of exposure of an organism to a substance. Response is the effect of a given dose on that organism. In studies of plant response to air pollutants, dose has been traditionally defined as the product of the concentration (c) of a pollutant and the duration (t) of exposure to that pollutant ('exposure dose'). The term 'uptake dose' may be used and refers to the amount of pollutant that enters the organism. However, we prefer exposure dose for use in studying the effect of air pollutants on plants and it is used interchangeably with dose in the remainder of this paper (Heck, Larsen and Heagle, 1981).

An understanding of the effects of various doses of any atmospheric chemical, or complex of chemicals, on plants is needed as a basis for understanding the effects of the chemical(s) on plants and as scientific criteria for setting ambient air-quality standards. The ingredients of dose–response designs (pollutant concentration, duration and frequency of exposure, response of the receptor) are also necessary for the development of predictive equations (models). These models can be used to predict yield reductions and thus fit into overall econometric models.

The dose–response experimental approach to develop predictive equations involves the study of plant responses across a range of pollutant concentrations, exposure durations or some combination of the two. Plant response data from dose–response designs are amenable to regression analysis or to the separation of points on a curve using mean separation tests. The complex experimental design requires a control (this can be a background, a subthreshold or a zero concentration of the pollutants(s)) and a series of pollutant concentrations that span expected ambient concentrations over various durations. Maximum pollutant doses should ensure at least a 50% response (modification of variable studied) in order to characterize the response curve (surface), unless such pollutant doses would be excessively high.

Dose–response information cannot be gained by using only ambient concentrations of a pollutant that occur in a given region. Neither can dose–response information be obtained by experiments using a single duration of exposure and a single concentration of pollutant. Useful dose–response information can be obtained when one of the two variables (time or concentration) is held constant and the second is varied over three or more levels.

Sensitive compared with resistant genotypes

The use of two genotypes (cultivars) within a species, that respond in differing degrees to a given pollution insult (sensitive versus resistant), can be used to study the mechanism of plant response to air pollutants. This approach has value whether the level of understanding is molecular, physiological or whole-plant.

The identification of the test genotypes requires the screening of a number of genotypes to ascertain whether the species has sufficient genetic variability to identify sensitive and resistant genotypes. The genotypes may or may not have a similar growth habit, but those with a similar growth habit would permit greater comparability of response factors. The screening should use a dose–response design (with only concentration as a variable). The response used in such a preliminary test could be foliar injury or some other simple response. Confirmation of acute screens should be made on selected genotypes using a chronic exposure design and growth or yield as the response measured.

The sensitive and resistant genotypes should first be studied in relation to pollution stress. The effects of other stress phenomena on the response of the genotypes to pollutants should then be determined. Information obtained from such studies would contribute to the development of mechanistic models. This approach is being used in our laboratory, using cultivars of snap beans, *Phaseolus vulgaris*. We have screened several hundred genetic selections of bean, with emphasis on the bush growth habit, and have identified a range of sensitivities to ozone. Two sensitive and two resistant genotypes were tested in field plots during a growing season and showed the same relative effects as from the acute screen. Subsequent mechanistic studies are being planned.

A similar approach is being used at the Beltsville laboratory, where the Bush Blue Lake 290 (BBL-290) cultivar of snap bean is the test organism (Jesse Bennett and Edward Lee, personal communication). The sensitivity of the BBL-290 is changed by the application of EDU (N-[2-(2-OXO-1-imidazolidinyl)ethyl]-N'-phenylurea) which is a chemical known to increase the resistance of some plants to ozone. This approach is similar to, and permits response comparisons which are similar to, the comparisons of sensitive and resistant genotypes.

Acute compared with chronic exposures

The use of acute or chronic exposures, or some combination of the two, is implicit in any research approach to the study of atmospheric chemicals. An acute exposure is generally exposure to a high concentration that will cause classic symptoms of acute injury on the test plants in a relatively short time. Chronic exposures are multiple exposures of plants to relatively low pollutant concentrations, which may fluctuate. Generally, chronic injury symptoms are more difficult to identify than classic acute symptoms. In ambient air, acute exposures are usually associated with point sources, or several point sources that are close together: chronic exposures are usually associated with multiple sources or regional sources. The response mechanism may be different in the two cases. Chronic exposures are more pervasive and more liable to produce effects on growth and yield without showing obvious symptoms of injury.

The exposure design used in controlled exposures must be determined after the basic objectives of the research are known. Both acute and/or chronic exposure designs should be considered, for both occur in nature. Most earlier studies stressed acute exposures, whereas most current studies tend to stress chronic exposures.

In general, we know that crops affected by pollutants in the field are exposed to some combination of acute and chronic exposures over the growing season. This is true for ozone and true for sulphur dioxide in some places. However, more and more we are dealing with chronic effects, except within several kilometres of major sources. In addition, plants in the field are probably exposed to *mixtures* of pollutants at acute or chronic exposure levels.

Effects of other external factors

The effects of air pollutants on agriculture cannot be fully understood until they are considered in relation to other factors that affect the plant. Research to date gives ample evidence that factors such as climatic and edaphic changes, and biotic pathogens, do affect the response of plants to air pollutants. These factors (*Figure 19.3*) must be studied independently and in combination, if we are to understand the mechanism of plant response to air pollutants. Results must be incorporated into modelling efforts whenever possible. These factors can best be studied by using the approaches discussed above.

Response of plants

Biological systems must be understood, to develop future directions and research approaches to studies of air-pollutant effects on plant systems. Effects of various stresses are often considered in terms of the level of organization at which the effects are studied (biochemical [subcellular], cellular, plant organs, whole plant or plant community). It is also important for the scientist to know which are the most important types of effects to study.

BIOLOGICAL ORGANIZATION

To increase our understanding of cause–effect relationships, it is necessary that research be directed more towards the simpler organizational level.

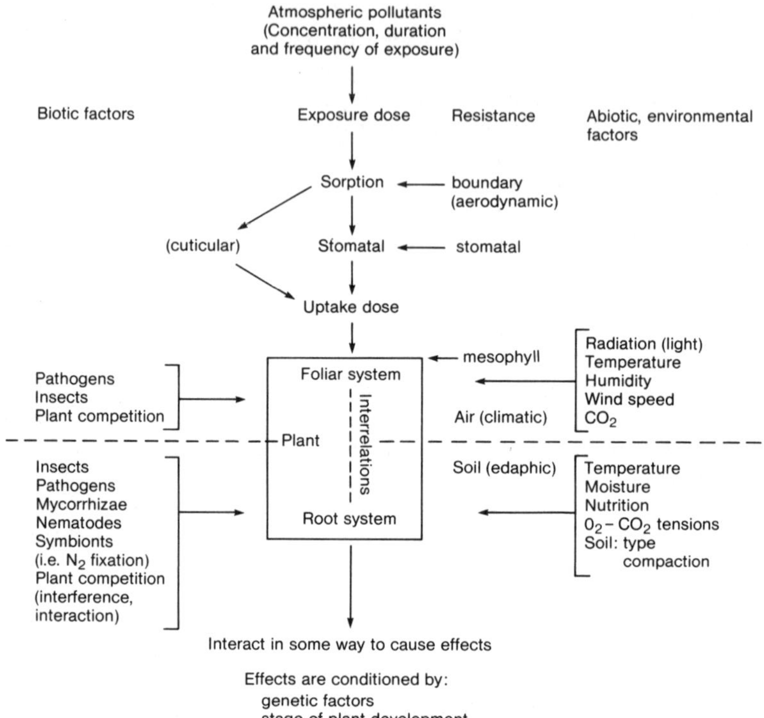

Figure 19.3 Biotic and abiotic factors that can affect the response of plants to an air-pollutant exposure dose

Thus, to understand completely a mechanism of action, we must know the effects on the specific chemical systems, within the plant, that control a given response. Through these studies we can understand the effects on metabolic pathways and thus the organelles containing these systems. This understanding can lead in turn to an understanding of cellular effects and eventually to effects on plant communities. The response system outlined by Heck (1973) is an overall view of the plant system and the foci within that system at which different effects may occur. A simplified representation of this scheme is shown in *Figure 19.4*.

Neither historically nor in practice have scientists followed the above research pattern. This is partly due to the fact that the effects of air pollutants have initially been associated with visible injury (chlorotic or

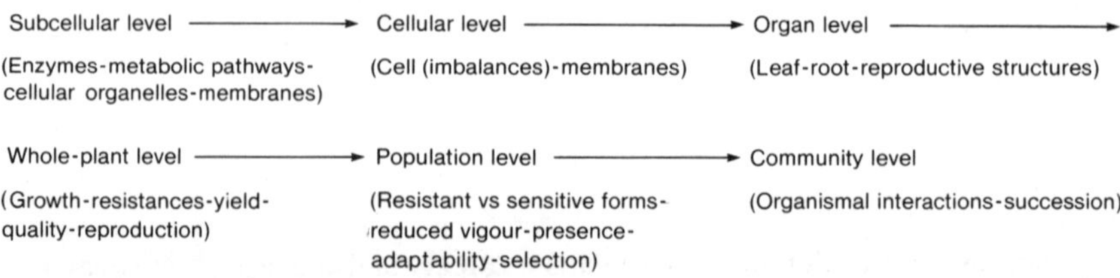

Figure 19.4 Levels in the plant system at which different effects may occur

necrotic foliar symptoms). This knowledge led to studies to identify cause–effect relationships between air pollutants and vegetation for foliar injury, and to attempt to describe this relationship in terms of the intensity of foliar injury. The next logical step in the cause–effect relationship was to understand the impact of air pollutants on plant growth and yield as a function of pollutant dose (concentration, duration, frequency of exposure). The determination of injury, growth and yield are empirical responses that permit some understanding of the nature and magnitude of the effects; thus, hypotheses could be developed on the mechanism of action. The empirical approach has shown the practical significance of air-pollution effects on an organ and whole-plant level.

Following these studies at the organ and whole-plant level, research emphasis moved to both simpler (cellular and subcellular) and more complex (species and community) levels of biological organization. Such studies are necessary for a complete understanding of cause–effect relationships but they are more difficult, and less sure of producing interpretable results. However, these results will support the development of mechanistic models that permit projection of results to plant systems that have not been intensively studied.

TYPES OF EFFECTS TO STUDY

Before determining which response to study, the primary objectives of air-pollution research should be reviewed. The ultimate goals are to understand and quantify effects in natural and managed biological systems. At present, only visible foliar injury can be identified conclusively as the result of ambient pollution under field conditions. Some research workers believe that foliar injury cannot be quantified and that it has little relation to more important plant responses such as biomass or yield. Such an attitude is detrimental to the identification and understanding of the effects of air pollution on plants. This is not to say that foliar injury should be the primary response of interest, except for certain types of preliminary studies. However, it should always be one of the responses determined because it is a visible manifestation that can be related to specific air pollutants, and can be used to identify potential areas of concern. The following major responses should be considered whenever air-pollution research is planned. Those chosen must be related to the research objectives.

1. *Visible foliar injury* The presence or absence of injury should always be determined, together with the initial site of injury on the leaf. Injury can be quantified and objective if it is carefully determined. The lack of classic injury symptoms should not be confused with the absence of visible foliar injury. We know that chronic injury can take many forms, including the apparent early senescence of leaf tissue. The researcher should make observations during the entire life cycle of the test plants, and take care before deciding that the results of a particular experiment do not show visible symptoms.
2. *Biomass* Changes in biomass are critical to assessment of long-term effects of pollutants on plants. Biomass measurements permit studies

of growth rates and correlations with injury, yield or other characteristics studied. They may be related to photosynthetic studies and are essential for many types of modelling.

3. *Yield*   Changes in yield, defined as the plant product of economic interest to man, is the final punch-line that is the overriding concern of the agriculturalist or the economist. Yield data can be used in the development of simple empirical models to assess the impact of air pollutants over a large region (Heagle and Heck, 1980). Implicit in the concept of yield is the quality of the plant product.

4. *Physiological processes*   These include gas-exchange phenomena (pollutant uptake, photosynthesis, respiration, transpiration), water relations, translocation, exudation, nitrogen fixation, etc. They are often determined over time and expressed as rates. These rate functions are essential in the development of certain types of models. They can often be studied as whole-plant processes and should be studied in this way whenever possible.

5. *Biochemical processes (reactions)*   These can involve both *in vitro* and *in vivo* studies but the thrust of such studies must be the latter and eventually must be shown in the intact plant. Many of these studies will also develop rate functions that can be used in process-model development.

These are the basic types of effects that must be understood in order to permit a wider application of data on pollution effects. All specific research

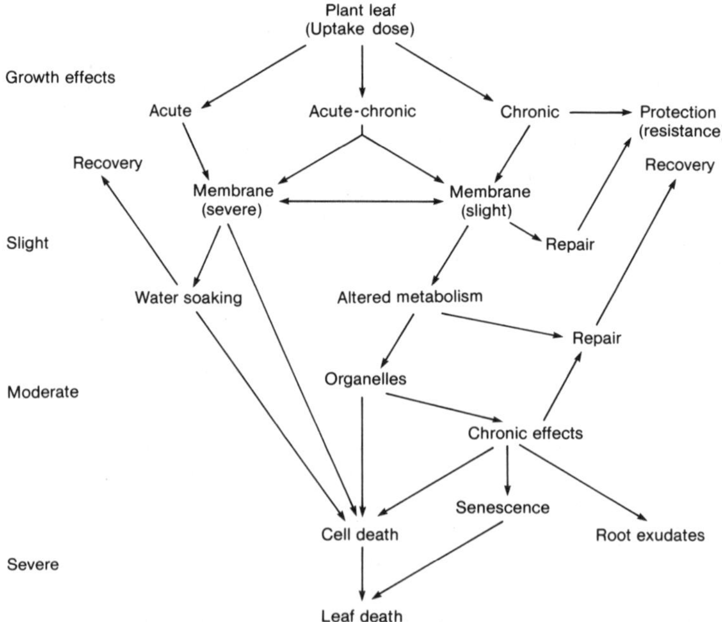

Figure 19.5   Functional relationships after the uptake dose enters the leaf. The uptake dose is that which enters the plant and sets off chemical reactions. This does not necessarily mean that measurable responses will be found. The level of expected effects on growth (left side) is associated with the degree of functional impairment: slight = < 10%, moderate = 10–20%, severe = > 20%)

can be studied in relation to these major types of effects. The effects can be studied using one or more of the approaches discussed earlier. *Figure 19.5* shows the functional relationships within the leaf that can be associated with different levels of an uptake dose of a given pollutant(s). It suggests that membranes are the initial site of reaction, that most other effects are secondary and that a different mechanism may exist for acute and chronic responses. Possible growth effects are shown to the left (*Figure 19.5*). The level of growth effect relates more to chronic than to acute effects. The levels should shift down for acute effects. *Table 19.2* lists the processes that

Table 19.2 PROCESSES AND OTHER FACTORS
THAT MAY BE AFFECTED BY AIR POLLUTANTS

Abiotic	*Population*
Precipitation	Mortality
Solar radiation	Abundance
Temperature	Biomass
Humidity	
	Community
Organisms	Diversity
Disease resistance	Dominance
Phenology	Spacial distribution
Nutrient uptake	
Nitrogen fixation	*Ecosystem*
Reproduction	Decomposition
Transpiration	Production
Photosynthesis	Water cycle
Respiration	Nutrient cycle
Translocation	Food chains, webs
Senescence	Resiliency
Decomposition	Resistance
Abscission	Toxins
Growth	Accumulation
Longevity	

can be affected by air pollutants; they should be considered in experimental designs.

It is not possible to study all pollutant–crop–environmental combinations. Thus, it is crucial that we conduct mechanistic studies that will eventually permit the application of results in a wider context.

Future research direction: priorities and management

The earlier sections have developed the basis for this section on recommendations for future directions in air-pollution research. Facility needs, approaches to research and types of effects to study have been discussed, but not details of specific research. General research priorities are developed here without actually making a list of specific research needs. A critique of current research management in air-pollution research is presented, with a plea for a search for new management approaches.

RESEARCH PRIORITIES

It is impossible to develop a simple list of priorities for research because too many factors derived from local, regional, or national needs are involved in deciding what should be done. However, a scheme for establishing research priorities, based on the objectives and goals of specific projects, is shown in *Table 19.3*.

Table 19.3 cannot be used to develop an empirical approach by administrators responsible for funding research. Each major heading has a list of priorities in research needs, that is based on a general need to know as much as possible as quickly as possible. However, every item listed in *Table 19.3* is an essential element of research for the next decade. The pollutants are listed in order of priority for the USA, but another order might be necessary for another country, or when specific problems are to be investigated. Dose–response should be an integral part of any research, and modelling concepts should be used as a basis on which all research should be planned. The varietal comparison should be used in studies to understand the mechanisms of response to a given pollutant(s). For the pollutants of regional concern, it is not possible to use the natural pollution loading; the controlled field exposures are therefore critical. For point-source problems the natural system may be preferable, if other factors are understood and not too variable. To develop mechanistic mathematical models, and to understand certain processes and reaction mechanisms, the controlled-environment facilities are essential. Greenhouse research is often a necessary adjunct for air-pollution studies.

Research designs should always consider obtaining at least some of the types of data shown in *Table 19.3*. Rate functions cannot be developed when data are taken only once. However, these types of data do require less manpower and are critical for short-term needs. Process studies are an integral part of the development of any mechanistic model and are essential if we are to understand how the pollutants affect plants.

All studies will be either empirical, mechanistic or a combination of the two. Field studies, by their very nature, will involve interactions but the research may not be designed to study these interactions. Greenhouse studies are similar but they may be designed to give some understanding of the interaction(s). Controlled-environment studies may be specifically designed either to remove the interactions or to study the interactions. Results of controlled-environment studies will help research workers to understand the potential confounding of results from field studies, caused by environmental factors.

The special studies are included but are generally of lower priority. For specific objectives and under certain circumstances these areas would take on added importance and should be studied.

1. *Biomonitoring* Basic studies on biomonitoring are essential if this area is to serve any long-range goals. At present, biomonitoring merely informs us of the presence of pollution and gives us some idea of the degree of impact of the pollutant on the biomonitor. Unless the relationship between effects on the monitor and effects (e.g. yield) on species of importance to man can be shown, the monitoring system will

Table 19.3 A SYSTEM FOR DEVELOPING RESEARCH PRIORITIES[1]

Pollutants[2]	Research approaches Approaches[3]	Facilities[4]	Data types[5] Single measures	Processes	Degree of understanding[6]	Interacting factors	Special studies
O_3	Modelling	Field, controlled ambient	Injury	Growth rates	Empirical	Pollutants	Biological monitoring
SO_2	Dose–response	Greenhouse	Biomass	Uptake, flux	Mechanistic	Climatic	Protection
NO_2	Sensitive–resistant	Controlled	Yield, quantity quality	Photosynthesis		Edaphic	Reproduction
Acidic precipitation				Respiration		Biotic	Evolution
HF			Biochemical	Biochemical			
CO_2				Other			
Others							

[1]Each list is generally in a priority order. The research objective(s) would set the specific priorities. All areas listed are in need of research effort during the coming decade.

[2]The pollutants are listed in priority order for the USA. This order would change depending upon the country and the specific problem. The 'others' category would include all specific source problems, agricultural chemicals, etc.

[3]The acute and/or chronic exposure designs are not listed because all research involves one or both of these exposure regimes.

[4]The field work should be stressed, but for some objectives the greenhouse or controlled environments must be used. Cost, time and the lack of an acceptable research approach may exclude ambient air studies.

[5]This is not meant to be an exhaustive list but it does cover the more important types of data.

[6]If the objective is to understand the system (mechanistic), this will change the priority ranking of other parts of the system.

be of little value except as a means of alerting the public to the presence of pollution. Eventually the monitor should be able to predict yield effects.

2. *Protection* Develop ways to protect sensitive species or selected varieties from pollution effects.

3. *Reproduction* Studies should focus on the direct effects of air pollutants on the reproductive structures (e.g. pollen [Feder, 1968]). This should include possible effects on the photoperiodic response of plants.

4. *Evolution* Studies should include the development of resistance in natural populations caused by selective pollution pressures (Roberts, Horsman and Bradshaw, 1980; Roose, Roberts and Bradshaw, this volume, Chapter 18) and the possible genetic effects of pollutants on plants exposed over several generations (Bruton, 1974).

Whenever possible, experimental designs should consider the direct economic effects of air pollutants on crop production. Cooperative work with economists should be encouraged, to consider both the direct and indirect losses in crop production.

For more discussion on specific research needs, Heck, Taylor and Heggestad (1973) should be consulted.

RESEARCH MANAGEMENT

Some changes in the management of air-pollution research are essential, if we are to progress in our understanding of air pollution and its effects on agricultural production. Two major concerns with present management practices (agency responsibility and funding methods) are considered and a suggested alternative management scheme is proposed.

Agency responsibility

As far as research into air pollution is concerned, the trend of governments is to place the research with the agency responsible for regulating air pollutants (in the USA this is the Environmental Protection Agency (EPA)). In the USA, some of the research responsibility has also been given to the Department of Energy (DOE), whose major mission is energy development. The DOE was given some responsibility in air-pollution research to permit them to determine the environmental impact of new energy technologies. We need to consider seriously the advisability of keeping research on the effects of air pollutants in regulatory agencies or in agencies whose primary mission is contrary to this type of research.

Initially it seems appropriate to ask the regulatory agency to carry out relevant research, in order to assure a rapid response to agency needs. This need for a rapid research response and the regulatory mission give some research workers grave doubts about the advisability of continuing the research arm of regulatory agencies. Regulatory agencies usually exist because the government and the public feel that industry has not been sufficiently sensitive to the public welfare. Thus, to show a need for continued existence, the agencies tend to look only at the negative aspects of effects. This is understandable and laudatory when the missions are understood. But the need to focus on negative aspects raises a serious conflict of interest with objective science. Thus, the regulatory agencies, historically and currently, support research aimed at furthering their mission of supporting regulatory controls and have not funded research that might show no impact or even positive effects. Other research agencies usually refuse to fund extramural research on the effects of air pollution, assuming that it is being done through the regulatory agency. This often results in biased research, directed at showing only deleterious effects. The bias of DOE may be the reverse, because their primary mission is the development of energy technologies and the environmental concerns are bound to be secondary.

The regulatory agency has to be responsive to unexpected occurrences that often rapidly alter their priorities. In most cases these changes alter the emphasis of the research programmes and much of the 'research' may become assessment and 'fire-fighting' in nature. Thus, the research often results in short-term projects to answer immediate needs. Usually the agency does not give long-term support to obtain answers that are critically needed for rational regulatory policies. This dichotomy of need is real and must be accepted. However, it has resulted in a lack of long-term research commitment, so that necessary research has not progressed as far as it should in the last decade. This situation could continue through the decade of the 1980s if research management stays with the regulatory or energy-development agencies.

Research management for air pollution should be transferred to those agencies, the primary mission of which is research, and not regulation. It should be done in such a way that close cooperation is ensured between research groups and agencies involved in the air-pollution system (*Figure 19.1*). Many believed that this could be done better if all parts of the system were within a single agency, as is currently the case in many countries. However, the conflict of interest between regulation and research, and the final responsibility for research direction by nonbiologists, have shown that keeping all parts of the air-pollution system within a single agency has not worked. Thus, the research management and regulatory functions should be placed so that one does not exert direct control over the other. This would permit more open development of research and the completion of long-term research projects that would support air-pollution regulations. The regulatory agency still needs the ability to carry out rapid research, and this should be associated with a research group, but this 'action group' should not have the responsibility of developing and directing research. This recommendation contains one problem that requires a solution. Many agricultural research agencies in different countries have not been willing to accept research on the effects of air pollution as part of their mission. This feeling is slowly changing in the USA, where the US Department of Agriculture is starting to accept this facet of their environmental responsibility.

Funding methods

A major departure from conventional research support is essential if we are to answer many of our current questions on the effects of air pollutants on crop production. It is apparent that the traditional grant–contract funding approach has not been successful in supporting air-pollution research. This is true in the United States, as well as in other countries where such research has been funded. In our attempt to be 'fair' and 'impartial', funds have been put into numerous programmes that have not had the critical mass (of personnel or funding) to develop the research necessary to answer questions on air pollution. Unless this approach is changed, we may open the decade of the 1990s little further advanced in our understanding than we are today.

Another problem with the grants and contracts is in the use of scientific time and research continuity. Several issues are included in this: the basic inequalities of 'peer'-reviewed grant and contract proposals; the lost scientific talents consumed in the frustrations of 'grantsmanship'; the relegation of the actual research work to younger staff and technicians; and, in the resulting attempt to survive, an inordinate amount of time spent on investigating where research has been, with correspondingly less time to spend on looking at where research should be going.

In order to answer problems concerned with air pollution, funding agencies should join hands with leading research groups to develop strong interdisciplinary approaches to research. Initial evidence of such approaches can be seen in the United Kingdom (UK) and the United States. The Natural Environment Research Council in the UK has a liaising Committee on Air Pollution Effects Research (CAPER) that could function in this way but still functions through the grants–contract route. (However, 'grantsmanship' is not as highly developed in the UK, and research support appears to rest more on scientific merit. In addition, those who are engaged in contract research often understand the need for research on long-term effects.) The NERC may use a sole-source approach by relying on the scientific expertise, persistence and facilities of certain groups. In the USA, the EPA has two programmes that are moving towards the use of leading research groups to develop strong interdisciplinary approaches to research on air-pollution effects. The EPA–Acid Precipitation Programme has placed research management with groups currently involved in research in acid precipitation. The EPA National Crop Loss Assessment Network (NCLAN) also involves several major laboratories that are cooperatively responsible for research direction, management and accomplishment. Other agencies within the USA are considering similar approaches. The NCLAN programme is being developed around the management system discussed below.

Research management system

Figure 19.6 is a stylized diagram showing a research management protocol that could be adopted as an approach to research into air pollution. Such an approach is important if we are really to understand the effects of air pollution on crop production.

The agencies with air-pollution research missions should identify the research needs, should develop cooperatively the research management committee (RMC) and assign co-project monitors (if agency funding

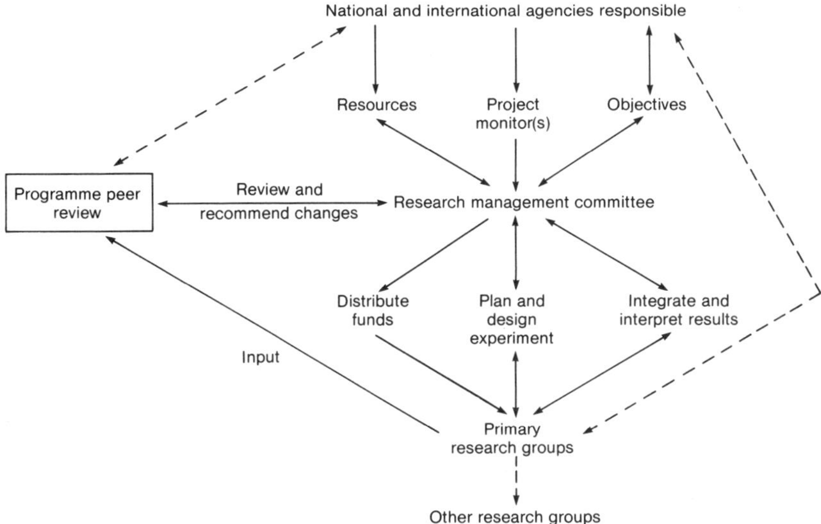

Figure 19.6 *Research management system:* the research management committee (RMC) is composed of representatives from the primary research groups asked to participate in this programme. *Primary research groups:* these research groups will perform the research, analyse and report results, and recommend additional research. They may ask the other research groups to work cooperatively with them on special projects

justifies such an assignment) who should be liaison members of the RMC. The RMC would be responsible for a full assessment of the problem area, development of the research approach, identification of additional research groups, and recommendations for distribution of the resources. These responsibilities should be carried out in close cooperation with the project monitor(s). The RMC would ultimately be responsible for integrating and interpreting the research results through working with the various research groups and the individual investigators. The research groups would perform the research, and analyse and publish their results. The programme peer review group would be composed of scientists outside the main line of the research programme. Their primary role would be to review the overall programme, assess the strengths and weaknesses, and recommend continuation or change of direction.

The RMC would be composed of senior scientists from the primary research groups that are asked to participate in the research programme. The groups chosen should have recognized ability in the area of research and should have sufficient talent and facilities available to cooperate in the research. The research groups should recognize the importance of international cooperation, should maintain open laboratories and should encourage both national and international visiting scientists to participate in their programme.

The development of such a management system could largely overcome many of the existing problems (agency responsibilities and funding methods) relating to research into the effects of air pollution on agricultural systems. However, these problems do exist and agency responsibilities should be changed.

References

BRUTON, C. (1974). *Environmental Influences on the Growth of* Arabidopsis thaliana. PhD thesis, North Carolina State University, Raleigh, NC

FEDER, W.A. (1968). *Science,* **160**, 1122

GALLOWAY, J.N., COWLING, E.G., GORHAM, E. and McFEE, W.W. (1978). *A National Program for Assessing the Problem of Atmospheric Deposition.* A report to the Council on Environmental Quality. National Atmospheric Deposition Program, National Resource Ecology Laboratory, Colorado State University, Fort Collins, Colorado

GUDERIAN, R. (1977). *Air Pollution: Phytotoxicity of Acid Gases and Its Significance in Air Pollution Control.* Ecological Studies 22. Springer-Verlag, Berlin

HEAGLE, A.S. and HECK, W.W. (1980). In *Proceedings of the E.C. Stakman Commemorative Symposium—Assessment of Losses Which Constrain Production and Crop Improvement in Agriculture and Forestry,* pp. 296–305. University of Minnesota, Misc. Publ. 7-1980, Agric. Exp. Sta./St Paul

HEAGLE, A.S., PHILBECK, R.B., ROGERS, H.H. and LETCHWORTH, M.B. (1979). *Phytopathology,* **69**, 15–20

HECK, W.W. (1973). In *Air Pollution Damage to Vegetation,* pp. 118–129 (Naegele, J.A., Ed.). Adv. Chem. Series 122. American Chemical Society, Washington, DC

HECK, W.W. and BRANDT, C.S. (1977). In *Air Pollution,* 3rd Edition, Vol. 2b, pp. 157–229 (Stern, A.C., Ed.), Academic Press, New York

HECK, W.W., KRUPA, S.V. and LINZON, S.N., Eds (1979). *Handbook on Methodology for the Assessment of Air Pollution Effects on Vegetation.* Air Pollution Control Association, Pittsburg, Pa

HECK, W.W., LARSEN, R.I. and HEAGLE, A.S. (1980). In *Proceedings of the E.C. Stakman Commemorative Symposium—Assessment of Losses Which Constrain Production and Crop Improvement in Agriculture and Forestry,* pp. 32–49. University of Minnesota, Misc. Publ. 7-1980, Agric. Exp. Sta./St Paul

HECK, W.W., MUDD, J.B. and MILLER, P.R. (1977). In *Ozone and Other Photochemical Oxidants,* pp. 437–585. National Academy of Sciences, Washington, DC

HECK, W.W., PHILBECK, R.B. and DUNNING, J.A. (1978). *A Continuous Stirred Tank Reactor (CSTR) System for Exposing Plants to Gaseous or Vapour Contaminants: Theory, Specifications, Construction, and Operation.* Series No. ARS-S-181. Agricultural Research Service

HECK, W.W., TAYLOR, O.C. and HEGGESTAD, H.E. (1973). *Journal of the Air Pollution Control Association,* **23**, 257–266

KRUPA, S.V. and LEGGE, A.H., Eds (1982). *Proceedings of an International Conference on Air Pollutants and their Effects on the Terrestrial Ecosystem.* Wiley, New York

MILLER, J.E. (1979). *An Open-air Fumigation System for Studies of SO_2 Effects on Crops.* Argonne National Laboratory, RERD, ERC-79-21

MUDD, J.B. and KOZLOWSKI, T.T., Eds (1975). *Response of Plants to Air Pollutants.* Academic Press, New York

NATIONAL RESEARCH COUNCIL (1977a). *Ozone and Other Photochemical*

Oxidants. Committee on Medical and Biological Effects of Environmental Pollutants. National Academy of Sciences, Washington, DC

NATIONAL RESEARCH COUNCIL (1977b). *Nitrogen Oxides*. Committee on Medical and Biological Effects of Environmental Pollutants. National Academy of Sciences, Washington, DC

NATIONAL RESEARCH COUNCIL (1979). *Sulfur Oxides*. Committee on Medical and Biological Effects of Environmental Pollutants. National Academy of Sciences, Washington, DC

NRIAGU, J.O., Ed. (1978). *Sulfur in The Environment: Part 1. The Atmospheric Cycle*. Wiley, New York

OSHIMA, R.J., POE, M.P., BRAEGELMANN, M.K., BALDWIN, D.W. and VAN WAY, V. (1976). *Journal of the Air Pollution Control Association*, **26**, 861–865

ROBERTS, T.M., HORSMAN, D.C. and BRADSHAW, A.D. (1980). In *Symposium on the Effects of Air-borne Pollution on Vegetation*, pp. 273–275. Economic Commission for Europe, Warsaw, Poland

SHINN, J.H., CLEGG, B.R. and STUART, M.L. (1977). *A Linear-Gradient Chamber for Exposing Field Plants to Controlled Levels of Air Pollutants*. Preprint UCRL-80411. Lawrence Livermore Laboratory, Livermore, California

20

TOWARDS AN UNDERSTANDING OF PLANT RESPONSES TO POLLUTANTS

F.T. LAST
*Institute of Terrestrial Ecology, Bush Estate, Penicuik, Midlothian,
Scotland*

Perspective

The meeting described in this volume contrasted sharply with that held
earlier in the year at Sandefjord, Norway (*see footnote*). Instead of dealing
with the ecological impact of acid precipitation and therefore being
primarily concerned with *particulate* pollutants, their removal by *wet
deposition* and their mainly *indirect effects* on freshwater, forest and other
natural/semi-natural terrestrial ecosystems, it focused on *gaseous* pollutants, their removal from the atmosphere by *dry deposition* and their
mainly *direct effects* on agricultural and horticultural crops. In short, the
two meetings had their separate identities, but none the less it should be
recognized that areas with gaseous pollutants are also usually subject to
acid precipitation and vice versa—the relative importance is a matter of
degree, with particulate SO_4^{2-} being of increasing importance as distances
from sources of SO_2 emissions increase, a change associated with the more
complete conversion (oxidation) of SO_2 to SO_4^{2-} as residence periods
lengthen.

The problems associated with acid precipitation, attributed to particulate pollutants, are mainly located in some of the areas where soils are
formed from slowly weathering granitic, porphyritic and gneiss bedrocks,
as in southern Norway, south west Scotland and north east USA (Last *et
al.*, 1980). At these locations the noticeable loss of fish, in particular
salmonids, seems to be causally related to the occurrence of toxic
concentrations of aluminium which are found increasingly as lakes change
from bicarbonate to strong acid/aluminium buffering systems (Henriksen,
1980); the cause(s) of concomitant changes in the assemblages of aquatic
plants and invertebrates remain unexplained. For a time it was thought
that the growth of trees had been adversely affected but it is now generally
agreed that the evidence does not warrant such a conclusion, although it
seems that acid precipitation can recognizably alter rates of some soil
processes which might be expected to affect plant growth adversely
(Abrahamsen, 1980).

Ecological impact of acid precipitation. The proceedings of this international conference at
Sandefjord, Norway from 11–14 March 1980 (Eds. Drabløs, D. and Tollan, A.) were
published (1980) by SNSF project, P.O. Box 61, 1432 Ås-NLH, Norway.

Rather surprisingly, the investigations associated with acid precipitation have compelled us to recognize that we still know far too little about the interrelations between soil properties and processes on the one hand and plant production on the other (Tamm, 1976). To what extent must soil properties be altered before growth is affected? Perhaps we should remind ourselves that agricultural soils are more sensitive to acid inputs than many forest soils. But, of course, acid pollutant inputs are relatively small compared with the acidifying effects of fertilizers, the more abundant use of which in agriculture has traditionally been offset by the addition of different forms of lime (*see* Bache, 1980). Had this not been so, more of our time (at the Conference described in this book) would have been devoted to effects on soil processes, cation exchange, etc.

There is some evidence to suggest that acidity has deleterious effects on foliage: Ferguson, Lee and Bell (1978) have presented strong circumstantial evidence indicating that the demise of many species of *Sphagnum* in the Pennine region of the UK can be attributed to bisulphite ions occurring in rain. Although reference is rarely made to the occurrence of bisulphite ions when discussing acid precipitation, largely because problems occur at sites remote from emission sources and where prolonged 'residence times' will have allowed the oxidation of bisulphite to sulphate, we should not exclude the possibility that damage done to foliage by precipitation, including possibly undesirable leaching and cuticle erosion, may exacerbate the effects of gaseous pollutants.

End-of-term report

I attended the Conference because I am attracted by the scientific fascinations of pollution research, also because I help to allocate research resources and am a member of society concerned with the activities of man on his environment. When listening to the different contributions I was, therefore, making assessments of different sorts and in doing so I was made aware of the contrast between pollution research concerned with the health of agricultural animals and that focused on the damage done to crop plants. Possibly because of the stimulus provided by obvious discomfiture and by the political pressures exerted by the 'dramatic' incidence of lameness in cattle, a considerable amount is known about the field incidence of fluorosis when herbage polluted with fluoride is grazed (*see* Davison, Blakemore and Craggs, 1979). While a similar amount may be known of the effects of atmospheric pollutants on the development of *individual plants*, or small assemblages of plants, much less is known of the effects on large populations of plants, *crops*. Regrettably, few effects studies have any direct relevance to conditions in the field. But this is not to suggest that the studies themselves lacked precision. Were experiments being done to elucidate mechanisms of pollutant damage or to determine likely yield losses? These are two perfectly valid, but not necessarily compatible, aims. I would very seriously question the purpose of many investigations, linking my enquiry with an examination of amounts of fumigant(s) and periods of exposure. Judging by the usually unrealistically prolonged exposure to large and unchanging concentrations, I am led to the conclusion that most experiments have been concerned with mechanisms. It could reasonably be argued, even when making allowances for technical difficulties, that this is putting the cart before the horse. Is it unreasonable for administrators to expect to have been given some idea of the damage done in the field by current concentrations of pollutants which, in many instances, because of the introduction in many countries of a variety of 'clean air acts' enforcing air-quality standards, are likely to be appreciably smaller than in earlier years?

To help develop my appraisal of our knowledge of gaseous pollutants, I thought I would subject my colleagues to an end-of-term assessment similar in format to school reports given by teachers, who know they cannot be called to account for their usually succinct and often derogatory statements until the start of the next term—a position of privilege. I have decided to examine three aspects:

1. *Defining the polluted atmosphere.* Assessment—*improving.*
2. *Physiological and biochemical responses to pollutants.* Assessment—*strong.*
3. *Air pollutants and the growth and quality of crops* (a subhead also including a consideration of interactions). Assessment—*good in parts, but usually weak.*

I now owe it to my colleagues to proffer some explanation of my terse, yet carefully considered, generalizations.

Defining the polluted atmosphere

There is now sufficient evidence to indicate that atmospheric pollutants occur mostly in mixtures and that the different components of the mixtures have their own distinctive patterns of diurnal and seasonal changes (Fowler and Cape, this volume, Chapter 1). These changes may, in some circumstances, be synchronous for considerable periods e.g. winter concentrations of SO_2 and NO_x near Edinburgh (Nicholson *et al.*, 1980). If realistic estimates of the effects of pollutants on crops are to be made, and if we hope to make predictions from relatively few experiments, then we must understand and relate the changing loads of pollutants to crop phenology. By analogy with the changing responses of ageing plants to attack by pathogens, a discussion of the effects of annual mean concentrations of pollutants on the development of winter-sown cereals is unlikely to be very instructive, knowing that concentrations of pollutants are predictably likely to be larger in winter, before ear emergence, than during spring and summer when ears and grain develop. Is the damage done to deciduous perennial plants, such as apples, attributable to the influence of winter mixtures of pollutants on bud development and/or to that of spring and summer mixtures, which may differ in composition and concentration (probably including more ozone), on leaf, fruit and shoot development? It is necessary to monitor the changing mixtures of atmospheric pollutants at many more locations: the principles have been established—details are now required.

Is it possible to assess loads of atmospheric pollutants without recourse to expensive equipment needing careful maintenance? Do biological indicators have a part to play and can they be used to help predict crop losses? As yet I think the answer is 'I don't know'. Whereas I can envisage that, in regions with different mixtures of pollutants, the different assemblages of lichen species (Hawksworth and Rose, 1976), and the occurrence of tar spot of sycamore *Acer pseudoplatanus* caused by *Rhytisma acerinum* (Bevan and Greenhalgh, 1976) and black spot of roses caused by *Diplocarpon rosae* (Saunders, 1966), may be loosely correlated with crop performance, I am not persuaded that the use of sensitive indicators which develop blemishes will materially help, except to indicate that the concentration of a particular pollutant sometimes exceeded a threshold related to the sensitivity of that indicator. Biological indicators can pinpoint a pollution episode of a particular intensity, whereas lichens and moss bags can integrate events over protracted periods. Whatever the position, use of indicators would add another stage in the process of yield-loss assessment. Not only would it be necessary to relate concentrations of atmospheric pollutants to the damage done to indicator plants, but also to associate the latter with the damage done to crops. But, do we know if the responses of indicator plants sensitive to ozone, NO_x or other substances, are modified by previous, concurrent and/or subsequent exposure to other pollutants? Would the responses of crop plants be similar or different? Although I am sure that ambiguities will have been eliminated from this book I suspect that many of us, perhaps unthinkingly, still assume that the relative sensitivities of different plants to concentrations of pollutants causing *chronic* damage are similar to those sensitivities to concentrations associated with *acute* damage. There is no justification for this assumption (Bell, this volume, Chapter 11).

The resistance analogue, including aerodynamic and surface components, was evolved to formalize the factors influencing the transfer of gases to plants and other surfaces. We now know something about the magnitude of the different resistances to the deposition of SO_2, including those attributable to stomata, cuticle and surface films of moisture (Fowler and Unsworth, 1979) but virtually nothing of their relative importance to the transfer of ozone, NO_x etc. This deficiency should be corrected.

Our improved understanding of the mechanisms involved in the dry deposition of pollutants has suggested a reason for the many apparently conflicting results of effects studies when plants were exposed, in chambers, to similar atmospheric concentrations of the same pollutant (Unsworth and Mansfield, 1980). As found by Ashenden and Mansfield (1977), the effects of pollutants are dependent upon leaf boundary-layer resistances, which largely reflect air movement. It seems that pollutants are less damaging to plants where the degree of atmospheric turbulence is insufficient to prevent the build-up of significant leaf boundary-layer resistances—a hazard that needs to be taken into account when devising closed fumigation chambers, but which is of little or no concern in field conditions. At its simplest, 'dose' should reflect the concentration and the duration of exposure (fumigation), in other words:

$$\text{Dose} = \text{Concentration} \times \text{Time (or } \int \text{(concentration)}dt)$$

But how should 'concentration' be defined and modified to take account of boundary resistances and, more important in most instances, periods of rapid dry deposition when stomatal resistances are small? For this purpose it is attractive to think of including a term describing canopy or stomatal conductance, always remembering that pollutants sometimes cause stomata to open, as happened when beans (*Vicia faba*) were exposed to SO_2 (Unsworth, Biscoe and Pinckney, 1972).

Physiological and biochemical responses to pollutants

To me, this is arguably the most straightforward part of pollution research. Much of the physiological and biochemical work has been aligned to very clear sets of objectives, whether attempting to synthesize the effects of pollutants from first principles, considering the adsorption of SO_2 and other pollutants by membranes or, starting from the other extreme, by identifying specific enzymes that are impaired when leaves are fumigated. Thirty years ago it seemed that much of the biochemical work was concerned with exceptional concentrations of pollutants which bore no relation to reality. However, despite technical difficulties, commendable attempts are now being made to experiment with concentrations typical of those occurring in the field. None the less, I am tempted to question whether the effects of SO_2, NO_x etc. should be tested against control plants exposed in pollution-free air or in ambient atmospheres from which the pollutant in question was excluded—a matter of objectives! More cynically, I wonder how long it will be before it is possible to extrapolate from events on membranes to potential losses of yield. But, realistically, I suspect we should be aiming at a better understanding of the detoxification of possibly obnoxious substances, gaining at the same time a better understanding of sulphur metabolism, nitrogen utilization, etc. In doing so, derangements may be identified that may prove, in the long term, to be useful indicators of incipient macroscopic damage (Malcolm and Garforth, 1977).

Air pollutants and the growth and quality of crops

Pathologists are usually confronted with a diseased plant or animal and are expected to find the cause of the derangement. This approach suited our understanding of the pollution problem, as related to plants, up to 30 years ago, when concentrations causing acute damage were still widespread. However with (1) the introduction of emission technologies that lessen the concentrations of pollutants and (2) the acceptance that yield losses can occur without blemishes, the nature of the problem has greatly changed; however, have we, in attempting to identify the apparently intangible, redefined our objectives? By definition, chronic damage is likely to have a relatively small effect on yields and is therefore likely to prove difficult to substantiate. Thus, are we concerned with the events in the relatively restricted areas of intense pollution, or should we be concerned, as I think we should, with events in the 60 million ha of North America and western Europe, where concentrations are smaller but still large enough possibly to cause chronic damage? At this stage I would like to draw an analogy to problems confronting those of my colleagues concerned with frost tolerance/sensitivity in trees. In the first instance, it is essential to obtain as much information as possible about the likely occurrence of frosts, their intensity and seasonality, in the locations for which frost-tolerant trees are required. With this information, it is then possible to determine the degree of selection that must be imposed in order to gain a worthwhile increase in frost tolerance, remembering that this must take into account different sensitivities at different stages of growth—bud break, shoot growth and approaching dormancy. Thus, risk and sensitivities should be considered in parallel. Reverting to the study of pollutants, it would therefore seem desirable to describe the mixtures of pollutants and their daily and longer-term changes in relation to crop phenology and, with this knowledge, to structure a series of experiments with different mixtures at different stages of growth, possibly enabling the detection of changes in crop sensitivity. While doubtless exaggerating, I gained the strong impression that many participants were unrealistically expecting to derive generalizations, from relatively few experiments, that could be applied to a greater range of crops. There is no reason *a priori* to expect different cultivars, let alone plant species, to behave in a similar manner. Pathologists learnt many years ago that each host/pathogen combination had to be examined separately and that it was desirable to make a series of sequential analytical observations, recognizing that observations restricted to one stage of cropping were likely to be of little value. Research workers concerned with pollutants must follow this example and embrace the concept of *'epidemiology'*. However, this, in most instances, will necessitate an intensification of effort with, as a result, fewer but more detailed experiments being done. I am sure that such experiments would aid the prediction of effects. While referring to a greater intensity of effort in fewer experiments, I would like to refer to the suggestion, made more than once during the Conference, that the results of fumigating individual plants might differ from those of fumigating swards. If there really is a difference, and if this difference is attributable to the selection of tolerant forms within the sward, and not to physical factors, then, surely, it is of sufficient importance to warrant careful investigation. In this connection I would like to argue against the ill-considered use of the words 'evolution' and 'resistant'. In the example just quoted, resistance would not have evolved; it already existed. Instead, the experimental conditions might have restricted the growth of sensitive seedlings and, in so doing, favoured the growth of tolerant specimens, the multiplication of which might therefore have been selectively enhanced. In other words, populations of tolerant plants may be 'built-up'. Now, turning to 'resistant': I doubt if resistant plants actually prevent (resist) the entry of pollutants. Instead they are, after entry, able to tolerate these substances—I suspect that they, like other plants that withstand toxic concentrations of heavy metals, have 'tolerance'.

At this stage I should address myself to administrator colleagues primarily, to plead for patience. Experience suggests that it is foolhardy to generalize about effects of pathogens without evidence from four or more cropping seasons. It is then possible to take note of the ways in which weather affects seasonal yields and possibly, at the same time, plant responses to, in our instance, pollutants. For example, are plants which are growing at different rates equally sensitive?

At an early stage I indicated that I intended to include 'interactions' as part of the section concerned with effects of pollutants on the growth and quality of crops. Following the clear indication of the interacting effects of sulphur dioxide and ozone to cause the formation of needle necroses in *Pinus strobus*, eastern white pine (Dochinger and Heck, 1969; Dochinger *et al.*, 1970), much of the work has been done with unrealistically large concentrations of pollutants. There has been much discussion of the meaning of additive effects, synergistic effects etc. with attention being turned in recent years to the interplay between SO_2 and NO_x. In this connection, the results presented by Ashenden and Mansfield (1978) are particularly timely as they also illustrate the inherent dangers of ill-advised generalizations. Thus, the effects of equal mixtures of SO_2 and NO_2 on the dry weights of *Poa pratensis* were additive (SO_2, NO_2 and $SO_2 + NO_2$ decreased weights by 46, 38 and 84% respectively), whereas on *Lolium multiflorum* they were synergistic (the comparable losses being 5, 10 and 52%). Bearing in mind that nearly identical concentrations of SO_2 and NO_x occur concomitantly at some seasons at some locations, I wonder how much damage ascribed to SO_2 should have been attributed to mixtures of these two substances? Farrar, Relton and Rutter (1977) found that the occurrence of Scots pine (*Pinus sylvestris*) in the industrial Pennines of Britain was negatively correlated with mean winter atmospheric concentrations of SO_2. But would the outcome of their analysis have been the same had they been able to relate growth to the prevailing concentrations of NO_x, or for that matter to mixtures of SO_2 and NO_x? Undoubtedly, there is need to reconsider our approach to the assessment of damage done in field conditions, more particularly to the assignation of cause. There are many approaches that could be adopted but, bearing in mind the complexity of the atmospheric environment, there is much to be said in favour of the use of filtered and unfiltered atmospheres. These can be used with a variety of chambers, including open-top chambers, but the use of chambers inevitably introduces an element of artificiality. To a large extent it is a matter of degree. Open-top chambers have a number of defects, two of which must surely be amenable to correction. *First*, without making special provision, foliage is likely to be without moisture films for longer periods in chambers than in the field and, as a result, surface resistances are likely to be larger, with a consequent decrease in the deposition of, at least, SO_2; the resistance of wet leaves to deposition is minimal (Fowler and Unsworth, 1979). But, because periods of surface moisture overlap to a considerable extent with periods of darkness, when stomata are closed, some may argue that the decreased rates of deposition to plants in dry chambers are of little biological significance. However, Fowler *et al.* (1980) found that ambient mixtures of atmospheric pollutants accelerate the degradation of epicuticular wax structures on needles of Scots pine (*Pinus sylvestris*), an effect associated with decreased cuticular resistances to water loss. *Second*, air flows within chambers differ from those in the field, but surely this problem could be resolved with the help of ventilation engineers? Is it not possible to countenance modifications whereby experimental atmospheres flow down through crop canopies, unlike most systems in use at present where air, introduced at or near ground level, permeates upwards, possibly losing some of its load of pollutants before the uppermost leaves are reached? My support for the use of filtered and unfiltered atmospheres should not be interpreted as being at the expense of other systems of exposure but, unless concentrations of gases can be varied, they would seem to be of more value for the elucidation of mechanisms. It may be argued that field exposure systems can be used for obtaining dose–response relationships, but so could chambers supplied with different mixtures of filtered and unfiltered air. Doubtless there is more than one solution to the problem, but this should not be an excuse for delay. Do the ambient concentrations of pollutants experienced in N.E. America and Western Europe damage crop yields? Bearing in mind the possible implications for the abatement of emissions, if this is to be considered rationally we must work with a sense of urgency but not of haste. Every so often I hear mention of 'difficulties'. All too often a plea of difficulty expresses an inability or unwillingness to dissect a multifaceted problem into its series of usually simple steps.

The future

With notable exceptions, research on the effects of pollutants has meandered. It is time to introduce greater objectivity and thrust. At the end of the Conference I think we may legitimately be asked if we are now nearer to knowing about the effects of pollutants on field crops. Regrettably my answer can be only a qualified yes. Recalling that atmospheric concentrations of pollutants are predictably likely to be larger in rural areas near to conurbations than in those at a distance, that there is some evidence to suggest that winter growth is less tolerant of pollutants than growth made at other times of the year, and that horticultural crops tend to be concentrated around centres of populations, it is a little surprising to find very little research being done on winter/spring maturing horticultural crops. Would there be any merit in re-acquainting ourselves with the elegant work done by Professor J.B. Cohen and his associates in and around Leeds more than fifty years ago? When examining the growth of lettuces and other crops, grown at different locations, they found a good but not perfect correlation with freedom from pollutants (*see* Crowther and Ruston, 1911/12; Mansfield and Freer-Smith, 1981).

To further our knowledge of the effects of gaseous pollutants on agricultural and horticultural crops much more needs to be known about:

1. The diurnally and seasonally varying *mixtures of atmospheric pollutants*, with particular emphasis on concentrations occurring in rural localities;
2. The factors, physical and biological, controlling the deposition of pollutants on to plants, there now being a greater need of information about oxides of nitrogen and ozone than SO_2;
3. With (1) and (2) it should be possible to estimate *dose* (= concentration × time, atmospheric concentrations being modified to take account of stomatal conductances and leaf boundary-layer resistances).

The better understanding of the agents of injury gained from the above list should enable *relevant series* of *effects studies* to be structured. Some of these should be concerned with 'mechanisms' whereas others, forming a very important group, should attempt to identify effects on crop growth in conditions as near as possible to those in the field.

1. Effects in the field
   (i)   To facilitate extrapolations and predictions, an intensive series of observations should be made relating the varying concentrations of pollutants to crop phenology.
   (ii) Although the use of filtered and unfiltered atmospheres, and mixtures of filtered and unfiltered air, has its attractions, technological advances are increasingly enabling the use of injection systems with programmable systems of concentration control. Equal effort should be devoted to methods of environmental control to ensure that air flows and periods of surface moisture mimic those in the field.

2. Mechanisms   To ensure that the fullest use will be made of the relatively few field studies that can be afforded, it is essential to sustain parallel series of mechanism studies. Four aspects, whether at the biochemical or physiological level, need to be taken into consideration:
   (i)   Because there is some evidence to suggest that the relative sensitivities of different plant species to small concentrations of pollutants, causing chronic damage, differ from those to larger concentrations, causing acute damage, it is essential to clarify experimental objectives. Are experiments being done to study chronic or acute damage?
   (ii) Because pollutants usually occur in mixtures, studies of responses to mixtures need to be emphasized.
   (iii) Because field crops exposed to gaseous pollutants will also be subject to acid precipitation, the interplay between gaseous and particulate pollutants must be explored.
   (iv) Experiments are required to reveal whether plants and their tissues are equally sensitive/tolerant to pollutants at all stages of development.

References

ABRAHAMSEN, G. (1980). In *Ecological Impact of Acid Precipitation*, pp. 58–63. Proceedings of an International Conference, Sandefjord, Norway, 11–14 March 1980. (Drabløs, D. and Tollan, A., Eds). SNSF project 1432 Ås-NLH, Norway

ASHENDEN, T.W. and MANSFIELD, T.A. (1977). *Journal of Experimental Botany*, **28**, 729–735

ASHENDEN, T.W. and MANSFIELD, T.A. (1978). *Nature*, **273**, 142–143

BACHE, B.W. (1980). In *Effects of Acid Precipitation on Terrestrial Ecosystems*, pp. 183–202. (Hutchinson, T.C. and Havas, M., Eds). Plenum Press, New York

BEVAN, R.J. and GREENHALGH, G.N. (1976). *Environmental Pollution*, **10**, 271–285

CROWTHER, C. and RUSTON, A.G. (1911/12). *Journal of Agricultural Science*, **4**, 25–55

DAVISON, A.W., BLAKEMORE, J. and CRAGGS, C. (1979). *Environmental Pollution*, **20**, 279–296

DOCHINGER, L.S. and HECK, W.W. (1969). *Phytopathology*, **59**, 399

DOCHINGER, L.S., BENDER, F.W., FOX, F.L. and HECK, W.W. (1970). *Nature*, **225**, 476

FARRAR, J.F., RELTON, J. and RUTTER, A.J. (1977). *Environmental Pollution*, **14**, 63–68

FERGUSON, P., LEE, J.A. and BELL, J.N.B. (1978). *Environmental Pollution*, **16**, 151–162

FOWLER, D. and UNSWORTH, M.H. (1979). *Quarterly Journal of the Royal Meteorological Society*, **105**, 767–783

FOWLER, D., CAPE, J.N., NICHOLSON, I.A., KINNAIRD, J.W. and PATERSON, I.S. (1980). In *Ecological Impact of Acid Precipitation*, p. 146. Proceedings of an International Conference, Sandefjord, Norway, 11–14 March 1980. (Drabløs, D. and Tollan, A., Eds). SNSF project 1432 Ås-NLH, Norway

HAWKSWORTH, D.L. and ROSE, F. (1976). *Lichens as Pollution Monitors*. Studies in Biology no. 66, Institute of Biology. Edward Arnold, London

HENRIKSEN, A. (1980). In *Ecological Impact of Acid Precipitation*, p. 68–74. Proceedings of an International Conference, Sandefjord, Norway, 11–14 March 1980. (Drabløs, D. and Tollan, A., Eds). SNSF project 1432 Ås-NLH, Norway

LAST, F.T., LIKENS, G.E., ULRICH, B. and WALLØE, L. (1980). In *Ecological Impact of Acid Precipitation*, p. 10–12. Proceedings of an International Conference, Sandefjord, Norway, 11–14 March 1980. (Drabløs, D. and Tollan, A., Eds). SNSF project 1432 Ås-NLH, Norway

MALCOLM, D.C. and GARFORTH, M.F. (1977). *Plant and Soil*, **47**, 89–102

MANSFIELD, T.A. and FREER-SMITH, P.H. (1981). *Biological Reviews*, **56**, 343–368

NICHOLSON, I.A., FOWLER, D., PATERSON, I.S., CAPE, J.N. and KINNAIRD, J.W. (1980). In *Ecological Impact of Acid Precipitation*, p. 144–145. Proceedings of an International Conference, Sandefjord, Norway, 11–14 March 1980. (Drabløs, D. and Tollan, A., Eds). SNSF project 1432 Ås-NLH, Norway

SAUNDERS, P.J.W. (1966). *Annals of Applied Biology*, **58**, 103–114

TAMM, C.O. (1976). *Ambio*, **5**, 235–238

UNSWORTH, M.H. and MANSFIELD, T.A. (1980). *Environmental Pollution*, *(Series A)*, **23**, 115–120

UNSWORTH, M.H., BISCOE, P.V. and PINCKNEY, H.R. (1972). *Nature*, **239**, 458–459

VI

Poster Session Abstracts

CONCENTRATIONS OF SULPHUR DIOXIDE, OXIDES OF NITROGEN AND OZONE IN AIR AT RURAL SITES IN GREAT BRITAIN AND IRELAND

A. MARTIN

CEGB, Scientific Services, Ratcliffe on Soar, Nottingham

Contours of annual mean sulphur dioxide values for rural sites in Great Britain and Ireland were presented, predicted from a mathematical model using estimated sulphur emissions and dispersion parameters. Rural sites were defined as at least 100 m from a single house, 1 km from a village, 10 km from a town. The predicted values agree well with values measured by specific methods at 20 rural sites throughout the UK. The contours range from $3\,\mu g\,SO_2\,m^{-3}$ curving across north-west Scotland and the western headlands of Ireland, to $40\,\mu g\,SO_2\,m^{-3}$ encircling Liverpool to Sheffield, Sheffield to Birmingham, the London area. Further details, including estimated rates of dry and wet deposition of sulphur (contours of total deposition range between 15 and $50\,kg\,S\,ha^{-1}\,y^{-1}$) are given in Martin (1980).

Values of sulphur dioxide, nitric oxide, nitrogen dioxide and ozone measured simultaneously using specific methods have been published, and summer mean values were shown for four rural sites in Great Britain and Ireland. The summer mean values in ppb of the four gases were respectively:

7, 2, 6, 35 at Harwell, Oxfordshire (Cox *et al.*, 1976)
0.7, 0.1, 0.4, 45 at Adrigole, Ireland (Cox, 1977)
7, <2, 10, 31 at Heysham, Lancashire (Harrison and McCartney, 1980)
10, 6, 7, 29 at Bottesford, Leicestershire (Martin and Barber, 1981).

Similar results for Devilla, Scotland are given by Fowler and Cape (this volume, Chapter 1).

Graphical relationships between the different site summer mean values were shown, i.e.

$$SO_2 \text{ (ppb, summer)} = 0.74 + 0.70\,(NO + NO_2, \text{ ppb}), \, r = 0.96$$
$$O_3 \text{ (ppb, summer)} = 41 - 0.94\,(NO + NO_2, \text{ ppb}), \, r = 0.99$$

Average diurnal variations in the concentrations at Bottesford were shown (from Martin and Barber, 1981), for July 1979 and November 1978. For the summer months, on average, the peak values of SO_2 and NO occurred in the hour ending 10.00 GMT, for O_3 at 16.00 GMT, for NO_2 at

450

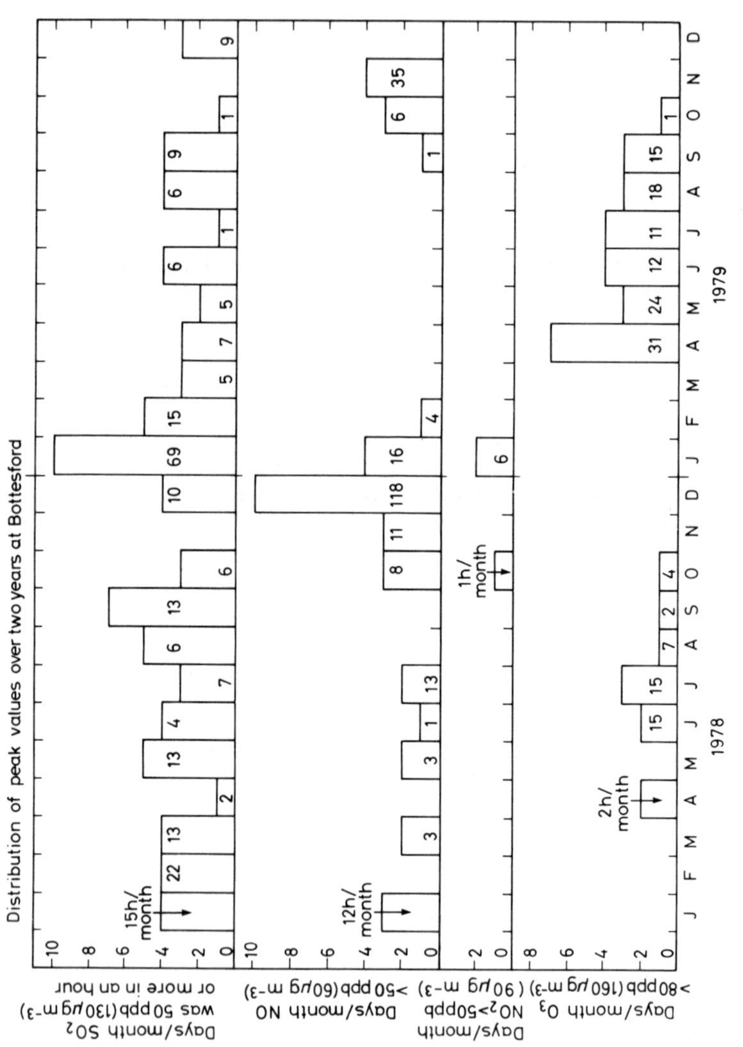

Figure 1   Distribution of peak values over two years at Bottesford

10.00 and 21.00 GMT with a trough between. For the winter month, the peak NO was at 12.00 GMT, peak O_3 at 13.00 GMT, and NO had a broad peak at 14.00–19.00 GMT. Further details and explanations are given by Martin and Barber (1981).

Figure 1 shows the distribution of peak hourly concentrations of the gases during two years' measurements at Bottesford. An hourly SO_2 concentration of 50 ppb or more occurred during almost every month, on average for 2 h during each of 3 days in any summer month and for 5 h during each of 3 days in any winter month (worst month, January). NO concentrations exceeding 50 ppb per hour occurred only during 6 days in two summers, but on average for 6 h during each of 3 days in any month in winter (worst month, December). NO_2 concentrations exceeding 50 ppb per hour occurred during only 3 days in two winters (worst month, January). O_3 concentrations exceeding 50 ppb per hour, occurred almost every month, but O_3 of 80 ppb per hour occurred only in summers, on average for 5 h during each of 3 days in any summer month (worst month, April). Other statistics are given by Martin and Barber (1981).

Sulphur dioxide has been said to act on plants by means of the sulphite ion in solution. Measured and calculated sulphite concentrations in equilibrium with levels of sulphur dioxide in rural air were shown to be unlikely to exceed 0.1 mM sulphite, rather less than many workers have used. The theoretical and measured relationship between concentrations of SO_2 in solution and in air is roughly linear at air concentrations up to 200 µg m^{-3} and then curves steeply. This was shown by Martin (1979) and Davies (1979). Taking the SO_2 concentrations from Bottesford, then both the annual mean there (30 µg m^{-3}) and the 99-percentile hourly value (140 µg m^{-3}) are equivalent to around 3 mg SO_2 ℓ^{-1} in solution as sulphite (0.05 mM). The absolute maximum hourly value in 2 years (1070 µg m^{-3}) is equivalent to 6 mg ℓ^{-1} (0.1 mM). A value of 1.0 mM would not be reached unless the air concentration was over 60 000 µg m^{-3} (20 ppm), a value normally only found in waste gases very close to a chimney mouth.

References

COX, R.A. *et al.* (1976). *AERE Report R8324*, Harwell, UK

COX, R.A. (1977). *Tellus,* **29**, 356–362

DAVIES, T.D. (1979). *Atmospheric Environment,* **12**, 1275–1286

HARRISON, R. and McCARTNEY, H.A. (1980). *Atmospheric Environment,* **14**, 223–244

MARTIN, A. (1979). In *SCI Symposium*, pp. 49–66. Society for Chemical Industry, London

MARTIN, A. (1980). *Environmental Pollution (Series B),* **1**, 177–193

MARTIN, A. and BARBER, F.R. (1981). *Atmospheric Environment,* **15**, 567 –578

A NEW METHOD FOR EXPOSURE OF FIELD CROPS TO SULPHUR DIOXIDE

P. GREENWOOD, A. GREENHALGH, C.K. BAKER and M.H. UNSWORTH
Department of Physiology and Environmental Science, University of Nottingham School of Agriculture

To evaluate the effects of sulphur dioxide on growth, development and yield of field crops, a method was developed for exposing an area of about $100\,m^2$ to raised SO_2 concentrations. Four independent line sources of SO_2 form a square, with sides about $20\,m$ long. Within the square, a central area $10\,m \times 10\,m$ is designated the main treatment area in which SO_2 concentrations are monitored sequentially at several points by a Meloy analyser. Under the direction of a minicomputer, SO_2 is released from upwind line sources so that it diffuses over the treatment area. The rate of

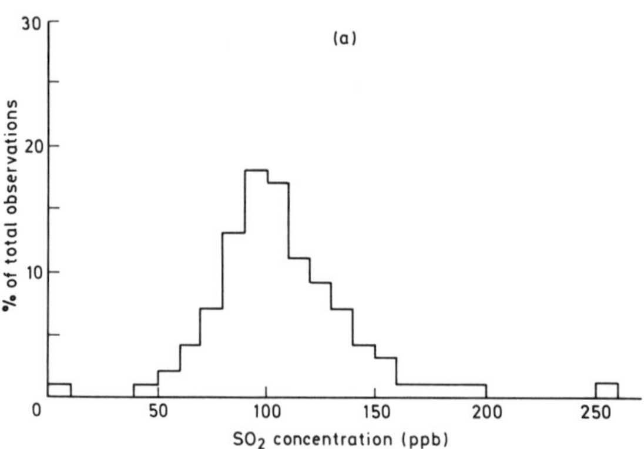

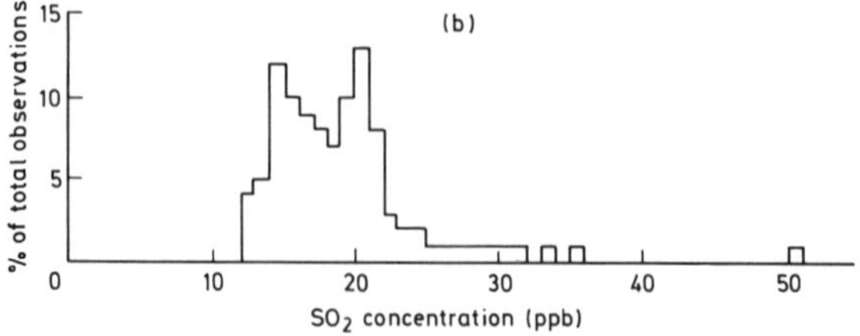

Figure 1   Frequency distributions of (a) the difference in SO_2 concentration between the 'treatment' and 'background' and (b) the background SO_2 concentration observed over winter wheat in the period 27 July–5 August, 1980. The system was operated to produce a mean treatment concentration of 100 ppb above background. Each observation was a 5-min average

release is adjusted by computer-controlled needle valves to maintain the desired SO_2 concentration in the treatment.

The system was operated from April to August 1980 at Sutton Bonington in a field of winter wheat. Concentrations in the treatment area were arranged to exceed ambient levels by about 100 ppb. Regular samples of plants were taken for analyses including dry weights, leaf area and aspects of development. Stomatal responses of fumigated plants were compared with those at control sites. *Figure 1* shows examples of the frequency distributions of the 'treatment' and 'background' SO_2 concentrations observed during a 10-day period of fumigation. Each observation was a 5-min mean concentration. The Figure shows that the mean treatment concentration was maintained close to the desired value—100 ppb above background—and that the distribution of peaks and troughs about the mean was similar to that occurring naturally.

This technique allows large areas of crops to be exposed to air pollutants, thus permitting regular destructive sampling of plants without severely modifying the plant density. In addition, plants from closer to the line sources can also be studied, to reveal responses to higher mean concentrations and to greater peaks. It is proposed to use the system in future seasons to expose winter wheat to raised SO_2 concentrations from emergence to harvest.

Acknowledgement

This work is supported by the Department of the Environment

DESIGN OF PLANT EXPOSURE RIG TO MEASURE PHYSIOLOGICAL RESPONSES TO NOXIOUS GASES FROM COMBUSTION SOURCES

N.M. DARRALL

Biology Section, Central Electricity Research Laboratories, Leatherhead

Chambers were of welded PVC lined internally first with silvered polyester film and then with transparent teflon film. A single layer of teflon film formed the roof of the chamber. A flow rate of $60\,\ell\,min^{-1}$ through the chambers gave one air change every 2 min and a small fan fitted with a rheostat was installed in each chamber to increase turbulent air flow. Three high-pressure mercury vapour lamps (400 W type HLRG) were installed above each chamber and the light intensity could be altered by varying the height of the bulbs above the chambers. A flowing waterbath between the lights and the fumigation chamber reduced heating within the chambers.

Shoot systems of whole plants were fumigated; pots in which they were grown were enclosed in teflon film and the shoots were sealed around the base with silicone rubber compound. Plants were acclimatized for up to 24 h before fumigation and mass flow meters were used to monitor the air flow to the chambers and control the injection of noxious gases at a predetermined ratio. SO_2 concentrations at the chamber inlet and outlet and also at plant height were measured by flame photometry. All air lines from the chambers were maintained at 40°C using self-compensating heating tape.

SO_2 depletion by *Vicia faba* was approximately half the inlet concentration when below 100 ppb. Rates of uptake did not increase at higher concentrations but remained at about $3.75\,mg\,m^{-2}\,h^{-1}$.

AN EXPERIMENTAL DESIGN FOR STUDYING THE EFFECTS OF FLUCTUATING CONCENTRATIONS OF SO₂ ON PLANTS

S.G. GARSED, P.W. MUELLER and A.J. RUTTER

Imperial College at Silwood Park, Ascot

It is now widely accepted that exposure to less than 0.1 ppm of SO_2 for several months can cause reductions in plant growth. However, it is difficult to extrapolate the results of experimental fumigations to predict crop losses in the field, because of the multitude of interacting factors. One particular difficulty is that experimental fumigations normally employ a constant concentration of the pollutant, whereas, in the field, the concentrations will fluctuate about the overall mean. This abstract describes an experiment designed to examine whether fluctuating concentrations of SO_2 modified the response of plants compared with those of plants exposed to a constant concentration.

Recent developments in the continuous monitoring of low concentrations of SO_2 in the field, and the discovery that the frequencies of the logarithms of daily mean concentrations are approximately normally distributed, provided a basis for simulating field fluctuations in controlled conditions. In this experiment, concentrations were varied by stepwise changes between the values 0, 50, 125, 300 and 750 μg m⁻³. The control system allowed a change of concentration each hour. The standard error of the hourly log concentrations around the daily mean in a given treatment was approximately the same as that of the daily log mean concentration around the overall mean. Treatment C (*Table 1*) simulates the distribution of daily mean concentrations observed at a semi-rural site near Nottingham with an annual mean of 100 μg m⁻³ and a 4% frequency of daily means exceeding 250 μg m⁻³. Treatment F is more extreme, in that daily means of 750 μg m⁻³ occur with 4% frequency. Treatments D and E were designed to complete (with C and F) a factorial arrangement in which two peak concentrations (300 or 750 μg m⁻³ as hourly means) were combined with two mean durations of peak (5 or 21 h). Each of the treatments C–F had an overall mean of 100 μg m⁻³ and peaks of short duration were, therefore, more frequent than those of long duration (*see Table 1*). The experimental design was completed with a treatment (B) of constant concentration 100 μg m⁻³ and a treatment (A) with clean air.

For each treatment there was a separate fumigation chamber, constructed of Perspex and illuminated naturally. The air supply to each chamber ensured two air changes per minute and SO_2 was bled into each air supply via a solenoid valve, capillary tube and flow-controller. The controller was set to give the maximum concentration required in that

Table 1 SUMMARY OF GAS TREATMENT IN EXPERIMENT WITH FLUCTUATING CONCENTRATIONS OF SO_2, TOGETHER WITH PRELIMINARY DATA FOR *PINUS SYLVESTRIS* HARVESTED AFTER 275 OR 650 DAYS' TREATMENT

Chamber	Treatment[1]	Mean SO_2 concn (μg m^{-3})	Max. hourly SO_2 concn (μg m^{-3})	% time* > 300 μg m^{-3}	Characteristics of peaks				Dry weight increment of Pinus sylvestris (g)	
					% time* > 750 μg m^{-3}	Mean duration (h)	Mean interval between (d)	% days with peaks	275 d	650 d
A	Clean air	0	0	0	0	–	–	–	28.8	118.5
B	Constant	100	100	0	0	–	–	–	30.4	101.9
C	SF	100	300	11.1	0	5	1	50	26.3	101.8
D	LO	100	300	11.1	0	21	22	12	24.0	92.8
E	SF	100	750	11.1	6.6	5	1	50	28.4	100.9
F	LO	100	750	11.1	6.6	21	22	12	25.3	91.2
									LSD = 5.9	LSD = 10.6

[1]SF = short, frequent peaks; LO = long, occasional peaks
*Includes hours >750 μg m^{-3}

treatment and lower concentrations were obtained by allowing the solenoid valve to open for a limited number of seconds in a 45-s cycle. The supply system was controlled by a microprocessor which also controlled sampling valves whereby the gas in each chamber could be sampled twice each hour and the SO_2 measured by a Meloy SA-285 SO_2 analyser.

The system operated almost continuously for 650 d. *Pinus sylvestris* (initially 3 years old) in the chamber was harvested after 275 and 650 d. No significant effects of treatment were found at the first harvest, but SO_2 significantly reduced growth in all treatments at the second (*Table 1*). There was no differential effect of peak height, but infrequent long peaks caused greater reductions in growth than frequent short peaks ($P < 0.05$). However, the main factor affecting growth was the overall mean SO_2 concentration.

EFFECTIVE POLLUTANT DOSE

G.E. TAYLOR, Jr., S.B. McLAUGHLIN and D.S. SHRINER
Environmental Sciences Division, Oak Ridge National Laboratory, Oak Ridge, Tennessee

The pollutant dose to which a plant is exposed (i.e. ambient dose) is not a direct quantitative measure of the dose that causes a physiological response (i.e. effective dose). The latter is a function of the rate at which pollutant/derivative molecules arrive at perturbation sites within cells of the leaf interior. This rate is controlled primarily by the conductivity of the gas-to-liquid pathway which, in turn, varies with the environment and genotype. Thus, only a fraction of pollutant molecules in ambient air react with foliage. Moreover, only a portion of the total flux is absorbed into the leaf interior, with the remainder being adsorbed to the leaf surface. We propose that total flux in general, and internal flux specifically ($mg\,m^{-2}\,h^{-1}$) can be a measure of the dose to which plants respond.

The importance of discriminating between ambient and effective dose can be demonstrated by experiments in which pollutant flux to vegetation was estimated by mass balance calculations from data derived from an open, gaseous exchange system (McLaughlin and Taylor, 1981). Three-week-old *Phaseolus vulgaris* L. (Bush Blue Lake 274) plants were exposed to gaseous pollutants in teflon-covered chambers and under the following environmental regime: $31 \pm 1°C$ air temperature, $370\,\mu E\,m^{-2}\,s^{-1}$ photosynthetically active radiation, and $70\% \pm 5\%$ relative humidity.

In the first study, fluxes of ozone and sulphur dioxide were measured in chambers containing either ozone alone (0.10 ppm) or in combination with increasing levels of sulphur dioxide (0–0.8 ppm). As steady-state flux was achieved with each increment of sulphur dioxide, estimates of internal leaf flux of each pollutant were calculated (McLaughlin and Taylor, 1981). Flux for each gas was reported as a ratio of internal flux to ambient pollutant concentration: if a proportional relationship existed between the two, the ratio would be constant. The ratio for ozone remained unchanged over the course of the 4-h exposure in both treatments (*Figure 1*). Neither continuous ozone exposure nor increasing levels of sulphur dioxide in the presence of ozone had an effect upon ozone flux into the leaf interior. In contrast, a threefold increase in the sulphur dioxide concentration did not result in a proportional rise in internal flux. The flux ratio declined from 3.0 at 0.2 ppm to 0.5 at 0.8 ppm. As the ozone ratio was unaffected, the absence of proportionality between sulphur dioxide flux and ambient concentration was not due to variable stomatal conductance. Therefore, a proportional relationship between ambient and effective pollutant dose did not exist.

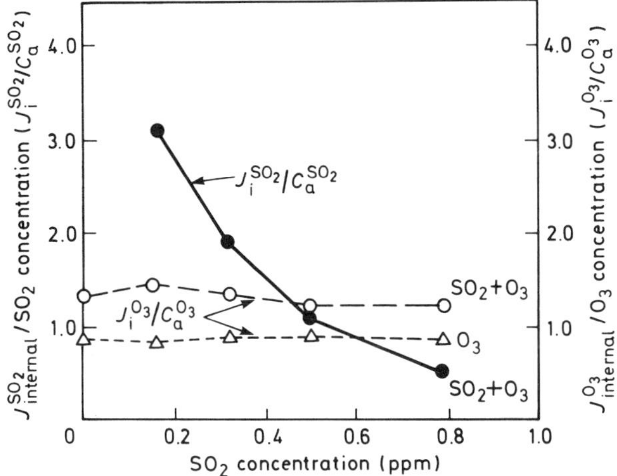

Figure 1   Ratio of pollutant flux to ambient concentration for each gas. Treatments were ozone alone and ozone plus inreasing levels of sulphur dioxide

In the second study, leaf flux of five sulphur-containing gases was measured. The gases are common emissions from coal conversion processes and were carbonyl sulphide, carbon disulphide, hydrogen sulphide, methyl mercaptan, and sulphur dioxide. Plants were exposed to $500\,\mu g\,m^{-3}$ of each gas alone. Among gases, total leaf flux of sulphur varied threefold from 1.0–$3.0\,mg\,m^{-2}\,h^{-1}$ (*Figure 2*). Solubility in water was positively correlated with pollutant flux ($r = +0.94$). This reflects the importance of pollutant movement into and through a water-dominated cell environment

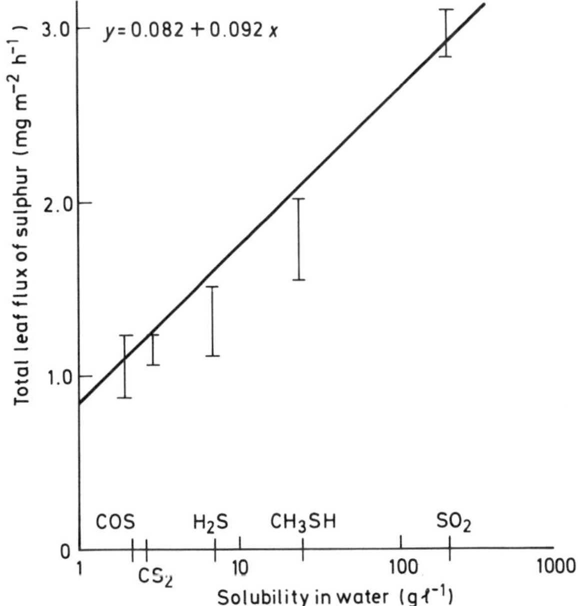

Figure 2   Total flux of sulphur to foliage of *Phaseolus vulgaris* for five coal-conversion gases at an ambient concentration of $500\,\mu g\,m^{-3}$: carbonyl sulphide, COS; carbon disulphide, CS_2; hydrogen sulphide, H_2S; methyl mercaptan, CH_3SH; and sulphur dioxide, SO_2

and suggests that the transfer of molecules from the gas-to-liquid phase is a rate-limiting process in pollutant flux.

The issue of ambient versus effective dose has implications for both applied and basic research efforts. Variation in flux is relevant to establishing acute and chronic rankings of phytotoxicity, because effective pollutant dose differs markedly among gases, even though ambient dose is equivalent. Moreover, pathway conductance can be significantly influenced by variations in environmental conditions during exposure (McLaughlin and Taylor, 1981) and by specific characteristics of leaf structure and function within a species. As for more basic concerns, such as physiochemical mechanism of injury, it is important to assess effective pollutant dose so that model experimental systems can be more effectively designed to represent field conditions.

Acknowledgement

Research performed under contract W-7405-eng-26 by the Department of Energy with Union Carbide Corporation at Oak Ridge National Laboratory, Oak Ridge, Tennessee.

Reference

McLAUGHLIN, S.B. and TAYLOR, G.E. Jr (1981). Relative humidity: Important modifier of pollutant uptake by plants. *Science*, **211**, 167–169

DEPOSITION OF NITROGEN OXIDES TO SCOTS PINE (*PINUS SYLVESTRIS* L.)

C. BENGTSON, PERINGE GRENNFELT and LENA SKARBY
Swedish Water and Air Pollution Research Institute, Goteborg, Sweden

Because of constantly increasing emissions the importance of nitrogen oxides (NO_x) as acidifying agents in terrestrial ecosystems has probably also increased. However, the environmental impact of NO_x has not yet been clarified. The aim of the investigation was to determine the uptake of NO_x in Scots pine and to study the effects on photosynthesis and transpiration. On the basis of deposition rates calculated from the uptake rates found, the yearly deposition of NO_x to a forest stand of Scots pine will be estimated.

In one experiment the deposition rates of nitrogen dioxide (NO_2) and nitric oxide (NO) to 1-year-old pine seedlings were determined at the laboratory by using a cuvette system. The seedlings were exposed for 6 h in light and 6 h in darkness to the following concentrations of NO_2 or NO: 40 ppb, 100 ppb, 200 ppb, and 400 ppb. The climate conditions were kept constant. In addition to uptake of NO_x, rates of photosynthesis, respiration and transpiration of the seedlings were determined.

In a similar experiment (carried out in August, 1979, at the field site of the Swedish Coniferous Forest Project) enclosed shoots of two 20-year-old Scots pines were exposed to NO_x. The gas exchange (NO_x, H_2O, CO_2) was registered continuously.

A preliminary evaluation of the results from the laboratory experiment shows that the uptake of NO_2 in light increased as the concentration was increased. On the other hand, the uptake rate of NO_2 in darkness was constantly low. The deposition rate of NO in light and darkness was essentially constant and low, independent of concentration of NO. The results indicate that neither NO_2 nor NO had any effect on rates of photosynthesis, respiration or transpiration in the concentration range used. At present the results of the field experiment mentioned above are being evaluated.

A QUANTITATIVE STUDY OF THE FATE OF SULPHUR DIOXIDE DEPOSITED ON TO PLANTS

S. ROWLATT, D.V. CRAWFORD and M.H. UNSWORTH
Department of Physiology and Environmental Science, University of Nottingham School of Agriculture

The cycling of sulphur within the atmosphere/wheat/soil system has been quantified. The major fluxes of sulphur were identified as follows:

1. Dry and wet deposition from the atmosphere
2. Root absorption
3. Plant leaching
4. Root loss
5. Soil leaching.

These fluxes have been modelled to describe sulphur cycling in a cereal crop. Further work has been carried out to assess the turnover of sulphur in the soil and the rates at which wheat required sulphur if it was to reach maximum yield.

During the last 100 days before harvest $5\,kg\,S\,ha^{-1}$ was deposited from the atmosphere to the crop: 90% of this entered the plants through the stomata, the remaining 10% being retained on the cuticle and subsequently washed off by rain. Before anthesis $21\,kg\,S\,ha^{-1}$ was absorbed by the roots: after anthesis $5\,kg\,ha^{-1}$ was lost from the roots to the soil and $3\,kg\,S\,ha^{-1}$ leached from the aerial parts of the plants by rain. These fluxes resulted in a transfer of $\sim15\,kg\,S\,ha^{-1}$ from the soil to the crop during the growing season. This flux was partly compensated by the deposition of $4\,kg\,S\,ha^{-1}$ in rain and $2\,kg\,S\,ha^{-1}$ returned to the soil after harvest if straw was burnt on site. Thus the net loss of sulphur from the soil over the experimental period was small ($\sim9\,kg\,ha^{-1}$) compared with the total soil sulphur reserves to 1 m depth of $8 \times 10^3\,kg\,ha^{-1}$.

Acknowledgement

This work is supported by the National Environment Research Council.

THE SULPHUR BALANCE OF AN AGRICULTURAL CATCHMENT

D. CRAGGS, D. HERBERT and S. ROWLATT
Department of Physiology and Environmental Science, University of Nottingham School of Agriculture

The sulphur balance of an agricultural catchment was determined. The catchment, situated 15 km east of the School of Agriculture, covers an area of 23 km^2. The soils are principally chalky boulder-clay types and the catchment is essentially watertight. Sulphur inputs (both wet and dry deposition) were assessed from data collected at the School of Agriculture and the Scientific Services Division of the CEGB. Sulphur output was determined from measurements of the sulphur content of a brook draining the catchment and from the total quantity of water leaving the catchment.

Removal of sulphur occurred mainly during the winter when streamflow was significant. The sulphur concentration in the brook was inversely proportional to flow rate during the winter (range 120–250 µg SO$_4$-S ml^{-1}). Sulphur concentrations were in general lower in summer than in winter. The input of sulphur from the atmosphere was comparatively uniform during the year. The total quantity of sulphur in the catchment therefore increased during the summer when plant demand for sulphur was at a maximum.

LEACHING OF PLANT NUTRIENTS BY RAIN

IOLA H. HUNTER
Department of Physiology and Environmental Science, University of Nottingham School of Agriculture

The leaching of nutrients from plants by rain was investigated using a purpose-built rain chamber. Both excised and attached leaves were exposed to artificial rain (distilled water, pH 5.3). Leaves were exposed to rain for 4–10 h, and the leachates analysed. It was found that actively growing young wheat plants lost approximately 3% of their potassium (K) and 9% of their sulphur (S). In contrast, senescent leaves lost 92% of their K and 54% of their S. The quantity of nutrients leached from mature leaves was proportional to the degree of yellowing.

Rain adjusted to values between pH 4–pH 8 had little effect on the quantities of nutrients leached, particularly from young leaves which retain their waxy cuticle.

The increased leaching of nutrients may be of particular significance in cereal crops, which are allowed to senesce before harvest.

LOSS OF FLUORIDES FROM GRASS SWARDS AND OTHER SURFACES

SULE TAKMAZ-NISANCIOGLU and ALAN DAVISON

Department of Plant Biology, University of Newcastle upon Tyne

Simultaneous sampling of air and grass fluoride in the vicinity of a major source has shown (Davison and Blakemore, 1976 and unpublished data) that grass fluoride follows fluctuations in air fluoride much more closely than expected. On days when the air fluoride increases, the grass fluoride rapidly follows, while a decrease in air fluoride is apparently followed by a decrease in grass fluoride. This latter observation implies that there is a continuous and often rapid loss (of the order of 30–100 μg F g^{-1} dry wt d^{-1}) from the sward which is greater than can be accounted for by the mechanisms of leaching, death of leaves and growth dilution. Possible mechanisms have been investigated by means of turf transplants in the field and by recording loss from model leaves.

Transfer of turf from the field site to an area where the air fluoride was consistently less than 0.1 μg F m^{-3} confirmed that rapid losses occurred and that they appeared to be unrelated to rainfall or growth rate. However, spatial variation of fluoride within the turf and the consequent sampling problems made interpretation of the experiments difficult, so in later work model leaves were used.

Paper models have shown that, where fluoride is applied as NaF solution, it is lost by desorption. The rate, which is apparently sufficiently rapid to explain most of the losses recorded in the field, depends on pH, surface wetness, concentration on the surface and the boundary-layer resistance of the model. The influences of temperature and relative humidity on fluorine loss are still being investigated.

Reference

DAVISON, A.W. and BLAKEMORE, J. (1976). In *Effects of Air Pollutants on Plants*, pp. 17–30 (Mansfield, T.A., Ed.), Society for Experimental Biology Seminar Series, Vol. 1. Cambridge University Press, Cambridge

THE USE OF INDICATOR PLANTS FOR PHOTOCHEMICAL OXIDANTS IN DENMARK

LISBETH MORTENSEN
*National Agency of Environmental Protection, Air Pollution Laboratory,
Risø National Laboratory, Roskilde, Denmark*

Phytotoxic ozone levels in Denmark were first demonstrated by means of indicator plants in the summer of 1977 (Ro-Poulsen *et al.*, 1981). Tobacco varieties Bel W3 and Bel-C were exposed at 14 stations in and around Copenhagen in urban, suburban and rural areas. Plants were significantly more damaged at rural sites than at urban sites. The occurrence of leaf damage was log-normally distributed, except at one site which was situated between two area sources, the city and the airport. It was found that plants at the sites west of Copenhagen must have been exposed to higher concentration episodes than the plants at sites in the city of Copenhagen.

This work is now followed up by a project in which tobacco-plant reactions are compared with physical–chemical ozone measurements. An

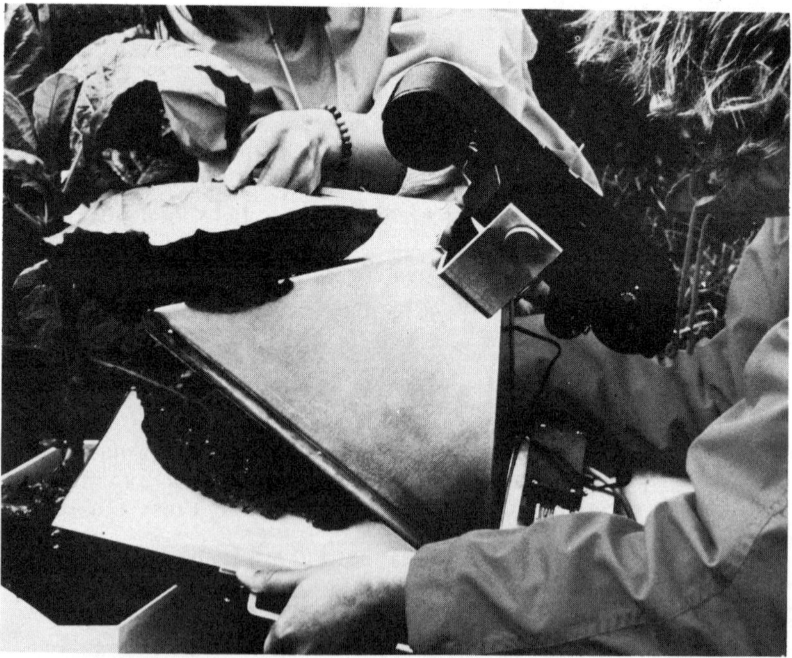

Figure 1   Photographing in the field

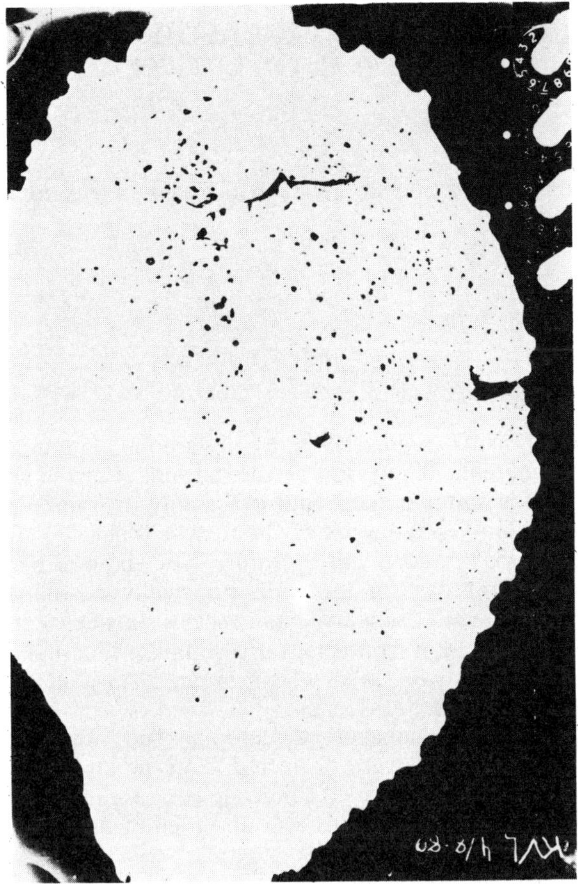

Figure 2   A copy of the negative which is scanned by the TV system

objective leaf-damage scoring system has now been developed. The scoring of the leaves is performed in two steps:

1. Photographing in the field.
2. Scoring of the negatives by a TV camera coupled to a microcomputer.

Each leaf is photographed (*Figure 1*) *in situ* with an integrated unit which illuminates the leaves from behind and keeps the camera in a fixed position relative to the leaf. By using a microfilm and a suitable minus-green filter, it is possible to obtain negatives in which the necrotic flecks appear as dark spots on a white leaf (*Figure 2*).

Reference

RO-POULSEN, H., ANDERSEN, B., MORTENSEN, L. and MOSEHOLM, L. (1981). Elevated ozone levels in ambient air in and around Copenhagen indicated by means of tobacco indicator plants. *OIKOS*, **36**, 171–176

ACUTE AND SUBACUTE DAMAGE TO VEGETATION BY AMMONIA FOLLOWING ACCIDENTAL POLLUTION

L. DE TEMMERMAN
Ministry of Agriculture, Institute for Chemical Research, Tervuren, Belgium

In an accident at Vilvoorde on 10 August, 1979, 8 t of gaseous ammonia were released in the atmosphere in 30 min. The accidental spill occurred at 3.15 p.m. on a slightly cloudy summer day, with practically no rainfall. During the spill there was a faint WNW breeze of about 12–13 km h^{-1}. The relative humidity was about 57% and air temperature was about 17°C. Within a distance of 300–400 m east of the spill, practically all plant species were heavily burned. The most sensitive plant species showed clear symptoms, even at a distance of 1 km from the source. Only a few plant species such as *Berberis vulgaris*, *Begonia tuberhybrida*, *Pelargonium zonale* and *Buxus sempervirens* were not damaged.

We could distinguish acute and subacute symptoms. The subacute effects were observed on plants which were protected from the ammonia cloud by other plants or obstacles and received a much lower dose than the surrounding vegetation. Subacute effects were also observed at distances of more than 500 m from the spill.

Most of the affected plants showed interveinal necrosis, sometimes localized at the leaf margin, as an acute effect (cauliflower, cucumber, tomato, rhubarb, clover, potato, sunflower, annual mercury, privet) or marginal necrosis progressing inwards between the larger veins (dwarf mallow, celery). For some plant species the necrotic flecks were bordered by a reddish-brown to violet-red margin (bean, rhubarb, red clover, lupin, sheep's sorrel, smartweed, knotgrass, dandelion, rose).

In addition, different forms of subacute effects have been observed, such as silvering and bleaching (cabbage, Brussels sprouts, leek), reddish-brown and violet-red discolorations (strawberry, rhubarb, violet) and bronzing (red cabbage, cucumber, pumpkin, scorzonera, tomato, potato, rhubarb, marigold, burdock, *Galinsoga parviflora*, mugwort).

Most of the observed discolorations appeared on the upper leaf surface of dicotyledonous plants. In some monocotyledonous plants, interveinal streaking has also been found (gladiolus and grasses). In some *Acer* species (sugar maple, sycamore) marginal necrosis developed 3–4 weeks after the incident.

There were no visible effects on fruits and vegetables such as apple, cucumber, tomato or bean. The most sensitive plant species, such as *Arctium tomentosum*, *Sonchus asper* and *Galinsoga parviflora* showed clear symptoms (bronzing) even at a distance of 1 km from the spill.

Most of the trees, shrubs and herbaceous plants were forming new leaves, which were symptomless. Other plants, which were at the end of their growing season, died (peas, beans, tomatoes, cauliflowers, potatoes). Further details and photographs of symptoms are available in de Temmerman (1980).

Reference

DE TEMMERMAN, L. (1980). Les dégâts aigus et subaigus occasionnés aux plantes par une décharge accidentelle d'ammoniac. *Revue de l'Agriculture,* **33**, 763–776

OZONE INJURY TO WHITE BEAN (*PHASEOLUS VULGARIS* L.) IN SOUTHWESTERN ONTARIO, CANADA: CORRELATION WITH OZONE DOSE, PAN EVAPORATION, PLANT MATURITY AND RAINFALL

E.I. MUKAMMAL and H.H. NEUMANN
Atmospheric Environment Service, Downsview, Ontario

G. HOFSTRA
Department of Environmental Biology, University of Guelph, Ontario, Canada

During July and August 1978, several fields in southwestern Ontario were periodically inspected for symptoms of injury by ozone and assessments were made of percentage leaf area injured. The observed injury, which ranged up to 53.7% leaf damage, was poorly correlated either with ozone concentration on the day before assessment or with accumulated ozone dose up to that day; for one site, weak negative correlations occurred. However by limiting analysis to one cultivar, Seafarer, for which several repetitive assessment visits were made to a few sites, it was possible to relate nearly all the variability of injury to a combination of environmental factors.

Sufficient data were available to perform single-station analysis at two sites. Assessments were made at intervals of 1–2 weeks: five assessments were available from a site near Wallacetown and four from a site near Tavistock, but one of the latter was eventually omitted as being of doubtful validity. Finally, when attempting to combine the single-station results, data from a site near Ridgetown consisting of two assessments, were also included.

The empirical relationship used to fit individual sites took the form

$$I = A + BD$$

where regression analysis was used to determine coefficients A, B relating injury, I to a combined dose parameter, D. The parameter D was determined from

$$D = CHU \times O/E$$

CHU was the parameter chosen to represent crop maturity at the assessment date. It was defined as the ratio of accumulated corn heat units (Brown, 1969) up to the assessment day relative to accumulated corn heat units required for crop maturity with accumulation beginning at planting date. Corn heat units are a modified form of the degree–day or heat unit concept, and incorporate both daily minimum and maximum temperatures in a nonlinear fashion. They were developed for climatological rating of hybrid corn (*Zea mays* L.) cultivars grown in southern Ontario but should also apply to white bean reasonably well (Brown, personal communication). The heat unit requirement for maturity of the white bean cultivar, Seafarer, was derived from the average of actual accumulated corn heat units at maturity for the years 1973–79 inclusive at test plots at the University of Guelph (personal communication, Prof. W. Beversdorf, Dept. of Crop Science).

The weighted accumulation daily ozone dose from 0700 to 2100 LST was 0 in ppb-hours and E in mm was the weighted accumulated daily 'lake' evaporation derived from measured Class A pan evaporation.

The accumulated values were formed by summing the product of the daily values multiplied by the weighting function f going backward in time, beginning the day before rating $(x=1)$ where $f=e^{-(x-1)/Tc}$ and Tc was an empirically determined time constant. Tc=4 days was found to give best fit for our data. This exponentially decreasing weighting function was introduced to account for the abscission of injured leaves, which was usually observed a few days after the day of injury, and also to allow for any emergence of new leaves which would result in an apparent decrease in percentage total leaf area injured.

Equivalent 'lake' evaporation was selected as a suitable factor because it would represent the evaporative demand of the atmosphere and hence reflect moisture stress on the plant. Moisture stress is known to reduce injury of plants by ozone.

The resultant regression lines fit the observations extremely well (r >0.998, SEE <1%) for the two sites treated individually, but regression coefficients between sites were significantly different. In order to coalesce the data into a single relationship, two additional parameters were introduced so that D was now given by

$$D = R \times CHU \times O \times M/E$$

where the relative rainfall factor R, which expressed the accumulated rainfall since planting relative to that at Wallacetown for the same CHU, was introduced to reflect differences in moisture availability while M, the minimum daily 'lake' evaporation observed for the site during July and August was used to normalize E. The combined data of Wallacetown and Tavistock gave a linear result with $r = 0.987$ and SEE = 3.4% and if the two points from Ridgetown were included then the fitted line became

$$I = -35.6 + 1.0 \times D$$
with
$$r = 0.966 \text{ and SEE} = 5.1\%$$

Reference

BROWN, D.M. (1969). *OMAF Fact Sheet on Corn Heat Units. AGDEX 111/31.* Ontario Ministry of Agriculture and Food, Toronto, Canada

RECENT RESEARCH ON RELATIONSHIPS BETWEEN AIR POLLUTION AND PLANTS IN CHINA

WANG CHIA-HSI (WANG JIAXI)
Jiangsu Institute of Botany, Nanjing, China

Effects of phytotoxic air pollutants on plants

Acute symptoms of injury from various pollutants have been observed and described. Several institutions have worked on the recognition and diagnosis of air-pollutant injuries (SO_2, HF, Cl_2, NH_3, ethylene and others) to vegetation in the field and in fumigation chambers and together have compiled a pictorial atlas of this project.

Anatomical features of various pollution injuries have also been studied for SO_2, HF and O_3.

The effects of SO_2 on growth, development and yield of crops and vegetables have also been studied. It has been shown that there are different degrees of reductions of output during distinct developmental stages at the same concentration of SO_2, and that the heading–flowering stage in cereal crops seems to be the most sensitive.

Foliar injury in species exposed to NH_3 and SO_2 separately as well as in combination has been studied. The combination of NH_3 and SO_2 caused less-than-additive foliar injury and neutralization may occur between these two gases during the process of exposure.

Some workers have determined the effect of SO_2 on the differential permeability of plasma membranes. When the concentration of SO_2 exceeded the injury threshold, ion leakage was closely related to visible injury which appeared later, and ethylene production increased markedly before visible symptoms of injury appeared.

Resistance or susceptibility of plants to air pollutants

Scientists studied differences in SO_2 resistance of various plants. Susceptibility of plants to injury depends on the developmental stage. Diurnal and seasonal changes in resistance of some woody plants to fluoride were studied. The most sensitive stage in the day appears to be between about 10.00 to 14.00 h. The annual fluctuation of resistance was characterized by a rapid decrease from the late spring, to a minimum in summer.

Selection of trees resistant to air pollutants for industrial areas

By means of fumigation techniques, field surveys, cultivation experiments with plants in polluted areas, and leaf-dipping tests with various chemicals, more than 100 species resistant to air pollutants have been selected and screened.

Using plants as indicators of air pollution

The use of plants as monitors of air pollution has been increasingly accepted. In sensitive species such as *Fagopyrum cymosum* and some cultivars of gladiolus, the degree of injury and the fluoride concentration in leaves are significantly correlated with HF concentrations in the atmosphere; these indicators have been used successfully to detect air pollution of fluoride in the vicinity of factories.

The role of metropolitan trees in protecting and improving the environment

The abilities of metropolitan trees and shrubs to protect the environment have been investigated. Effects include: reduction of gaseous pollution (e.g. measured loss of atmospheric SO_2 and HF respectively over a forest belt or a green area); reduction of atmospheric particulates; suppression of bacteria; noise abatement; and reductions of summer temperature.

Absorption, accumulation and translocation of air pollutants in plants and ecosystems

A great amount of work has been done on pollutant analysis in leaves, demonstrating that many species are able to absorb and accumulate SO_2, HF and Cl_2 from the air. Distribution of $^{35}SO_2$ in seedlings was studied soon after exposure: more than 70% of the ^{35}S was in the form of inorganic sulphate.

Scientists have described the translocation and migration of fluorides in ecosystems, particularly in the food chain 'air–forage–cattle'. Wang Chia-Hsi *et al.* (1980a,b) have investigated the effects of fluorides on mulberry and silkworm and their migration in the mulberry–silkworm ecosystem. The results showed that polluted mulberry leaves containing more than 30 ppm fluorides (dry wt) may cause acute damage in silkworms.

References

WANG CHIA-HSI, LI ZHENG-FANG, QIAN DA-FU, GAO XU-PING, MA DE-HE and LI RONG-QI (1980a). *Environmental Quarterly*, **(1)**, 22–25
WANG CHIA-HSI, QIAN DA-FU, LI ZHENG-FANG, GAO XU-PING, MA DE-HE and LI RONG-QI (1980b). *Environmental Quarterly*, **(2)**, 33–38

THE RELATIVE SENSITIVITIES OF CONIFER POPULATIONS TO SO₂ IN SCREENING TESTS WITH DIFFERENT CONCENTRATIONS OF SULPHUR DIOXIDE

S.G. GARSED and A.J. RUTTER
Imperial College at Silwood Park, Ascot

The reliability of tests using short exposures to high concentrations of SO_2 for predicting the relative sensitivity of plants to long-term low levels of SO_2 has been increasingly questioned. We have studied the relative sensitivities of eight conifer populations to different SO_2 treatments.

In the first year *Pinus sylvestris*, *Pinus nigra* ssp. *maritima* and three populations of *Pinus contorta* (Clearwater, Terrace and Long Beach) were used. In the second year, three spruce populations were studied: *Picea abies* and two origins of *Picea sitchensis* (Forks and East Graham Island), together with *P. sylvestris* and *P. nigra* ssp. *maritima* to allow comparison with the previous year's work.

Three concentrations of SO_2 were used (*Table 1*). In Treatment (a) ($200 \, \mu g \, m^{-3}$) damage was measured as percentage reduction in relative growth rate (RGR) compared with clean-air controls. Treatment (c) was an acute fumigation ($8000 \, \mu g \, m^{-3}$) in which damage was measured as percentage necrosis on the needles. Treatment (b) ($750 \, \mu g \, m^{-3}$) was near the borderline between acute and chronic injury. Cumulative scoring of acute injury for 67 days showed classic acute injury symptoms (banding and tip-burn) on several individuals in the first part of the experiment, but a tendency to rapid senescence of whole needles in the second part. Individuals damaged during the early stages were not necessarily affected by the senescence in the later stages. Thus, different orders of sensitivity were obtained after 35 days and after 67 days.

The results (*Table 1*) show that the order of sensitivity differs greatly between treatments. Duplicated experiments (not shown here) gave consistent results. In both series of experiments note that the order obtained at $8000 \, \mu g \, m^{-3}$ was virtually the reverse of that at $200 \, \mu g \, m^{-3}$.

We conclude that:

1. The relative sensitivity of the populations depended almost entirely on the concentration and duration of exposure, and we emphasize that short-term fumigations at high SO_2 concentrations cannot be used to predict responses to long-term exposure to SO_2 in the field.
2. Many of the populations showed a wide range of sensitivity, indicating the need for substantial replication and full statistical analysis in any screening work.
3. The different responses to the high and low concentrations of SO_2 suggest that more than one physiological mechanism is involved in resistance.

Table 1 MEAN SCORES OF TREE POPULATIONS IN VARIOUS TESTS OF SENSITIVITY TO SO_2, TOGETHER WITH RANKINGS IN ORDER OF SENSITIVITY[1]

Tree population	Treatment							
	(a) $200\,\mu g\,m^{-3}$ $(n = 14)^2$ 11 months		(b) $750\,\mu g\,m^{-3}$ $(n = 14)^3$ 35 days		67 days		(c) $8000\,\mu g\,m^{-3}$ $(n = 16)^3$ 6 hours	
	Value & sig.[2]	Rank	Value & sig.[3]	Rank	Value & sig.	Rank	Value & sig.[3]	Rank
Pinus sylvestris	96.1 n.s.	1	1.4 a	3=	3.0 a	2	7.4 a	5
P. nigra ssp. *maritima*	95.2 n.s.	2	2.5 b	5	5.7 cd	4	6.6 ab	4
P. contorta Clearwater	89.9 n.s.	3	0.4 d	1	2.2 a	1	4.8 cd	2
P. contorta Terrace	89.7 n.s.	4	0.8 acd	2	4.1 bc	3	5.5 bce	3
P. contorta Long Beach	86.2*	5	1.4 ac	3=	5.9 bcd	5	4.1 de	1
SECOND SERIES								
P. sylvestris	90.8 n.s.	2	3.2 a	4			7.4 a	5
P. nigra ssp. *maritima*	102.0 n.s.	1	3.5 a	5			6.6 a	4
Picea abies	[4]		3.1 a	3	Data not		1.3 b	3
Picea sitchensis Forks	84.9**	3–5	0.3 b	1	available		0.6 b	1
P. sitchensis East Graham Island			0.9 b	2			0.8 b	2

[1]Rankings are from 1–5 according to the mean values. 1 = most resistant, 5 = most sensitive. Conclusions as to the validity of such rankings may be drawn from the statistical tests.

[2]Values are $\dfrac{\text{RGR in } SO_2}{\text{RGR in control}} \times 100$

* and ** indicate differences from control treatment significant at $P < 0.05$ and $P < 0.01$ respectively.

[3]Values are the mean score for acute injury on current year's needles for each tree. 0 = undamaged, 1 = 1–9% damage, 2 = 10–19% etc., 10 = 100%. Values with the same letter are not significantly different ($P < 0.05$).

[4]Several of the *Picea* sp. were attacked by aphids and discounted. The results for the remaining trees of all three populations have been pooled for statistical analysis.

STUDIES OF FLUORIDE POLLUTION IMPACT ON CONIFEROUS FORESTS

JEAN-PIERRE GARREC
CEA/Centre d'Etudes Nucléaires de Grenoble, DRF/Laboratoire de Biologie Végétale, Grenoble Cedex, France

Relationship between fluoride concentration and forest growth

In polluted coniferous forests, studies of fluoride concentrations in needles of spruce, and growth reduction based on the increment core rings measurements have shown a positive correlation between these two parameters (Garrec and Audigier, 1980). It appears that there is a logarithmic relationship between these two factors, and in the case of the coniferous forests studied (Maurienne, Savoie, France), the relation is

$$y = -10.7 + 11.02 \ln x \quad r = 0.80$$

where y = growth loss in %
   x = fluoride concentration in spruce needles in ppm.

This relationship shows that in polluted areas, low levels of fluoride have a real effect on forest growth, and that above 100 ppm of fluoride in needles, growth reduction is very important (more than 40%).

Relationship between precipitation and fluoride accumulation

In the same polluted forests, studies on fir trees from 1970 to 1979 have shown that the fluoride accumulation in the needles is highly significantly correlated to the fluoride emissions from aluminium smelters and precipitation (Garrec and Plebin, 1981). The best estimation of this relationship was given by the regression equation

$$F = 2.92 \, E^{1.14} \, P^{-0.51} \quad r = 0.73$$

where F is fluoride concentration in the needles in ppm, E is fluoride emissions in tons f and P is rainfall in mm.

This relationship shows that fluoride accumulation is directly related to the fluoride emissions and that high precipitation will be related to a small decrease of the fluoride level of the trees, in comparison to the marked increase of this level resulting from low precipitation amounts.

References

GARREC, J.P. and AUDIGIER, C. (1981). *European Journal of Forest Pathology*, **10**, 432–438

GARREC, J.P. and PLEBIN, R. (1981). *European Journal of Forest Pathology*, **11**, 129–136

SELECTION OF PLANTS RESISTANT, ABSORPTIVE AND SENSITIVE TO AIR POLLUTANTS

WANG CHIA-HSI, QIAN DA-FU, LI ZHENG-FANG, GAO XU-PING, TANG SHU-YU and PAN RU-GUI
Jiangsu Institute of Botany, Nanjing, China

By means of fumigation, field surveys, cultivation experiments with plants in polluted areas, leaf-dipping tests with various chemicals and analysis of pollutant concentration in leaves, dozens of species resistant and absorptive to air pollutants have been selected and screened.

Some important resistant species are: *Ailanthus altissima, Broussonetia papyrifera, Canna indica, Celtis sinensis, Distylium racemosum, Euonymus japonicus, Firmiana simplex, Juniperus chinensis* var. *kaizuca, Ligustrum lucidum, Nerium indicum, Pittosporum tobira, Trachycarpus fortunei, Viburnum odorastissimum, Yucca gloriosa.*

At the same time, several sensitive species such as *Fagopyrum cymosum, Antenoron neofiliforme, Hosta plantaginea, Hemerocallis fulva, Calystegia hederacea* and some varieties of gladiolus have been found to be very sensitive to HF and other pollutants.

INFLUENCE OF AERIAL POLLUTION ON CROP GROWTH AND YIELD

ANNE H. BUCKENHAM, M.A.J. PARRY and C.P. WHITTINGHAM
Rothamsted Experimental Station, Harpenden

Since 1972, Rothamsted Experimental Station has been conducting experiments to determine the effects of ambient concentrations of pollutants on crop growth. Experiments have been confined to the Marston Valley, the centre of the UK brick-making industry. Sulphur dioxide and fluoride are the main pollutants.

Modified open-topped chambers were used to compare the growth of plants in ambient field air that had been cleaned by charcoal filters. Pollutant concentrations were reduced by 60–70% in the filtered chambers compared with the outside, and the concentration in the unfiltered chambers was reduced by 12%. Spring barley (Magnum) was grown inside eight open-topped chambers and on eight outside plots.

Sulphur dioxide was monitored using a Meloy SA285 continuous analyser and by Warren Spring bubblers. Gaseous and particulate fluorides were measured using impregnated filter papers which were analysed using a specific ion electrode (*Table 1*).

Table 1 AMBIENT POLLUTANT CONCENTRATIONS

Year	Mean F conc. ($\mu g\,m^{-3}$)		Highest daily F conc. ($\mu g\,m^{-3}$)		Mean SO_2 conc. ($\mu g\,m^{-3}$)	Highest daily SO_2 conc. ($\mu g\,m^{-3}$)
	Particulate	*Gaseous*	*Particulate*	*Gaseous*		
1979	–	–	–	–	48	224
1980	0.02	0.13	0.11	1.27	52	182

Measurements to record the growth and the development of the plants were made at 2- to 3-day intervals. Destructive harvests were taken at anthesis and maturity to determine leaf and stem areas, partitioned dry weights, fluoride and sulphur contents.

At anthesis the most marked effect was that plant development in the chambers was eight days ahead of that outside; in addition, the chambers substantially reduced tiller and ear numbers. At maturity the same effects were apparent and the decreased ear number in the chambers reduced grain yields. Cleaning the air had little effect on tiller production or ear number, but increased both grain and straw yields (*Table 2*).

Table 2   THE EFFECT OF OPEN-TOPPED CHAMBERS AND FILTRATION ON YIELDS OF SPRING BARLEY CV. MAGNUM IN THE MARSTON VALLEY IN 1979 AND 1980

Environment	Ear number (m^{-2})		Grain dry weight $(g\,m^{-2})$		Straw dry weight $(g\,m^{-2})$		1000-grain weight (g)		Leaf fluoride content (ppm wt/wt)	
	1979	1980	1979	1980	1979	1980	1979	1980	1979	1980
Filtered chamber	541[a]	715[a]	489[a]	560[a]	470[a]	569[a]	43.8[a]	42.4[a]	27.7[a]	28.7[a]
Unfiltered chamber	488[a]	709[a]	308[b]	518[a]	345[b]	475[b]	37.6[b]	42.0[a]	53.6[b]	42.8[b]
Outside	977[b]	955[b]	559[a]	631[b]	597[c]	675[c]	31.7[c]	35.0[b]	89.4[c]	110.6[c]

Means followed by the same letter within a column are not significantly different at 95% probability level

Because the filtration was nonselective, the injurious agent was not identified. These results indicate that the ambient concentrations of pollutants in the Marston Valley are reducing the yield of some cereal crops.

FIELD TRIALS AND OPEN-TOP CHAMBER STUDIES OF THE INFLUENCE OF SULPHUR DIOXIDE ON THE GROWTH OF RYEGRASS

KAY E. COLVILL and DAVID C. HORSMAN
Department of Botany, University of Liverpool

Field trials were set up to compare the growth of a range of ryegrass populations in polluted, i.e. urban and urban fringe, and clean, i.e. rural, areas in N.W. England. Populations from urban and rural areas, as well as selected tolerant and sensitive clones (as determined by wind-tunnel fumigations) of commercial varieties were compared. It was envisaged that the tolerant urban populations and the tolerant clones would act as controls in the trials, as they would be little affected by sulphur dioxide (SO_2). The trials were started during the summer of 1976 and although there was no marked effect of the urban environment on the growth of all the material during the first summer, winter and following spring, there was a significant decline in that summer's yield, particularly for the rural populations and sensitive clones. A similar pattern has been observed in subsequent trials. However, there has been no significant effect of the urban environment on the growth of the material taken as a whole, or on the urban and rural populations differentially. Wind-tunnel fumigations have demonstrated that the populations do differ in their sensitivity to SO_2. The lack of an effect of the urban environments can therefore most easily be explained by the general decline in mean SO_2 levels, particularly overwinter in urban areas: for example in Manchester, winter SO_2 levels have declined from $181\,\mu g\,SO_2\,m^{-3}$ in 1976/77 to $73\,\mu g\,SO_2\,m^{-3}$ in 1979/80, so that SO_2 is not now having an effect in field conditions.

Open-top chamber studies have been used to complement the field trials, as they allow plant growth to be studied under controlled air-pollutant conditions but in an otherwise similar environment—thus eliminating 'site' factors in obtaining different SO_2 levels. Overall, filtering the air in an urban environment has caused an increase in the yield of ryegrass over winter, but effects during the rapid growth period in the summer have been much smaller. This suggests that there are still some pollutants in urban air which affect plant growth, but whether or not SO_2 alone is the causative factor cannot be proved in such an experiment. However, at a control site, open-top chambers have been used for adding SO_2 to rural air. Mean levels of 150–$240\,\mu g\,SO_2\,m^{-3}$ over the last two winters have produced yield losses in the spring harvests. Summer levels of $<100\,\mu g\,SO_2\,m^{-3}$ have had no comparable effect.

In summary, concentrations of $>150\,\mu g\,SO_2\,m^{-3}$ under field conditions have produced reductions in the yield of ryegrass, particularly over winter, but these have not been evident at concentrations of $<100\,\mu g\,SO_2\,m^{-3}$, especially in summer.

OZONE EFFECTS ON POLLEN-TUBE GROWTH *IN VIVO* AND *IN VITRO*

W.A. FEDER and G.H.M. KRAUSE
Suburban Experiment Station, University of Massachusetts, Waltham, USA

B.H. HARRISON and W.D. RILEY
Department of Biology, University of Massachusetts, Boston, USA

Tube elongation of germinated pollen of the ozone-sensitive cultivar Tiny Tim of tomato (*Lycopersicum esculentum*) was retarded *in vitro* by concentrations of ozone as low as 0.15 ppm for 15 min. Fluorescent microscopy and scanning electron microscopy were used to evaluate pollen-tube elongation and stigmal penetration *in vivo*. Ozone at the abnormally high concentration of 0.8 ppm for 6 h had no effect if pollen was treated immediately after deposition on the untreated stigma, or if the stigmal surface was treated with ozone and then pollinated with pollen which had not been treated with ozone. When pollen was incubated for 2 h *in vitro* and then transferred to the stigma and exposed to 0.5 or 0.8 ppm ozone for 6 h, subsequent tube elongation was markedly inhibited. The results indicate that the stigma affords protection for pollen-tube elongation under ozone stress. There is evidence that tomato pollen tubes respond to ozone stress by changing their interaction with the stigmal surface.

SOME EFFECTS OF LOW CONCENTRATIONS OF SO_2 AND/OR NO_2 ON THE GROWTH OF GRASSES AND POPLAR

MARY E. WHITMORE, P.H. FREER-SMITH and TERESA DAVIES
Department of Biological Sciences, University of Lancaster

The grass *Poa pratensis* L. (Monopoly) was sown in four outdoor fumigation chambers in October 1979. Pollutants were added to three chambers for 5 days a week, and charcoal-filtered air was added for the remaining 2 days. The mean weekly pollutant concentrations were 60 ppb SO_2, 60 ppb NO_2, and 60 ppb SO_2 + 60 ppb NO_2. A fourth chamber was supplied continuously with charcoal-filtered air (control). Plants were harvested at intervals until May 1980. *Figure 1* shows the increase in the plant dry weight in each of the four treatments over the fumigation period. At the mid-February and mid-March harvests, when the seedlings were still very small, the combined fumigation caused 50% and 60% reductions in plant dry weight, respectively, compared with the controls, while the individual pollutants had no effect. By the time of the three later harvests, however, the SO_2 and the NO_2 treatments both caused ~50% reductions in dry weight and the combined fumigation caused a 75% reduction. Both the

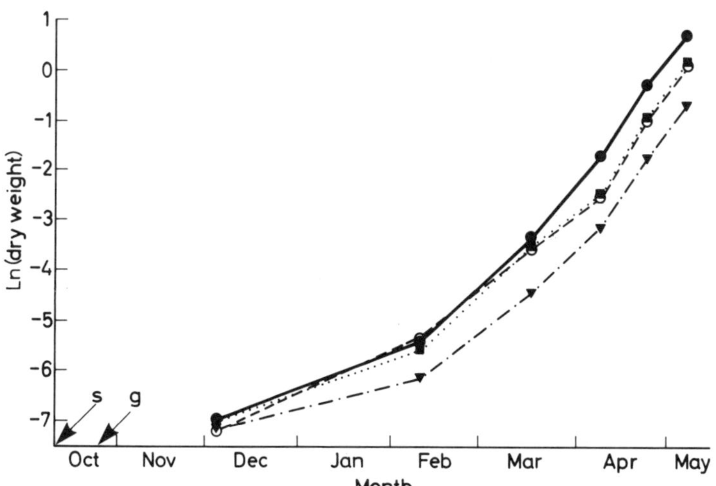

Figure 1 Change in dry weight of *Poa pratensis* L. (g, log scale) with time under natural lighting, ambient temperature, when exposed to clean air (●———●), 60 ppb NO_2 (■·····■), 60 ppb SO_2 (○--○) or 60 ppb SO_2 + 60 ppb NO_2 (▼·−·▼); s, date sown; g, date of germination

"

SO$_2$ alone and the SO$_2$ + NO$_2$ treatments caused increases in the shoot:root ratio at every harvest from February onwards, whereas the NO$_2$ had no significant effect.

Twelve first-year cuttings of *Populus nigra* L. (Italica) were fumigated for eight weeks during June and July 1980 in each of the treatments described above. At the end of this period, only the SO$_2$ + NO$_2$ treatment caused a significant reduction in mean dry weight compared with the control. These plants showed a 37% reduction in growth. However, morphogenic effects were observed in all treatments. Both SO$_2$ and NO$_2$ caused reductions in leaf area and in the number of side shoots, but stimulated increases in the height of the plants. Such increases were greatest in plants exposed to SO$_2$ alone, while side-shoot development was inhibited most in those plants exposed to SO$_2$ and NO$_2$ in combination.

Previous research at Lancaster has shown that the grass *Phleum pratense* L. is most sensitive to SO$_2$ pollution under conditions of low light stress (Davies, 1980). When growth was followed in the laboratory throughout a 40-day exposure period under conditions of low photon flux density (220 μE m^{-2} s^{-1}), an early interference from SO$_2$ was observed. Ten-day-old seedlings of *Phleum pratense* L., S48, exposed to 120 ppb SO$_2$, showed a 21% reduction in total yield after only 10 days. *Figure 2* shows the

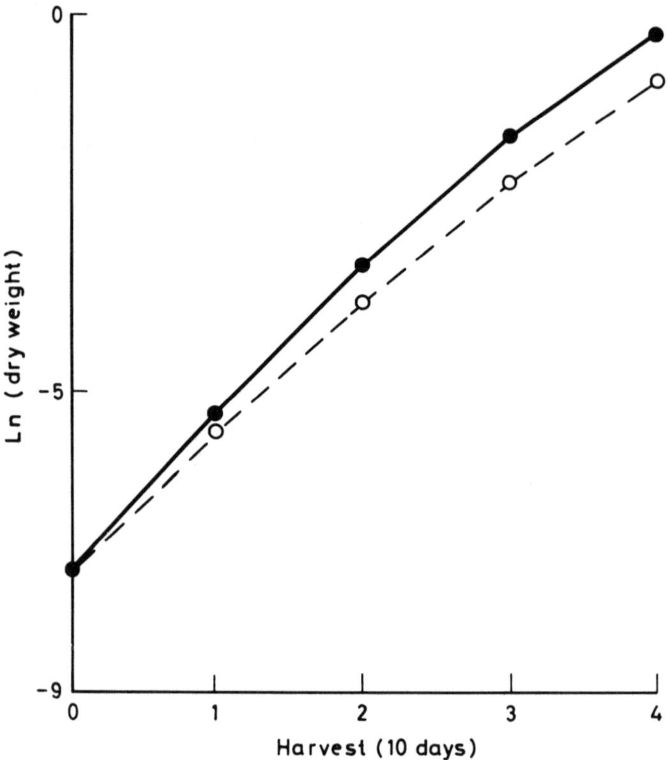

Figure 2   Change in dry weight of *Phleum pratense* L. (g, log scale) with time under low photon flux density, short daylength, when exposed to clean air (●——●) or 120 ppb SO$_2$ (O--O)

increase in plant dry weight with time, with and without SO_2. The initial interference was in root growth, which was reduced by 40% during the first 10 days. After 40 days' exposure to SO_2, 51% and 62% reductions in total shoot and root weight, respectively, were observed. During the first 20 days, SO_2 caused large reductions in the relative growth rates of all plant parts. After this period, the low photosynthetic capacity produced by SO_2 was partly compensated by increased partitioning to the leaves, and an increase in leafiness. However, these changes failed to alleviate the combined stresses of low photon flux density and SO_2, and marked growth reductions were observed.

Acknowledgement

This work was supported by grants from the Natural Environment Research Council.

Reference

DAVIES, T. (1980). *Nature,* **284**, 483–485

AN INVESTIGATION OF THE EFFECTS OF FLUORIDE ON SELECTED MOSS SPECIES

D. PATERSON and J.B. KENWORTHY
Department of Botany, University of Aberdeen, Scotland

Vegetation downwind from an aluminium smelter at Fort William has been compared with that of a control area in the vicinity. The results show that fewer species are found in the neighbourhood, and it is postulated that the distribution of vegetation in the smelter area is determined by the output of fluoride from the smelter. Levels of up to 4.1 ppm fluoride (F^-) in rainfall have been measured in the immediate vicinity of the source. Species growing next to the smelter appear to be more resistant to fluoride than those found further away, and vice versa. *Sphagnum* spp., when transplanted into the polluted area, accumulate up to 200 ppm F^- within 2 weeks. There is also visual evidence of damage, accompanied by a 40% loss in chlorophyll. The order in which species enter the community with increasing distance from the smelter can be used to predict their relative sensitivities to fluoride. From field observations it can be postulated that *Polytrichum commune*, for example, is more resistant than *Sphagnum* spp. from the same site.

Using these two mosses, and others, the uptake of fluoride and the resultant effects on various physiological processes have been studied in the laboratory. Extensive experimentation has produced a technique which involves placing mosses in bathing solutions with sodium fluoride concentrations containing between 1.0 and 100.0 ppm F^- for periods of 4 h, after which time uptake had ceased. Uptake was monitored continuously with a selective ion electrode. Even at low concentrations of fluoride (below 10 ppm) the presence of fluoride in the bathing solutions caused a significant increase in respiration and decrease in photosynthesis (*Figure 1*), a breakdown of chlorophyll and increased release of potassium from the cells. In all cases, *Sphagnum* was more severely affected than *Polytrichum*.

The results indicate that the proportion of fluoride accumulated in the cell walls is consistently higher, over a wide range of fluoride concentrations, in *Polytrichum* than in *Sphagnum* (*Figure 2*). Significantly, potassium leakage from cells, attributable to the breakdown of membranes, is much greater in the sensitive *Sphagnum* species. It is suggested, therefore, that a partitioning of fluoride between cell walls and cell contents may determine the sensitivity of these mosses to low concentrations of fluoride. A consequence of this is that sites of resistance to fluoride pollution may be found either in cell walls and/or at the membrane level.

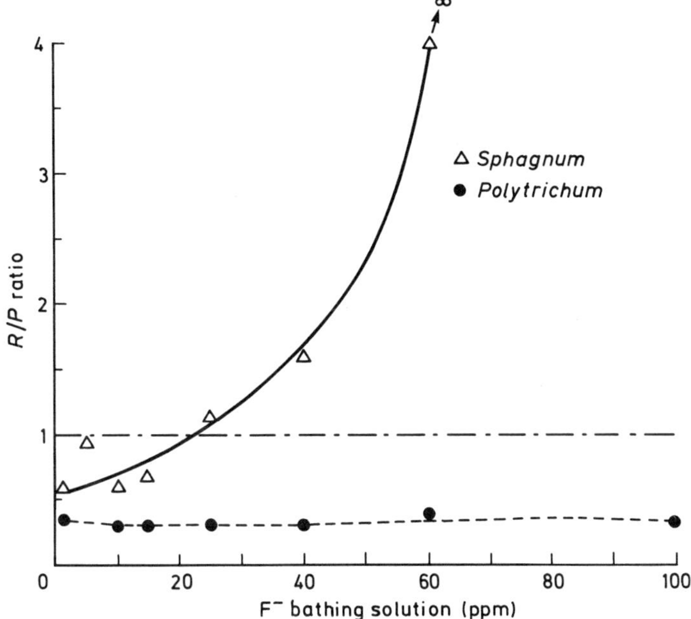

Figure 1   The respiration/photosynthesis ratio for *Sphagnum* and *Polytrichum* after immersion in a range of fluoride concentrations. When the R/P ratio exceeds 1, then a carbon imbalance will result

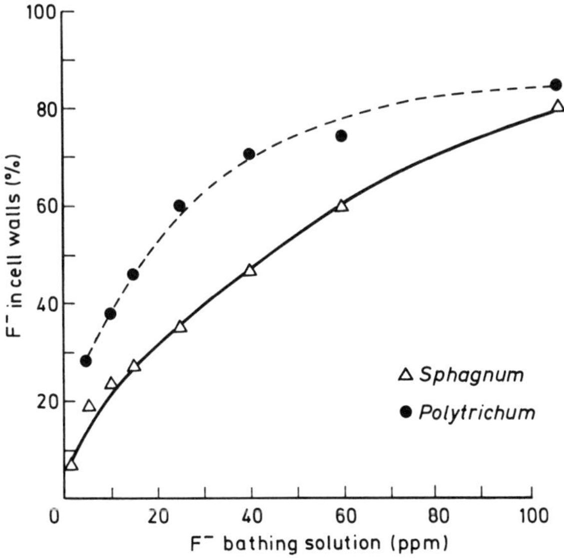

Figure 2   Proportion of fluoride fixed in cell walls of *Sphagnum* and *Polytrichum* with different levels of bathing-solution fluoride

Few workers have demonstrated that sites of sensitivity or resistance in bryophytes function at levels of organization below that of the membrane, but data presented here clearly demonstrate that physiological differences exist between species defined as sensitive or resistant by other methods. The differences found in the responses of these two species to fluoride pollution in the laboratory relate to their distribution downwind from the smelter at Fort William.

THE INFLUENCE OF ELEVATED ATMOSPHERIC CO_2 AND SO_2 ON THE GROWTH AND PHOTOSYNTHESIS OF EARLY SUCCESSIONAL SPECIES

ROGER W. CARLSON

Institute for Environmental Studies, University of Illinois, Urbana

Plant growth may be reduced by SO_2 and increased by exposure to high CO_2. High CO_2 may ameliorate the effect of SO_2. The environmental variables of relative humidity, temperature, light intensity, soil moisture, and soil nutrient level may alter the response of plants to SO_2 and CO_2. Elucidation of these interactions is necessary in order to determine appropriate regulatory levels for atmospheric SO_2.

Reported here are the results of a study in which six early successional annual species (*Chenopodium album* L., *Datura stramonium* L., *Polygonum pennsylvanicum* L., *Amaranthus retroflexus* L., *Setaria faberii* Herrm. and *Setaria lutescens* (Wiegel) F.T. Hubb.) were exposed for up to 4 weeks to 0.0 or 0.25 ppm SO_2 at 300, 600, or 1200 ppm CO_2. The primary pathway of CO_2 fixation for *Chenopodium*, *Datura*, and *Polygonum* is that of the reductive pentose phosphate cycle (C_3) while *Amaranthus* and the two *Setaria* species utilize the C_4-dicarboxylic acid cycle (C_4). This difference in carbon metabolism coupled with other differences between C_3 and C_4 plants could lead to differences in response to fumigation with air pollutants.

Plants of six species were grown from seed in a glasshouse for 28 d and transferred to growth cabinets maintained at either 300, 600, or 1200 ± 15 ppm CO_2. Air temperature was maintained at 27 ± 3°C during 14-h days and 18 ± 3°C at night with relative humidity at 75 ± 5%. The growth cabinets were exposed to ambient summertime sunlight supplemented with illumination from high-intensity metal halide lamps. SO_2 at 0.25 ppm was administered for 8 h each day for 5 d during both the 2nd and 4th weeks of CO_2 treatment. Photosynthesis and transpiration were measured under nonfumigated conditions at 300, 600, and 1200 ppm for two plants of each species from each treatment regimen on days 14 and 28 after the start of CO_2 treatment. Shoot weight, root weight, and leaf area were determined at the end of the experiment. The total amount of water transpired by each plant was measured during the second 5 d of fumigation with SO_2. This was done by placing the root volume of each plant in a plastic bag to prevent evaporation from the pot, weighing each plant daily, and watering to replace transpired water.

Results

GROWTH

Total plant growth increased with CO_2 for *Chenopodium, Datura*, and *Polygonum* (C_3 species) at both 0.0 and 0.25 ppm SO_2. Growth of SO_2-fumigated plants was significantly less than that of nonfumigated plants at 300 ppm, but not significantly different for plants grown at 600 or 1200 ppm CO_2. Total plant growth of *Amaranthus* did not increase with an increase in CO_2 concentration. Growth of *Setaria faberii* was greater at 1200 ppm and growth of *Setaria lutescens* was greater at 600 ppm than that obtained at the other CO_2 concentrations. Total biomass of fumigated *Amaranthus* and *Setaria* species declined with increasing CO_2 and was significantly less than that of nonfumigated plants at 1200 ppm CO_2.

SHOOT:ROOT RATIO

The shoot:root ratio of nonfumigated plants was constant across all CO_2 concentrations for the C_3 species, decreased with increasing CO_2 for *Amaranthus*, and increased slightly with increasing CO_2 for both *Setaria* species. In general the shoot:root ratio of plants fumigated with SO_2 was similar to that of nonfumigated plants, except for a marked increase at 1200 ppm CO_2 for *Amaranthus*.

LEAF AREA

Leaf area increased only slightly with increasing CO_2 for all nonfumigated plants. Fumigation with SO_2 caused a larger reduction in the leaf area of C_3 species at 300 ppm CO_2 than at the higher CO_2 concentrations. For *Amaranthus* and both *Setaria* species (C_4 plants) leaf area was reduced more by fumigation with SO_2 at the higher concentrations of CO_2 than at 300 ppm.

Leaf area ratio (LAR), the ratio of photosynthetic to respiring material within a plant, declined in nonfumigated plants with increasing CO_2 for C_3 species and did not change across CO_2 concentrations for C_4 species. For C_3 species, LAR was greater for fumigated plants at 300 ppm than that of nonfumigated plants. There was no difference in LAR between fumigation treatments at the higher CO_2 concentrations for all C_3 species and both *Setaria* species. LAR of fumigated *Amaranthus* was slightly higher than that of nonfumigated plants at all CO_2 concentrations.

PHOTOSYNTHESIS

Photosynthesis increased with an increase in the concentration of CO_2 during the measurement of photosynthesis for all species. However, the rate of photosynthesis at each measurement concentration was not different between growth concentrations. In general, the rate of photosynthesis for fumigated plants was slightly less than that of nonfumigated plants.

TRANSPIRATION

Transpiration declined with increasing CO_2 for all species. At each measurement concentration the rate of transpiration was not significantly different between plants grown at different concentrations of CO_2. Transpiration of fumigated plants was similar to that of nonfumigated plants.

Leaf water use efficiency, the rate of photosynthesis divided by the rate of transpiration for individual leaves, increased with increasing CO_2 for all species, was not different between plants grown at different concentrations of CO_2, and was not significantly different between fumigation treatments.

Water uptake of nonfumigated plants decreased with increasing CO_2 for all C_3 species and was not significantly different between CO_2 concentrations for the C_4 species. Only for *Datura* at 300 ppm CO_2 was there a significant difference in water uptake between fumigation treatments, with more water taken up by the fumigated plants. At other CO_2 concentrations for *Datura*, and at all CO_2 concentrations for the other species, there was no difference in water uptake between fumigated and nonfumigated plants.

Acknowledgement

Research supported by DOE Contract EE-77-S-02-4329.

INJURIOUS EFFECTS OF LOW CONCENTRATIONS OF SO$_2$ AND O$_3$ ON THE LEAF SURFACES AND STOMATAL CONDUCTANCES OF FIELD BEAN

C.R. BLACK, V.J. BLACK[1] and D.P. ORMROD[2]
Department of Physiology and Environmental Science, University of Nottingham School of Agriculture

Plants of field bean (*Vicia faba* L., cv. Dylan) were exposed to known concentrations of SO$_2$ and O$_3$ applied singly or in mixtures in controlled-environment chambers. Exposure to 17.5–175 ppb SO$_2$ increased stomatal conductance and transpiration by about 20%. Measurements with a diffusion porometer showed that the stomatal conductances of upper and lower leaf surfaces responded similarly to SO$_2$.

In order to study the cellular basis of the stomatal response, epidermal strips were removed from both control and polluted plants. The strips were incubated in a medium containing neutral red, a vital stain, for approximately 30 min and subsequently examined microscopically to assess survival and visible injury in the guard cells and adjacent epidermal cells.

Strips from control plants contained a consistently high proportion of viable epidermal cells, although the percentage survival figures were invariably slightly lower on the abaxial surface (about 75–80% and 65–75% respectively). However, the viability of the adjacent epidermal cells was reduced in plants exposed to SO$_2$ concentrations between 17.5 and 175 ppb for as little as 2 h. The reduction in epidermal-cell survival increased with SO$_2$ concentration and ranged between 5–35% and 25–50% on the upper and lower leaf surfaces respectively. The reduction in epidermal cell survival was invariably larger on the lower epidermis. This response probably results from more rapid pollutant fluxes to the lower epidermis caused jointly by the larger abaxial stomatal conductances and the greater intercellular volume adjacent to the lower epidermis. Injury to guard cells was observed infrequently, unless SO$_2$ concentrations exceeded 175 ppb or prolonged exposures were used.

Using this technique it is impossible to assess how far the increased proportion of dead epidermal cells observed in strips from polluted plants originates from damage during stripping to membranes already weakened by exposure to SO$_2$. Membrane injury is known to be an important early symptom of SO$_2$ toxicity. However, examination of intact leaf samples from control and polluted plants, using. scanning electron microscopy,

[1]Present address: Ecology Section, Department of Human Sciences, Loughborough University of Technology
[2]Permanent address: Department of Horticulture, University of Guelph, Guelph, Ontario, Canada

supported the evidence obtained using epidermal strips, and confirmed that exposure to SO_2 causes extensive loss of turgor in the epidermal cells. These results support the view that the stomatal opening induced by low SO_2 concentrations is caused by preferential injury and loss of turgidity in epidermal cells adjacent to the stomata, thus permitting the stomatal pores to increase in size. In contrast, the stomatal closure observed at higher SO_2 concentrations ($\geqslant 1$ ppm) appears to result from injury to the guard cells themselves.

Exposure to gaseous pollutants such as SO_2 at low concentrations has several damaging consequences:

1. Net photosynthesis may be reduced by as much as 20–25%.
2. Transpiration is increased to a similar extent.
3. Water-use efficiency must be adversely affected by these responses.
4. The increased stomatal conductance facilitates entry of SO_2 and other gaseous pollutants into leaves and may increase injury in the mesophyll and other tissues.

Brief exposures to low levels of O_3 and to mixtures of O_3 and SO_2 also induced cellular injury and depressed net photosynthesis (*see abstract by Black, Ormrod and Unsworth, page 494*). Exposure to 60 ppb O_3 for 2 h was sufficient to cause a small reduction in epidermal-cell viability, which increased sharply in severity as O_3 levels were raised to about 120 ppb. There was no evidence of further increases in epidermal cell injury after exposure to 150–300 ppb O_3. Mixtures containing 40 ppb SO_2 generally caused more severe damage than ozone alone, particularly at O_3 concentrations below 200 ppb. The survival of guard cells was not affected significantly by O_3 levels below about 200 ppb, but extensive disorganization and collapse of guard-cell protoplasts occurred after exposure to 300 ppb for 2–4 h. Once the threshold for visible injury was exceeded, the severity of the injury to guard cells caused by ozone appeared much greater than that induced by comparable levels of SO_2.

EFFECTS OF OZONE AND OZONE/SULPHUR DIOXIDE ON THE GAS EXCHANGE OF *VICIA FABA* L.

D.P. ORMROD[1], V.J. BLACK[2] and M.H. UNSWORTH
Department of Physiology and Environmental Science, University of Nottingham School of Agriculture

Investigations of plant responses to air pollutants are usually made using exposures to single gases. However, in the atmosphere, air pollutants rarely occur singly. In many areas of the UK, Europe and North America, sulphur dioxide and ozone commonly may be found in combination. Unfortunately the majority of reports of plant responses to this pollutant combination deal solely with the effects of unrealistically high concentrations of the gases, and in particular, with the appearance of visible injury symptoms (e.g. Hofstra and Beckerson, 1981).

In this study, the physiological responses of plants to the low concentrations of SO_2 and O_3 found in areas of Britain were investigated. In rural or semi-rural areas of Central England, for instance, a low background of SO_2 (10–50 ppb) is supplemented on 10–30 d y^{-1} by concentrations of O_3 which exceed 80 ppb for at least 1 h. Maximum hourly ozone concentrations of 130–260 ppb have also been reported (*see* Martin, this volume, page 449).

Vicia faba L. cv. Dylan (field bean) plants were grown in controlled-environment rooms. When they were 3 weeks old or comprised three pairs of fully expanded leaves, they were transferred to 250 ℓ exposure chambers (*described fully by* Black and Unsworth, 1979a) and allowed to acclimatize for 24 h exposed to clean air. The next day, plants were either exposed for 4 h to clean air or to air containing 50–300 ppb O_3 or to 50–300 ppb O_3 supplemented with 40 ppb SO_2. SO_2 was supplied from a cylinder and ozone was produced by passing air or oxygen over an ultraviolet source. Pollutant concentrations in the air stream entering or leaving the chambers were monitored using Meloy sulphur and Dasibi ozone analysers. Exit pollutant concentrations were considered to approximate closely to exposure concentrations. Chamber conditions were maintained at $22 \pm 2°C$, 280 W m^{-2} PAR, 0.9 ± 0.2 kPa vapour-pressure deficit and 320 ± 15 ppm CO_2. Air flow through the chambers was 20–25 ℓ/min and internal fans minimized boundary-layer resistance.

In each experiment conducted (18 with O_3 alone and 21 with $O_3 + SO_2$), photosynthetic rates of plants exposed to pollutants were compared with

[1]Permanent address: Department of Horticulture, University of Guelph, Guelph, Ontario, Canada
[2]Present address: Ecology Section, Department of Human Sciences, Loughborough University of Technology

those of control plants exposed to clean air. Carbon dioxide and water vapour exchange were monitored by infrared gas analysis and dew point hygrometry.

The main conclusions of this study were:

1. Net photosynthesis was depressed significantly by 40 ppb SO_2 alone, an observation reported previously (Black and Unsworth, 1979b);
2. Net photosynthesis was depressed significantly by ozone above a threshold of 50 ppb. The magnitude of depression of photosynthesis and the rate at which the depression occurred increased as O_3 concentration increased, reaching a maximum depression of 15–20% at 300 ppb O_3;
3. The addition of SO_2 to O_3 further increased the inhibition of net photosynthesis, especially at low O_3 concentrations where inhibition was greater than additive. A maximum inhibition of 20–30% was observed at 300 ppb O_3 + SO_2, but was less than additive;
4. Visible injury to leaves was observed only after plants had been exposed to O_3 above 200 ppb, whether or not SO_2 was present.

Clearly, these results suggest that photosynthesis, which is depressed by SO_2 at concentrations commonly occurring in Britain, would be depressed further by the superimposition of ozone generated photochemically.

Acknowledgement

This work was supported by the Natural Environment Research Council.

References

BLACK, V.J. and UNSWORTH, M.H. (1979a). *Journal of Experimental Botany,* **30**, 81–88

BLACK, V.J. and UNSWORTH, M.H. (1979b). *Journal of Experimental Botany,* **30**, 473–483

HOFSTRA, G. and BECKERSON, D.W. (1981). *Atmospheric Environment,* **15**, 383–389

EFFECTS OF OZONE AND SULPHUR DIOXIDE ON SOYBEANS

G.C. PRATT and S.V. KRUPA
Department of Plant Pathology, University of Minnesota, St Paul

Soybean (*Glycine max* (L.) Merr. cv. Hodgson) plants were fumigated under controlled conditions with ozone (0.08 to 0.25 ppm—160 to 500 µg m^{-3}), sulphur dioxide (0.10 to 0.40 ppm—266 to 1066 µg m^{-3}), and combinations of the two pollutants. Fumigations were of 0.5–4 h duration as single or multiple (5 consecutive days) events. Following the fumigations, unifoliate and first trifoliate leaves were evaluated for percentage visible injury, chlorophyll concentration, and sulphur concentration.

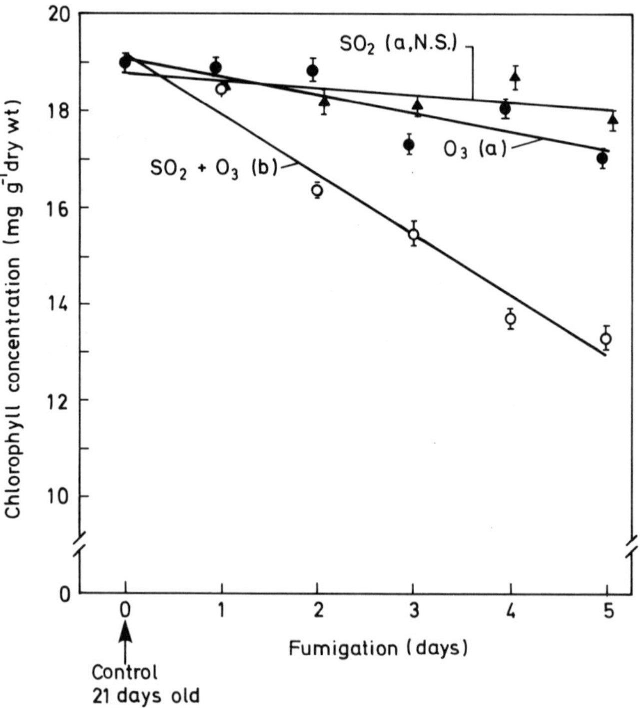

Figure 1   Relationship between chlorophyll concentration of first trifoliolate leaves of soybean cv. Hodgson and successive days of fumigation with ozone (0.10 ppm, 2 h d^{-1}), sulphur dioxide (0.40 ppm, 2 h d^{-1}), or the two pollutants simultaneously (same concentrations, 2 h d^{-1}). (N.S.: regression line with slope not significantly different from null, different lower case letters indicate significantly different slope at $P = 0.05$ by the F-test)

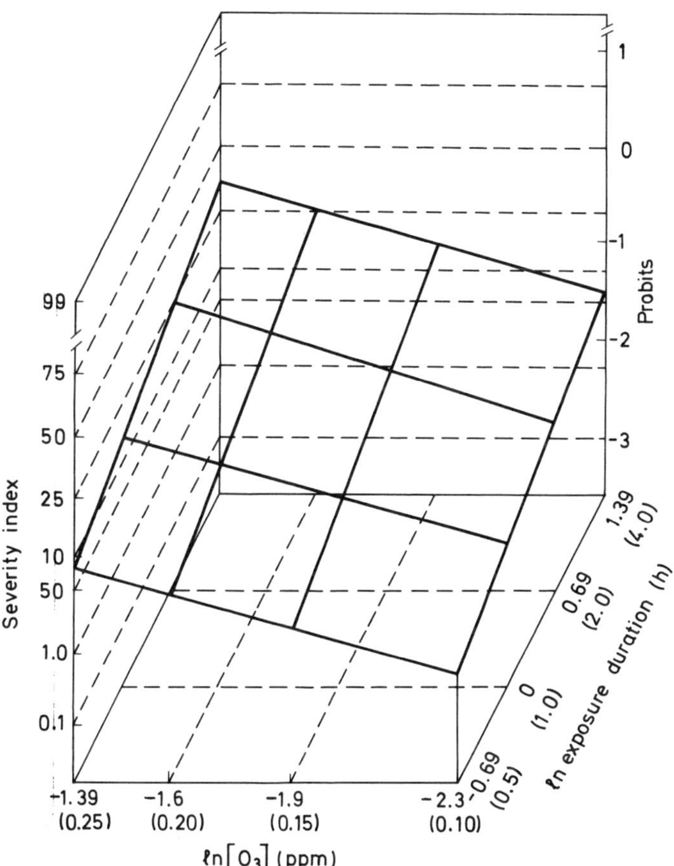

Figure 2   Response plane generated using a log-probability model for first trifoliolate leaves of soybean cv. Hodgson exposed to ozone. The vertical axis is a normal probability scale and the other axes are natural log scales

Visible symptom development was well correlated with reduction in chlorophyll concentration. A reduction in chlorophyll concentration was also measured in leaves fumigated with low concentrations of ozone and showing few or no visible symptoms (*Figure 1*). The relationships between percentage visible injury and acute ozone dose were modelled using the method of Larsen and Heck (1976). The data could be satisfactorily explained using this procedure (*Figure 2*). An injury threshold (1 per cent leaf surface injured) for a 1-h exposure to 0.09 ppm ozone is predicted.

Fumigation with sulphur dioxide alone resulted in no visible symptoms and no significant reduction in the chlorophyll concentration at any of the concentrations used. Exposure to sulphur dioxide resulted in an increase in foliar sulphur concentration.

Single-event fumigations with combinations of ozone and sulphur dioxide (0.08 or 0.10 ppm ozone and 0.40 or 0.20 ppm sulphur dioxide) of up to

4 h duration resulted in minimal visible symptoms. There was also no significant reduction in chlorophyll concentration. The sulphur concentrations of plants fumigated with the pollutant combinations were significantly lower than the sulphur contents of plants exposed only to sulphur dioxide.

Exposures of 2 h per day for 5 consecutive days to the pollutant combinations (0.08 or 0.10 ppm ozone and 0.40 or 0.20 ppm sulphur dioxide) resulted in more-than-additive visible injury. The reduction in chlorophyll concentration of leaves exposed to the pollutant combinations was significantly more than in leaves exposed to a single pollutant (*Figure 1*). In addition, the increase in sulphur concentration of leaves exposed to the pollutant combination for 5 consecutive days was less than that of leaves exposed to 5 days of sulphur dioxide alone.

These results indicate that soybean cultivar Hodgson is relatively sensitive to ozone. In addition, exposures to combinations of ozone and sulphur dioxide resulted in more-than-additive injury compared with the effect of single pollutants. Sulphur uptake did not occur to the same extent after fumigation with both ozone and sulphur dioxide as with sulphur dioxide alone. The acute visible injury response of soybean cultivar Hodgson to ozone can be explained with a log-probability model.

Reference

LARSEN, R.I. and HECK, W.W. (1976). *Journal of the Air Pollution Control Association*, **26**, 325

COMBINED EFFECTS OF SALT (NaCl) AND AIR POLLUTION (SO₂) STRESS ON *CUCUMIS* PLANTS

P.R. VAN HASSELT and M.J. WASSEN
Department of Plant Physiology, University of Groningen, The Netherlands

It is anticipated that the concentration of SO_2 in the atmosphere will increase in the near future as a result of the replacement of oil and natural gas by coal, which contains more sulphur. Such an increase can be expected in the highly industrialized west part of Holland. In this area salt (NaCl) in the soil water causes serious problems in horticulture (Sonneveld and Voogt, 1978). The effect of a combined stress of NaCl and SO_2 on plants is unknown. We therefore studied the effect of SO_2 on photosynthesis and dark respiration of salt-stressed plants. In preliminary experiments transpiration and peroxidase activity were also measured.

Cucumis plants were grown in a greenhouse in pots using a drip culture system. A series of salt stresses was obtained by adding 0, 5, 25 and 50 mM NaCl to the nutrient solution. The shoot of an 18–21-day-old plant was enclosed in a transparent chamber and was fumigated with SO_2–air mixtures in the light for 10 min (total dose 0.17 mg SO_2), 20 min (total dose 0.14 mg) or 30 min (total dose 0.12 mg). Apparent photosynthesis and dark respiration were measured with an IR gas analyser before and after fumigation. Photosynthesis decreased in proportion to the dose of SO_2 to which the plants were exposed. After exposure to 0.17 and 0.14 mg, the photosynthesis of plants subjected to a mild salt stress (5 mM) decreased less than the photosynthesis of plants grown without NaCl. Plants subjected to a high salt stress (25 and 50 mM) showed more inhibition of photosynthesis than plants grown without salt. No significant changes in dark respiration were observed after fumigation.

Reports on the effect of SO_2 on stomatal conductance are contradictory (Majernik and Mansfield, 1970; Biscoe, Unsworth and Pinckney, 1973). Therefore, in some experiments the transpiration of plants before and after fumigation was measured. Transpiration was inhibited by SO_2 fumigation, both in salt-stressed and in control plants. Photosynthesis and transpiration changed in different ways, indicating a primary inhibitory effect of SO_2 on the mechanism of photosynthesis, e.g. the activity of ribulose diphosphate carboxylase (Ziegler, 1972).

Peroxidase activity is reported to increase after fumigation with SO_2 in some plant species (Posthumus, 1978). In preliminary experiments the peroxidase activity of *Cucumis* leaves before and after fumigation was studied. Peroxidase activity of salt-stressed plants was not changed compared with that of control plants. SO_2 caused a decrease of peroxidase activity in control plants, while it had no significant effect in salt-stressed plants.

References

BISCOE, P.V., UNSWORTH, M.H. and PINCKNEY, H.R. (1973). *New Phytologist,* **72**, 1299
MAJERNIK, O. and MANSFIELD, T.A. (1970). *Environmental Pollution,* **1**, 149
POSTHUMUS, A.C. (1978). *VDI-Berichte,* **314**, 225
SONNEVELD, C. and VOOGT, S.J. (1978). *Plant and Soil,* **49**, 595
ZIEGLER, I. (1972). *Planta,* **103**, 155

EFFECTS OF H₂S ON WATER-EXTRACTABLE SULPHYDRYL CONTENT OF CROP PLANTS

L.J. DE KOK
University of Groningen, The Netherlands

C.R. THOMPSON
University of California, Riverside, USA

J.B. MUDD
MSU/DOE Plant Research Laboratory, Michigan State University, East Lansing

Fumigation of crop plants with low concentrations of H_2S (30–300 ppb) resulted in growth inhibition. At the same time an increased water-extractable sulphydryl-compound content was observed. Sulphydryl content increased more than fourfold at 300 ppb H_2S. The H_2S-induced increase was noticeable during the entire fumigation period, even though ageing of the plants was responsible for a decrease in the actual sulphydryl content. The increase in sulphydryl content was not due to an increased sulphide content of the tissue. It is suggested that the sulphydryl compound was the product of a disturbed sulphur metabolism.

THE EFFECT OF SULPHITE ON LIGHT ACTIVATION OF FRUCTOSE 1,6-BIS-PHOSPHATASE IN SOYBEAN CHLOROPLASTS

RUTH ALSCHER-HERMAN
Boyce Thompson Institute, Cornell University, Ithaca, NY

SO_2 at levels below those which cause visible injury has a reversible inhibitory effect on apparent photosynthesis. Data from other laboratories suggest the light modulation (LEM)[1] system of several enzymes of carbon fixation as one possible site for reversible SO_2 action in the chloroplast (Ziegler, Marewa and Schoepe, 1976; Anderson and Duggan, 1977). Two cultivars of soybean, Hark and Beeson, have different SO_2 susceptibilities in the field with respect to yield (Amundson and Weinstein, personal communication). The relative susceptibilities of the LEM systems for alkaline FbPase[2] to sulphite *in vitro* in Hark and Beeson have been investigated in the work reported here, with a view to determining whether metabolic differences in SO_2 susceptibility between the two cultivars correlate with these field data. FbPase was chosen for this study as it has been demonstrated to be a key enzyme in the regulation of rates of carbon fixation (Heldt *et al.*, 1977). Chloroplasts were isolated from the leaves of soybean plants (5th–7th trifoliate stage) which had been subjected to a 5-d subdued-light regime (16-h photoperiod, $80 \, \mu E \, m^{-2}$) before harvest. The isolation method was that of Nakatani and Barber (1977) with an added filtration step through 15 nm precision-woven nylon mesh to remove the whole cells which commonly contaminate chloroplast preparations from soybean. Light activation was carried out *in vitro* according to the method of Anderson and Avron (1976). Alkaline FbPase activity was either measured as inorganic phosphate released (Anderson, 1974) or by the coupled assay described by Preiss and Greenberg (1961). Protein was determined by a dye-binding method, utilizing Coomassie Blue.

The effect of the presence of 5 mM sulphite on light activation of FbPase present in stromal extracts prepared from the Hark cultivar of soybean is demonstrated in *Table 1*. Reconstituted Hark chloroplasts showed a two-fold light-mediated stimulation of FbPase activity. This stimulation was abolished when illumination was carried out in the presence of 5 mM sulphite. *Table 2* summarizes the results of several experiments in which 1 and 5 mM sulphite were tested for their effect on light activation of FbPase in Hark and in Beeson chloroplasts. Results are presented as the ratio of FbPase activity obtained after illumination in the presence of sulphite, to

<hr>

[1]LEM = light effect mediator
[2]FbPase = fructose 1,6-bisphosphatase

Table 1   EFFECT OF SULPHITE ON LIGHT ACTIVATION OF CHLOROPLAST FbPase

| Sample | Condition | | Light |
| | Light | Dark | |
	FbPase activity (nmol P$_i$ produced[a] /100 µg Chl)		Dark
Hark membranes + stroma	490	228	2.15
Hark membranes + stroma + 5 mM sulphite	243	257	0.95

[a]P$_i$ is inorganic phosphate

Table 2   EFFECT OF THE PRESENCE OF SULPHITE DURING ILLUMINATION ON SUBSEQUENT ACTIVITY OF CHLOROPLAST FbPases. STROMA + MEMBRANES WERE PRESENT IN EACH REACTION MIXTURE. (Hark and Beeson)

Cultivar	Sulphite concentration	Activity + sulphite / Activity − sulphite
Hark (SO$_2$-susceptible)	1 mM	0.41
	5 mM	0.47
Beeson (SO$_2$-resistant)	1 mM	1.22
	5 mM	0.51

that obtained after illumination when sulphite was absent from the reaction mixture. The presence of sulphite at either 1 or 5 mM during illumination resulted in a decrease in the subsequent activity of chloroplast FbPases from Hark, whereas the presence of 1 mM sulphite during illumination appears to produce the opposite effect in Beeson.

This influence of sulphite on the light activation of chloroplast FbPase appeared to be confined to stromal enzyme, because membrane-bound

Table 3   EFFECT OF THE PRESENCE OF SULPHITE ON MEMBRANE-BOUND FbPase FROM SOYBEAN LEAVES

Sample	Conditions	FbPase activity (nmol Pi produced/100 µg Chl)
Hark	Dark	24.62
Hark	Dark, 5 mM sulphite	108
Hark	Light	170.4
Hark	Light, 5 mM sulphite	158.6
Beeson	Dark	20.6
Beeson	Dark, 5 mM sulphite	157
Beeson	Light	179
Beeson	Light, 5 mM sulphite	151

Table 4   EFFECTS OF THE PRESENCE OF SULPHITE ON THE ACTIVITY OF FbPase IN STROMAL EXTRACTS FROM HARK AND BEESON

| Sample | Control | FbPase activity (nmol F6P produced min^{-1} mg protein^{-1}) | |
		1 mM sulphite	5 mM sulphite
Hark stroma	356		359
Beeson stroma	408	408	450

FbPase in both cultivars reacted quite differently to sulphite, as is illustrated in *Table 3*. In both Hark and Beeson, 5 mM sulphite appears to stimulate membrane-bound FbPase, and this stimulation took place equally well in the light and in the dark. Sulphite at 1 or 5 mM had no effect on the activity of the stromal enzyme in either Hark or Beeson, as is shown in *Table 4*. Thus, sulphite must be exerting its effects on the light activation process *per se*. As the light-activation process involves electron transport (Anderson, 1979) it must, of necessity, involve an association of the enzyme with the membrane. But bound enzyme does not exhibit this same sulphite susceptibility. It seems, therefore, that the inhibitory effect of sulphite is being exerted on the stromal enzyme during the binding to the thylakoid membrane, which must take place before the enzyme can be

Table 5   EFFECT OF LIGHT ACTIVATION ON THE AMOUNT OF FbPase ACTIVITY PRESENT IN THE FREE (STROMAL) FORM. MEMBRANES AND STROMA WERE ILLUMINATED TOGETHER. AFTER THE INCUBATION, THE MEMBRANES WERE REMOVED BY A BRIEF (2 min) CENTRIFUGATION AND ENZYME ACTIVITY IN THE SUPERNATANT WAS DETERMINED

Sample	*FbPase activity* (nmol F6P mg protein^{-1} min^{-1})	
	Light	*Dark*
Membranes + stroma	2061	1546
Supernatant (contains free enzyme)	294	443

activated. The data of *Table 5* show that illumination does bring about the binding of the enzyme to the membrane. Binding may be affected by sulphite. Differences between Hark and Beeson with regard to sulphite sensitivity might reside in differences in the binding process, such as variations in the nature or in the number of binding sites for FbPase on the membrane.

References

ANDERSON, L.E. (1974). *Biochemical and Biophysical Research Communications*, **59**, 907–913

ANDERSON, L.E. (1979). *Encyclopedia of Plant Physiology, New Series, 6,* pp. 271–281 (Gibbs, M. and Latzko, E., Eds). Springer Verlag, New York

ANDERSON, L.E. and AVRON, M. (1976). *Plant Physiology,* **57**, 209–213

ANDERSON, L.E. and DUGGAN, J.X. (1977). *Oecologia,* **28**, 147–151

HELDT, W.W., CHON, C.J., LILLEY, R.M. and PORTIS, A. (1977). *Proceedings of the 4th International Congress on Photosynthesis*, pp. 469–478. The Biochemical Society, London

NAKATANI, H.Y. and BARBER, M. (1977). *Biochimica Biophysica Acta,* **461**, 510–512

PREISS, J. and GREENBERG, E. (1961). *Methods in Enzymology,* **23**, 691–696

ZIEGLER, I., MAREWA, A. and SCHOEPE, E. (1976). *Phytochemistry,* **15**, 1627–1632

THE CORRELATION BETWEEN SO_2 SUSCEPTIBILITIES OF PLANTS AND pH OF LEAF SAP

YU SHU-WEN, LIU YU, LI ZHEN-GUO, YANG WEI-DONG and
WU YOU-MEI
Shanghai Institute of Plant Physiology, Academia Sinica, Shanghai, China

Different plants vary in their resistance to SO_2. Wu stated that the susceptibility of plants to SO_2 was related to the pH of their leaf sap: plants with lower pH values are more susceptible, while those with pH values around 7 are resistant. Therefore the pH of leaf sap may serve as a physiological index in assessing the resistance of plants to SO_2. We present here the results of a study on the mechanism of the correlation between pH and SO_2 susceptibility.

Segments of wheat seedlings were treated with H_2SO_3 or $NaHSO_3$ (5 mM) solution at various pH values. The production of ethane, the leakage of K^+ and the destruction of chlorophyll were measured and taken as indices of the extent of injury by H_2SO_3 or $NaHSO_3$ solution. Solutions with different pHs were prepared, using citrate–phosphate buffer.

The amount of ethane produced by the action of H_2SO_3 itself (effect of (citrate phosphate buffer + H_2SO_3, 5 mM) − effect of citrate phosphate buffer alone) varied with pH, attaining a maximum at pH 3–4. The relative amount of the bisulphite ion (HSO_3^-) present in the solution of H_2SO_3 was also dependent upon the pH value of the medium. It attained a maximum at pH 3–4 also. The curve of ethane production by H_2SO_3 and the pH-dependence curve of the proportion of HSO_3^- are similar, i.e. the curve of injury by H_2SO_3, measured as ethane production, ran almost parallel to the curve of proportion of HSO_3^-.

The curve of injury measured as changes of permeability (leakage of K^+) also ran parallel to the curve of proportion of HSO_3^-, the heaviest injury being at pH 3–4; measurement of chlorophyll content showed that the injury curve measured as the destruction of chlorophyll was also similar to the curve of proportion of HSO_3^-.

SO_2 in the atmosphere enters the leaves through the stomatal apertures, dissolves in the water bathing the cell wall, then injures leaf cells. The pH of the cell-wall solution is in equilibrium with the pH of cell sap, the action of which is the same as that of the pH of the incubation medium. The results of the present study therefore provide the experimental evidence for the relationship between the SO_2 injury and pH of leaf tissue.

The effect of pH on the SO_2 injury comprises two facets: 1. *The direct destructive action of acidity.* SO_2 is an acidic gas: when plants absorb it, the pH of leaf sap becomes more acidic, hence some injury is induced. However, this effect is apparent only at strong acidity ($<$pH 3), and within the range of physiological acidity is not significant. 2. *The indirect effect due to the influence of pH on the partition among the three species existing in the solution: HSO_3^-, SO_3^{2-} and undissociated H_2SO_3 molecules.* According to the results of the present study, this is the main effect. Within the range of pH 2–5, where HSO_3^- is predominant, tissue injury occurs heavily. At pH > 5, where the major component in the solution is SO_3^{2-}, injury is slight. At pH <2, where undissociated molecules of H_2SO_3 become the principal component, serious damage is caused by strong acidity, but not by H_2SO_3 itself. HSO_3^- is the major factor, whereas SO_3^{2-} and undissociated H_2SO_3 molecules are less toxic. In the past it was believed that because undissociated H_2SO_3 molecules penetrated readily into cells, they were more toxic. But this view perhaps had made no distinction between the toxicity of H_2SO_3 itself and the destructive effect of strong acidity on which the existence of undissociated H_2SO_3 molecules was dependent.

STUDIES ON THE MECHANISM OF SO_2 INJURY IN PLANTS

YU SHU-WEN, LIU YU, LI ZHEN-GUO, TAN CHANG and YU ZI-WEN
Shanghai Institute of Plant Physiology, Academia Sinica, Shanghai, China

Experiments were carried out by exposing potted plants to SO_2 in dynamic fumigation cabinets in a phytotron or by treating segments of leaves with HSO_3^- as a model system. The effect of SO_2 on the differential permeability of plasma membranes was studied. Permeability changes were followed by measurement of electrolyte leakage, especially K^+ leakage from leaves. It was shown that the permeability changes were closely related to SO_2 injury. When the concentration of SO_2 exceeded the threshold of injury, the ion leakage increased markedly. The more the ion leaked, the more serious the visible injury which appeared later. When the concentration of SO_2 did not exceed the threshold value, little change in permeability was observed. The destruction of differential permeability indicates damage of plasma membrane. Evidence suggests that the membrane lipid is affected by SO_2. Three kinds of products of peroxidation of membrane lipids, especially their component polyunsaturated fatty acids, were detected in plants during the injury process: 1) Ethane was produced, the amount of which increased with increasing concentration of SO_2 or HSO_3^-. 2) Production of thiobarbituric acid-reactive substances (TBA-RS), mainly malondialdehyde, was increased. 3) Fluorescent substances accumulated in leaf tissue with an excitation maximum at 360 nm and a fluorescence emission maximum at 470–490 nm. We therefore conclude that injury by SO_2 is closely related to the peroxidation of membrane lipids.

It was well known that peroxidation of membrane lipids could be induced by free radicals which might be generated during the oxidation of sulphite to sulphate. Hence we proposed that SO_2 injury is mediated probably by the reaction of free radicals generated in SO_2 oxidation. In order to verify this supposition, five kinds of free-radical scavengers (diphenyl amine; sodium benzoate; butylate hydroxytoluene; tocopherol (vitamin E) and propyl gallate), were used to spray plants or treat segments of leaves before exposing them to SO_2 or HSO_3^-. The results showed that plants or segments of leaves pretreated with free-radical scavengers were indeed less injured. At the same time the production of ethane, of TBA-RS and of fluorescent substances decreased correspondingly. Since all of the five free-radical scavengers tested have a more-or-less protective effect against SO_2 injury, it can be concluded that free radicals do take part in the process of injury. This provides another proof of the hypothesis that the peroxidation of membrane lipids is an important part of SO$_2$ injury. The interrelation can be summarized as follows:

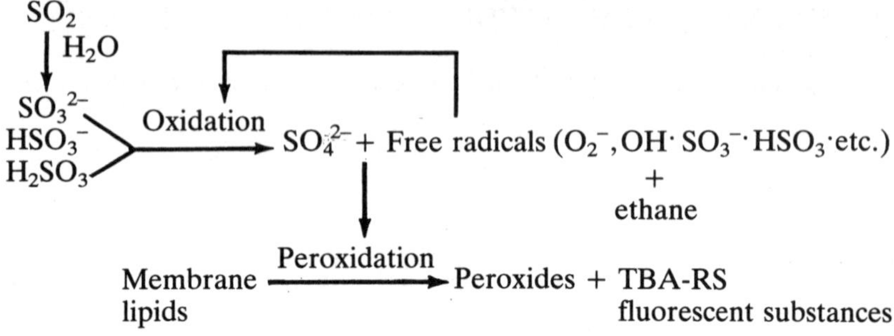

APPENDIX 1 LATIN AND COMMON NAMES USED IN THE TEXT AND THEIR COMMON AND LATIN NAME EQUIVALENTS

Abies fir
Acacia acacia
Acacia georginae no common name: an acacia
Acer maple, sycamore etc.
Acer pseudoplatanus sycamore, sycamore maple
Acer rubrum red maple
Acer saccharum sugar maple, rock maple
Agrostis palustris creeping bentgrass
Agrostis stolonifera fiorin, a native bentgrass
Agrostis tenuis common bentgrass, colonial bentgrass
Ailanthus altissima tree-of-heaven
alfalfa *Medicago sativa*
Allium onion, leek etc.
Allium cepa onion
Allium porrum leek
Alnus glutinosa black alder
Alnus incana European alder
Alternaria solani early blight fungus
Amaranthus retroflexus red-root pigweed
Anthoxanthum odoratum sweet vernal-grass
Antirrhinum snapdragon
Aphelenchoides fragariae a nematode of strawberry
apple *Malus sylvestris*
apricot *Prunus armeniaca*
apricot, Chinese *Prunus armeniaca*
Arachis hypogaea peanut, groundnut
Arctium tomentosum burdock
Armillaria mellea honey fungus, a root fungus
ash, green *Fraxinus pennsylvanica*
ash, white *Fraxinus americana*
aspen, trembling *Populus tremuloides*
Atriplex confertifolia saltbush
Atriplex sabulosa an atriplex
Atriplex triangularis an atriplex
Avena sativa oat

barley *Hordeum vulgare*
bean, bush, navy, pinto, snap, white *Phaseolus vulgaris*
bean, field *Vicia faba*
bean, lima *Phaseolus limensis* or *P. lunatus*
beet *Beta vulgaris*
Begonia tuberhybrida tuberous begonia
Belonolaimus longicaudatus a nematode
bentgrass, creeping *Agrostis palustris*
bentgrass, common or colonial *Agrostis tenuis*
Berberis Darwinii a barberry
Berberis vulgaris common barberry
Beta vulgaris beet, mangel, sugarbeet, Swiss chard
Betula pendula silver birch, European white birch
birch *Betula* spp.
black spot of roses *Diplocarpon rosae*
bluegrass, Kentucky *Poa pratensis*
Botrytis cinerea grey mould fungus
broccoli *Brassica oleracea* var. *botrytis*
bromegrass *Bromus* spp.
Broussonetia papyrifera paper mulberry
Brussels sprouts *Brassica oleracea* var. *gemmifera*
buckwheat *Fagopyrum esculentum*
burdock *Arctium* spp.
Buxus sempervirens common box

cabbage *Brassica oleracea* var. *capitata*
Calystegia hederacea a bindweed
camellia *Camellia* spp.
Canna indica Indian shot
Capsicum annuum red pepper, sweet bell pepper
carnation *Dianthus caryophyllus*
carrot *Daucus carota*
Carthamus tinctoria safflower
castor oil plant *Ricinus communis*
cauliflower *Brassica oleracea* var. *botrytis*
celery *Apium graveolens*

Celtis sinensis   a hackberry
Chamaenerion angustifolium   rosebay willow-herb, fireweed,
Chenopodium album   fat hen, common pigweed, lambs-quarters
cherry   *Prunus* spp.
Chlorella   a green alga
Chlorella sorokiniana   a green alga
Citrus limon   lemon
Citrus paradisi   grapefruit, pomelo
Citrus sinensis   sweet orange
clover   *Trifolium* spp.
clover, Egyptian   *Trifolium alexandrinum*
clover, red   *Trifolium pratense*
cocksfoot   *Dactylis glomerata*
Commelina communis   day flower
corn, field   *Zea mays*
corn, sweet   *Zea mays*
Cornus alba   tatarian dogwood
Cornus florida   flowering dogwood
Coronilla varia   crown vetch
Corynebacterium nebraskense   a bacterial disease agent
Cotoneaster   cotoneaster
cotton   *Gossypium* spp.
Crataegus monogyna   English hawthorn
Cronartium fusiforme   fusiforme rust
cucumber   *Cucumis sativus*
Cucumis sativus   cucumber
Cucurbita pepo   squash, field pumpkin

Dactylis glomerata   orchard grass, cocksfoot
dandelion   *Taraxacum officinale*
Datura stramonium   thorn apple, Jimson weed, Jamestown weed
Diplacus aurantiacus   a monkey-flower
Diplocarpon rosae   black spot fungus of roses
dogwood, flowering   *Cornus florida*

Euglena   a green alga
Euonymus japonicus   a spindle-tree

Fagopyrum cymosum   a buckwheat
Fagopyrum esculentum   common buckwheat
Fagus sylvatica   European beech
fat hen   *Chenopodium album*
fescue, meadow   *Festuca pratensis*
fescue, red   *Festuca rubra*
Festuca rubra   red fescue
Festuca pratensis   fescue, meadow
fir   *Abies* spp.
fir, Douglas   *Pseudotsuga taxifolia* or *P. menziesii*
fir, silver   *Abies alba* or *A. amabilis*
Firmiana simplex   Chinese parasol tree, Phoenix tree
flax   *Linum usitatissimum*
Fragaria   strawberry
Fragaria virginiana   common wild strawberry

Fraxinus   ash
Fraxinus americana   white ash
Fraxinus excelsior   European ash
Fraxinus pennsylvanica   green ash

Galinsoga parviflora   galinsoga
geranium   *Geranium* spp. or *Pelargonium* spp.
Geranium carolinianum   Carolina crane's-bill
gladiolus   *Gladiolus hortulanus*
Glomus fasciculatus   a mycorrhizal fungus
Glomus macrocarpus   a mycorrhizal fungus
Glycine max   soybean
goatweed   *Hypericum perforatum*
gooseberry   *Ribes uva-crispa*
Gossypium hirsutum   upland cotton
grapefruit   *Citrus paradisi*
grape   *Vitis* spp.
grass, crested wheat   *Agropyron cristatum*
grass, annual meadow   *Poa annua*
grass, Japanese lawn   *Zoisia japonica*
grass, Kentucky blue   *Poa pratensis*
grass, knot   *Paspalum distichum*
grass, orchard   *Dactylis glomerata*
grass, herds, timothy   *Phleum pratense*

Helianthus annuus   sunflower
Helminthosporum maydis   maize blotch and blight fungus
Hemerocallis fulva   common orange day lily
Heterobasidion annosum   a fungus
Heterodera glycines   a nematode of soybean
hemlock, western   *Tsuga heterophylla*
Heteromeles arbutifolia   tollon, toyon, Christmas-berry
hickory   *Carya* spp.
Holcus lanatus   Yorkshire fog, velvet grass
honey fungus   *Armillaria mellea*
horseradish   *Amoracia rusticana* or *Cochlearia armoracea*
Hordeum   barley
Hosta plantaginea   fragrant plantain-lily
Hypericum   St. Johnswort

Ilex aquifolium   English holly

Jerusalem cherry   *Solanum pseudo-capsicum*
Juniperus chinensis   Chinese juniper

knotgrass   *Paspalum distichum* or *Polygonum aviculare*

larch, European   *Larix decidua*
Larix   larch
Larix decidua   European larch
Larix leptolepsis   Japanese larch
leek   *Allium porrum*
Leguminosae   pea or pulse family
lentil   *Lens culinaris* or *Ervum lens*
Lepidium virginicum   pepper grass

lettuce   *Lactuca sativa*
Ligustrum lucidum   a privet
Ligustrum ovalifolium   California privet
lilac   *Syringa* spp.
Lilium   lily
lily   *Lilium* spp.
Liquidambar styraciflua   sweet-gum
Liriodendron tulipifera   tulip-tree,
   whitewood
Lolium multiflorum   Italian ryegrass
Lolium perenne   perennial ryegrass,
   English ryegrass
lucerne   *Medicago sativa*
lupin   *Lupinus* spp.
Lupinus bicolor   a lupin
Lycopersicon esculentum   tomato

Mahonia aquifolium   holly mahonia,
   barberry
maize, grain   *Zea mays*
mallow, dwarf   *Malva rotundifolia*
Malus pumila   paradise apple
mangel   *Beta vulgaris*
maple, red   *Acer rubrum*
maple, sugar   *Acer saccharum*
marigold   *Tagetes* spp.
Medicago sativa   alfalfa, lucerne
Meloidogyne hapla   a nematode
Microsphaera alni   a fungus
milo maize   *Sorghum vulgare*
mugwort   *Artemisia vulgaris*
mulberry   *Morus* spp.
mustard   *Brassica* spp.

Nerium indicum   oleander
nettle, small   *Urtica urens*
Nicotiana glutinosa   a tobacco
Nicotiana tabacum   common tobacco

oak, black   *Quercus velutina*
oak, willow   *Quercus phellos*
oat   *Avena sativa*
onion   *Allium cepa*
orange   *Citrus* spp.
orange, Mandarin   *Citrus reticulata*
orange, sweet   *Citrus sinensis*
orchard grass   *Dactylis glomerata*
Oregon grape   *Mahonia nervosa*
Oryza sativa   rice
Oxydendrum arboreum   sour-wood, sorrel
   tree

Papaver somniferum   opium poppy
parsley   *Petroselinum crispum*
pea   *Pisum sativum*
peach   *Prunus persica*
peanut   *Arachis hypogaea*
pear   *Pyrus communis*
Pelargonium × *hederaefolium*   a hybrid
   geranium
Pelargonium × *hortorum*   a hybrid
   geranium

Pelargonium hortorum   fish geranium
Pelargonium zonale   zonal or horseshoe
   geranium
pepper, sweet   *Capsicum frutescens*
perilla   *Perilla* spp.
petunia   *Petunia hybrida*
Petunia hybrida   petunia
Petunia nyctaginiflora   petunia
Phaseolus vulgaris   bean, snap bean, white
   bean, navy bean, kidney bean, pinto bean
Phleum bertolonii   cat's-tail
Phleum pratense   timothy, herdsgrass
Phytophthora infestans   potato and tomato
   blight fungus
Picea abies   Norway spruce
Picea excelsa   Norway spruce
Picea pungens   Colorado spruce
Picea sitchensis   sitka spruce
pine   *Pinus* spp.
pine, eastern white   *Pinus strobus*
pine, loblolly   *Pinus taeda*
pine, lodgepole   *Pinus contorta*
pine, ponderosa   *Pinus ponderosa*
pine, red   *Pinus resinosa*
pine, Scots   *Pinus sylvestris*
pine, white   *Pinus strobus*
Pinus   pine
Pinus contorta   lodgepole pine
Pinus echinata   shortleaf pine
Pinus montana   Swiss mountain pine
Pinus nigra   Corsican pine, Austrian pine
Pinus ponderosa   western yellow pine
Pinus resinosa   red pine
Pinus strobus   eastern white pine
Pinus sylvestris   Scots pine
Pinus taeda   loblolly pine
Pisum sativum   garden pea
Pittosporum tobira   Japanese pittosporum
plane   *Platanus occidentalis*
Plantago lanceolata   ribwort, ribgrass,
   English plantain
Platanus occidentalis   buttonwood,
   American plane tree
plum   *Prunus* spp.
Poa   meadow grass, bluegrass
Poa annua   annual meadow grass, annual
   bluegrass
Poa pratensis   smooth-stalked meadow
   grass, Kentucky bluegrass
Polygonum affine   a knotweed
Polygonum pennsylvanicum   Pennsylvania
   persicaria
poplar   *Populus* spp.
poplar, black   *Populus nigra*
poppy   *Papaver somniferum*
Populus deltoides   cottonwood, northern
   cottonwood
Populus deltoides × *Populus trichocarpa*   a
   hybrid poplar
Populus euramericana   a poplar
Populus nigra   black poplar
Populus tremuloides   trembling aspen

potato   *Solanum tuberosum*
powdery mildew   a fungus disease
Pratylenchus penetrans   a nematode
Primula ioessa   a primrose
privet   *Ligustrum* spp.
Prunus   stone fruits
Prunus avium   sweet cherry
Pseudomonas glycinae   bacterial blight of
   soybean
Pseudomonas phaseolicola   bacterial blight
   or halo blight of bean
Pseudotsuga menziesii or *P.
   taxifolia*   Douglas fir
Puccinia graminis   rust fungus
pumpkin   *Cucurbita pepo*

Quercus   oak
Quercus alba   white oak
Quercus borealis   red oak
Quercus phellos   willow oak
Quercus velutina   black oak

radish   *Raphanus sativus*
rape   *Brassica napus*
Raphanus sativus   radish
Rhizobium   root nodule bacterium
rhubarb   *Rheum rhaponticum*
Rhytisma acerinum   sycamore tar spot
   fungus
rice   *Oryza sativa*
ring spot   a plant disease
rose   *Rosa* spp.
rosebay willow-herb   *Chamaenerion
   angustifolium*
Rubus hispidus   groundberry
Rumex acetosa   garden sorrel
rust of wheat, black   *Puccinia graminis*
rust, fusiform   *Cronartium fusiforme*
rutabaga   *Brassica napobrassica*
rye   *Secale cereale*
ryegrass, annual or Italian   *Lolium
   multiflorum*
ryegrass, perennial   *Lolium perenne*

Saccharomyces cerevisiae   bread yeast
safflower   *Carthamus tinctorus*
Salix fragilis   brittle willow, crack willow
Sambucus nigra   European elder
Scirrhia acicola   a fungus disease
Scorzonera   a salsify
Senecio laxifolius   groundsel
Setaria faberii   a millet or foxtail-grass
Setaria lutescens   a millet or foxtail-grass
smartweed   *Polygonum* spp.
snapdragon   *Antirrhinum* spp.
Solanum tuberosum   potato
Sonchus asper   spiny sow-thistle
Sorbus aria   white beam-tree
Sorbus aucuparia   European mountain ash,
   rowan
sorghum   *Sorghum vulgare*
sorrel, sheep's   *Rumex* sp.

soybean, soyabean   *Glycine max*
Sphagnum   a moss
spinach   *Spinacia oleracea*
Spinacia oleracea   spinach
spruce   *Picea* spp.
spruce, white   *Picea glauca* or *P. alba*
squash   *Cucurbita pepo* or *C. maxima*
Stachys lanata   lamb's ears
strawberry   *Fragaria* spp.
strawberry, wild   *Fragaria virginiana*
sunflower   *Helianthus* spp.
sugar beet   *Beta vulgaris*
sugar cane   *Saccharum officinarum* or *S.
   officinale*
Swiss chard   *Beta vulgaris*
sycamore   *Acer pseudoplatanus*
Symphoricarpus ribularis   snowberry,
   waxberry

Tagetes   marigold
tea   *Thea sinensis* or *Camellia sinensis*
tar spot of roses   *Diplocarpon rosae*
Tiarella cordifolia   foam flower, codwort
tobacco   *Nicotiana tabacum*
tomato   *Lycopersicon esculentum*
Trachycarpus fortunei   windmill palm
Tradescantia paludosa   a spiderwort
Trifolium alexandrinum   Egyptian clover
Trifolium repens   white clover
Triticum aestivum   bread wheat
Tsuga heterophylla   Western hemlock
tulip   *Tulipa gesneriana*
turnip   *Brassica rapa*

Ulmus americana   American elm
Ulmus glabra   Scotch elm, wych elm
Uromyces phaseoli   bean rust fungus
Urtica urens   small nettle

vetch, crown   *Coronilla varia*
Viburnum odorastissimum   a native shrub
   viburnum
Viburnum opulus   European cranberry
   bush
Vicia   vetch
Vicia faba   field bean, broad bean
violet   *Viola* spp.
Vitis labrusca   fox grape
Vitis vinifera   grape

wheat   *Triticum aestivum*

Xanthomonas alfalfae   a bacterial disease of
   alfalfa
Xanthomonas fragariae   a bacterial disease
   of strawberry
Xanthomonas phaseoli   bacterial blight of
   bean

Yucca gloriosa   Spanish dagger

Zea mays   maize, Indian corn, field corn,
   sweet corn
Zoysia japonica   Japanese lawn grass

APPENDIX 2 UNITS, PHYSICAL PROPERTIES AND CONVERSION FACTORS

Concentrations of pollutant gases in air are commonly expressed either on a mass/volume basis (e.g. microgrammes per cubic metre, $\mu g\,m^{-3}$) or on a volume/volume basis. Many pollutant gases in the atmosphere occur at concentrations between one part in 10^6 by volume and one part in 10^9. For convenience, in this book we use the American billion, 10^9, and so volumetric concentrations are usually expressed in parts per billion, ppb. Authors have been encouraged to use either ppb or $\mu g\,m^{-3}$ in their papers. There are benefits in expressing air pollution concentration in air on a volume/volume basis because the value is usually independent of temperature and pressure, whereas mass/volume depends on these factors. On the other hand, for accumulation of pollutant elements or for evaluation of fluxes a mass basis is more appropriate. To avoid repetition, conversion factors between the two types of unit are listed below.

Conversion factors from $\mu g\,m^{-3}$ to ppb and vice versa depend on molecular weight M, temperature T and pressure P. The appropriate expressions, assuming that all gases obey the ideal gas laws at temperatures above their boiling points are:

$$\mu g\;m^{-3} = \frac{ppb \times M \times 10^{-3}}{V_o} \times \frac{T_o}{T} \times \frac{P}{P_o}$$

$$ppb = \frac{\mu g\,m^{-3} \times V_o \times 10^{-3}}{M} \times \frac{T}{T_o} \times \frac{P_o}{P}$$

where T_o and P_o are standard temperature and pressure, 273 K and 101.3 kPa (1 atmosphere) respectively and V_o is the molar volume of ideal gas ($22.4 \times 10^{-3}\,m^3\,mole^{-1}$ at standard temperature and pressure). These equations are adequately accurate for the usual range of temperature and pressure found in laboratory or field experiments. Conversion factors given in *Table 1* have been calculated at reference temperature $T_R = 293$ K (20 °C) and pressure $P_R = 101.3$ kPa. Factors at other temperatures and pressures can be calculated from the above equations.

Table 1 CALCULATED CONVERSION FACTORS

Gas	Molecular weight (g mol^{-1})	Boiling point (°C)	Density (kg m^{-3})	Diffusivity in air (mm^2 s^{-1})	To convert from ppb to μg m^{-3}, multiply ppb by	To convert from μg m^{-3} to ppb, multiply μg m^{-3} by
Ammonia NH$_3$	17.0	−33.4	0.72	25	0.71	1.414
Chlorine Cl$_2$	70.9	−34.6	2.99	12	2.95	0.339
Ethylene C$_2$H$_4$	28.0	−103.8	1.17	19	1.16	0.859
Fluorine F$_2$	38.0	−187.0	1.58	16	1.58	0.633
Hydrogen chloride HCl	36.5	−85.0	1.53	17	1.52	0.659
Hydrogen fluoride HF	20.0	19.4	0.83	23	0.83	1.202
Hydrogen sulphide H$_2$S	34.1	−59.6	1.43	18	1.42	0.705
Nitric oxide NO	30.0	−151.0	1.25	18	1.25	0.801
Nitrogen dioxide NO$_2$	46.0	21.3	1.88*	15	1.91	0.523
Ozone O$_3$	48.0	−112.0	1.99	15	2.00	0.501
Peroxyacetyl nitrate (PAN) CH$_3$COONO$_2$	105.0	**	**	10	4.37	0.229
Sulphur dioxide SO$_2$	64.1	−10.0	2.73	12	2.67	0.375

*At 25°C, 101.3 kPa (note boiling point is 21.3°C).
**Values not known

The density ρ of a gas varies with temperature and pressure. For ideal gases the relation is

$$\rho = MP/RT$$

where R is the universal gas constant, $8.314\,\text{kJ}\,\text{mol}^{-1}\,\text{K}^{-1}$. Values in *Table 1* are at $T_R = 293\,\text{K}$ (20°C), $P_R = 101.3\,\text{kPa}$, and should be adjusted for other temperatures and pressures assuming that

$$\rho(T,P) = \rho(T_R,P_R)(T_R/T)(P/P_R)$$

The molecular diffusivity D of a gas in air depends on temperature, pressure and atomic size (Jarvis, 1971). The temperature and pressure dependences are reasonably well represented by

$$D(T,P) = D(T_R,P_R)(T/T_R)^{1.75} (P_R/P)$$

where temperatures are in degrees Kelvin. Diffusivities in *Table 1* have been calculated for $293\,\text{K}$, $101.3\,\text{kPa}$ relative to the diffusivity of water vapour ($D_W = 24\,\text{mm}^2\,\text{s}^{-1}$) assuming that

$$D(M) = D_W(M_W/M)^{1/2}$$

where M is molecular weight (Unsworth, Biscoe and Black, 1976). Measured diffusivities in air for many of the gases in *Table 1* are apparently not published, but these calculations agree reasonably well with the few measured values listed by Andrussov (1969), who also lists measured diffusivities of a large number of other inorganic and organic gases.

References

ANDRUSSOV, L. (1969). In *Zahlenwerte und Funktionen aus Physik, Chemie, Astronomie, Geophysik, Technik*, Edition 6, Vol. 2, Part 5a, pp. 551–557 (Landolt–Börnstein, Eds). Springer Verlag, Berlin
JARVIS, P.G. (1971). In *Plant Photosynthetic Production: Manual of Methods*, pp. 566–570 (Sestak, Z., Catsky, J and Jarvis, P.G., Eds). Junk, The Hague
UNSWORTH, M.H., BISCOE, P.V. and BLACK, V.J. (1976). In *Effects of Air Pollutants on Plants*, pp. 5–16 (Mansfield, T.A., Ed.). Cambridge University Press, Cambridge

LIST OF PARTICIPANTS

Alscher-Herman, Dr Ruth	Boyce Thompson Institute, Cornell University, Ithaca, NY 14853, USA
Baker, Dr C.K.	Nottingham University School of Agriculture, Sutton Bonington, Loughborough LE12 5RD, UK
Bateman, Mrs H.	Nottingham University School of Agriculture, Sutton Bonington, Loughborough LE12 5RD, UK
Bell, Dr J.N.B.	Imperial College Field Station, Silwood Park, Ascot, Berks, UK
Bengtson, Dr C.	Swedish Water and Air Pollution Research Institute, PO Box 5207, S-40224, Goteborg, Sweden
Black, Dr C.R.	Nottingham University School of Agriculture, Sutton Bonington, Loughborough LE12 5RD, UK
Black, Dr V.J.	Nottingham University School of Agriculture, Sutton Bonington, Loughborough LE12 5RD, UK
Bleasdale, Prof. J.K.A.	National Vegetable Research Station, Wellesbourne, Warwick CV35 9EF, UK
Bonte, Dr J.	Ministère de L'Agriculture, I.N.R.A., Laboratoire D'Etude de la Pollution Atmospherique, Montardon 64160 Morlaas, France
Borka, Dr G.	Agricultural University, Keszthely, Deak F. u. 16. H-8361, Hungary
Bowler, Mr G.K.	Technical and Research Section, London Brick Co. Ltd, Stewartby, Bedford, UK
Brown, Dr D.H.	Botany Department, University of Bristol, Bristol BS8 1OG, UK
Bucher, Dr J.B.	Swiss Federal Institute of Forestry Research, CH-8903 Birmensdorf, Switzerland
Callander, Dr B.	Nottingham University School of Agriculture, Sutton Bonington, Loughborough LE12 5RD, UK
Cape, Dr J.N.	Institute of Terrestrial Ecology, Bush Estate, Penicuik, Midlothian, UK
Carlson, Prof. R.W.	Institute for Environmental Studies, University of Illinois, Urbana, Illinois 61801, USA
Colvill, Dr K.E.	Botany Department, University of Liverpool, PO Box 147, Liverpool L69 3BX, UK
Cowling, Mr D.W.	Grassland Research Institute, Hurley, Maidenhead, Berks, UK

Crawford, Dr D.V.	Nottingham University School of Agriculture, Sutton Bonington, Loughborough LE12 5RD, UK
Crump, Mr T.J.	Crump Scientific Products Ltd, 166A/168A High Street, Rayleigh, Essex, UK
Davies, Miss T.	Biological Sciences Department, University of Lancaster, Bailrigg, Lancaster, UK
Davison, Dr A.	Plant Biology Department, Ridley Building, University of Newcastle upon Tyne, Newcastle upon Tyne NE1 7RU, UK
Deeley, Ms S.	Butterworth & Co. (Publishers) Ltd, Borough Green, Sevenoaks, Kent TN15 8PH, UK
De Kok, Mr L.J.	Plant Physiology Department, Biological Centre, University of Groningen, PO Box 14, 9750 AA Haren, The Netherlands
De Temmerman, Dr L.	Institute for Chemical Research, Museumlaan 5, B-1980 Tervuren, Belgium
Feder, Prof. W.A.	Suburban Experiment Station, University of Massachusetts, 240 Beaver Street, Waltham, MA 02154, USA
Fletcher, Mr J.	ICI Agricultural Division, PO Box 6, Billingham, Cleveland, UK
Ford, Mr J.D.	Environmental Biology Department, University of Guelph, Guelph, Ontario, Canada
Fowler, Dr D.	Institute of Terrestrial Ecology, Bush Estate, Penicuik, Midlothian, UK
Freer-Smith, Mr P.H.	Biological Sciences Department, University of Lancaster, Bailrigg, Lancaster, UK
Furukawa, Dr A.	Division of Environmental Biology, National Institute for Environmental Studies, Yatabe Tsukuba, Ibaraki 305, Japan
Gardner, Mr D.W.	Plant Pathology and Physiology Department, Virginia Polytechnic Institute and State University, Blacksburg, Virginia 24061, USA
Garrec, Dr J-P.	CEA Centre D'Etudes Nucleaires de Grenoble, Laboratoire de Biologie Vegetale, 38 Grenoble, France
Garsed, Dr S.G.	Imperial College Field Station, Ascot, Berks SL5 7PY, UK
Godzik, Dr S.	Polish Academy of Sciences, Institute of Environmental Engineering, Zabrze, Poland
Greenwood, Mr P.	Nottingham University School of Agriculture, Sutton Bonington, Loughborough LE12 5RD, UK
Hand, Dr D.W.	Glasshouse Crops Research Institute, Worthing Road, Rustington, Littlehampton, West Sussex BN16 3PU, UK
Hashimoto, Prof. Y.	Agricultural Engineering Department, Ehime University, Tarumi, Matsuyama 790, Japan
Heagle, Dr A.S.	USDA-SEA/AR, Plant Pathology Department, North Carolina State University, Raleigh NC 27607, USA
Hebblethwaite, Dr P.D.	Nottingham University School of Agriculture, Sutton Bonington, Loughborough LE12 5RD, UK
Heck, Dr W.W.	Botany Department, North Carolina State University, Raleigh, NC 27609, USA
Hunter, Mrs I.H.	Nottingham University School of Agriculture, Sutton Bonington, Loughborough LE12 5RD, UK
Jacobson, Dr J.S.	Boyce Thompson Institute, Cornell University, Ithaca, NY 14853, USA

Jakobsson, Mrs C.K.	Plant and Forest Protection Department, Swedish University of Agricultural Sciences, Box 7044, 75007 Uppsala, Sweden
Jensen, Mr K.F.	USDA Forest Service, Box 365, Delaware, OH 43015, USA
Jensen, Mrs C.K.	317 S. Sectionline Road, Delaware, OH 43015, USA
Kluczewski, Mr S.M.	Botany Department, University of Nottingham, University Park, Nottingham, UK
Koziol, Mr M.J.	Botany School, University of Oxford, South Parks Road, Oxford OX1 3RA, UK
Krupa, Prof. S.	Plant Pathology Department, University of Minnesota, St. Paul, Minnesota 55108, USA
Kup, Miss S.	Nottingham University School of Agriculture, Sutton Bonington, Loughborough LE12 5RD, UK
Kvist, Dr K.	Plant and Forest Protection Department, Swedish University of Agricultural Sciences, Box 7044, S-75007 Uppsala, Sweden
Lane, Mr P.I.	Imperial College Field Station, Silwood Park, Ascot, Berks SL5 7PY, UK
Last, Prof. F.T.	Institute of Terrestrial Ecology, Bush Estate, Penicuik, Midlothian EH26 0QB, UK
Law, Mr R.M.	Biological Sciences Department, University of Lancaster, Bailrigg, Lancaster LA1 4YQ, UK
Le Sueur-Brymer, Mrs P.	Atmospheric Environment Service, 4905 Dufferin Street, Downsview, Ontario M3H 5T4, Canada
Lockyer, Mr D.R.	Grassland Research Institute, Hurley, Maidenhead, Berks, UK
Lorenzini, Dr G.	Institute of Plant Pathology, University, Via Del Borghetto 80, 56100 Pisa, Italy
McGowan, Dr M.	Nottingham University School of Agriculture, Sutton Bonington, Loughborough LE12 5RD, UK
McLeod, Mr A.R.	Biology Section, Central Electricity Research Laboratory, Kelvin Avenue, Leatherhead, UK
Mansfield, Prof. T.A.	Biological Sciences Department, University of Lancaster, Bailrigg, Lancaster LA1 4YQ, UK
Martin, Mr A.	Central Electricity Generating Board, Scientific Services, Ratcliffe on Soar, Nottingham NG11 0EE, UK
Mellanby, Prof. K.	Monks Wood Experimental Station, Huntingdon PE17 2LS, UK
Mitchell, Miss M.J.	Laud House Room 314, Central Electricity Generating Board, 20 Newgate Street, London EC19 7AX, UK
Monteith, Prof. J.L.	Nottingham University School of Agriculture, Sutton Bonington, Loughborough LE12 5RD, UK
Moriarty, Dr F.	Monks Wood Experimental Station, Huntingdon PE17 2LS, UK
Mortensen, Mrs L.	National Agency of Environmental Protection, Air Pollution Laboratory, Risø National Laboratory, DK-4000, Roskilde, Denmark
Moult, Dr F.	The Analytical Development Co. Ltd, Pindar Road, Hoddesdon, Herts EN11 0AQ, UK
Mudd, Dr J.B.	MSU/DOE Plant Research Laboratory, Michigan State University, East Lansing, Michigan 48824, USA
Murray, Mr A.J.S.	Biological Sciences Department, University of Lancaster, Bailrigg, Lancaster LA1 4YQ, UK

Neumann, Dr H.H.	Atmospheric Environment Service, 4905 Dufferin Street, Downview, Ontario, Canada
Nicholson, Mr I.A.	Institute of Terrestrial Ecology, Hill of Brathens, Banchory, Kincardineshire AB3 4BY, UK
Nisancioglu, Mr S.	Plant Biology Department, University of Newcastle upon Tyne, Newcastle upon Tyne NE1 7RU, UK
Ormrod, Prof. D.P.	Horticultural Science Department, University of Guelph, Guelph, Ontario, Canada
Parry, Mr M.A.J.	Rothamsted Experimental Station, Harpenden, Herts, UK
Paterson, Mr D.	Botany Department, University of Aberdeen, St. Machar Drive, Aberdeen AB9 2UD, UK
Posthumus, Dr A.C.	Research Institute for Plant Protection, Binnenhaven 12, Wageningen, The Netherlands
Potter, Mr E.	Delta-T Devices, 128 Low Road, Burwell, Cambridge CB5 0EJ, UK
Pratt, Mr G.C.	Plant Pathology Department, University of Minnesota, St. Paul, Minnesota 55108, USA
Rajagopal, Prof. R.	Plant Physiology and Anatomy Department, Royal Veterinary and Agricultural University, Thorvaldsensvej 40, DK-1871 Copenhagen, Denmark
Rogers, Mr F.S.M.	Warren Spring Laboratory, Department of Industry, Gunnels Wood Road, Stevenage, Herts SG1 2BX, UK
Roose, Dr M.L.	Botany Department, University of Liverpool L69 3BX, UK
Ro-Poulsen, Dr H.	University of Copenhagen, Institute of Plant Ecology, Øster Farimagsgade 2D, DK-1353, Copenhagen K, Denmark
Rowlatt, Dr S.	Nottingham University School of Agriculture, Sutton Bonington, Loughborough LE12 5RD, UK
Rutter, Prof. A.J.	Imperial College Field Station, Silwood Park, Ascot, Berks, UK
Sardi, Dr K.M.	Agricultural University, Keszthely, Deak F. u. 16 H-8361, Hungary
Saunders, Dr P.J.W.	Natural Environment Research Council, Polaris House, North Star Avenue, Swindon SN2 1EU, UK
Saxe, Dr H.	Plant Physiology and Anatomy Department, Royal Veterinary and Agricultural University, Thorvaldsensvej 40, DK-1871 Copenhagen, Denmark
Schultz, Dr M.G.	Natural Environment Research Council, Polaris House, North Star Avenue, Swindon SN2 1EU, UK
Scott, Dr N.M.	Macaulay Institute for Soil Research, Craigiebuckler, Aberdeen, UK
Taylor Jr., Dr G.E.	Environmental Sciences Division, Oak Ridge National Laboratory, Oak Ridge, TN37830, USA
Taylor, Mr E.G.	Environmental Sciences Division, Oak Ridge National Laboratory, PO Box X, Oak Ridge, TN 37830, USA
Thiel, Dr W.R.	Verein Deutscher Ingenieure, Kommission Reinhalt ung der Luft, Postfach 1139, D-4000 Dusseldorf 1, FRG
Tingey, Dr D.T.	US Environmental Protection Agency, 200 SW 35th Street, Corvallis OR 97330, USA
Unsworth, Dr M.H.	Nottingham University School of Agriculture, Sutton Bonington, Loughborough LE12 5RD, UK

Unwin, Mr R.J.	Ministry of Agriculture, Fisheries and Food, Great Westminster House, Horseferry Road, London SW1P 4AE, UK
Usher, Ms S.	Botany and Plant Technology Department, Imperial College, Prince Consort Road, London SW7, UK
Van Hasselt, Dr P.R.	Rijks Universiteit Groningen, Department of Plant Physiology, Biologisch Centrum, Postbus 14, 9750 AA Haren, The Netherlands
Van Zinderen Bakker, Dr E.M.	Kananaskis Centre, University of Calgary, Calgary, Alberta T2N 1NE, Canada
Wang, Dr C.H.	Ecology and Environmental Protection Department, Jiangsu Institute of Botany, Nanjing, China
Weinstein, Dr L.H.	Boyce Thompson Institute, Cornell University, Ithaca, NY 14853, USA
Wellburn, Dr A.R.	Department of Biological Sciences, University of Lancaster, Bailrigg, Lancaster LA1 4YQ, UK
Wilson, Mr R.B.	Department of the Environment, Room 629 Becket House, Lambeth Palace Road, London SE1, UK
Wilson, Mr G.B.	Imperial College Field Station, Silwood Park, Ascot, Berks, UK
Whitmore, Mrs M.E.	Biological Sciences Department, University of Lancaster, Bailrigg, Lancaster LA1 4YQ, UK
Whittington, Prof. W.J.	Nottingham University School of Agriculture, Sutton Bonington, Loughborough LE12 5RD, UK
Yu, Prof. Shu-wen	Shanghai Institute of Plant Physiology, Academia Sinica, 300 Fonglin Road, Shanghai, China 200032

INDEX OF SPECIES WITH POLLUTANTS

Well-known agricultural and horticultural species are listed under their common names; others are listed under Latin names. Appendix 1 gives Latin and common name equivalents.

Abies spp., 143, 145, 146, 154
Acer spp.
  F, 153
  gas mixtures, 312, 321
  O_3, 117, 129, 387, 388
  SO_2, 334, 343
Agrostis spp.
  heavy metals, 381–383
  O_3, 288
Alfalfa (*Medicago sativa*)
  F, 142, 152, 269, 277, 283, 285
  gas mixtures, 80, 309–311, 313, 319, 321
  H_2S, 365, 367
  NO_x, 183
  O_3, 121, 129, 295, 341, 388
  SO_2, 29, 50, 73, 79, 252, 253, 353, 363–367
Alder, *see Alnus*
Alnus, 153
Alternaria solani, 334, 339, 346
Amaranthus retroflexus, 489–491
Antirrhinum, 262
Anthoxanthum odoratum, 383
Aphelenchoides fragariae, 335, 338
Apples (*Malus sylvestris*)
  F, 208, 214, 215
  gas mixtures, 310, 311, 318, 321, 326
  SO_2, 252
Apricot (*Prunus armeniaca*), 141, 157, 208, 283
Armillaria mellae, 333
Ash, *see Fraxinus* spp.
Atriplex spp., 69, 79, 119

Barley (*Hordeum vulgare*)
  F, 152, 209, 278, 357–358
  gas mixtures, 285–286, 312, 479
  O_3, 336, 358
  SO_2, 78, 82, 248, 253, 255, 256, 353

Bean (*Phaseolus vulgaris*), *see also* Bean, bush, etc.
  and acid rain, 346
  carbonyl disulphide, 459
  carbonyl sulphide, 459
  F, 156, 211, 284, 339, 346, 357, 358, 391
  gas mixtures, 312–313, 386, 391
  H_2S, 459
  methyl mercaptan, 459
  NO_2, 369
  O_3, 117–119, 123–124, 129, 344, 336, 359, 361, 362, 386, 389, 393, 394, 458–459
  SO_2, 69, 70, 77, 253, 256, 334, 343, 390
Bean, broad, *see Vicia faba*
Bean, bush
  F, 154
  O_3, 295, 298
  SO_2, 249–250
Bean, field, *see Vicia faba*
Bean, kidney, 258, 339–340, 346
Bean, lima, 297
Bean, navy, 69, 340
Bean, pinto, 69, 312, 314, 319, 341
Bean, snap, 311, 326, 422
Bean, white
  gas mixtures, 309, 311, 312, 319, 326
  O_3, 340, 470
  SO_2, 69
Beet (*Beta vulgaris*), 248, 249, 252, 358, 359, *see also* Beet, sugar; Chard
Beet, sugar (*Beta vulgaris*), 365, 367
Begonia
  gas mixtures, 311, 313–314, 338
  O_3, 338
  SO_2, 262, 335
Belonolaimus longicaudatus, 335, 338
Betula spp., 285
Birch, *see Betula*
Botrytis cinerea, 336, 337, 344